ELEMENTS OF
Radio Servicing

WILLIAM MARCUS

The combination of technical skill, practical experience, teaching ability, and skill in writing has given William Marcus eminence as a teacher and as an author. He has taught radio theory and practice at the Abraham Lincoln High School and at the Melville Radio School and he has written technical manuals for the Navy at the Hazeltine Electronics Corporation. William Marcus is currently the principal of a Bayside school.

ALEX LEVY

Practical experience as a radio serviceman, as a sound technician, as the owner of a radio service shop, and as a writer of technical manuals on electronic equipment for the armed forces well qualifies Alex Levy as a writer on radio servicing. He is a teacher of radio mechanics and electronics at the Thomas A. Edison Technical and Vocational High School in New York City. Alex Levy is the author of the syllabus for Radio and Television Mechanics for the vocational high schools in the city of New York.

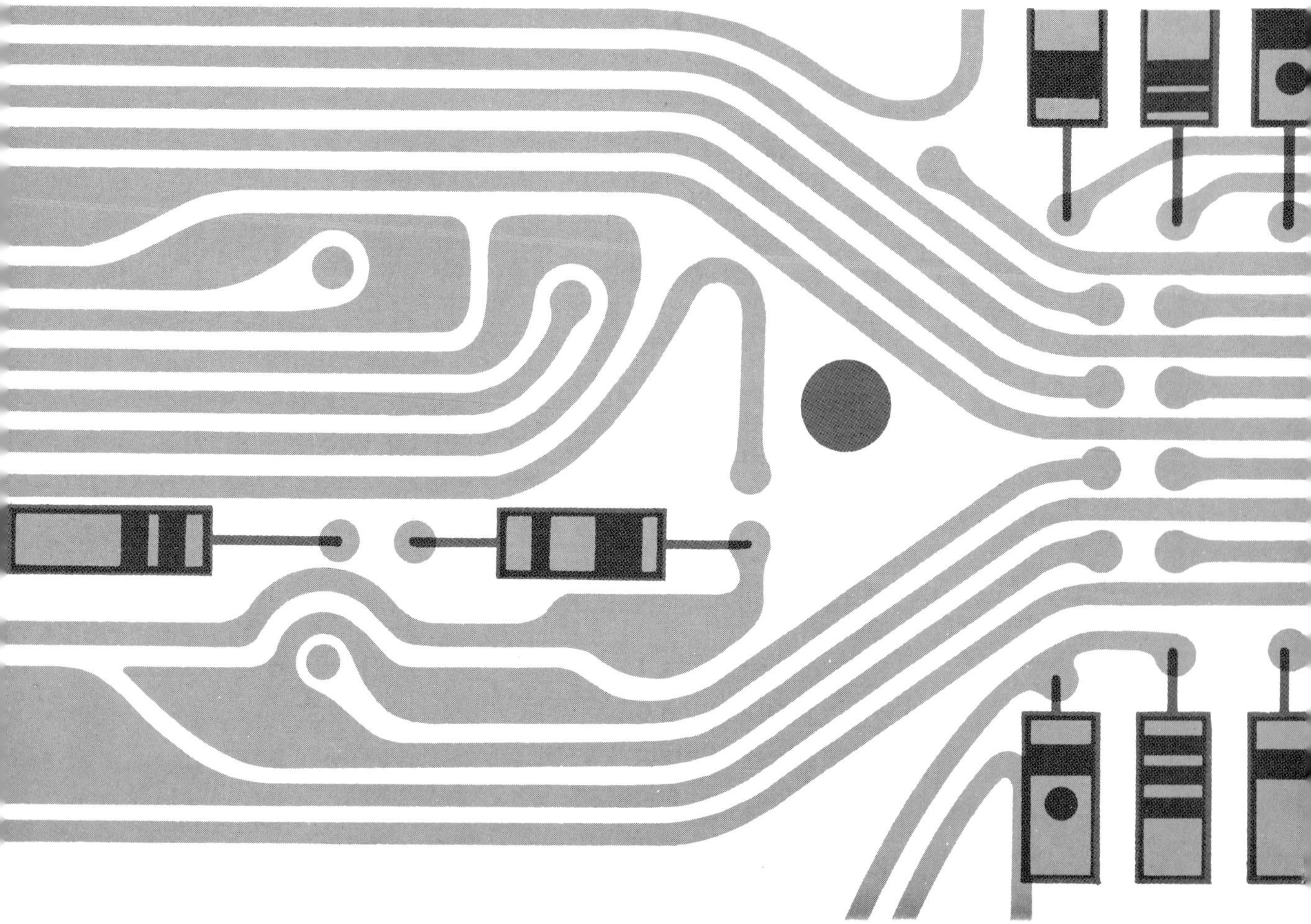

third edition

ELEMENTS OF Radio Servicing

Marcus and Levy

McGRAW-HILL BOOK COMPANY
ew York • St. Louis • San Francisco • Toronto • London • Sydney • Mexico • Panama

ACKNOWLEDGMENTS

The authors wish to acknowledge their indebtedness to the following companies for their cooperation in furnishing photographs, line drawings, schematic diagrams, and information.

Cornell Dubilier Electric Corp.
Emerson Radio & Phonograph Corp.
Essex Electronics
Galvin Mfg. Corp. (Motorola)
General Electric Company
Hytron Radio and Electronic Corp.
Jensen Manufacturing Co.
P. R. Mallory & Co., Inc.
Meissner Mfg. Div., Maguire Industries, Inc.
Philco Corp.
Pilot Radio Corp.
Precision Apparatus Co., Inc.
Quam-Nichols Co.
Radio Corp. of America
Radio City Products Co., Inc.
Howard W. Sams & Co., Inc.
Stromberg-Carlson Co.
Thordarson Electric Mfg. Div., Maguire Industries, Inc.
Weston Electric Instrument Corp.
Zenith Radio Corp.

ISBN 07-040290-6

910 VHVH 76

PREFACE

The field of radio servicing still offers many opportunities for establishing a source of lucrative income, even though its companion field, television, continues its great expansion. The mastery of radio servicing, a profitable field in its own rights, is the inevitable prerequisite for the study of television servicing.

As in the case of the previous editions of this book, basic and necessary information are presented together with a dynamic technician's procedure. The many tributes paid to the previous editions have indicated to the authors the sound basis of their pedagogical methods.

It is assumed that the reader has acquired an elementary background of electronic theory prior to delving into service work. Nevertheless, elementary theory is presented in this book wherever it serves to make a particular procedure more clear.

Design theory has been eliminated since it is felt that such theory does not fall within the province of the repairman. It is axiomatic that the technician must never redesign a receiver brought in for repair unless so advised by the manufacturer.

Since the publication of the second edition of this book, new practices and circuits have been developed. Therefore new chapters have been added to update the information. A new chapter is devoted to the popular and growing field of hi-fi phonograph assemblies. An addition has been made to the chapter on f-m radio to include multiplexing and f-m stereo receivers. The chapter on auto radio servicing has been completely rewritten in line with current practices. The three-way portable radio receiver servicing chapter has been replaced with one on the servicing of transistor portable sets. All other chapters have been updated to include current practices. Information on printed circuits has been merged into several chapters to give greater proficiency in general servicing. New information has been added to the appendix to make it more useful.

The authors feel that this edition of the book is the most comprehensive treatment on radio servicing found in any one book. Indeed, it has passed from basic elements to a far more extensive treatment of radio servicing.

A book of this limited size could not possibly cover all the individual variations found in radio receivers. So the authors have confined their survey to the most widely used practices of the last ten years. It is felt that with this background, the technician will be able to figure out the other variations with greater ease.

We are sure that old friends of the previous editions will find this third edition sufficiently new to be of extreme value to them. May we thank the many friends who wrote to us through the past years in praise of what they had learned from the previous editions.

WILLIAM MARCUS
ALEX LEVY

CONTENTS

Introduction 1

Functional servicing. Thinking, especially in the solving of problems, involves the application of random bits of information to a particular situation. Two distinct elements are involved in this procedure. The first is that sufficient information to draw from is available. The second is that the information necessary for the solution is applied to the particular problem. The first element is a static one; the information may be compiled in a book for continuous reference. The second element is a dynamic one and cannot be assumed to develop from the first element unless specific exercise is provided.

Too many servicing manuals and books are organized on the premise that servicing skills can be developed if only enough bits of information are presented. In this respect, they fail to develop functional skills. The learner finds his path a slow and uncertain one.

The purpose of this book is to apply the psychology of learning to radio servicing. Basic information is presented at all times. In addition, the information is so organized that it develops whole dynamic procedures for application to specific radio troubles.

Scope of the book. It would be impossible to present in any small book procedures for servicing all types of radio receivers, as well as all the variations of each type. For this reason, the scope is restricted to the most widely used receiver—the superheterodyne.

All the individual variations could not be given. Therefore, a standard circuit, based on the most widely used practices, is presented as the basis for study. This circuit is shown in Fig. 1-1. In all probability, there is no receiver that incorporates all the features indicated; but for study purposes, such a standard circuit will be found invaluable. Throughout this book, the standard circuit is broken down and analyzed by stages, in accordance with the plan described in the following section.

All modern practices could not possibly be indicated in one schematic diagram. Therefore, a section on widely used variations in design is included in each chapter of stage analysis. It is felt that enough information will be obtained from the standard circuit and the variations sections to understand and service any other variation.

To promote understanding of the basic circuits, the text describes and discusses the use of such test instruments as the signal generator, the multimeter, the sweep generator, and the cathode-ray oscilloscope. These are the recommended instruments for basic radio servicing.

Several chapters are devoted to the a-c transformer power supply, the a-c/d-c transformerless power supply, and their variations. For in the last analysis, the signal circuits of most receivers are basically similar superheterodyne circuits with either of the two main types of power supplies.

One chapter is devoted to home audio systems. These refer to hi-fi phonograph equipment and its servicing. Stereo systems are

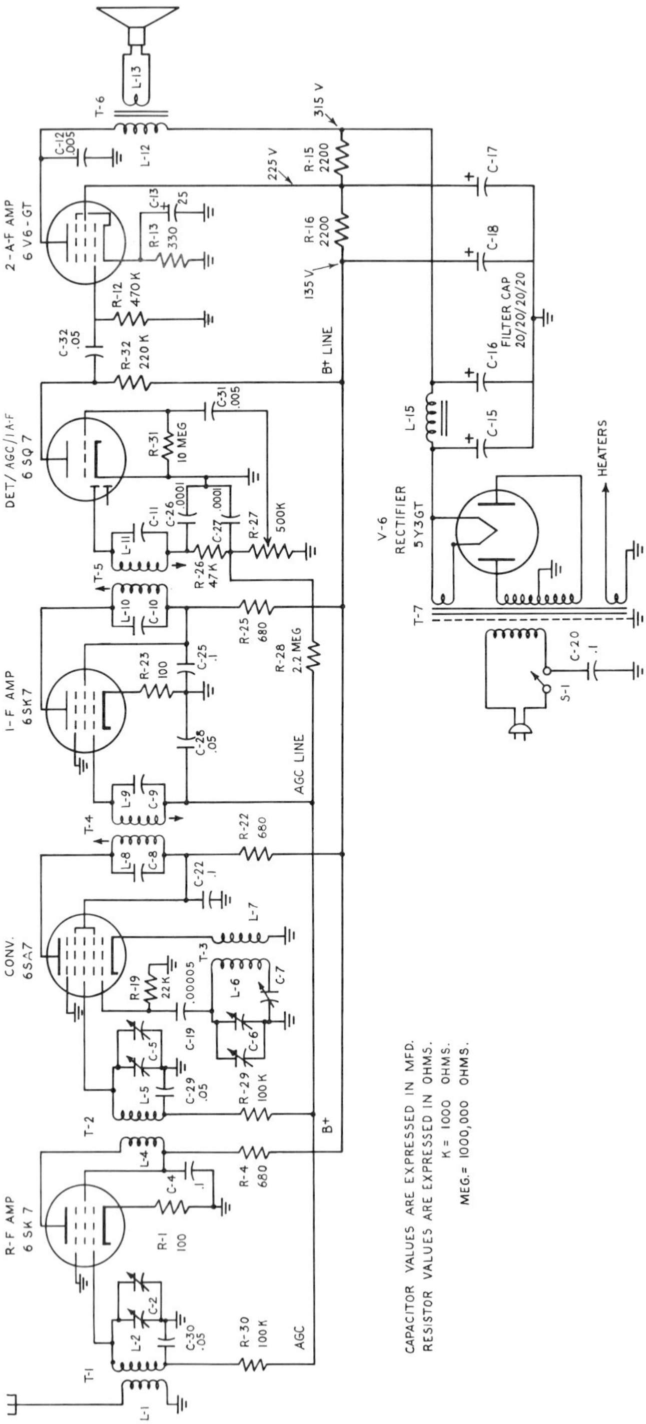

FIG. 1-1. Standard receiver circuit used for basic study.

described in this chapter. Another chapter is devoted to alignment of receivers. This procedure involves the proper adjustment of tuned circuits for best reception of station signals and for tracking with tuning-dial settings.

The remainder of the book is concerned with the servicing of transistor portable receivers, auto radios, and f-m receivers. These are explained in separate chapters, more quickly than the analysis of the basic receiver of the book, since they could be considered as variations of the basic superheterodyne receiver, with special power supplies and adjustments. The final chapter is a relatively rapid overall summary of the causes of specific defects and of servicing procedures.

Organization of dynamic material. In order to make the material of this book dynamically functional, information is presented in the sequence that it would be used practically in servicing a superheterodyne receiver. Instead of proceeding from stage to stage in the order that a radio signal would pass from the antenna to the speaker, the stages are presented in the order that a technician would investigate a defective receiver. Standard radio-servicing procedures are given for each stage. In addition, simple practical tests performed by technicians on the bench are presented. These tests are based on years of practical servicing experience.

Each stage is presented in a similar manner. The following topics are discussed as appropriate:

---

1. Quick check for normal functioning of the stage
2. Function of the stage
3. Typical or basic circuit schematic
4. Function and common value for each component part
5. Normal test data for the stage
6. Common troubles encountered in the stage
   - How they are found
   - Special problems involved in replacement of components
7. Variations from the typical stage that are frequently used; special trouble-shooting procedures in these variations
8. Summary of tests including outline of procedure to be followed in tracing various symptoms to their cause

---

The organization of the information, as outlined above, is the method by which the material information will become quickly functional. A little practice in its use will assure a quick practical approach to radio servicing problems.

It should be understood that this book is not intended to be an encyclopedia of radio servicing. Once the method of attack is mastered, reference to service notes distributed by radio-receiver manufacturers will be more useful than before. Where an unusual circuit is encountered, such notes will prove to be of great value.

Superheterodyne Receivers 2

Block diagram of a superheterodyne receiver. Before the stage analysis of the superheterodyne receiver is presented, it is advisable for the technician to have an overview concept of how it works. This picture will be obtained readily from a block diagram. Each block represents a stage that will be shown later in schematic and more detailed form. The accompanying wave forms or pictures of the types of electric currents show how each stage alters the signal entering it. It will be seen later that some of these stages may be omitted or that two stages may be combined into one. The block diagram of the superheterodyne receiver is given in Fig. 2-1.

How the superheterodyne receiver works. An analysis of the block diagram shown will clarify this matter. The antenna picks up the modulated r-f carrier signals of all stations within the receiving area of the set. In the broadcast band, they vary from 550 to 1,600 kc. Before passing through the r-f stage, one station is selected by tuning and its signal is passed on. The modulated r-f carrier signal is a high-frequency wave modulated or varied by a lower frequency wave known as the "audio modulation." The audio modulations represent the useful component that will eventually drive the speaker.

The r-f stage merely amplifies the station to which we are tuned and passes the amplified signal with its audio modulation on to the mixer. The audio modulation retains the same wave form as the signal received at the antenna.

The mixer and local oscillator work together as a team. Often the two stage functions are performed by one tube, which is called a "converter." The local oscillator is a generator of unmodulated r-f waves, automatically adjusted to a frequency of about 455 kc above that of the received station r-f frequency. When the output of the local oscillator is mixed with the r-f station frequency in the mixer stage, the resulting output of the mixer is at a frequency of 455 kc, with the same audio modulations as that of the original signal that came down the antenna.

The 455-kc signal is then fed into the i-f stage, which is fixed-tuned to about 455 kc. Here the signal is amplified and fed into the detector. The audio modulations still retain the original wave form.

The detector stage removes the 455-kc r-f component from the audio modulation component and passes the latter into the first audio stage. This detector is frequently referred to as the "second" detector, and the mixer or converter is called the "first" detector.

The audio component enters the first audio stage, where its voltage is amplified. It still retains the same wave form as that of the original audio modulation on the station carrier.

The second audio stage amplifies the audio signal even more, developing sufficient power to drive the speaker, which is a power-driven device. The audio signal still retains its same wave form at the input to the speaker. The speaker response is a series of sound waves.

Power for the entire receiver is usually obtained from a 110-volt, 60-cycle a-c source or a 110-volt d-c source. The power supply will rectify the a-c supply, where such power is supplied, and will filter the rectified voltage to obtain a fairly smooth direct current, which now becomes our *B* supply. Where 110-volt d-c power is furnished, the power supply will merely filter it to obtain the *B* supply. In battery portable sets the supply is obtained directly from batteries, which do not require additional filtering.

Using the block diagram. It is important that the block diagram shown in Fig. 2-1 be committed to memory before going on. Where test instruments are used, the input and the output waves of each stage will determine how to make proper settings. This is especially important in signal-substitution methods where a signal generator is used.

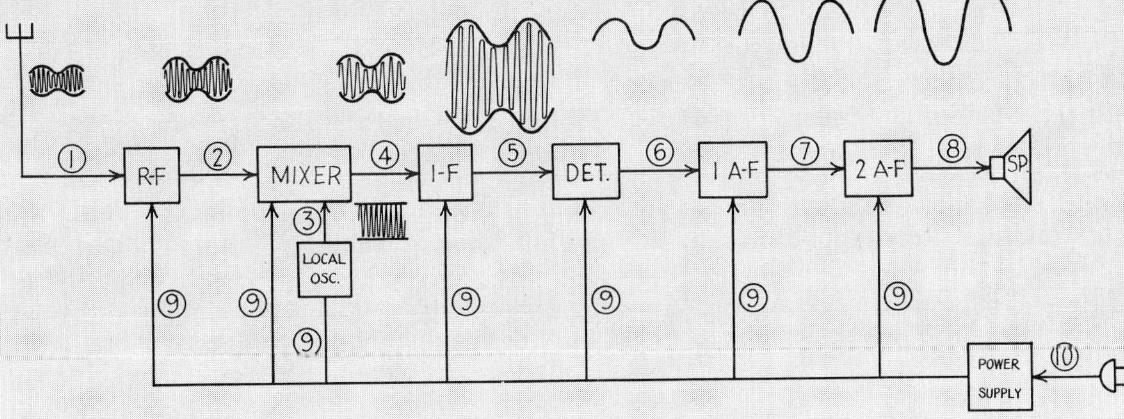

1. R-f (550 to 1,600 kc)—modulated at radio frequencies.
2. Tuned and amplified r-f (550 to 1,600 kc)—modulated at audio frequencies.
3. Unmodulated r-f (r-f one + 455 kc).
4. I-f (455 kc)—modulated at audio frequencies.
5. Amplified I-f (455 kc)—modulated at audio frequencies.
6. Audio frequencies (50 to 10,000 cycles) for high-fidelity receivers.
7. Amplified audio frequencies (50 to 10,000 cycles) for high-fidelity receivers.
8. Amplified audio frequencies (50 to 10,000 cycles) for high-fidelity receivers.
9. D-c *B*— supply.
10. 110-volt, 60-cycle a-c supply or 110-volt d-c supply.

FIG. 2-1. Block diagram of a superheterodyne receiver with associated wave forms.

Servicing Procedure

3

Receiver servicing systems. When a radio receiver is brought in for servicing, the demand made of the technician is that he put the set back into normal operation. The means is of relatively no importance to the customer. Although this end also becomes the aim of the technician, he is confronted with a more immediate goal. What method shall he follow in locating the defect?

The various techniques that he uses can be grouped into a few systems of procedure, which are listed below:

1. Reliance on sight, touch, smell, and past experiences with the same type of receiver.
2. Part-substitution method.
3. Voltage measurements across components.
4. Point-to-point resistance measurements.
5. Electrode-current checking.
6. Signal substitution.
7. Dynamic-signal tracing with a vacuum-tube voltmeter and oscilloscope.

The first system is a self-evident one. Wherever a component appears to be broken or burned, or smells as if it has been overheated, or feels too hot, the assumption might reasonably be made that it is defective and should be replaced. Similar difficulties previously experienced with the same type of receiver might guide the technician. Unfortunately, too many defects will not result in extremes of breakdown. Also, the defective component is not disclosed as the cause of the receiver failure or the result of some other defect. Finally, experience as the guide can at most be a helpful rather than an infallible aid.

The second system involves the substitution of a part, known to be good, for a similar part that seems to be defective in the receiver. The weakness in this procedure is that it is too time-consuming by itself and may be useless where the trouble involves a number of defective components.

The third system is one in which voltage measurements are taken across various components. When the observed values are compared with normal voltage data, defective components are readily found. There are several weaknesses in this system when used alone. The time required to make all voltage checks in a modern complex receiver makes it extremely inefficient. At the very best, it may be used alone for making routine checks. In addition, many defects will not alter voltage readings to an extent that would indicate where the defects may be found.

The fourth system is similar to the third, except that resistance measurements are taken with an ohmmeter across the various components, rather than voltage measurements with a voltmeter. Used alone, this system has the same weaknesses as the voltage test.

The fifth system is one in which current measurements are made in various portions of the receiver to locate deviations from normal values. It is not often used, because it involves either the opening of circuits to insert ammeters in series, or the use of special adapters.

The sixth system is a popular one. A signal, similar to the one normally encountered in operation, is fed into the input of a stage, and the result at the output is then observed and compared with normal expectations. It is not suitable when used alone, since it primarily locates a defective stage without indicating the defective component.

The last system is one that involves expensive equipment and complex techniques. Commercial instruments are of various types, but most attempt to analyze the stages of the receiver under actual working conditions. Basically, all are combinations of vacuum-tube voltmeters, capable of making measurements without loading the circuits tested, and are excellent for measuring weak signals in the order of microvolts. The signal indicators are of various types: oscilloscopes, electron-ray tubes, loudspeakers, meters, etc. These instruments readily indicate loss of gain of stages, distortion, intermittents, regeneration, oscillation, noise, and other conditions. However, they still require supplementation by the multimeter and the signal generator.

Which servicing procedures shall we use? No one of the servicing systems referred to in the above section can be used with speed and efficiency when taken alone. Experience has shown that it is most efficient first to determine the defective stage by means of a signal check and then to carefully analyze that stage for defective components.

This book assumes that the intelligent and combined use of the first four systems listed, plus the signal-substitution system, will give a highly efficient trouble-shooting procedure. Reference to the stage analysis in later sections will give great facility in the proper combined use of the suggested systems.

What instruments should the technician have? To follow the suggestions that are recommended, a voltmeter, an ohmmeter, and a signal generator are required. Two of these are combined in one popular instrument called a "multimeter," which combines a voltmeter, ammeter, and ohmmeter in one unit, with a switching device to obtain the desired function as well as the proper range.

Order of use of instruments. The advantages of the recommended procedures will become evident with use. The general rule to be followed in servicing a receiver is, first, to use the signal generator in order to locate the defective stage or interstage components. The voltmeter and ohmmeter are then applied in order to close in for the kill, that is, the determination of the actual defective components.

The latter part of this book breaks down a typical superheterodyne receiver into its stages and gives procedures for testing the normal operation of each one. For each stage, typical test voltage and resistance measurements are listed for comparison with those actually found in the defective receiver. In addition, where possible, practical methods of testing stages are listed.

Finally, the order of presentation of the stages analyzed is, in general, the order in which a technician would be expected to subject the defective receiver to analysis. It is felt that in this way he will use this book with a more functional approach to his problem.

The question might arise at this time as to the place of a tube tester in a service shop, since many receiver defects may be due to faulty tubes alone. A word with regard to this matter will explain the lack of emphasis placed on that instrument.

There are two types of tube testers: the mutual-conductance type of tester and the emission tester. In the first, a small designated change of grid voltage is applied to the tube.

The resulting change of plate current determines whether to call the tube good or bad. In the emission tester, the current flow or emission that results when the filaments are heated and a fixed voltage is placed on the plate determines whether to call a tube good or bad. Emission decreases with the age of the tube. In addition, both types of testers have circuits for determinating whether there is leakage or a short between the tube elements.

The tube tester is suitable for testing rectifier tubes. However, for other tubes, it does not measure their operation under the same dynamic conditions that they encounter in actual operation. Tubes that test good in it may be poor in actual receiver operation. A far better check for the technician is to hook up the signal generator and an output meter to the receiver and observe the output. Then substitute a good tube for the one believed to be bad and compare the two outputs. Of course, where the customer brings only his tubes for testing, the tube tester is the instrument to use, its limitations being understood.

4 Multimeters

A typical multimeter. The multimeter is one of the radio technician's constant companions. It is the instrument that finally localizes troubles in the receiver after the defective stage is found. A typical multimeter is shown in Fig. 4-1. Its purpose is primarily to make voltage, current, and resistance measurements throughout the receiver.

To perform its functions, the multimeter is a milliammeter, voltmeter, and ohmmeter combined in one case. In addition, it is designed to furnish various ranges of current, voltage, and resistance measurements. To select a particular function and a particular range from the instrument, a front-panel selector switch is provided. Each position of the switch is labeled for that purpose.

In describing the components of the multimeter, it is better to treat the voltmeter and ohmmeter as though they were separate. Nothing will be said about the milliammeter as a current measuring device, since few technicians will make such measurements without adapters. The only principle to be kept in mind, when currents are measured, is to be sure to be on the correct range. A good policy is to start at the highest range and switch down to lower ones until the correct one is reached.

General principles of the voltmeter. The purpose of a voltmeter is to indicate the potential difference or voltage between two points of a circuit. This is accomplished by connecting the two input terminals of the voltmeter to the two points to be tested in the circuit. The placement of the voltmeter, in parallel with the circuit to be measured, brings up some interesting factors that will be described later.

Essentially, the voltmeter is a D'Arsonval galvanometer in series with a fairly high-ohmage resistor. The latter is commonly called the "multiplier." Figure 4-2 shows a basic voltmeter. The size of the multiplier determines the range of the voltmeter. A brief analysis will make this point clear.

Begin with a galvanometer that gives full-scale deflection at 1 ma (0.001 amp). Such an instrument is usually called a "one-mil milliammeter." Assume that it has an internal resistance of 30 ohms. What must be the resistance of the multiplier to convert it into a 0 to 1 voltmeter? When so converted, 1 volt placed across the milliammeter and multiplier will drive 1 ma through it to give full-scale deflection, as shown in Fig. 4-3. Using Ohm's law, determine the resistance that will give this condition.

$$R = \frac{E}{I} = \frac{1}{0.001} = 1{,}000 \text{ ohms}$$

Since the milliammeter has a resistance of 30 ohms, the multiplier R_m must have a resistance of 1,000 minus 30, or 970 ohms. An instrument of this sort is called a 1,000-ohms-per-volt voltmeter, because 1 volt is applied across 1,000 ohms: (1,000/1) = 1,000. This designation is an indication of its sensitivity.

Suppose that it was desired to convert the same milliammeter into a voltmeter of 0 to 100 volts. What must be the resistance of the multiplier? By similar reasoning, 100 volts now placed across the milliammeter and multiplier will drive 1 ma through it to give full-scale deflection, as shown in Fig. 4-4. Using Ohm's law for the total resistance,

$$R = \frac{E}{I} = \frac{100}{0.001} = 100{,}000 \text{ ohms}$$

Again subtracting the milliammeter resistance from the total resistance, we find that the multiplier must have a resistance of 100,000 minus 30 = 99,970 ohms. Its sensitivity is still found to be 100,000/100, or 1,000 ohms per volt. A switch is usually provided on the multimeter to give a voltmeter of different ranges by cutting in different multipliers. Such a switching device is shown in Fig. 4-5.

In using various voltmeters, the technician may be surprised when he measures the voltage across two points of a circuit and obtains two different readings. His first impulse might be to say that one of the instruments is inaccurate. Yet they may both be right, and the technician must interpret his results more carefully.

The explanation for this condition lies in the different sensitivities of the voltmeters. The example given above was for a 1,000-ohms-per-volt voltmeter. The current tendency in radio servicing is to use voltmeters whose sensitivity is 20,000 ohms per volt. To illustrate, let us assume that a galvanometer requires 50 microamperes (0.00005 amp) for full-scale deflection. What must be the size of the total resistance to give a voltmeter with a range of 0 to 1 volt? From Ohm's law,

$$R = \frac{E}{I} = \frac{1}{0.00005} = 20{,}000 \text{ ohms}$$

The sensitivity of this voltmeter is 20,000/1, or 20,000 ohms per volt. Similarly, with the same basic movement, we could convert it into a voltmeter with a range of 0 to 120 volts. From Ohm's law,

$$R = \frac{E}{I} = \frac{120}{0.00005} = 2{,}400{,}000 \text{ ohms}$$

The sensitivity is still 2,400,000/120, or 20,000 ohms per volt. Now, consider the following circuit in Fig. 4-6, across which 60 volts are dropped. We shall use voltmeters on an assumed 100-volt range.

Since $R_1 = R_2$, the voltage dropped across each is equal and is 30 volts. If the 1,000-ohms-per-volt voltmeter is connected across R_2,

FIG. 4-1. The Simpson Model 260 multimeter.

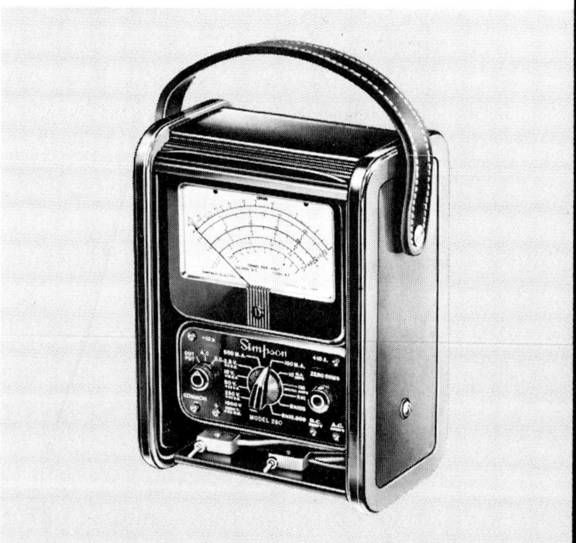

FIG. 4-2. A basic voltmeter.

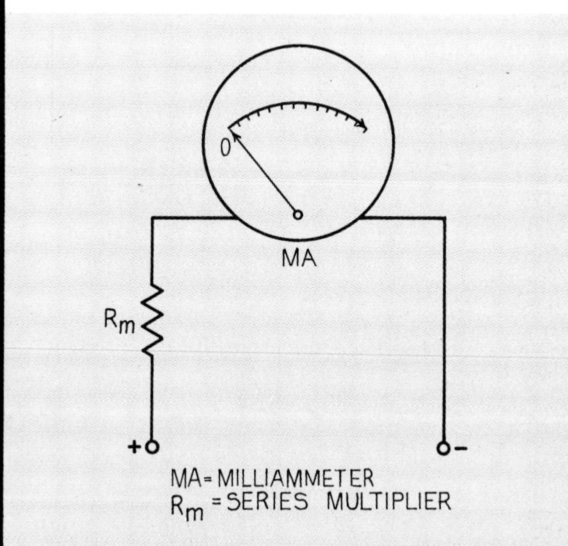

we have the condition indicated in Fig. 4-7. The voltmeter and R_2 are equal in resistance and in parallel. The combined resistance of the parallel branch is now 50,000 ohms, and the circuit now appears as in Fig. 4-8. Since the two resistors are now not equal, the voltage divides differently, (100,000/150,000) × 60, or 40, volts across R_1, and (50,000/150,000) × 60, or 20, volts is dropped across R_2 and the voltmeter. The voltmeter reads 20 volts. If the 20,000-ohms-per-volt voltmeter is substituted for the 1,000-ohms-per-volt voltmeter, the condition indicated in Fig. 4-9 prevails. The combined resistance of the voltmeter and R_2 is about 95,238 ohms. The circuit now appears as shown in Fig. 4-10. Across R_1, (100,000/195,238) × 60, or about 30.7 volts are dropped. Across R_2 and the voltmeter are dropped (95,238/195,238) × 60, or about 29.3 volts. The voltmeter reads 29.3 volts. In both cases above, the effects of change of current when the voltmeters were connected have not been taken into consideration because the relative results would still exist.

Which voltmeter was correct in its reading? If interpreted properly, both gave correct results. The technician may use either of the two voltmeters of different sensitivities, but at all times he must interpret his results. Generally, it is true that the voltmeter of the higher ohms-per-volt sensitivity will give a more accurate reading when the voltage across a high-resistance circuit is measured. Since radio-receiver circuits are generally of high resistance, voltmeters intended for measuring them have a sensitivity of at least 20,000 ohms per volt. Such voltmeters can be used for every radio voltage measurement, except where there is a low-voltage drop across a very high resistor. For example, agc voltage is developed across a high resistor of about 2.2 megohms. On a 12-volt range, the 2.4 megohms of the voltmeter across the 2.2 megohms of the resistor gives an overall resistance of about 1 megohm. This would reduce the low-voltage reading to a value too low to be read. To get a proper agc reading, use is made of a vacuum-tube voltmeter. It has a high input impedance of about 12 to 15 megohms and will therefore load the circuit to a lesser degree.

The voltmeter section of the multimeter is usually designed for various ranges of a-c

FIG. 4-3. Voltmeter with a 0- to 1-volt range at full-scale deflection.

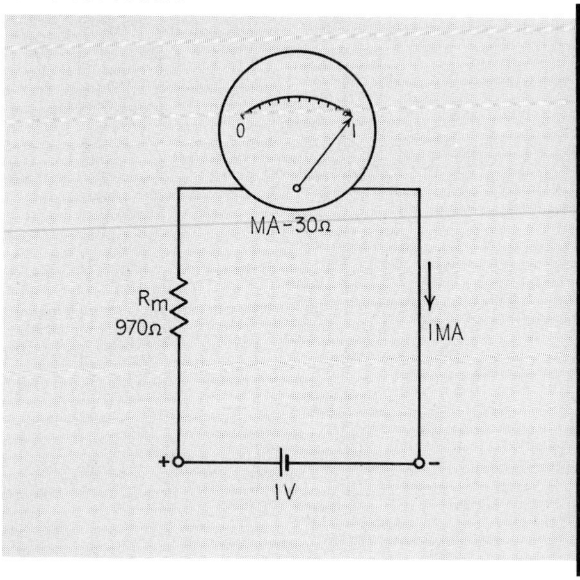

FIG. 4-4. Voltmeter with 0- to 100-volt range at full-scale deflection.

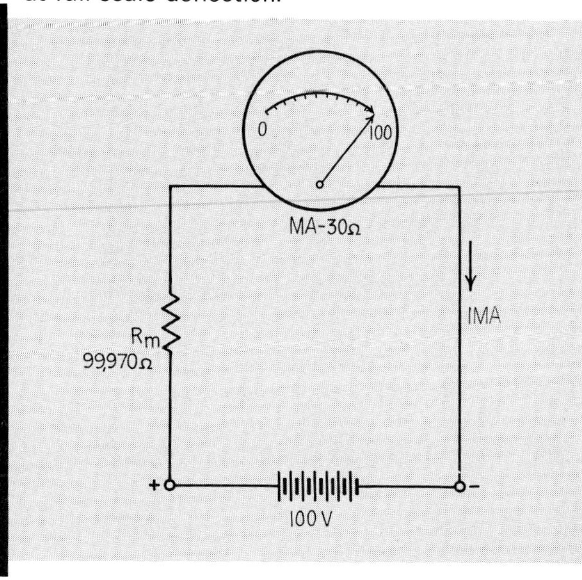

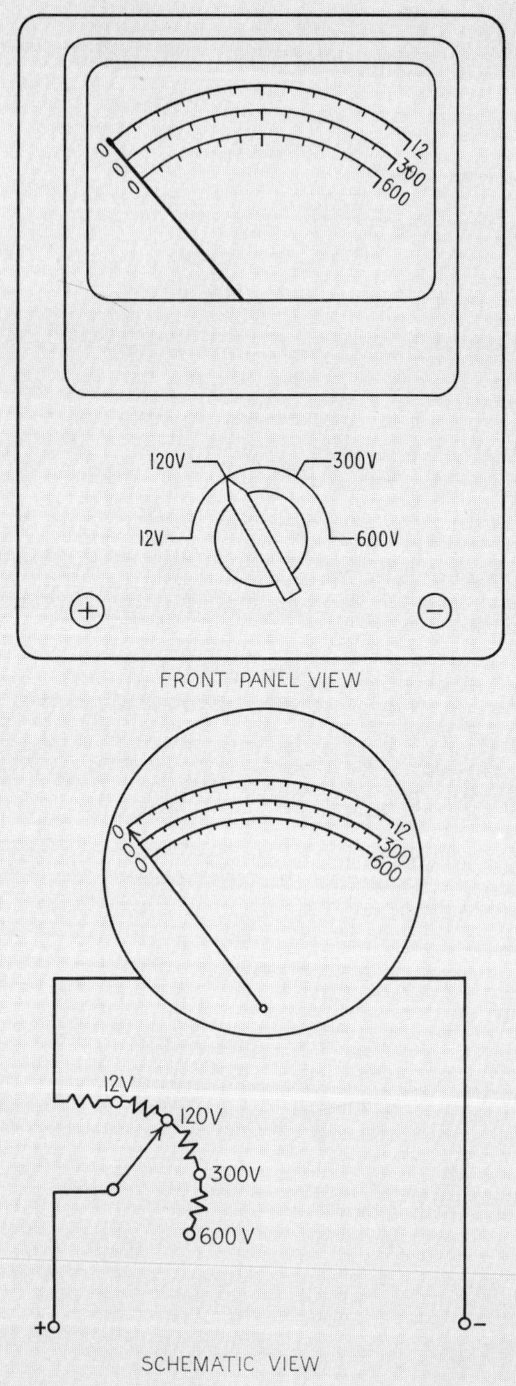

FIG. 4-5. Multirange voltmeter showing range switch.

as well as d-c voltage measurements. The voltmeter is converted into an a-c meter by placing a rectifier in the circuit, as shown in Fig. 4-11. The rectifier converts the alternating current to direct current, which is then read on the d-c meter. Different ranges of a-c voltage may be measured by use of the range switch, as was done for the d-c voltmeter. The rectifier is switched in and out of the circuit by a separate switch, one position of which is marked A-C and the other D-C.

Several considerations must be kept in mind when using the voltmeter. When used as a d-c voltmeter, polarity must be observed. There is a positive terminal and a negative terminal. The test leads are color-coded, one red and the other black. The usual convention is to connect the red lead to the positive terminal and the black lead to the negative terminal. It is advisable to clip the black, or negative, lead to the *B*– of the power supply or the chassis, and to tap the red, or positive, lead to points to be tested in the receiver. This latter step should be done with one hand to avoid severe shocks.

When using the instrument as an a-c voltmeter, such polarity need not be observed. Either terminal may be connected to any point. The rectifier takes care of the polarity required by the voltmeter itself.

A final important precaution to remember is that a high voltage, applied across the voltmeter when it is switched to a low-voltage range, will burn out the meter. With an a-c voltmeter, a similar error will burn out the rectifier. Good practice is to switch to the highest range and then to decrease the range by steps until the proper one is attained. Of course, voltmeter readings in the receiver are always taken with the power from the mains turned on. Voltage measurements on a receiver are usually taken with the volume control turned full on, and the tuning dial in an off-station position.

General principles of the ohmmeter. The ohmmeter is an instrument indicating the

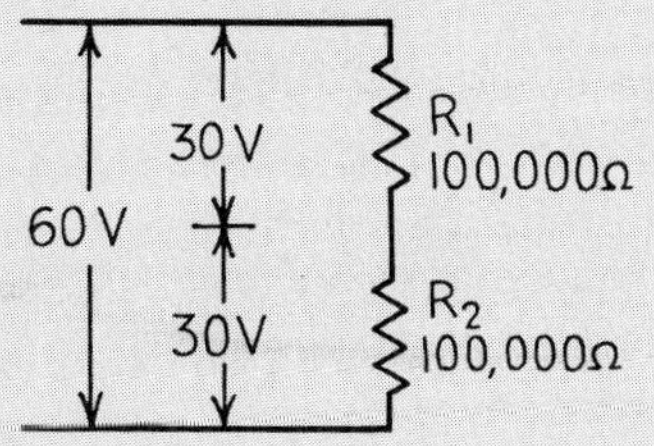

FIG. 4-6. Voltage distribution across two equal resistors.

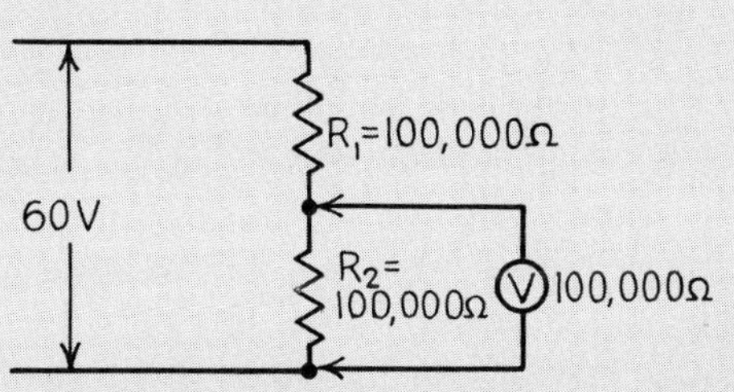

FIG. 4-7. Measuring voltage with a 1,000-ohms-per-volt voltmeter.

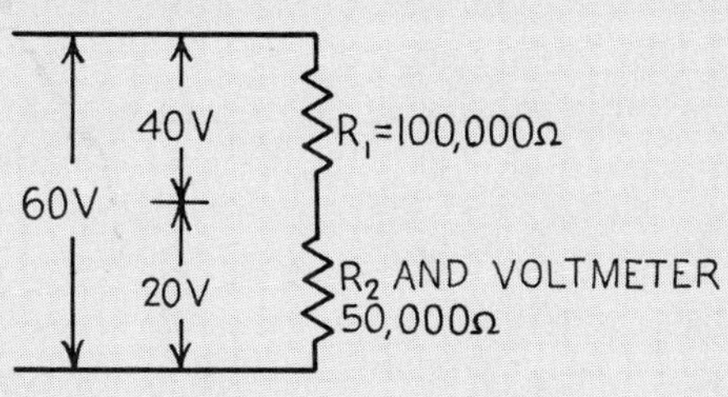

FIG. 4-8. Voltage distribution resulting from loading the circuit with a 1,000-ohms-per-volt meter.

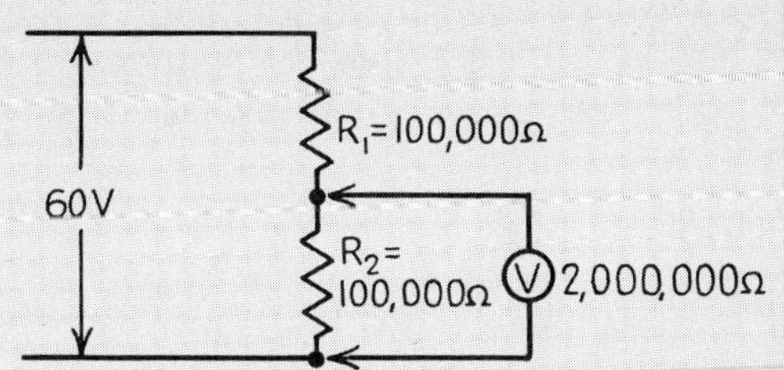

FIG. 4-9. Measuring the voltage with a 20,000-ohms-per-volt meter.

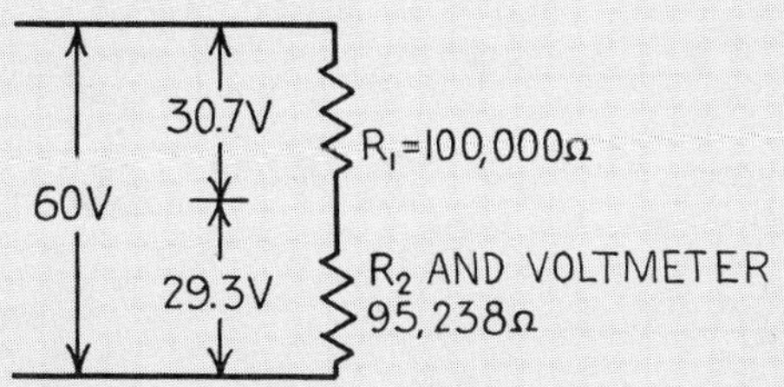

FIG. 4-10. Voltage distribution resulting from loading the circuit with a 20,000-ohms-per-volt meter.

amount of resistance that a component offers to the flow of a direct current. When used to make such measurements in a radio receiver, the power must be shut off if we do not wish to ruin the ohmmeter by placing an external voltage across it.

Basically, the ohmmeter is a milliammeter that requires current to energize it. Since the power in the receiver is off, another driving source of voltage is required. A battery is included in the instrument itself for this purpose. To compensate for any change in battery voltage as time goes on, a zero-adjusting rheostat is included. A basic circuit for an ohmmeter is shown in Fig. 4-12. The component to be measured is placed across the points marked $x-x$ in the figure. If the component has practically no resistance, the milliammeter will be fully deflected. The higher the unknown resistance, the less the amount of current through the milliammeter, and the less the deflection. For this reason, the zero of the ohmmeter scale is at the right and the scale increases toward the left.

Unfortunately, the scale is not linear; that is, the units are not equal. Values of resistance at the upper, or left, end of the scale are very crowded and hard to read. For this reason, a switching device is included to give various ranges. In some ohmmeters, the switch markings and scales present a problem in reading. For this reason, an example in reading would be of great value. Figure 4-13 shows the ohmmeter scale of a typical multimeter, with the meter needle indicating a particular reading. Note that the left end of the scale shows 2K. The letter K stands for 1,000. Unfortunately, owing to previous practice, the letter M is often used for 1,000. This latter practice leads to confusion. For example, 2M ohms equals 2,000 ohms, while 2Mc equals 2,000,000 cycles. It therefore becomes necessary for the technician to interpret the meaning of M in schematics. This book will use K for 1,000 and M for 1,000,000.

The ranges of such an ohmmeter are 0 to 2,000 ohms; 0 to 200,000 ohms; and 0 to 2 megohms. The switch ranges are indicated in either of two ways by multimeter manufacturers. These are shown in Fig. 4-14.

The switch designation in Fig. 4-14*B* is more convenient, since it tells directly by what value the scale reading must be multiplied in order to get the true reading for each range. It is suggested that, if the ohmmeter of the technician has scale indications as indicated in Fig. 4-14*A*, he paste multipliers similar to those at *B* over the ranges.

Now what is the reading if the switch of our meter is at *R*? Here the scale is read directly as 43 ohms, approximately. If the switch is at the $R \times 100$ range, the reading is 43×100, or 4,300 ohms. When the switch is at the $R \times 1{,}000$ range, the reading is $43 \times 1{,}000$, or 43,000 ohms.

A good rule to follow is to select the range

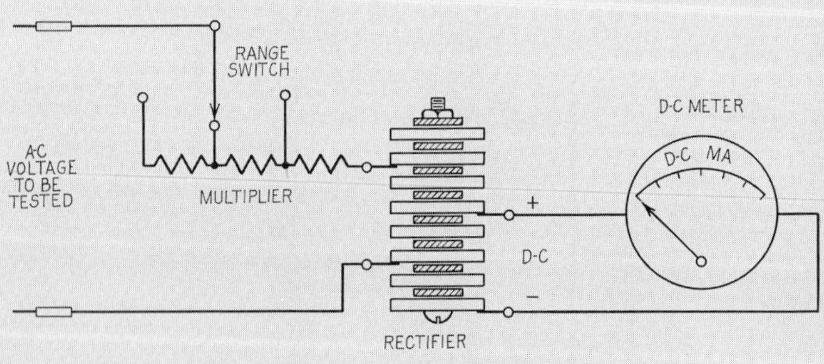

FIG. 4-11. Typical multirange a-c voltmeter.

that gives the resistance reading about the middle of the scale. Of course, since a battery is included in the ohmmeter and is properly polarized, no polarity need be observed when the resistance of components is measured.

A final word must be said about making resistance measurements in the receiver with an ohmmeter. The technician must be sure that there is no parallel branch across the component that he is measuring. Reference to a schematic of the receiver being tested will aid in such determination. When in doubt, disconnect one terminal of the component under test. The technician will also encounter difficulty where an electrolytic capacitor is in parallel with a tested unit. Normally, capacitors are practically infinite in resistance to direct currents. But electrolytic capacitors have a fairly low leakage resistance (from 1 to 50 megohms). The rule to follow, where such is the case, is to measure the resistance of the component, then reverse the ohmmeter prods, and measure again. This is done because the polarized electrolytic capacitor will show less leakage in one direction than in the other. Use the higher of the two readings obtained as the reading for the unit being tested. If there is any doubt, disconnect one terminal of the component, as for parallel resistors.

Electronic volt-ohmmeters. Reference was made previously to the vacuum-tube voltmeter with its high-input impedance. These instruments have in recent years become relatively inexpensive. Their popularity has also increased, since the function of ohmmeter has been added. They are referred to as "electronic volt-ohmmeters." Figure 4-15 shows such an electronic meter, the RCA Junior Voltohmyst.

The principle of operation of the vacuum-

FIG. 4-12. Basic ohmmeter circuit.

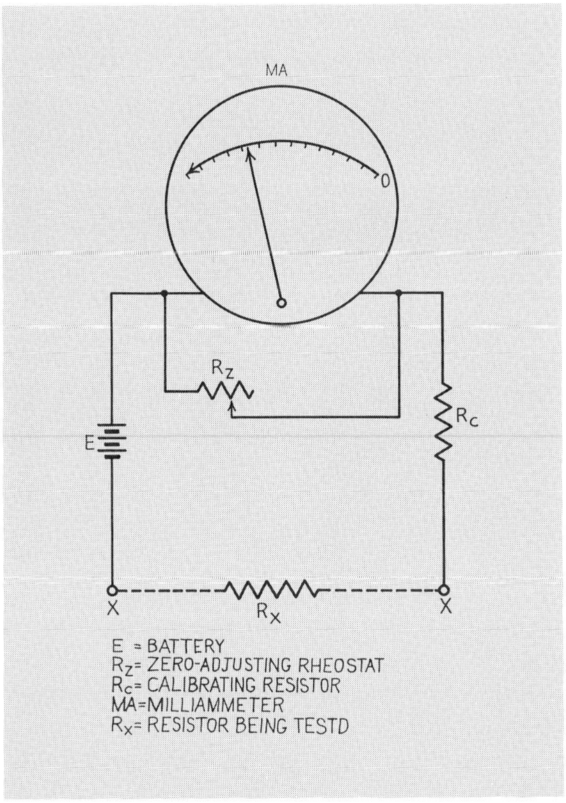

FIG. 4-13. Typical ohmmeter scale.

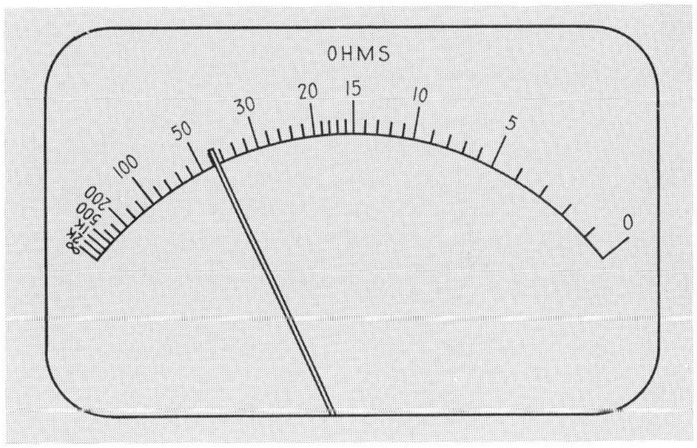

FIG. 4-14. Typical ohmmeter range switches.

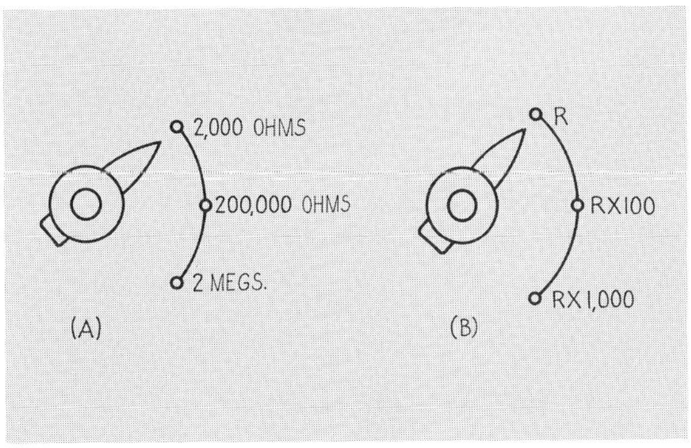

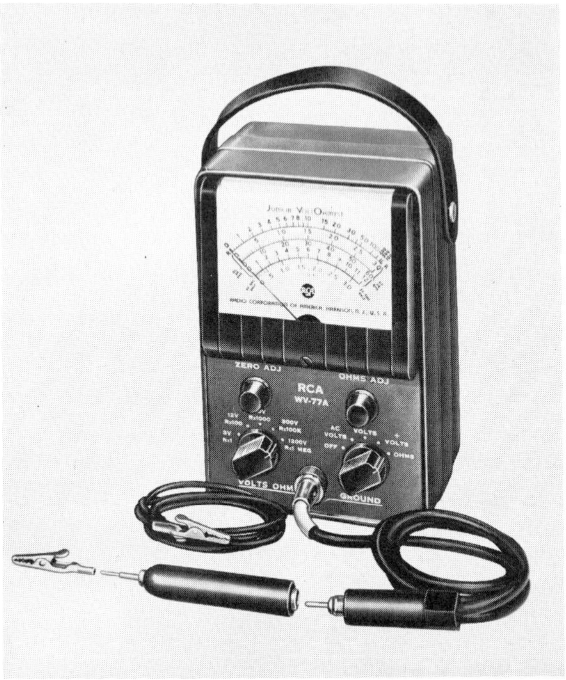

FIG. 4-15. RCA Junior Voltohmyst.

tube voltmeter is fairly simple. Examine the schematic of a basic d-c electronic voltmeter shown in Fig. 4-16. The voltage to be tested is applied across two test prods and is then fed to grid and cathode of a tube. The grid return to cathode is through a high resistance, giving the instrument its high-impedance input. The test voltage applied to the tube grid increases the plate current more or less, depending on the magnitude of the applied voltage. The increase of plate current is then indicated on the d-c current meter. The latter is calibrated in volts necessary to produce the amount of plate current increase. The range switch is similar in function to that in galvanometer instruments.

The electronic meter, then, is basically a d-c voltmeter. To measure low-frequency alternating voltage, a built-in rectifier is switched ahead of the basic d-c instrument, as shown in Fig. 4-17. Usually a diode tube is used for the rectifier. This changes the alternating current to direct current, which is then measured on the d-c vacuum-tube voltmeter.

The capacitance losses of the test leads make the above type of a-c voltmeter unsuitable for testing high-frequency a-c voltage. To overcome this difficulty, many instrument manufacturers supply an auxiliary high-frequency test probe. Here the rectifier (usually a germanium crystal) is enclosed in the probe, making for very short, low-loss r-f leads. The d-c output of the crystal in the probe is then measured by the basic d-c vacuum-tube voltmeter. The probe setup is shown in Fig. 4-18. This special probe enables the instrument to measure a-c signal voltages at frequencies as high as 250 megacycles.

The last function of the electronic instrument is that of an ohmmeter, shown in Fig. 4-19. In this function, the instrument includes a self-contained battery and calibrating resistor R, connected in series with the test prods, as shown in the diagram. The placement of an unknown resistor, R_x, across the prods, produces a closed series circuit, and battery current flows. This develops a voltage across the unknown resistor, of a magnitude dependent on its resistance. This voltage is then measured by the basic vacuum-tube voltmeter.

One of the scales on the instrument is calibrated directly in ohms. Note that the circuit is so arranged that zero ohms is at the left end of the scale, as shown in the top scale of Fig. 4-15. This is opposite to that of a moving-coil ohmmeter.

The general rules previously given for moving-coil meters apply equally to electronic meters. When measuring ohmage, get the reading around the center of the scale by choosing the proper range. Also, be sure that no parallel circuits exist across components being measured. Again, the same precaution exists for both instruments when measuring leakage resistance of electrolytic capacitors. Reverse the test prods, and take the higher reading as the leakage resistance. When measuring voltage, begin on a high-voltage range and work down for a good reading.

Both the moving-coil and vacuum-tube multimeters are extremely useful instruments for the technician. Many prefer the moving-coil type because of its greater mobility and greater stability during use. Others prefer the vacuum-tube type because of its ability to make all measurements, including those of low voltages across high resistances.

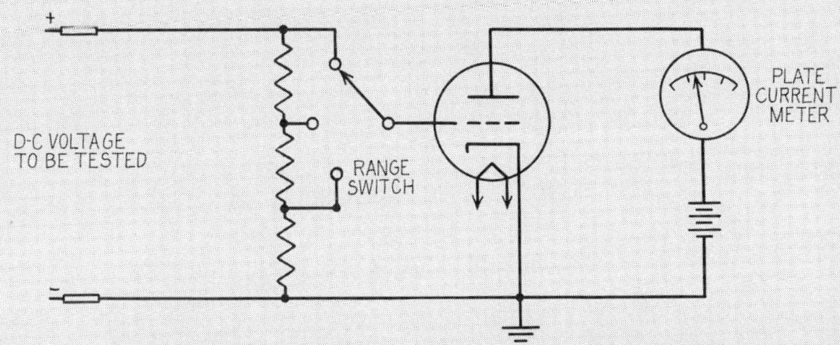

FIG. 4-16. Basic d-c electronic voltmeter.

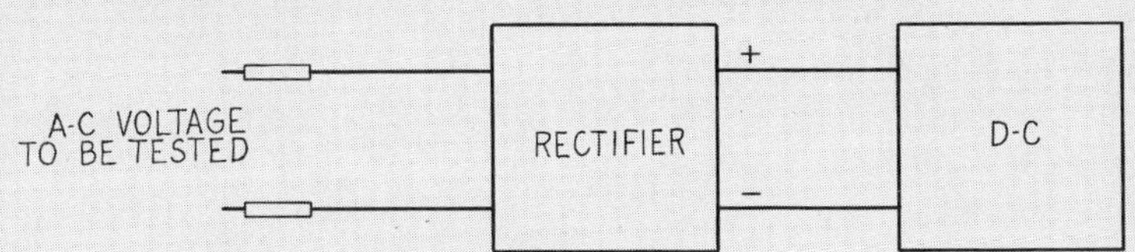

FIG. 4-17. Basic electronic a-c voltmeter.

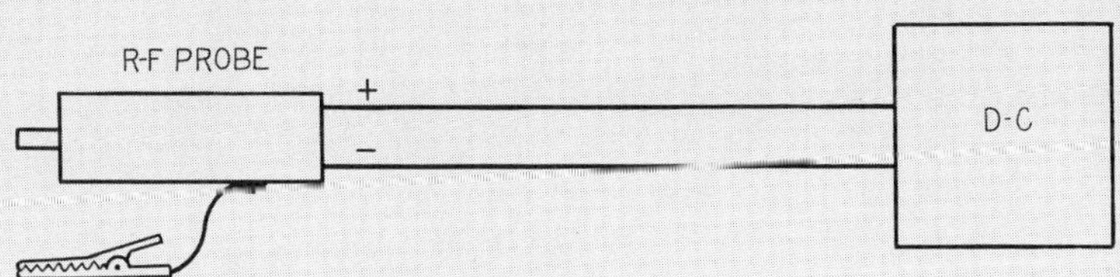

FIG. 4-18. Basic electronic r-f voltmeter.

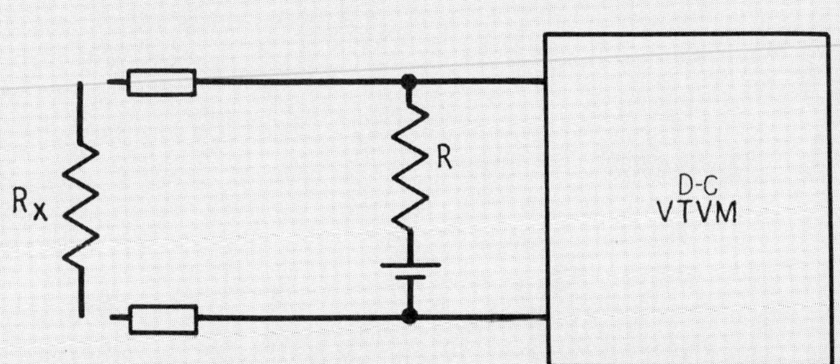

FIG. 4-19. Basic electronic ohmmeter.

Signal Generator— Introductory

5

Fundamentally, the signal generator is a device for placing into the input of a stage a signal similar to that of the input signal, when the set is receiving a station. In this way, it can be determined if a stage is operating normally. By placing the signal from the generator at various strategic points, interstage coupling components can also be tested for breakdown. Finally, the signal generator is an invaluable aid in receiver alignment.

Types of currents. A better understanding of the use of the signal generator will be obtained if time out is taken for a review of the various types of currents. The simplest type is the pure direct current. It is a flow of electrons at a steady rate in one direction through a circuit. Such a current would result from the use of a battery as a power source. The build-up and steady flow of such current could be represented as shown in Fig. 5-1. The fact that the current is steady is shown by the horizontal current line. The fact that the current flows in one direction is shown by the fact that the current line (graph) is always above the zero base line, in the plus direction.

Another type of current is the pulsating or varying direct current. Here, the electrons always flow in one direction but at a *varying* rate. Such a current would result from a varying voltage source or from a varying resistance in the circuit. Figure 5-2 represents the varying direct current resulting in a circuit that includes a flasher button which changes the resistance from that of the lamp alone to an infinite (open) resistance. Notice that the direct current flows only in one direction, as shown by the fact that the graph is always above the base line.

A third important type of current is the pure alternating current. This current continually changes in magnitude and periodically reverses in direction. An a-c generator as a power source would produce such a current, often called a "sine-wave current." Figure 5-3 represents a pure alternating current. That the magnitude is constantly changing is shown by the fact that every point of the current curve is different in value from every point adjacent to it. That the direction of electron flow is regularly changing is shown by the fact that the current curve regularly rises above and dips below the zero base line, first in the plus direction and then in the minus direction.

Alternating and direct currents need not be mutually exclusive. They may be mixed and combined in a single circuit. Figures 5-4 and 5-5 show two such combinations. In Fig. 5-4, a pure direct current from a 3-volt battery and an alternating current (1-volt peak) are mixed in a circuit. The result is a varying direct current, whose average is 3 amp, varying 1 amp above and below the average at the same rate as the alternating current. In Fig. 5-5, two alternating currents from two generators of different outputs and different frequencies are mixed. Sometimes their phase relationships are such as to add to each other; at other times, they oppose each other. The result is the regularly recurring a-c wave form in the diagram that is like neither of the two pure sine-wave components.

Types of alternating currents. Alternating currents present many interesting aspects that require explanation. Refer again to Fig. 5-3. The complete movement of electrons back and forth through the circuit is called one "cycle." The figure shows one cycle completed in 1 second. Hence the frequency of the current through the circuit is said to be one cycle per second. It is possible to have currents of any frequency, even up to millions of cycles per second.

On the basis of different frequencies and therefore use, alternating currents are divided into various categories. The first are the power frequencies, which are the alternating currents used to deliver power to lamps, radios, electrical appliances, etc. The most common frequency in this group is 60 cycles per second. Other power frequencies are 25 and 40 cycles per second.

The second category makes up the audio frequencies (a-f). These are alternating currents of frequencies from 20 to 20,000 cycles per second. They are characterized by the fact that, when fed into a reproducer like a pair of earphones or a speaker, they produce an audible sound.

A third category makes up the radio frequencies (r-f). These are alternating currents of frequencies above 20,000 cycles per second. Currents of such high frequencies have two important characteristics. If fed into a pair of earphones, they will not produce an audible sound. Also, they tend to radiate energy, in the form of radio waves out into space, from the circuit in which the current is flowing.

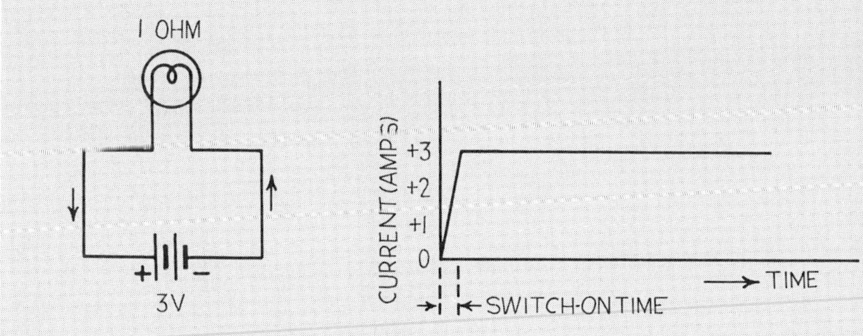

FIG. 5-1. Circuit and wave form for a pure direct current (d-c).

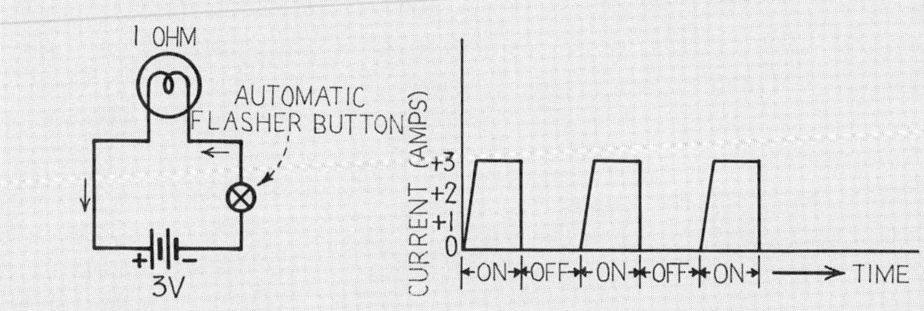

FIG. 5-2. Circuit and wave form for a varying direct current.

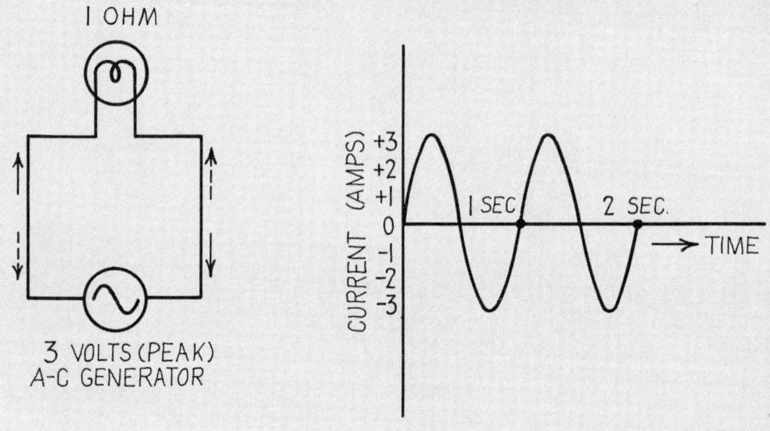

FIG. 5-3. Circuit and wave form for a pure alternating current (a-c).

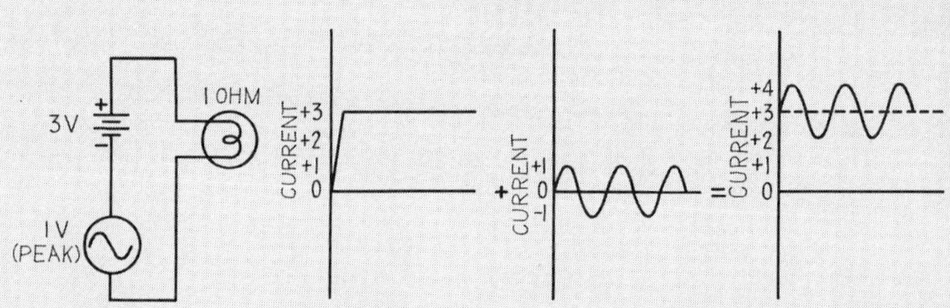

FIG. 5-4. Circuits and wave forms for mixture of d-c and a-c currents.

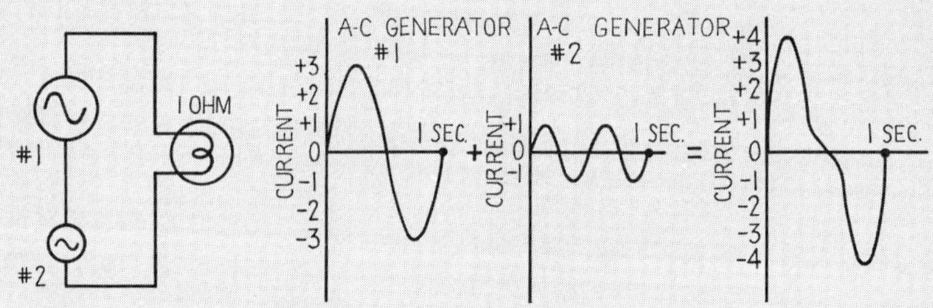

FIG. 5-5. Circuits and wave forms for mixture of two a-c currents.

Audio frequencies. Sound, as it comes to our ears, consists of nothing more nor less than vibrations of the air particles. However, our ears are limited to a relatively small range of vibration frequencies, about 20 to 20,000 vibrations per second. Anything below or above that range will not be heard; within it, different vibration rates will produce sounds of different pitch.

When a sound falls on our eardrums, it causes them to vibrate at the same frequency as that of the sound itself. Similarly, when it falls on a microphone, it sets up vibrations at the same frequency as the sound. A microphone is designed to produce alternating currents at the same frequency as the mechanical vibration produced by the sound. If these alternating currents are amplified and fed into a reproducer, like a loudspeaker, they make it vibrate mechanically at a frequency equal to that of the currents. This mechanical vibration of the speaker makes the air around it vibrate at the same frequency, and the original sound is reproduced. This sequence is illustrated in

Fig. 5-6. If the sound is complex instead of one-frequency, the electrical currents produced will also be complex as a result of the combination of various alternating currents. The end result will be the same.

Radio frequencies. The problem confronted by a broadcasting station is to radiate into space energy that will eventually result in sound at the reproducer of the radio receiver. Unfortunately, a-f currents will not radiate into space to any great extent. When we get up to currents of frequencies above 20,000 cycles per second, the radio frequencies, radiation of energy into space as radio waves becomes efficient. Unfortunately, the radio frequencies will not produce sound at the receiver reproducer.

To obtain the desired results, the sound-producing audio frequencies must be combined with the radiating radio frequencies. In this combination the radio frequency is called the "carrier" and the audio frequency the "modulating currents." The combined current is called a "modulated carrier." This relationship is shown in Fig. 5-7. The carrier is shown as a pure sine current at 1,000 kc (1,000,000 cycles per second). The audio current is shown as a pure sine current at 400 cycles per second. The modulated carrier is an r-f current whose peaks (envelope) vary at the audio rate (400 cycles per second).

This type of modulation of a carrier wave is known as "amplitude modulation" (abbreviated a-m), since the amplitude of the carrier wave is made to increase and decrease at the same rate or frequency as the modulating or audio signal.

Another type of modulation of a carrier wave is known as "frequency modulation" (abbreviated f-m). In this system, the audio signal does not alter the amplitude of the carrier but alters the *frequency* instead, at a rate equal to the frequency of the audio signal. For example, if a 400-cycle audio note were modulating an r-f carrier whose frequency is 88 megacycles per second, the carrier would be made to shift above and below 88 megacycles 400 times each second. A graph of the f-m system is shown in Fig. 5-8.

The branch of f-m receivers is a system by itself. In this book, we shall first describe the a-m superheterodyne receiver. Later chapters will describe the f-m receiver. The signal generator described at this time produces an a-m signal for a-m receivers.

Nature of an electric current. The question of the nature of an electric current should be cleared up at this point. Too much confusion has arisen from comparing different books. About 1765, Benjamin Franklin evolved a theory of electricity that became widely accepted. He believed that electricity (whatever it was) flowed in an electric circuit. By convention, he and many others assumed that electricity flowed from the + pole to the − pole. This conventional current flowing from + to − was described in technical literature for many years after, and still leads a virile life.

However, in 1897, J. J. Thomson discovered

FIG. 5-6. Basic sound system.

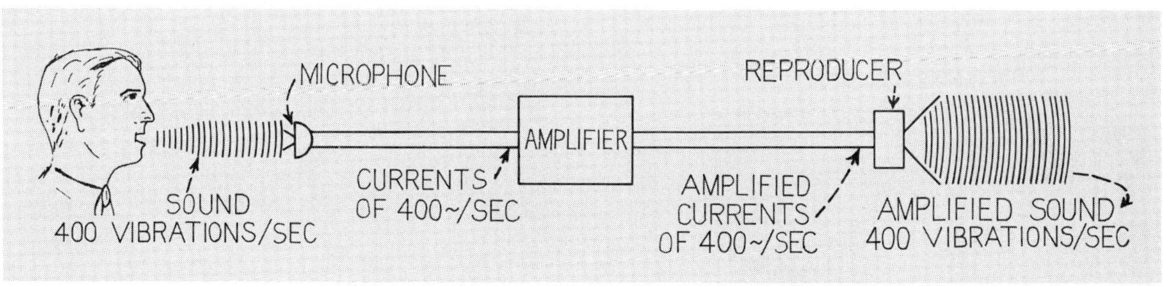

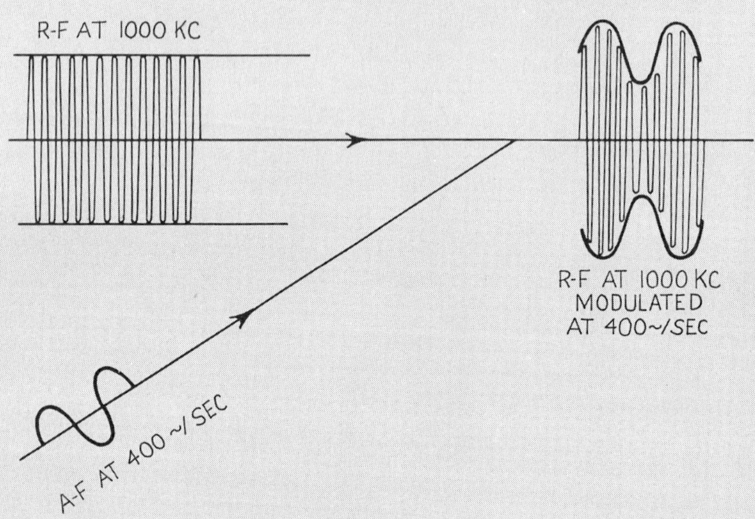

FIG. 5-7. R-f carrier (1,000 kc) modulated by 400-cycle audio note.

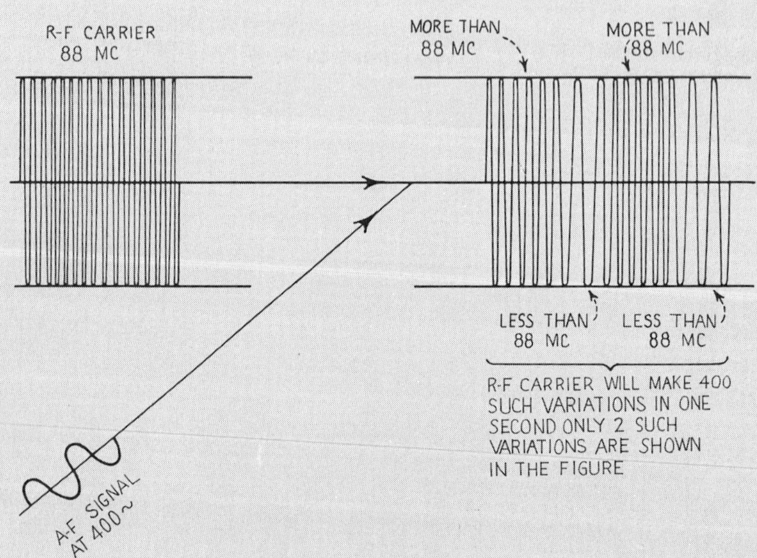

FIG. 5-8. R-f carrier (88 megacycles) frequency-modulated by 400-cycle audio note.

the electron, and the true nature of an electric current in a circuit became known. An electric current is the flow of negatively charged electrons through a circuit. Hence, the electrons must always flow from − to +, an idea opposite to that of the conventional theory.

The confusion arises because many authors do not define which concept they have in mind when referring to current. As a result, many beginning students confuse the two ideas and erroneously assume that when we say current flows from + to − (Franklin's convention), we mean that electrons flow from + to −. On the contrary, when we say current flows from + to −, we should forget all about electrons. Franklin did not know that they existed when he adopted that convention. When we say current flows from − to +, we are up to date and talking about electrons. Throughout this book, the authors will use the newer concept of the current; a flow of electrons from − to +.

Signal-generator output. The description given above will make the signal output from the signal generator more meaningful. Figure 2-1 shows the block diagram and wave forms of the superheterodyne receiver. Various types of currents are encountered. Modulated radio frequency enters the aerial and produces modulated r-f currents up to the mixer. The local oscillator produces pure unmodulated r-f currents. From the mixer to the detector stage, modulated r-f currents at a lower frequency (called "modulated intermediate frequencies," or i-f) are encountered. From the detector to the reproducer, the signal is at audio frequencies.

It is the function of the signal generator to generate all of the above current types to simulate regular receiver signals for testing. Figure 5-9 shows the output voltages and currents obtained from most generators.

The unmodulated radio frequency of the signal generator is an alternating current or voltage of a frequency anywhere above about 75,000 cycles per second (usually written 75 kc). Any frequency above that lower limit is selected by means of the various controls. Audio frequencies are alternating currents or voltages ranging from about 20 up to 20,000 cycles per second. Most signal generators have a fixed-frequency audio output of about 400 cycles per second, which is the standard test frequency. Another important output from the signal generator is a mixture of the radio frequency and the 400-cycle audio. This is known as "400-cycle modulated radio frequency." It simulates a modulated r-f radio signal. Means are often provided for mixing the r-f with an external a-f signal. This gives an output on the signal generator known as "externally modulated radio frequency." The r-f output signal is amplitude-modulated in either case.

FIG. 5-9. Output voltages and currents from a signal generator.

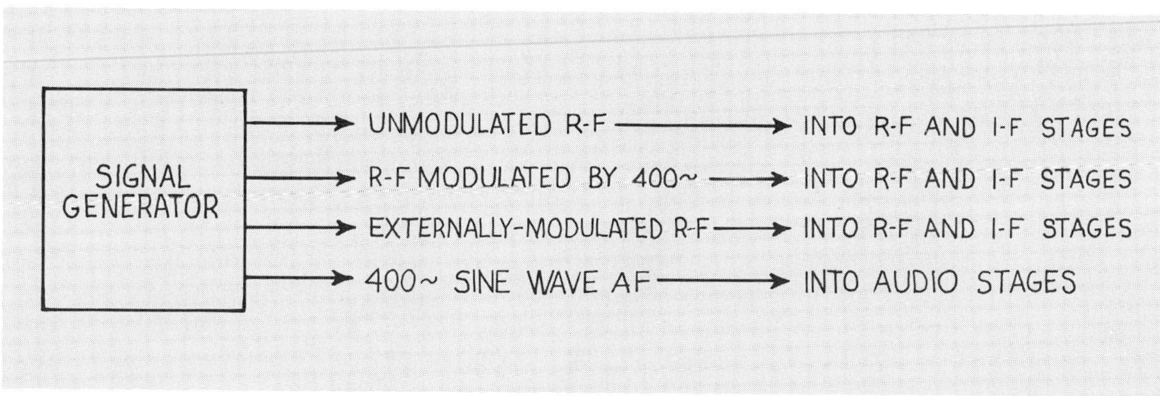

Setting Up the Signal Generator

6

Block diagram of the signal generator. There are various differences in detail between one signal generator and another; basically, they are very similar. A block diagram will show to best advantage the elements that make up an average signal generator (Fig. 6-1).

The r-f oscillator generates an r-f voltage with a range of about 75 kc to 60 megacycles. This range includes the intermediate frequencies of any standard receiver. The output from the oscillator itself is unmodulated.

The a-f oscillator, as its name implies, generates a voltage at an audio frequency, which is usually the audio test frequency of 400 cycles per second. On some signal generators, the audio output is variable from approximately 100 to 10,000 cycles per second. The a-f oscillator is used to modulate the r-f voltage generated by the r-f oscillator. In addition, most signal generators provide front-panel terminals where the a-f output is independently available. This independent a-f output may vary in voltage up to several volts. It is used to check the a-f stages in the receiver.

The modulation switch shown in Fig. 6-1 enables the operator to modulate the r-f with the a-f signal. The usual practice is to have 30 percent modulation at an audio modulating frequency of 400 cycles. The 30 percent modulation means that the r-f voltage is made to dip and rise 30 percent below and above its peak value, as shown in Fig. 6-2. Many signal generators make provision for modulating the r-f voltage with an external a-f signal of any frequency.

The strength of signals at various test points throughout the receiver will vary greatly, beginning at the antenna and ending at the loudspeaker. Since the signal generator must substitute signals comparable to the actual signals, it must have a great range of output. This function of variable output is taken care of by an attenuator that breaks the complete range of output into steps and then gives smooth variation within each step. For the most part, the output readings obtained from the attenuator primarily furnish a value to any setting of the output, rather than give an exact microvolt output for radio-servicing procedures. Later chapters in this book will make this statement more significant, especially in stage-gain measurements.

Up to this point, the description of the signal generator has been generalized to give an overview picture. A more detailed discussion of the actual controls will give greater skill with the instrument. Of course, there is great variation in the control designations. Some common ones will be described and should be sufficient to aid the technician in understanding any other variations. The manufacturer's instructions for all signal generators should serve as the final guide for specific operation.

A typical signal generator. To get a better understanding of the various signal generators in existence today, it might help to synthesize a typical front panel of such an instrument and study its controls. Of course, there probably is no generator that has this exact make-up. Figure 6-3 shows the signal generator that would be constructed. On the left center is found the POWER switch to energize the signal generator when it is to be used. On the right center is the OUTPUT jack from which the various outputs for application to various test points in the receiver are taken.

To determine the nature of the output, there is an OUTPUT SELECT switch for obtaining pure r-f, modulated r-f, or audio signals. This instrument is of the usual fixed a-f type with an audio output at 400 cycles per second. Therefore, when the OUTPUT SELECT switch is in the MOD. R-F position, the output is an r-f signal modulated approximately 30 percent by a 400-cycle audio note.

The entire r-f coverage is accomplished by the large tuning dial in the center. This frequency range of r-f output is quite large and could not be covered in one sweep of the tuning dial. Therefore, a band selector switch (BAND SW.) is provided to divide the complete coverage into bands. The complete swing of the tuning dial will therefore cover only one band. Four distinct bands are shown in our typical signal generator. They are labeled *A*, *B*, *C*, and *D*, each with a different range. Figure 6-3 shows band *C* chosen for coverage.

The output level is controlled by the two dials marked MICROVOLTS and MULTIPLIER. The first of these controls gives the number of microvolts from 0 to 10. It is usually a potentiometer control. The second is a 5-point switch for a step attenuator and determines by what value to multiply the reading from the MICROVOLTS dial to get the output level. The multiples shown are 1, 10, 100, 1K (1,000), and 10K (10,000). For example, the reading shown in Fig. 6-3 would be 6 × 1K, or 6,000 microvolts. The caution given in the previous section about the true value of this reading should be kept in mind.

The general information given above is important because the technician should see in what ways all signal generators are alike. However, each specific instrument will have its own variations, and the service manual supplied by the manufacturer should serve as the guide. The next section will describe a good commercial signal generator, to show how the controls should be operated to get the various outputs and output levels that are required in service work.

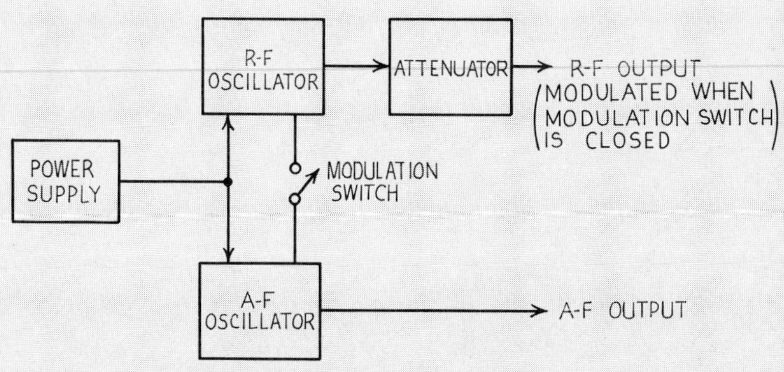

FIG. 6-1. Block diagram of an a-m signal generator.

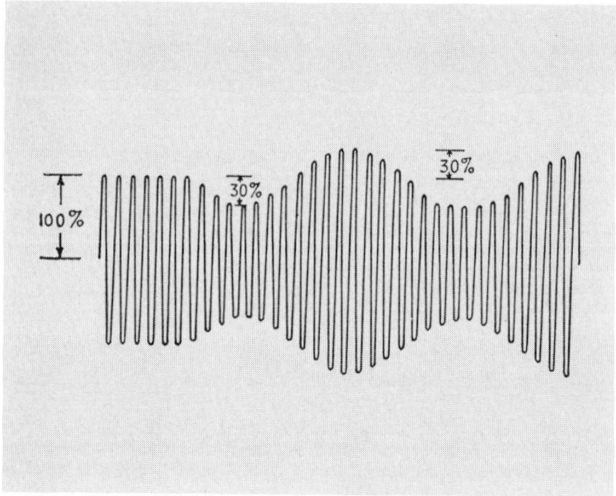

FIG. 6-2. A 30 percent modulated r-f signal.

The Precision E-200-C signal generator. In the Precision signal generator, the usual tuning dial is found in the upper center part of the front panel (see Fig. 6-4). Frequency coverage from 88 kc to 110 megacycles is performed in eight bands, indicated as *A, B, C, D, E, F, G,* and *H*. The BAND SELECTOR switch is located at the lower left end of the panel. The frequencies covered by each band are as indicated below.

A: 88–230 kc
B: 220–600 kc
C: 550–1700 kc
D: 1.60–5.0 mc
E: 5.0–15.5 mc
F: 15–29 mc
G: 29–55 mc
H: 55–110 mc

R-f output is taken from two jacks above the BAND SELECTOR switch. When large output is desired, the jack labeled HIGH is used; when low output is required, the jack labeled LOW is used. From these two jacks are obtained either unmodulated r-f signals or r-f signals that are modulated by the audio oscillator signal.

The type of output is determined by the setting of the control at the lower right end of front panel. The settings of this dial are R-F UNMOD., MOD. R-F, EXT. MOD., and 400 ~ AUDIO, giving unmodulated R-F, modulated r-f, externally modulated r-f, and 400-cycle audio, signal, respectively. The audio signal for the last-named position is obtained from two jacks labeled AUDIO SIGNAL under this control.

The level of the audio output is determined by the setting of a control at the upper right end of the panel. This is labeled MODULATION CONTROL. The setting of this dial also determines the percentage modulation of the r-f signal when the output type control is in the MOD. R-F position. The a-f output is very high, sufficient to operate a high-impedance speaker directly without an intervening amplifier.

Attenuation of the r-f output signal is accomplished by two controls at the upper left end of the panel. They are labeled R-F CONTROL–1 and R-F CONTROL–2. Each of these dials is arbitrarily divided into 10 main units. R-F CONTROL–1 delivers increasing outputs at each position as the knob is turned clockwise. The outputs in these various positions are not calibrated but are relative. R-F CONTROL–2 is a decimal multiplier. Thus, if the first dial is in position 3 and dial 2 is in position 7, it means that 7/10 of the total available output for position 3 of dial 1 is available. If dial 2 is turned to position 9, it means that 9/10 of the maximum available output for position 3 of dial 1 is delivered. If dial 2 is at position 10, then 10/10, or all, of the available output for position 3 of dial 1 is available. To get more output, return dial 2 to zero and set dial 1 in position 4. The greatest available output is delivered when dial 1 is at position 10 and dial 2 is also at position 10. In other words, dial 1 sets the limit of output and dial 2 tells us how many tenths of that limit are being delivered. Note, again, that the two dials give no actual output reading but merely arbitrary positions for any output obtained.

A final control on this signal generator is one marked AGC CONTROL. It determines the level of steady agc voltage delivered to two

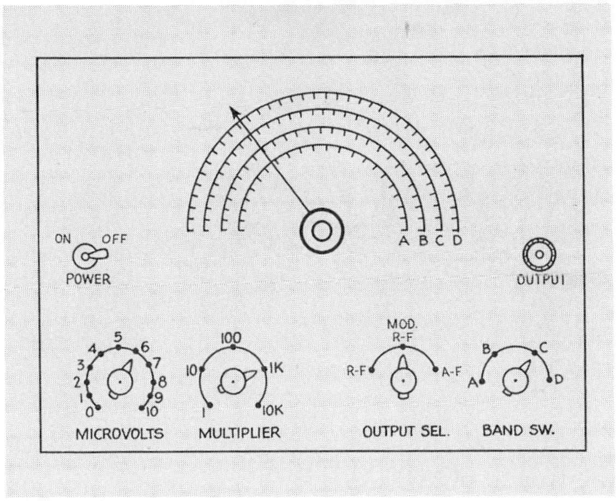

FIG. 6-3. Front panel, showing controls of a typical signal generator.

jacks marked AGC VOLTAGE beneath it. This agc voltage is used for checking agc operation in receivers, and in aligning receivers with agc control.

Checking signal-generator calibration. It is important that the frequencies of the signal generator should be accurately calibrated and regularly checked. To make such a check, it is necessary to have a standard for comparison that is accurate. The frequencies of the broadcast stations are valuable in this respect, since each station is assigned a fixed carrier frequency from which it deviates to a negligible degree.

It is not necessary to check the frequency calibration of the signal generator all over the dial. In radio service work a few test frequencies are important. These are 455, 600, 1,000 and 1,500 kc. The instrument will be extremely useful if these frequencies are accurately determined on the dial.

Let us see how we could make the check suggested above. Suppose that it is desired to see if 600-kc output from the generator is obtained when the frequency dial is set at 600. The output lead from the instrument should be connected through a 0.00025-mfd/600-volt capacitor to the antenna of a broadcast receiver. The generator ground and receiver ground should be commonly connected to a good ground.

If there is a station whose carrier frequency is exactly 600 kc, the check will be quite simple. We first tune our receiver sharply to that station. Then set the output selector switch of the signal generator to unmodulated r-f output. As we tune the frequency dial close to 600 kc, a high-pitched whistle is heard. This effect is due to a phenomenon known as "beats." For example, if the signal generator were producing an output at a frequency of 605 kc, it would mix with the station signal of 600 kc and produce a beat note of 5 kc—the difference between the two signals. Since 5 kc is in the audio frequencies, it would be heard in the receiver as a whistle. As the generator output approaches the station frequency, the difference becomes less, producing a lower and lower pitched sound in the speaker, since the beat frequency becomes less. When the two frequencies are identical or very nearly so, the beat note tends to disappear. At that position we have tuned for zero beat. As we tune the frequency dial past zero beat, we again begin to get the beat note. At zero beat, we could safely assume that the signal generator is at the same frequency as the station; namely, 600 kc.

It is not always possible to find a broadcast station with the exact frequencies that we wish to check. Such would be the case in the metropolitan New York area. Suppose the technician in that vicinity wanted to check 600 kc on his signal generator. The nearest stations to that frequency are WMCA at 570 kc and WVNJ at 620 kc. To check the signal generator at 600 kc, tune it for zero beat with WMCA, the station to which the receiver is sharply tuned. At that position, the output of the generator is 570 kc. Suppose its tuning dial reads 560 kc. We can then assume that it is 10 kc off and that therefore an output of 600 kc would be obtained when the generator tuning dial is at 590 kc. To verify, tune for zero beat with WVNJ at 620 kc and note whether it too is 10 kc off in the same direction.

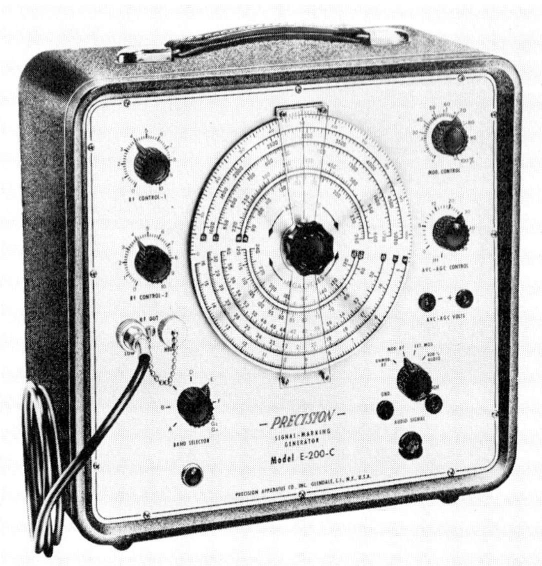

FIG. 6-4. The Precision signal generator, E-200-C.

Similarly, tuning-dial positions on the generator should be found for 1,000 kc and for 1,500 kc. The stations to use for 1,000 kc might be WJRZ at 970 kc and WINS at 1,010 kc. The stations to use for 1,500 kc might be WHOM at 1,480 kc and WQXR at 1,560 kc.

Determining the true setting for 455 kc requires a different analysis, because it is outside the broadcast band. At first, it would seem impossible to check until we realize that, when a signal generator oscillator is set at 455 kc, it is not only producing an output of 455 kc or thereabouts but also whole-number multiples thereof. Therefore, there would be concurrent signals at frequencies of $455 \times 2 = 910$ kc, $455 \times 3 = 1{,}365$ kc, $455 \times 4 = 1{,}820$ kc, etc. These simultaneous multiple signals are known as "harmonics." The fundamental frequency of 455 kc is often known as the "first harmonic," 455×2 as the "second harmonic," 455×3 as the "third harmonic," etc. Now, if we use the second harmonic of 455, or 910 kc, we find that it falls in the broadcast band. Therefore, set the signal generator up as before, but tune on the band including 455 kc. The two stations for comparison near 910 kc are WCBS at 880 kc and WJRZ at 970 kc. If we are tuning for zero beat with WCBS, our generator tuning dial should be at 440 kc, since we are using the second harmonic. If we obtain zero beat at 445 kc, the signal generator is off 5 kc. An output of 455 kc will then be obtained at a dial position of 460 kc. Again, this fact should be verified by beating the second harmonic of 485 kc from the signal generator with station WJRZ at 970 kc.

A special precaution is required when checking calibration in the i-f band. If the check receiver employs an i-f amplifier tuned to 455 kc, a confusing double beat may be obtained, since the signal-generator output may beat with the signal in the i-f amplifier as well as with the test station. However, if the receiver is equipped with an r-f stage and an i-f wave trap, there is little likelihood of the signal generator's output beating with the signal in the i-f amplifier, and it may be used. Another way of avoiding this effect is to use a receiver whose i-f amplifier is tuned to a frequency quite different from the signal being tested, such as an auto radio at 260 kc. Furthermore, a TRF receiver, if available, could be used for calibration purposes, since it has no i-f amplifier.

The proper settings for the important test frequencies should be recorded in some manner by the technician for later use. The same technique may be used for regions other than the metropolitan New York area by similarly choosing local stations close to the test frequency points.

Signal-Generator Applications 7

Uses of the signal generator. Throughout this text, various purposes will be served by means of the signal generator. First, the instrument will be used to determine if a stage and its associated coupling circuits are functioning properly. By placing the "hot" lead at various points in the radio receiver, this fact can easily be determined. This system of servicing is known as the "signal substitution" method and will receive more elaboration throughout the text.

Another use to which the signal generator may be put is that of receiver alignment. For most receivers brought into the technician's shop, this will not be a usual procedure. Where alignment is necessary, it is advisable to follow instructions given by the radio manufacturer. However, a generalized procedure will be given for those cases where the manufacturer's notes are not available.

A third use of the signal generator is to determine if each stage is giving proper gain. In this respect, a standard output will be measured by means of an output meter. Then the settings of the output of the generator will be compared with those necessary for each stage on a known good receiver, to obtain the abovementioned standard output.

How to connect the signal generator to a receiver. The output from the signal generator is fed to the receiver being tested through a coaxial cable or a shielded connector cable. In either case, the external conductor is grounded within the generator and the center, or hot, lead is connected to the receiver test points. The hot lead is usually coded red, and the ground lead is either black or bare braiding.

Both the signal generator and the receiver should be at the same ground potential. This condition may be obtained by connecting the ground lead of the signal generator to the receiver chassis, which in turn should be connected to a good ground. In a-c/d-c receivers, where the chassis is connected directly to one side of the power line, there is danger of a short circuit in following this direction. This danger may be overcome by connecting a capacitor of about 0.1 mfd/400 volts in series with the ground lead.

Where the hot lead is to be connected to an inductor like an antenna coil, it is advisable to use the Institute of Radio Engineers (I.R.E.) standard dummy antenna in series with the lead. This is shown in Fig. 7-1.

Under normal circumstances in using the signal generator for signal substitution service work, it is necessary only to connect a capacitor in series with the hot lead. This prevents high d-c potential points of the receiver from ruining the test instrument. In each case, the manufacturer's instructions should be fol-

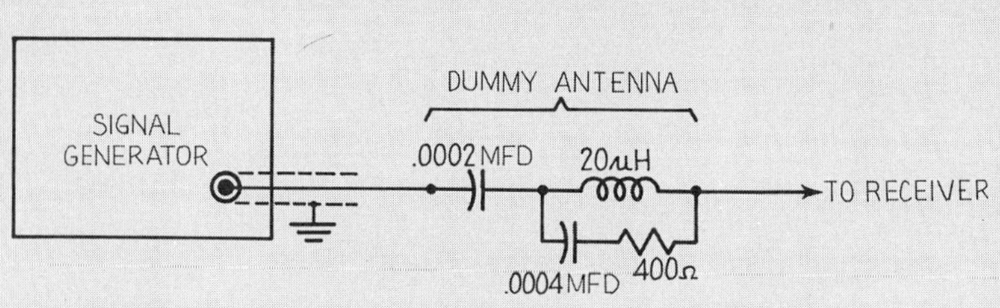

FIG. 7-1. The I.R.E. standard dummy antenna, connected to signal generator and receiver.

lowed. Generally, a 0.1-mfd/600-volt capacitor should be used where i-f and a-f signals are delivered to the set. Where r-f signals are delivered to the receiver, a 0.00025-mfd/600-volt capacitor may be used. When short waves (high-frequency r-f signals) are fed to the receiver, a 400-ohm resistor is used.

Signal-substitution method of servicing. The signal generator, as used through the remainder of this book, will primarily concern itself with signal substitution for servicing receivers. At various test points in the receiver it will introduce a signal, similar to the one received in normal broadcast reception, and the results will be observed. Where observed results are not normal or typical, trouble is indicated.

A brief description will serve at this time to set down the outline of testing to check that each stage is operative. Figure 7-2 shows a simplified diagram of a superheterodyne with strategic points indicated by the ballooned numbers. Above each number is indicated the type of signal input for testing the applicable stage. The sequence of the numbers is the order in which to make the test.

Point ① tests the speaker itself. The test cannot be made unless a signal generator with a high level of a-f output is available. Where such is the case, the audio note should be heard in the speaker.

Point ② checks the operation of the second a-f stage, once the speaker has been found to be in good shape. Because of the stage amplification, a lower level a-f signal is required at the input. If operation of the stage is normal, the audio signal should be heard clearly.

Point ③ is the test point for operation of the first a-f stage, if the preceding tests check perfect. Once again a lower level a-f input signal is required. Normal operation would result in a strong, clear audio note in the speaker.

Point ④ is the test point for operation of the detector stage. It should be remembered, as always, that all previous checks have shown proper stage operation. A modulated i-f signal introduced at this test point should produce a clear modulation note in the speaker. The intermediate frequency, of course, is that for the particular receiver.

Point ⑤ is the test point for the i-f amplifier. A modulated i-f signal from the signal generator, at the i-f for the particular receiver, should produce a clear modulation note in the speaker. The level of this signal input should be less than that for point ④, because of the gain of the i-f amplifier.

Point ⑥, the signal grid of the mixer, is the test point for the mixer and oscillator. A modulated r-f signal injected at this point should produce the modulation note in the speaker if the oscillator and the mixer are both operative. If no note is heard, then introduce a modulated i-f signal at this point. If the note is now heard, then the mixer is functioning and the

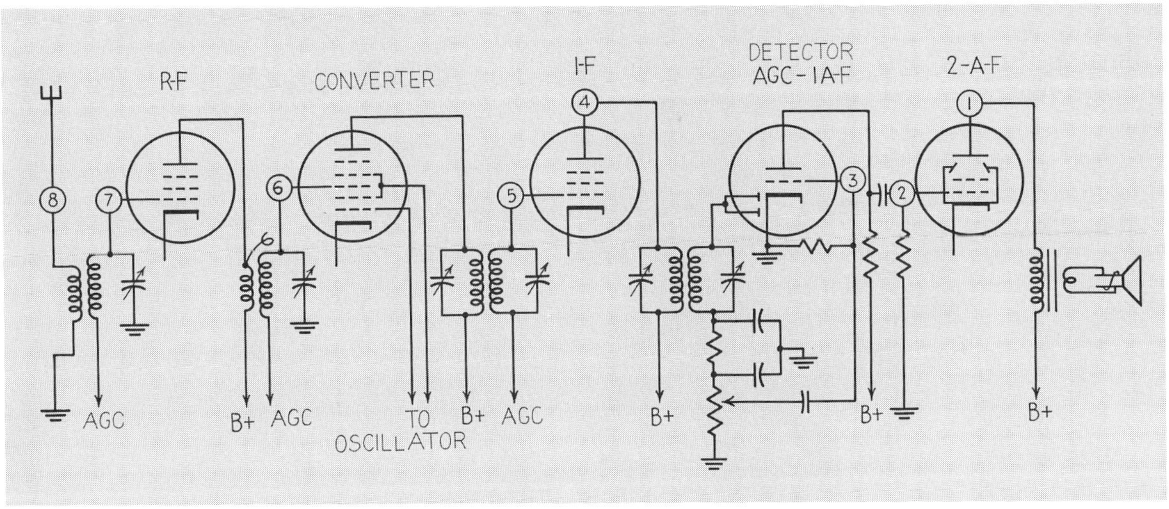

FIG. 7-2. Signal chain of a superheterodyne receiver showing test points.

oscillator may be assumed to be inoperative.

Point ⑦ is the grid of the r-f amplifier tube. A modulated r-f signal is introduced at this point to check the operation of the r-f stage. Again, it should require less input signal at point ⑦ than was needed at point ⑥, the converter grid, because of the gain of the r-f tube.

Point ⑧ is the test point for the antenna coil. A modulated r-f signal at a lower level than for point ⑦ should produce a clear modulation note in the speaker, if all else is well.

The check procedure presented briefly here will be elaborated in the stage analyses given later in the book. It should be noted that, where coupling devices are to be checked, introduction of the proper signal at the input and the output of the coupling device should produce modulation notes in the speaker. If the note is heard at the output but not at the input, then the device or its associated circuit is presumed to be defective.

Using the signal-substitution method of servicing. An example of how to use the signal-substitution method in localizing a defect will make clear its value. Refer to the receiver whose schematic is shown in Fig. 7-3. We assume a defect and try to localize it. Suppose i-f trimmer capacitor *C*-14 is shorted. The receiver is brought in with the complaint that it does not work.

Voltage analysis will not disclose the defect, because the d-c resistance of parallel coil *L*-6 is quite low, and the d-c voltage drop across it is very small. Ohmmeter analysis of the receiver would be too lengthy if used by itself.

Let us proceed by the signal-substitution method. An audio signal from the signal generator is delivered to the signal grid of the output tube. It is heard clearly in the speaker. This stage is considered to be all right. The audio signal is then introduced to the grid of the type 14B6 tube. Again the audio note is heard in the speaker and the first audio amplifier is assumed to be good. A modulated i-f signal is now introduced on the signal grid of the i-f amplifier. The modulation note is heard clearly in the speaker and the detector, and i-f stages need no further investigation. Now, when a modulated i-f signal is introduced on the signal grid of the type 14Q7 converter, the modulation note is not heard. This indicates that the trouble is between the converter signal grid and the i-f amplifier grid. Then a modulated i-f signal is introduced on the plate of the converter, and still no modulation note is heard. This localizes the defect between the plate of the converter and the signal grid of the i-f amplifier. Thereafter, a simple ohmmeter check across the primary and the secondary (*L*-6 and *L*-7, respectively) of the first i-f transformer will show the short across *L*-6.

Receiver alignment. The average superheterodyne receiver has seven or more tuned circuits, each one of which has to be in resonance at its proper frequency for best operation of the receiver. The procedure for bringing these circuits to resonance at their operating frequencies is called "alignment."

The signal generator is an invaluable tool in receiver alignment, since it is used to feed the proper aligning frequency to each circuit. The procedure consists essentially in connecting an output-measuring device across the speaker, which is the output of the receiver; feeding a voltage at the proper frequency to the circuit being aligned; and adjusting the variable component, usually trimmer capacitors provided for the purpose, to a maximum deflection of the output meter.

Alignment is necessary when one of the components of any tuned circuit becomes defective and is replaced. Alignment will also perk up a receiver where, owing to natural aging of the components with time and moisture, the tuning-circuit parts change in value.

FIG. 7-3. A Stromberg-Carlson a-c/d-c receiver.

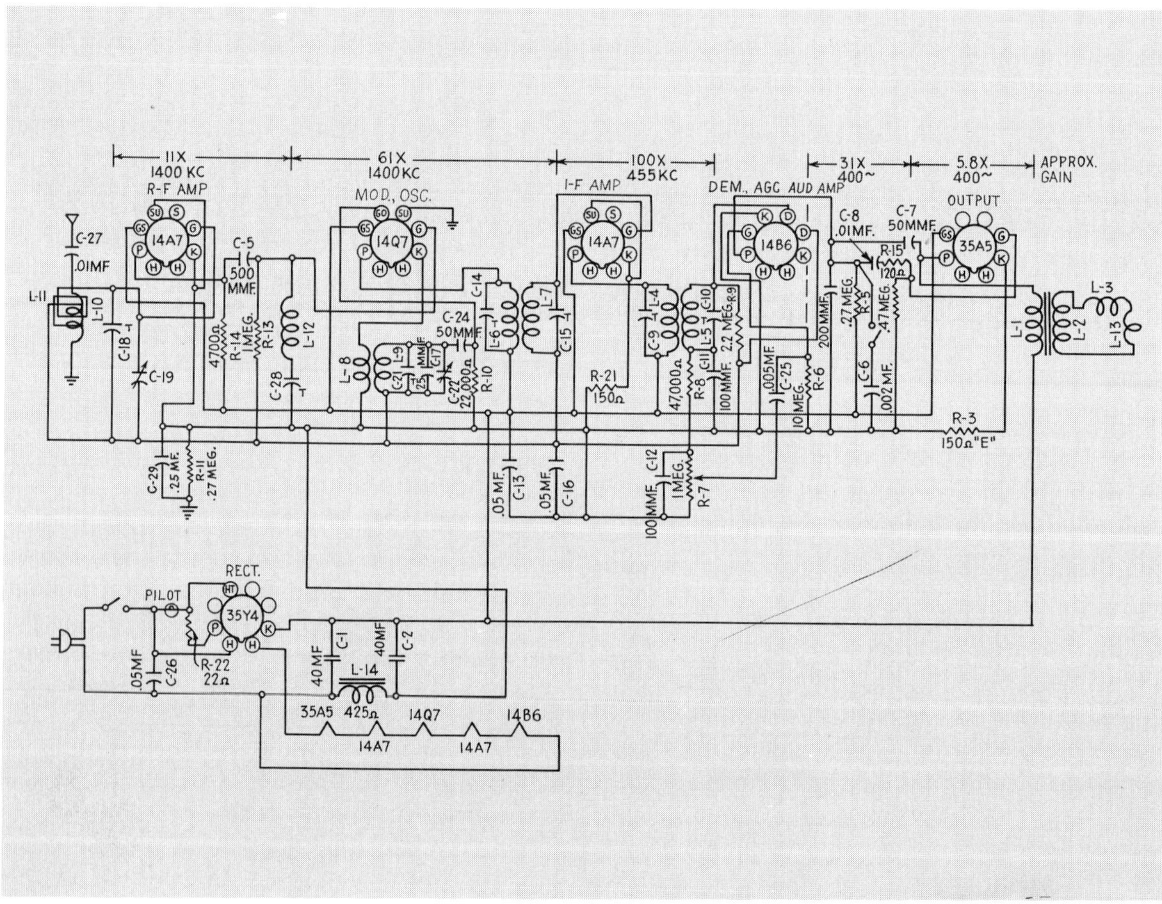

Stage-gain measurements. In a superheterodyne receiver, each stage, except the diode detector, amplifies the signal before it passes it on to the next stage. When the technician has an idea of the approximate amplification or gain that may be expected from each stage and is equipped to measure it while making a signal check of the receiver, he has a powerful service tool for quickly determining the location of many troubles.

For example, assume an open cathode bypass capacitor in a stage of a receiver that is perfect in all other respects. The receiver would produce a weak output. In servicing such a receiver by the old methods, tubes would check good, voltage measurements would be normal, and a routine ohmmeter check would also show nothing. The technician would then proceed to substitute parts, more or less at random, until he came to the defective capacitor.

With the aid of stage-gain measurements, he would be examining the defective stage in a matter of minutes. Although he would still be confined to the substitution of parts, he would be doing so for the components of only one stage found to be defective.

Accurate stage-gain measurements, as made in engineering laboratories, would require a considerable outlay in the matter of test equipment. However, for servicing purposes, great accuracy is not necessary since the offending stage will usually be far below normal when the receiver is brought in as defective. Adequate stage-gain measurements can be made with the equipment that the technician has on hand—a signal generator and an a-c voltmeter.

The theory underlying stage-gain measurements is quite simple. The receiver is held at all times during the check at one output, known as "standard" output. A signal from the generator is fed into the input of a stage, and the voltage of that signal, necessary to produce standard output, is noted. Then the signal is fed into the output of the stage. The voltage level of the signal is increased until standard output is again obtained. By dividing the second voltage by the first we obtain the gain of the stage. This sequence is illustrated in Fig. 7-4.

Let us take an example to illustrate the point. If 1 volt of signal at the input of a stage gives standard output, and the signal level must be increased to 10 volts to maintain the standard output when it is connected to the

FIG. 7-4. Sequence of measurements to obtain the gain of a stage.

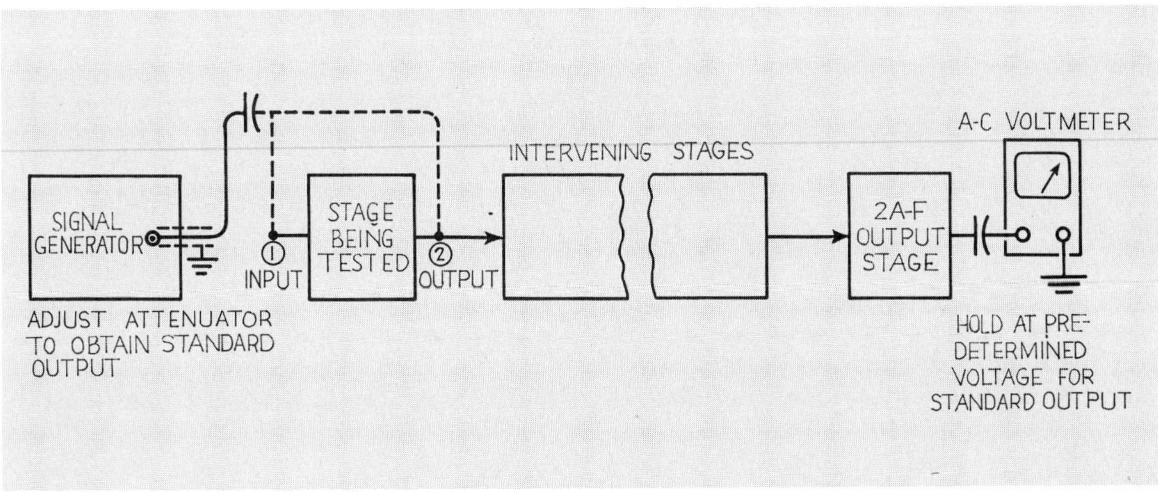

output of the stage being tested, then the gain of the stage is 10 to 1, or 10.

The standard output used in stage-gain measurements has been set by the I.R.E. at 50 mw of signal power fed into the speaker. The output power may be measured by connecting an a-c voltmeter across the speaker voice coil or, more conveniently, across the primary of the output transformer. In stage-gain measurements, the signal input level is adjusted to keep the output meter at the proper fixed value. This value corresponds to approximately 16 volts across the output transformer primary for most receivers. During stage-gain measurements, the agc system must be inoperative, or it will invalidate results. For this reason, the receiver output is maintained at the low level of 50 mw so that input signals necessary to attain that level will be too weak to activate the agc system.

The measurement points in the receiver for stage-gain checking are usually taken from one grid to the next. The amount of signal necessary to give standard output from any point in the receiver is often called the "sensitivity" of the receiver from that point on. When a signal of 3,500 microvolts is required at an i-f amplifier grid to give standard output, the sensitivity of the receiver at the i-f amplifier grid is said to be 3,500 microvolts.

For the practical technician, exact sensitivity measurements are not necessary. Comparative sensitivity measurements will serve as well. These may be obtained by actually making sensitivity measurements from various points in receivers known to be in perfect operating condition. In each case, the attenuator reading of the signal generator necessary to give standard output should be recorded. When completed, the readings for each point are averaged. As a result, the technician will have comparative data for determining proper sensitivity from various points for any receiver brought in. For example, if the attenuator position varies greatly at the grid of the i-f amplifier of an unknown receiver from the average setting just obtained, a defect in the i-f amplifier stage is indicated, if all later stages check perfect.

On the average, the sensitivity of radio receivers from various points may be summarized in the accompanying table. The diode detector is omitted because its purpose is not amplification but rather demodulation.

Sensitivity, average input	Generator frequency set at	Generator hot lead connected to	Output from the receiver
5–12 microvolts	600 kc	Antenna terminal	Standard
50 microvolts	600 kc	Modulator grid	Standard
3,500 microvolts	455 kc (or other I-f)	I-f grid	Standard
0.032 volt	400~	First A-f grid	Standard
1.6 volts	400~	Second A-f grid	Standard

After having obtained the attenuator setting at various points to give standard output, the technician may assume that the input values are those given in the table. Thereafter, he may make due allowance if he has service literature from the receiver manufacturer giving sensitivity at various points. For example, if the service data indicate that, for a particular receiver, the sensitivity at the i-f grid is 3,000 microvolts to give standard output, he knows that he must turn the attenuator up to give less than his comparative output, which is presumed to be 3,500 microvolts.

Stage-gain measurements are readily obtained from sensitivity measurements. Suppose that the signal generator delivered an output of 50 microvolts to the converter signal grid to develop standard output. The sensitivity of the receiver from that grid would be 50 microvolts. Now, suppose that the generator delivered an output of 3,500 microvolts to the grid of the next i-f amplifier to develop standard output. The sensitivity of the receiver from the i-f grid would be 3,500 microvolts. The gain of the converter stage would then be found by dividing the latter sensitivity by the former. It is found to be 3,500/50, or 70.

Gain per stage varies in different receivers; therefore a small range of figures rather than a

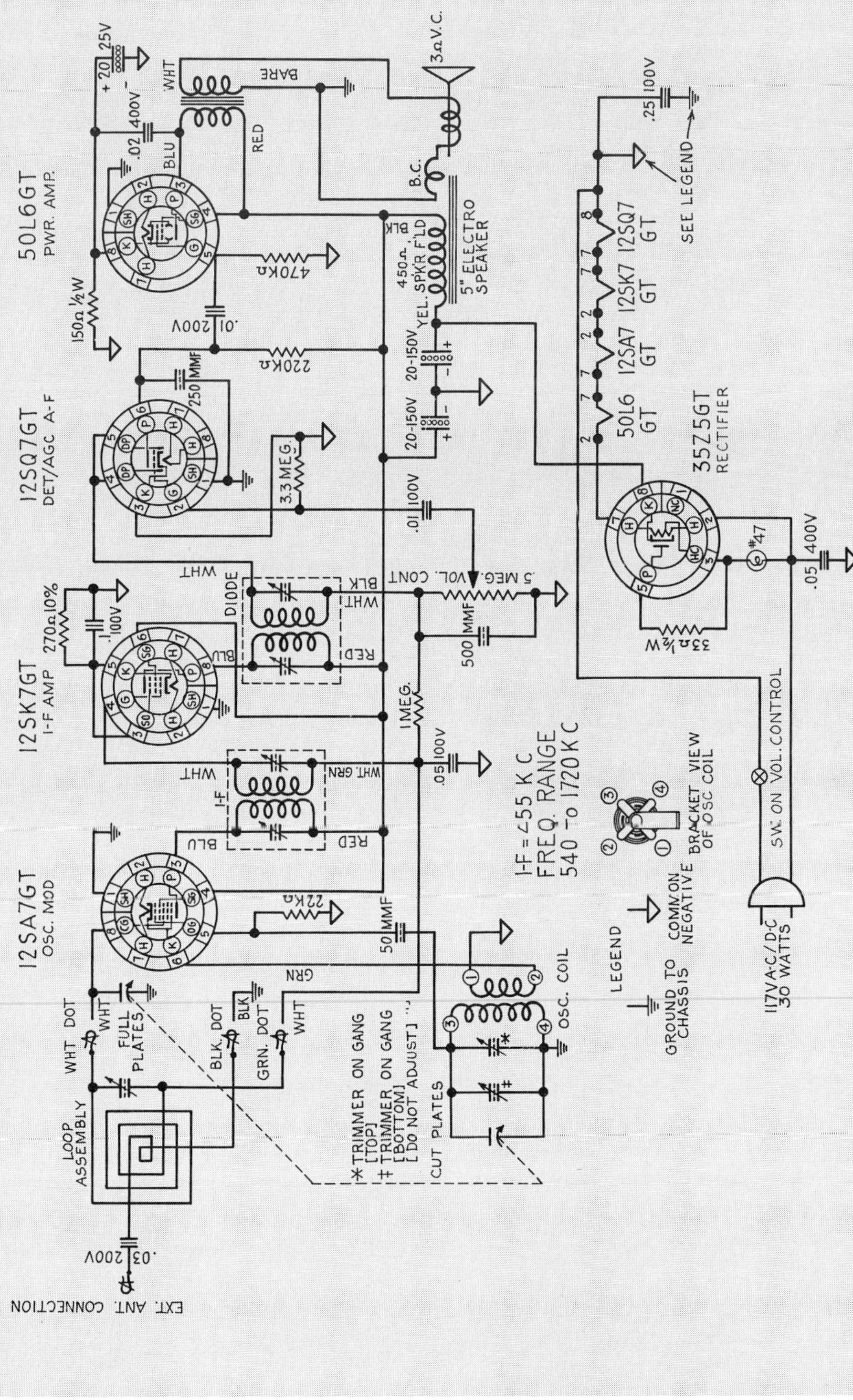

FIG. 7-5. A Motorola a-c/d-c receiver.

single figure would be desirable for comparative work. The accompanying table lists the various stages of a superheterodyne receiver, gives the test frequencies to the input of each, the ranges of gain for many receivers, and an average gain used in this book. For specific receivers, gain data furnished by the manufacturer in his service notes should be followed, if available.

Stage	Test frequency	Range of gain	Average gain
Second *A-f*	400~	5–15	10
First *A-f* (high-mu)	400~	40–60	50
I-f	455 *kc*	80–120	100
Converter	600 *kc*	60–80	70
R-f	600 *kc*	20–40	25

Examination of the service notes of a typical receiver will now show the value of this stage-gain technique. Figure 7-5 shows the schematic diagram for the receiver. Service notes given by the manufacturer give the data shown in the accompanying table. The dummy antenna capacity indicates values to be connected in series with the hot lead of the signal generator. In each case, the input signal is given which results in standard output. From the data given, it is seen that, from antenna to modulator grid (at the same modulated r-f frequency), there is a voltage gain of 55/15, or approximately 3.7. From modulator grid to i-f grid (at the same modulated i-f frequency) there is a voltage gain of 3,700/50, or 74. Any wide variations from these gain measurements would result in an indication of a defective stage.

Average microvolt input	Generator set at	Generator feeder connected to	Dummy antenna capacity
3,700	455 kc	I-f grid	0.1 mfd
50	455 kc	Modulator grid	0.1 mfd
55	600 kc	Modulator grid	0.1 mfd
15	600 kc	Antenna terminal	400 ohms

Another method of indicating stage gain is shown in Fig. 7-3. Here, stage gain is indicated between specified points. Beneath the stage-gain value is indicated the frequency to which the signal generator must be set in making the check.

The data may be analyzed as follows. The level of input signal from the signal generator at a modulated 1,400-kc frequency should be 11 times as great at the signal grid of the converter tube as it is at the r-f tube signal grid, to give standard output. This means that there is a voltage gain of 11 due to the amplification of the r-f tube. The level of input signal at a modulated 1,400-kc frequency should be 61 times as great at the signal grid of the i-f amplifier as it is at the signal grid of the converter tube, to give standard output. The level of input signal at a modulated 455-kc frequency at the detector plate should be 100 times as great as it is at the signal grid of the i-f amplifier, to give standard output. The level of input signal at 400 cycles per second should be 31 times as great at the signal grid of the output tube as it is at the signal grid of the first audio amplifier, to give standard output. And finally, the level of input signal at 400 cycles per second should be 5.8 times as great at the plate of the output tube as it is at the signal grid of the same tube, to give standard output.

When service notes do not include sensitivity figures or stage-gain data, the technician can still use these ideas for working on receivers with inadequate volume or sensitivity. As was stated previously, he should establish average attenuator settings on his signal generator based on results with several good receivers. For example, when checking from the i-f grid, the average attenuator setting turns out to be 20×100 to give standard output. Then, on a defective set, up to the i-f grid, normal attenuator settings produce standard output. But at this point, the attenuator has to be advanced to 80×100 for standard output. He then knows that a defective condition exists in the i-f stage.

A-C Transformer-Type Power Supply

8

Quick check. If all the tubes in the receiver light, there is no sign of overheating, the hum level is normal, and the $B+$ voltage measures 200 to 300 volts, the power supply is probably functioning properly, and the trouble shooter proceeds to check the next stage.

Function of power-supply stage. The power supply furnishes *A*, *B*, and *C* voltages for the rest of the receiver. The *A* supply lights the filaments of the tubes, the *B* supply furnishes the necessary d-c voltage to operate the plate circuit of the tubes, and the *C* supply furnishes d-c grid voltage for the tubes.

The power-supply stage can be a set of batteries, as is the case in portable and emergency equipment. Usually, the lighting mains are employed to furnish the power. The power-supply stage, therefore, converts the 110-volt lighting supply into the necessary *A*, *B*, and *C* voltages for the receiver.

Two main types of power supplies will be considered: the transformer-type power supply for use on a-c mains, and the so-called a-c/d-c type which permits receivers to be plugged into either a-c or d-c mains. The a-c/d-c power-supply stage will be treated in a later chapter.

Theory of operation of a-c power supplies. The basic parts of the power supply are shown in the block diagram of Fig. 8-1.

The power transformer, by stepping voltage up and down, supplies high voltage for the rectifier in the *B* supply, and low voltage for the tube filaments. The low-voltage windings of the power transformer are all that is needed for the *A* supply.

The rectifier allows current to flow only in one direction. Its output, therefore, is pulsating direct current.

The filter circuit smooths the pulsating direct current from the rectifier into unvarying direct current, for use as the *B* supply.

Standard circuit. See Fig. 8-2.

Functions and values of component parts. Transformer *T*-7 is the power transformer. It operates on the principle of electromagnetic induction. Current in the primary sets up a magnetic field in the iron core. Since the primary current is alternating, the magnetic field is constantly changing in magnitude and direction: building up, collapsing, building up in the opposite magnetic direction, collapsing, etc., with each change in the alternating current. A changing magnetic field induces voltage in any winding that is exposed to it, and the greater the number of turns, the greater will be the induced voltage. At this point, the inability of transformers to operate on direct current can be easily seen. Direct current sets up a steady magnetic field, and voltage will not be induced in the windings.

Power transformers for radio work are usually designed to operate at 2 to 4 turns per volt.

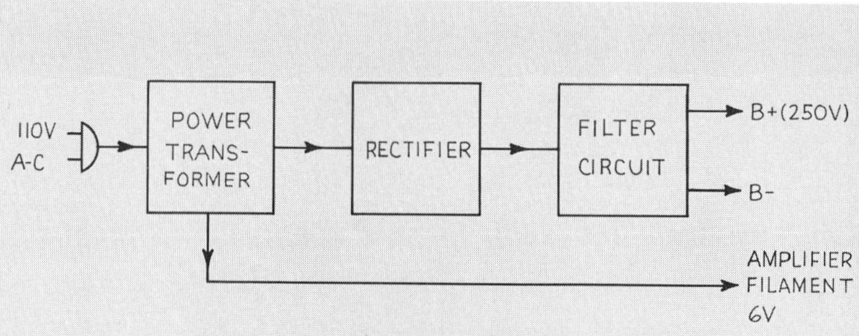

FIG. 8-1. Block diagram of a transformer-type a-c power supply.

Assume a 2-turns-per-volt transformer. Then the 120-volt primary will be wound with 240 turns. (Although the lighting mains are usually called "a 110-volt line," line voltage will actually measure more nearly 120 volts. Design work assumes a line voltage of 117.) Each 2 turns of secondary winding will have 1 volt induced in it. The 5-volt winding for the rectifier filament will be wound with 10 turns, and the wire will be comparatively heavy to carry the 2 amp that the rectifier filament draws. The high-voltage winding, usually 540 volts, will be wound with 1,080 turns. This will be fine wire, since the radio requires only about 70 ma (0.07 amp) of *B* current. Remember that 540 volts is dangerous. Care must be exercised in handling and measuring the high-voltage leads.

The filament winding for the other tubes in the receiver will be wound with 12 turns for 6 volts, and the wire will be heavy enough to carry the current drain of several tubes.

The high-voltage winding is always center-tapped for use in the full-wave rectifier circuit. The other windings are sometimes also tapped: the primary at the 220th turn, for use

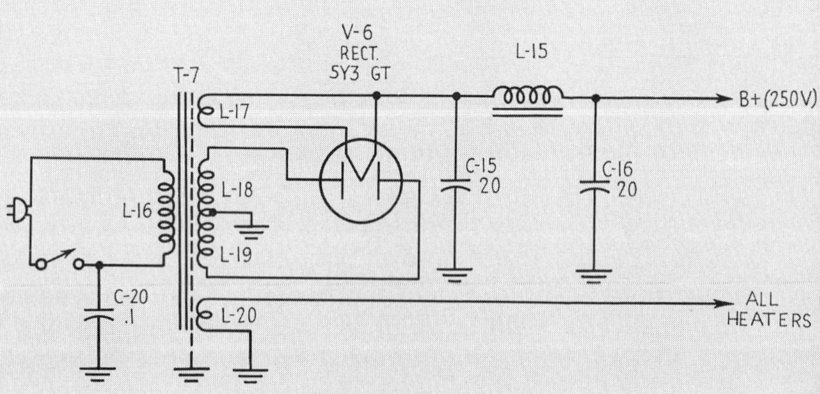

FIG. 8-2. Standard circuit of a transformer-type a-c power supply.

in areas where line voltage is low. The amplifier and rectifier filaments may also be tapped in the center.

Small receivers generally have fewer tubes and therefore draw less current than larger receivers with more tubes. Transformers for these smaller receivers will be correspondingly smaller in physical size. Although the secondary voltage ratings may be the same for both types, the heavier current drain of the larger radios requires heavier wire in the windings, resulting in an increase in overall size.

The rectifier is a conventional full-wave circuit. Tube *V*-6 is a 5Y3GT. In large receivers, where the plate current drain is heavy, the rectifier is a 5U4GB or a 5AR4. The full-wave rectifier, operating from a 60-cycle-per-second source, delivers 120 pulses per second to the filter circuit.

The filter circuit consists of *L*-15, *C*-15, and *C*-16. *L*-15 consists of a large number of turns of wire, wound on an iron core. Its action in the filter circuit is that of an inductor or choke. An inductor acts to retard any change in current through it in the following way. Any change in current will produce a change in the magnetic field. The changing magnetic field will induce voltage in any winding exposed to it, as it does in the case of the transformer. In the case of the choke, where there is only one winding, the voltage will be induced in that winding. Since the induced voltage is opposite in direction to the original source, it will always tend to oppose any change in current in the coil due to the varying magnetic field. The choke, therefore, has a high opposition to any change in current (alternating current or pulsating direct current), while its opposition to direct current (unchanging magnetic field) is comparatively low. Since the choke is connected in series with the power-supply output circuit, it tends to keep pulsations out of the output.

Capacitors *C*-15 and *C*-16 are connected across the power-supply output, one on each side of the choke. The action of a capacitor in a circuit containing pulsations is to stabilize the voltage across it. When the voltage across a capacitor is exceeded by the momentary peak from the rectifier, the capacitor charges and absorbs the peak. During the lull between peaks from the rectifier, when the voltage would drop, the capacitor discharges and maintains the voltage. Capacitors *C*-15 and *C*-16 are high-capacitance, high-voltage electrolytic capacitors. Often they are in the same container, which is called a "filter-capacitor block." A common size would be labeled "20–20 mfd-450 volts D-c–Surge voltage 525." Sometimes the block contains three capacitors, such as the one pictured in Fig. 8-3.

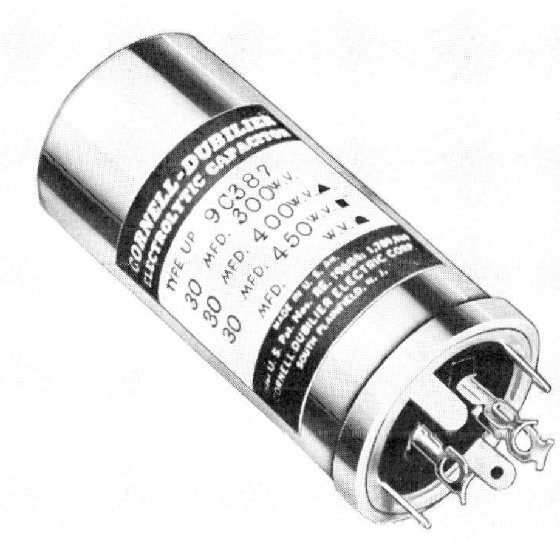

FIG. 8-3. A filter-capacitor block.

Switch *S*-1 is the on-off switch for the radio. It is often ganged with the volume control. Switch replacement notes will be found together with volume control replacement notes in Chap. 13 on the first a-f stage.

Capacitor *C*-20 is the line filter. Its action is to remove various r-f line disturbances, such as those caused by sparking brushes on electric motors, from entering the radio. The value of *C*-20 is not critical. Values ranging from 0.002 to 0.5 mfd are found in various radios.

NORMAL TEST DATA FOR THE POWER-SUPPLY STAGE

Check for normal stage operation.

All tubes light or heat.

No sign of overheating.

Voltage check, *B*+ to chassis, 200 to 300 volts.

Hum level—normal.

Most receivers normally have a slight hum, since it is rather costly to remove the last traces. This is known as "residual" hum, and the technician must have some way of determining whether the amount present is normal or excessive. A good check is to place the ear close to the speaker with no station tuned in. If the hum is just discernible, call it normal. This small amount will not be objectionable when the ear is at its usual distance from the speaker and a station is tuned in. If noises from the r-f amplifier interfere with the test, the r-f end of the receiver can be made inoperative by removing the i-f amplifier tube. If the test is being made with the speaker out of its cabinet, as is usual at the bench, the technician should remember that the cabinet baffle accentuates low-frequency response and, since 120-cycle hum is low-frequency, he should allow accordingly.

If the quick check indicates trouble in the power supply, disconnect the line plug and, before proceeding to further tests, discharge the filter capacitors by shorting them. The filter capacitors may retain a charge, with subsequent danger of shock or damage to test equipment.

Normal resistance data. Normal resistance data are given in the accompanying table.

Plug, prong to prong	2–15 ohms
Chassis to rectifier plates	35–200 ohms
Rectifier filament to *B*+, across choke *L*-15	40–400 ohms
Chassis to rectifier filament	100,000 ohms

The normal reading of 2 to 15 ohms across the prongs of the plug is a measure of the resistance of the primary winding of the power transformer. The range of resistance is caused by the different sizes of transformers. Large sets with large transformers draw heavy current through windings with heavy wire. Such transformers will show a primary resistance of about 2 ohms. Smaller transformers with finer wire drawing lower current will range up to 15 ohms.

The reading from the chassis to the rectifier plates measures each half of the high-voltage winding. Again, the larger transformers will show the lower reading. Similarly, when measuring choke *L*-15, the lower reading may be expected in sets with a larger current drain.

The reading from ground to rectifier heater is given as 100,000 ohms. This reading shows the leakage resistance of filter capacitors *C*-15 and *C*-16. When making this test, reverse the ohmmeter prods after taking the reading and read again. The higher of the two readings should be taken as the correct one.

Normal voltage data. Normal voltage data are given in the accompanying table.

Rectifier filament to filament	5 volts a-c
Across other tube heaters	6 volts a-c
Chassis to rectifier plate	200–350 volts a-c
Chassis to rectifier filament	220–385 volts d-c
Chassis to *B*+	200–360 volts d-c

Small receivers tend toward the lower *B* voltages. Large receivers tend toward the higher *B* voltages. The measured voltage from chassis to rectifier plate is the RMS or effective value. The rectifier voltage, measured from chassis to rectifier filament, is usually a little higher than the a-c input owing to the action of capacitor *C*-15, which maintains the rectified voltage at more nearly the peak value.

COMMON TROUBLES IN THE POWER SUPPLY

All the component parts in the power supply are common sources of trouble. Even the rectifier-tube socket is not immune. In the case of the socket, dirt between the rectifier plate pins causes the high voltage to arc across, burning up the socket material. This is found by inspection, and the cure is obvious: replacement of the socket. The power transformer should be carefully checked, since the heavy drain may have damaged it.

Troubles common to power transformers. The power transformer develops many ills, the chief cause of which is overheating due to overloads within the transformer or to external shorts. The ohmmeter check is not entirely reliable. For example, a few shorted turns in the high-voltage winding will not affect the ohmmeter reading to any great extent, while it will cause a heavy drain from the primary and consequent overheating. In a case like the above, even though the voltage would be considerably reduced, the radio would keep on playing, and it might not be brought in for repairs until the overload had caused the primary finally to open or the owner had become concerned about the smell from his radio. Incidentally, the smell from a burned transformer is unmistakable, and the technician need only follow his nose to the trouble. When the trouble has been determined, it is wise to check for external shorts before replacing the transformer. As an example of the necessity for this, assume a partial short in the dial-light wiring of a radio. The radio continues to play, and finally the overload causes the transformer primary to open. The technician quickly finds the open transformer; replaces it; checks the radio, which appears to operate satisfactorily; returns it to the customer; and, before long, the new transformer is burned owing to feeding current to the partial short that is still in the dial-light wiring.

FIG. 8-4. The power transformer.

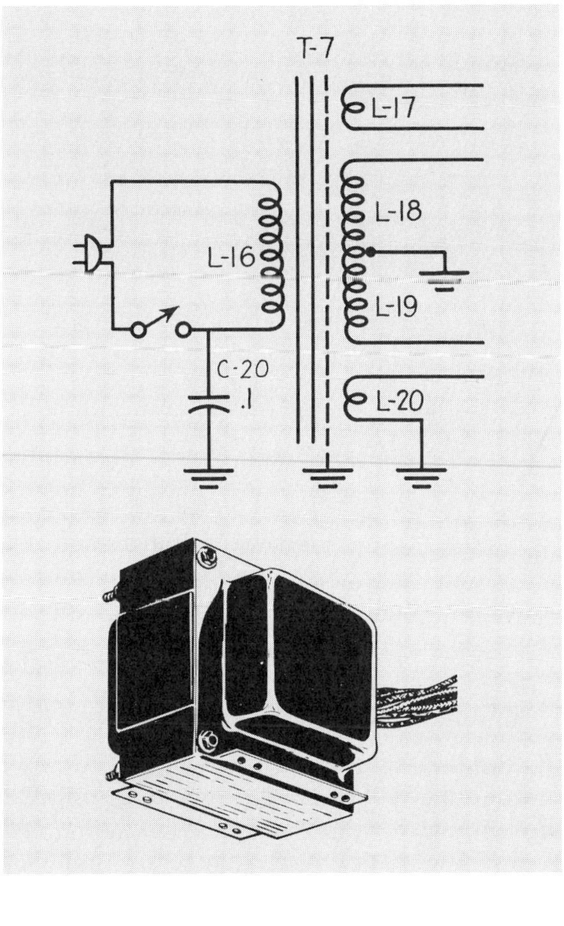

How to check the power transformer. The best check for normal operation of the power transformer is a wattmeter, or a-c ammeter, connected in the primary circuit. The technician's multitester, however, rarely includes scales and ranges that are suitable for this purpose. A good check with inexpensive equipment can be made as follows:

1. Remove all tubes from the radio.
2. Plug the radio into an outlet that contains an ordinary 25- or 40-watt lamp in series with the line, as shown in Fig. 8-5.
3. A good transformer will cause the lamp to glow dimly.
4. Any short that is present will cause the lamp to glow brightly.
5. If a short is present, remove the transformer secondary leads from their connection points, one winding at a time, to determine whether the short is internal or external; in the latter case, to determine which circuit contains the short.

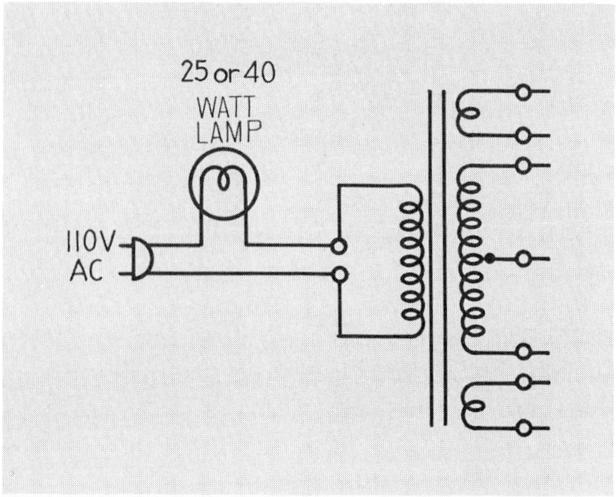

FIG. 8-5. Checking the power transformer.

To interpret the above checks, it might be well to give some more transformer theory. With all the tubes removed, the secondaries are not drawing current, and consequently, the primary should not be drawing current. This would be true if the transformer were 100 percent efficient. Since this is not so, the primary will draw a small amount of current to overcome the hysteresis and eddy-current losses in the iron core. With the average radio power transformer, this small amount of current is sufficient to cause the series 25-watt lamp to just glow. This is the test for a good transformer.

Now, assume some shorted turns, or a short in the 6-volt amplifier-filament wiring. The primary must furnish the power that this short consumes. The added primary drain causes more current to flow through the series 25-watt lamp, and the lamp glows more brightly. Now, suppose that we disconnect the 6-volt transformer leads. If the lamp brightness drops to just a glow, we must inspect the receiver filament circuit for a short. If the lamp filament continues to glow brightly, even after all circuits have been opened, the short is within the transformer.

When a power transformer is replaced, an exact duplicate is to be preferred. If this is unobtainable, the technician is beset by a number of questions. What size shall I use? Which winding is which? How can I tell the windings apart? What shall I do with the extra leads?

What size of replacement power transformer should be used? Replacement transformers are usually rated in the voltages and currents obtainable from the various secondary windings. A partial listing of general replacement transformers from the Stancor catalogue is shown in Fig. 8-6. These data must be compared with the calculated requirements of the tubes in the receiver being serviced. For example, checking the requirements of our standard receiver with the tube manual, we obtain the information shown in the accompanying table.

Tube complement	A requirements		B requirements	
	Volts	Amp	Volts	Total cathode current, ma
5Y3-GT	5	2		
6V6-GT	6.3	0.6	315	36.2
6SQ7	6.3	0.3	100	0.4
6SK7	6.3	0.3	100	13
6SA7	6.3	0.3	100	12.3
6SK7	6.3	0.3	100	13
Total	5 6.3	2 1.8	315 volts at 75 ma	

We should remember that transformers are rated in RMS values, and that the action of the input filter capacitor provides a maximum *B*+ voltage which is closer to the peak value of the input voltage. Therefore, a transformer rated at 260 or 270 volts RMS will easily provide the 315 volts at the maximum *B*+ point. The replacement transformer for the standard receiver should have secondary windings with the following ratings:

High voltage	270–0–270 volts at 90 ma
Rectifier filament	5 volts at 2 amp
Amplifier heaters	6.3 volts at 2 amp

The high-voltage winding in some catalogues is labeled "540V CT." The CT indicates that the high-voltage winding is center-tapped, giving 270 volts on each side of the tap.

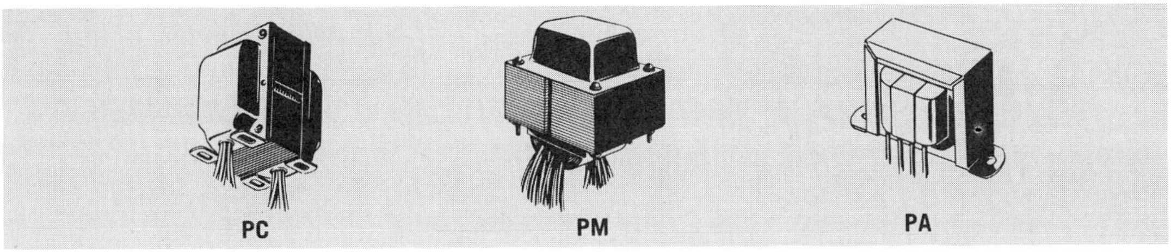

POWER TRANSFORMERS to Provide Approximately 260 Volts D.C. to Condenser Input Filter

Catalog No.	Plate Supply A. C. Volts	Plate Supply D.C. Ma.	Rectifier Fil. Volts	Rectifier Fil. Amps.	Other Windings Volts	Other Windings Amps.	Mtg. Type	Height Overall	Base Area
PC8401 / PM8401	235-0-235	40	5.0	2.0	6.3 CT	2.0	PC / PM	3 3/16 / 2 5/8	2 5/8 × 2 5/8 / 2 1/2 × 3
PC8402 / PM8402	240-0-240	55	5.0	2.0	6.3 CT	2.0	PC / PM	3 3/16 / 2 3/4	2 5/8 × 2 3/4 / 2 1/2 × 3
PC8403 / PM8403	250-0-250	70	5.0	2.0	6.3 CT	2.5	PC / PM	3 3/16 / 3 1/8	2 5/8 × 3 1/8 / 2 1/2 × 3
PC8404 / PM8404	260-0-260	90	5.0	2.0	6.3 CT	3.0	PC / PM	3 5/8 / 3 5/8	3 × 3 1/2 / 2 7/8 × 3 3/8
PC8405 / PM8405	270-0-270	120	5.0	3.0	6.3 CT	3.5	PC / PM	4 / 3 1/2	3 1/4 × 3 1/2 / 3 1/8 × 3 3/4

POWER TRANSFORMERS for Use with Choke Input Filter, VR-Tube Regulated Supply, Speaker Field in Filter, or Higher Voltage with Condenser Input Filter

Catalog No.	Plate Supply A. C. Volts	Plate Supply D.C. Ma.	Rectifier Fil. Volts	Rectifier Fil. Amps.	Other Windings Volts	Other Windings Amps.	Mtg. Type	Height Overall	Base Area
PM-8423	300-0-300	90	5.0	2.0	6.3 CT	3.5	PM	3	2 13/16 × 3 3/8
PC8406 / PM8406	325-0-325	40	5.0	2.0	6.3 CT	2.0	PC / PM	3 3/16 / 2 3/4	2 5/8 × 2 3/4 / 2 1/2 × 3
PC8407 / PM8407	325-0-325	55	5.0	2.0	6.3 CT	2.0	PC / PM	3 3/16 / 3 1/8	2 5/8 × 3 1/8 / 2 1/2 × 3
PC8422 / PM8422	325-0-325	150	5.0	3.0	6.3 CT	5.0	PC / PM	4 / 3 3/4	3 1/4 × 3 7/8 / 3 1/8 × 3 3/4
PC8408 / PM8408	340-0-340	70	5.0	2.0	6.3 CT	2.5	PC / PM	3 5/8 / 3 1/2	3 × 3 3/8 / 2 7/8 × 3 3/8
PC8409 / PM8409	350-0-350	90	5.0	2.0	6.3 CT	3.0	PC / PM	3 5/8 / 3 3/4	3 × 3 5/8 / 2 7/8 × 3 3/8
PC8410 / PM8410	360-0-360	120	5.0	3.0	6.3 CT	3.5	PC / PM	4 / 3 3/4	3 1/4 × 3 3/4 / 3 1/8 × 3 3/4
PC8411 / PM8411	375-0-375	150	5.0	3.0	6.3 CT	4.5	PC / PM	4 5/16 / 3 7/8	3 5/8 × 4 / 3 1/2 × 4 1/8
PC8412 / PM8412	400-0-400	200	5.0	3.0	6.3 CT	5.0	PC / PM	4 3/4 / 3 7/8	4 × 4 / 3 3/4 × 4 1/2

POWER TRANSFORMERS for Use with 6AX5, 6X4, 6X5, or Selenium Rectifiers

Catalog No.	Plate Supply A. C. Volts	Plate Supply D.C. Ma.	Rectifier Fil. Volts	Rectifier Fil. Amps.	Other Windings Volts	Other Windings Amps.	Mtg. Type	Height Overall	Base Area
PS8415	125 1/2-wave	15			6.3	0.6	PS	2	2 3/8 × 1 3/8
PS8416	125-0-125	25			6.3	1.0	PS	2 5/16	2 7/8 × 1 3/4
PA8421	125 1/2-wave	50			6.3	2.0	PA	2 1/4	3 3/4 × 2 1/8
PC8417	220-0-220	50	6.3	0.6	25.2	0.5	PC	3 3/16	2 5/8 × 2 5/8
PC8418 / PM8418	230-0-230	50			6.3	2.5	PC / PM	3 3/16 / 2 5/8	2 5/8 × 2 5/8 / 2 1/2 × 3
PC8419 / PM8419	240-0-240	70			6.3	3.0	PC / PM	3 3/16 / 2 7/8	2 5/8 × 2 7/8 / 2 1/2 × 3
PC8420 / PM8420	260-0-260	90			6.3	4.0	PC / PM	3 1/2 / 3 1/2	3 × 3 1/2 / 2 7/8 × 3 3/8
P-6358	300-0-300	65			6.3 CT	2.7	C	3 1/8	2 5/8 × 2 1/2

FIG. 8-6. A partial listing of general replacement transformers from the Stancor catalogue (Stancor Electronics, Inc.).

Reference to Fig. 8-6 shows that the Stancor PC8404 or PM8404 may be used as a replacement for the power transformer in the standard circuit. The next step is to compare the overall height and the base area of the original transformer with the figures shown in the catalogue. A good rule to follow is that of selecting a replacement transformer with about the same physical size as the original one.

Power-transformer color code. Most transformer manufacturers color their leads in accordance with the Electronic Industries Association (E.I.A.) color code. This can be used to advantage for replacement and is given in Fig. 8-7.

How to identify leads of an uncoded transformer. In case the manufacturer does not follow the code, the leads can be determined with an ohmmeter and voltmeter as follows:

1. Pair up the winding leads by means of an ohmmeter.
   a. First connect the ohmmeter to any lead and check for continuity with all the other leads, as shown in Fig. 8-8*A*. The lead that shows continuity is the other end of that winding or a tap. In the case of a tapped winding, three leads will show continuity.
   b. Separate these two or three leads, as the case may be, and repeat to find the other windings, as shown in Fig. 8-8*B*.
2. Read the resistance of each winding, as shown in Fig. 8-9.
   a. The primary will show a resistance of 2 to 15 ohms (240 turns).
   b. The high-voltage winding will show a resistance of 80 to 400 ohms (1,080 turns) for the entire winding.
   c. The filament windings will show a reading of less that 1 ohm (10 or 12 turns).

   There will be no mistaking the high-voltage winding. Tape the leads so there will be no danger of shock.
3. Connect the primary to the a-c line, and check the voltage of the filament windings to determine which is the amplifier and which the rectifier filament winding (Fig. 8-10).

What to do with unused leads. The replacement transformer often has leads that are not used in the original wiring diagram of the receiver. The filament center taps, for example, may not be used. If this is the case, tape the unused leads so that they will not short and dress them neatly in the receiver

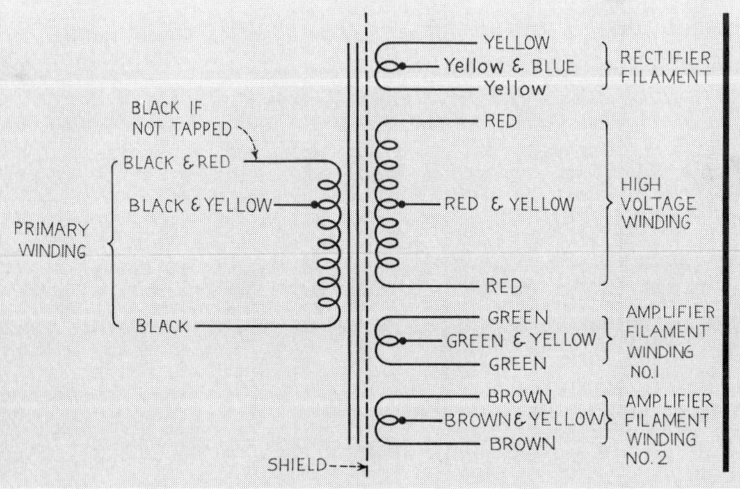

FIG. 8-7. Power-transformer color code.

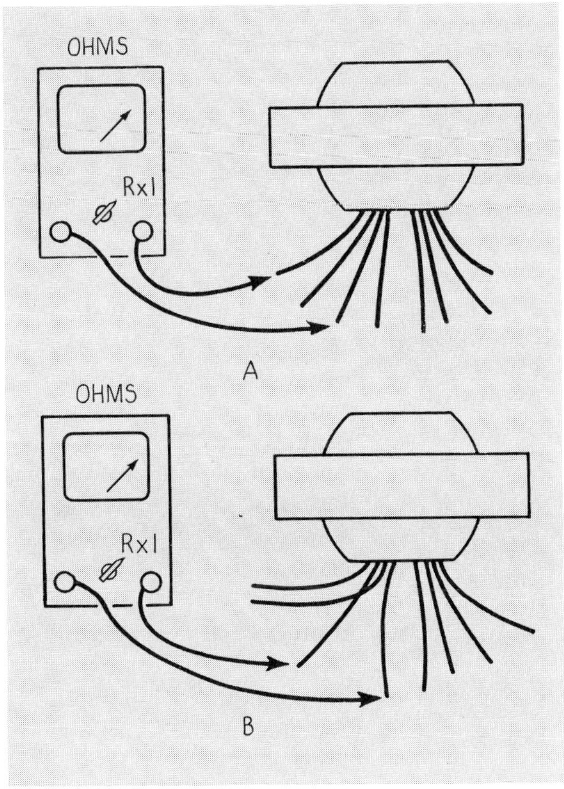

FIG. 8-8. Pairing the transformer leads.

chassis. If the unused center tap is of the type that has two separate wires in a single piece of spaghetti, solder these two wires together before insulating with tape.

Sometimes the replacement transformer has an uncoded lead that does not show continuity to any of the other leads. This lead will be the connection to a noise-reducing Faraday shield, between the primary and the secondary windings. If the transformer has such a lead, connect it to a chassis soldering lug.

General replacement notes. Before concluding this section of replacement notes on power transformers, the authors would like to remind the technician that it is a sign of good workmanship always to be careful of wiring and soldering and that this is especially important when replacing the power transformer. A poor connection or resin joint can cause much trouble when it is in the low-voltage high-amperage filament circuit. Poor insulation and sloppy soldering can also cause a messy recall job from flashovers in the high-voltage circuit. Of course, the line cord should be examined for frays, the grommet should be examined for breaks, and the knot should be in place behind the grommet on the inside of the chassis.

Troubles common to the rectifier tube. Rectifier tubes usually have a long life. The 5Y3-GT, for example, is rated at 125 ma of output current. This is rarely exceeded or even reached by the typical receiver; when it is, a larger tube, the 5U4-GB, is usually employed. As the tube ages, it gradually loses its emission, with a consequent loss in output voltage. Tube checkers are reliable in indicating this condition. Another check is a comparison of output voltage with another rectifier tube that is known to be good. Occasionally, rectifier tubes become gassy and glow with a purplish light. In this case, the receiver will not operate at all, or its speaker might

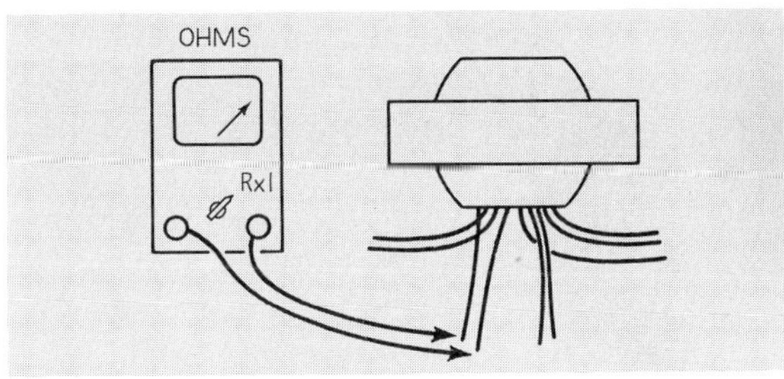

FIG. 8-9. Checking resistance of each winding.

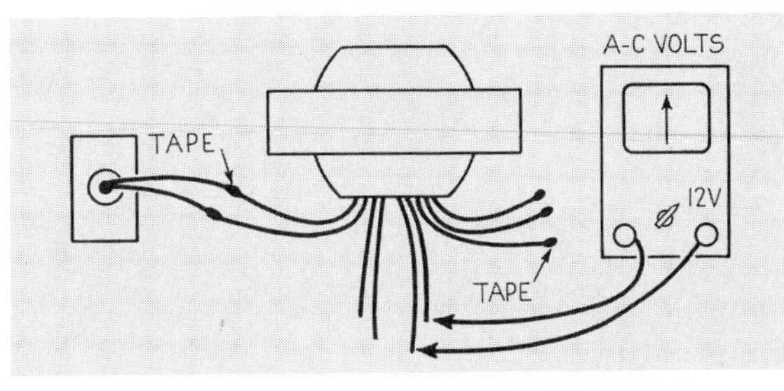

FIG. 8-10. Identifying the amplifier heater winding.

emit only a low tearing growl. Replacement of the tube is the answer. The above applies only to high-vacuum rectifiers like the 5Y3-GT, 5U4-GB, etc. It is normal for a glow to appear in gas rectifiers like the OZ4-G and in mercury-vapor rectifiers like the 82 and 83.

Troubles common to the filter choke. The common fault with filter choke *L*-15 is that the winding opens. This will be found on check, by no voltage at *B*+ and abnormally high voltage at rectifier filament. When he finds this condition, before checking to make sure that the choke is open, the technician should pull the receiver plug and discharge the filter capacitors. Input filter capacitor *C*-15 remains at full charge, since there is no discharge circuit when the choke is open.

Another trouble that sometimes occurs with filter chokes is that the insulation between the winding and the iron core breaks down, making the winding short to the core. Since the core is generally connected directly to the chassis, the plate power supply is shorted out. When this is the case, the receiver shows zero volts at *B*+, and the rectifier plates glow at red heat in their effort to feed the heavy current demanded by the short.

When the filter choke opens or develops a short, a replacement must be obtained. If an exact duplicate is not available, a general replacement type may be used. A partial listing of general replacement chokes from the Stancor catalogue is given in Fig. 8-11. When choosing a suitable choke, the important ratings to look for are the current-carrying capacity (DCMA) and the physical size (height and base area). For example, we have calculated a current drain of 75 ma for our standard receiver. The replacement choke should be rated at 75 ma or more. Of the two chokes listed at 75 ma, we would choose the one that is closest to the original in physical size. Note that the insulation rating is sufficiently high for any radio application. The inductance and d-c resistance ratings are generally not critical in the filter circuit.

Troubles common to the input filter capacitor. The input filter capacitor *C*-15 is the most common cause of trouble in the power-supply stage. It is a high-voltage, high-capacity electrolytic capacitor of either the wet or the dry type. With time, electrolytic capacitors lose capacitance and open. When this is the case, the *B*+ voltage will be low and the receiver will hum. The defect is confirmed by bridging the capacitor with a good one of similar capacitance and noting the improvement.

Capacitor *C*-15 also has the highest d-c voltage in the receiver across it. In addition, there are large surges in voltage across it. As a result, it is subject to voltage breakdown and shorting. When this happens, the *B*+ voltage is zero, and the rectifier-tube plates become red-hot from the heavy drain of current into the shorted *C*-15.

How to check an electrolytic capacitor. The handiest check for an electrolytic capacitor is a resistance measurement on the high-resistance range of the ohmmeter. This check is illustrated in Fig. 8-13. When the capacitor is checked, the meter pointer will kick up and then drop. The meter test prods are then reversed. The meter pointer should kick up further and then drop again. The surge of current, indicated by the kick, is caused by the capacitor's being charged by the battery in the ohmmeter. When the test prods are reversed, the charged capacitor adds its voltage to the battery in the ohmmeter, causing an increased surge of current, as indicated by the increased kick. An open electrolytic capacitor will show very little of this charge-and-discharge current.

Electrolytic capacitors normally have leakages, which will be different, depending on the polarity of the ohmmeter connections and that of the capacitor. Definite values cannot be assigned to the ohmmeter readings of this leakage resistance, owing to differences in capacitors as well as in ohmmeters. An approximation for capacitor *C*-15 is 50,000 ohms with the test prods connected one way, and

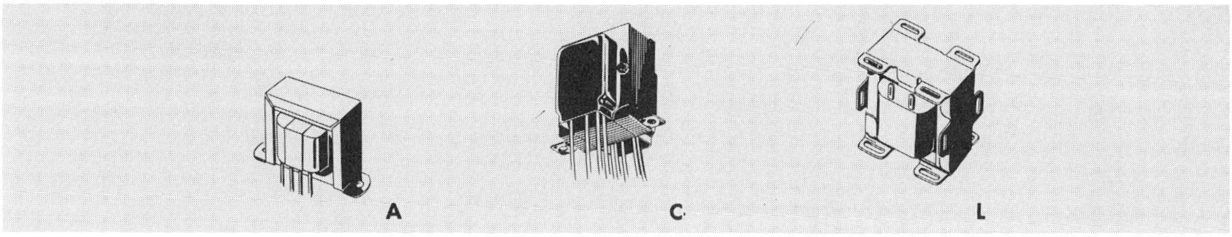

SMOOTHING CHOKES—For D.C. Power Supplies
Inductance values rated at 10 Volts R.M.S., 60 cycles

Catalog No.	Rating Induc. at DCMA	D.C. Res. in Ohms	R.M.S. V. Insul.	Mtg. Type	Height Overall	Base Area
C-1706	4.5 hy. at 50 ma.	300	1500	A	1⅜	2⅜ × 1⅜
C-1707	7.0 hy. at 50 ma.	550	1500	A	1⅜	2⅜ × 1⅜
C-1003	16.0 hy. at 50 ma.	580	1500	A	2	3¼ × 1¾
C-1708	13.0 hy. at 65 ma.	500	1500	A	2	3¼ × 1¾
C-1355	8.0 hy. at 75 ma.	290	1500	L	2¹⁄₁₆	2⁵⁄₁₆ × 1¾
C-1002	15.0 hy. at 75 ma.	400	1500	A	2¼	3¾ × 2¼
C-1420	16.0 hy. at 80 ma.	360	1500	C	3³⁄₁₆	2⅝ × 2⅝
C-1709	8.0 hy. at 85 ma.	250	1500	A	2	3¼ × 2
C-2704	9.0 hy. at 125 ma.	250	1500	A	2¼	3⅝ × 1⅞
C-2303	2.5 hy. at 130 ma.	100	2000	A	2	3¼ × 1¾
C-1421	7.0 hy. at 140 ma.	165	3000	C	3³⁄₁₆	2⅝ × 2⅝
C-2304	2.3 hy. at 150 ma.	60	1500	A	2	3¼ × 1¾
C-2309	3.0 hy. at 150 ma.	90	2000	A	2¼	3¾ × 2¼
C-1710	7.0 hy. at 150 ma.	200	1500	A	2⅝	4 × 2¼
C-2325	2.0 hy. at 200 ma.	60	1500	A	2¼	3¾ × 2¼
C-1646	5.0 hy. at 200 ma.	90	5000	C	4	3¼ × 3⅜
C-2705	10.0 hy. at 200 ma.	150	2500	C	3⁹⁄₁₆	3 × 3½
C-1412	4.0 hy. at 250 ma.	60	3000	C	3⅝	3 × 3½
C-2326	1.0 hy. at 300 ma.	43	1500	A	2¼	3¾ × 2¼
C-2334	2.8 hy. at 300 ma.	60	1500	A	2⅝	2¼ × 4
C-2308	8.0 hy. at 300 ma.	80	3000	C	2⅝	4 × 3⅞

FIG. 8-11. Stancor replacement chokes and a partial listing from the Stancor catalogue (Stancor Electronics, Inc.).

500,000 ohms on reversal. The difference is due to the fact that the capacitor is polarized. Capacitor C-15 must be disconnected from the circuit for this test, since other circuits are connected in parallel with it. The above explains the general rule when making resistance tests in a circuit bridged by an electrolytic capacitor: Reverse the test prods and take the higher reading.

Replacement of the input filter capacitor. When filter capacitor C-15 is replaced, the capacitance and voltage rating of the original should be used. A lower capacitance may cause hum; a lower voltage rating may soon cause breakdown. Correct polarity must be observed since, if polarity is reversed, the capacitor will overheat and possibly explode.

Sometimes, input-filter replacement capacitors continually break down. This is due to high surge voltage and is found in large receivers. The high surge voltage is due to the fact that, when the receiver is turned on, the filament-type rectifier furnishes high voltage almost immediately while the cathode-type amplifiers, which constitute the load, have not yet warmed up and are not drawing current. During the period of no load or low load as the amplifier tubes warm up, the voltage output of the power supply is high. Normally, in the average receiver, this is of no consequence, since the surge voltage developed from a 270–0–270 high-voltage winding is

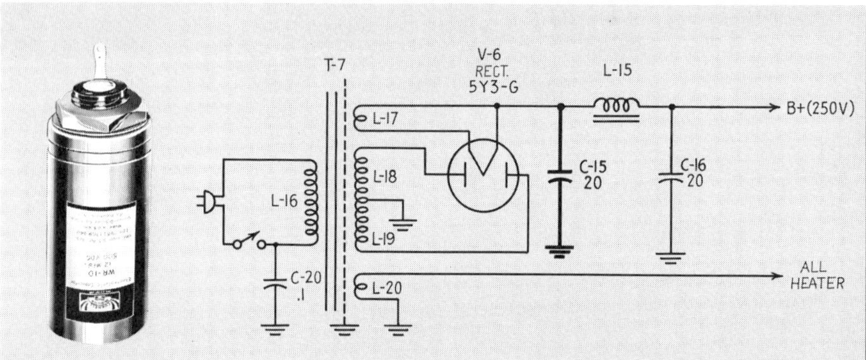

FIG. 8-12. A typical input filter capacitor and its position in the power-supply circuit.

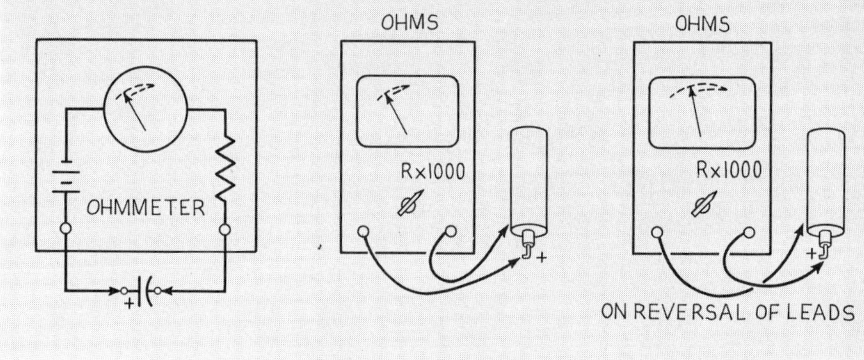

FIG. 8-13. Checking an electrolytic capacitor with an ohmmeter.

approximately 375 volts, well under the 525 surge-voltage rating of an electrolytic capacitor. In large receivers, however, where the tube complement includes a 5U4-GB and two or four 6BQ5 or 7189 tubes, the high-voltage winding may deliver higher voltage, and the voltage across *C*-15 may be 550 volts until the output tubes warm up. Where this is the case, there will be repeated breakdowns of capacitor *C*-15.

Surge voltage is easily checked. Simply allow the receiver to cool down, connect the voltmeter across capacitor *C*-15, turn the receiver switch on, and watch the voltmeter. If the voltmeter goes up to 400 or 425 volts when the switch is first turned on, and then settles back to about 300 volts as the tubes warm up, there is little likelihood of trouble from surge voltage. If the surge voltage climbs above 525, the safest procedure is to replace capacitor *C*-15 with two capacitors in series, as shown in Fig. 8-14. Capacitors *C*-15*A* and *C*-15*B* should each be twice the capacitance of capacitor *C*-15, since two equal capacitors in series have a total capacitance of half of one of them. The resistors should be 1 watt, 1 megohm (1,000,000 ohms) apiece. Their purpose is to equalize the voltage across capacitors *C*-15*A* and *C*-15*B*. Each capacitor, therefore, will have half of the total voltage across it. A circuit of this type, employing capacitors of the same voltage rating, will withstand any surge.

When a shorted input filter capacitor is replaced, it is advisable to check the rectifier tube to make sure that it was not damaged by the heavy overload.

Troubles common to the output filter capacitor. Output filter capacitor *C*-16 is usually similar to the input capacitor *C*-15 and is subject to the same troubles; it opens and shorts. When it opens, there is no effect on the *B*+ voltage, but there may be excessive hum, squeal, or motorboating, or a combination of

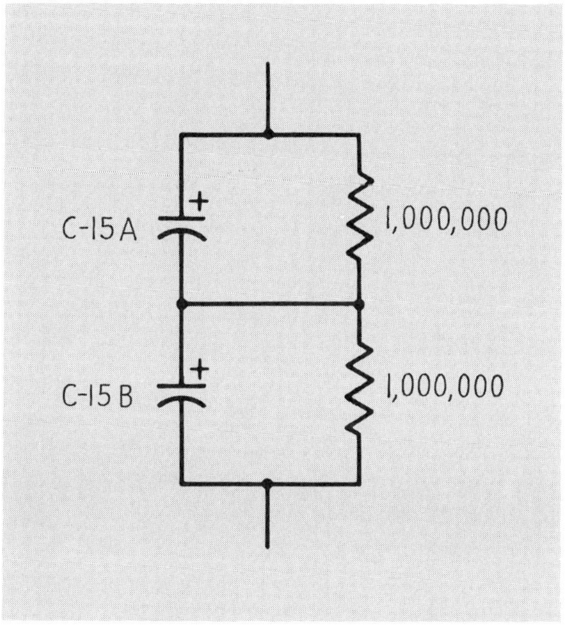

FIG. 8-14. Connecting two capacitors to increase voltage rating.

all three. Substituting another capacitor to see its effect is the fastest check. When it shorts, *B*+ voltage is zero, and the rectifier tube overheats, but not to the point of red plates.

Before condemning capacitor *C*-16, the technician should look for even a small *B*+ voltage. In parallel with capacitor *C*-16 is the plate circuit of every tube in the radio, and the short may very well be elsewhere. Figure 8-15 is a skeleton diagram of the receiver, showing only the plate and *B*+ circuits. If, for example, capacitor *C*-12 were shorted, *B*+ voltage would be low, the voltage at the rectifier filament would be almost normal, and the plate voltage of the second a-f tube, *V*-5, would be zero. It would be a good idea, therefore, to check all plate voltages before going further. Another good indication as to the location of the short would be an overheated resistor. Resistor *R*-4, *R*-22, or *R*-25 would be badly overloaded if capacitor *C*-4, *C*-22, or *C*-25 were shorted. If these methods do not locate the short, it would be necessary to open *C*-16 as well as the rest of the *B*+ circuit, one wire at a time, and hunt for the short with an ohmmeter. When the short is located, if it is an item other than capacitor *C*-16, replacement notes will be found for it in the chapter dealing with its particular stage.

When replacing capacitor *C*-16, the technician must be careful to observe polarity. Also, when replacing an open output filter capacitor, he should be careful to remove the connection from it when, for one reason or another, the original capacitor is left physically on the chassis. Even though the soldering lug might be handy for the replacement capacitor, leaving the old one connected in

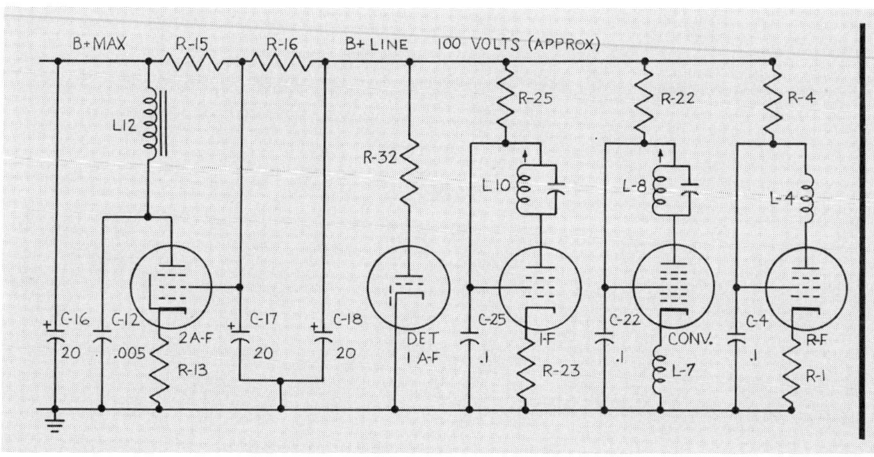

FIG. 8-15. Skeleton diagram of the standard receiver, showing the *B* circuit.

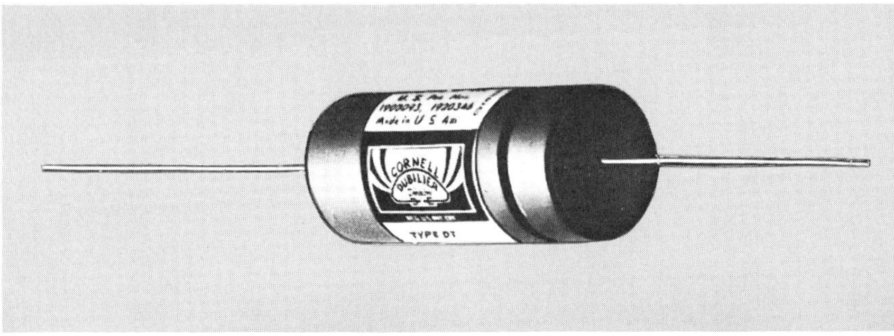

FIG. 8-16. Paper tubular capacitor, commonly used for line filter C-20.

the circuit is a potential source of trouble. Output filter capacitor C-16 is not nearly so susceptible to high surge voltage as input filter capacitor C-15, and the usual surge voltage rating of 525 volts is adequate.

Finally, capacitors C-15 and C-16 are often contained in one dual-section filter block. The fact that one capacitor has proved defective is no indication that the other cannot still give long, satisfactory service. Whether to replace the single unit or the entire block is up to the individual technician. Usually, it is preferable to replace the block.

Troubles common to the line filter capacitor. Line filter capacitor C-20 is a paper tubular capacitor, whose usual capacitance is 0.1 mfd. With the usual rating of 400 volts, voltage breakdowns are unknown. The capacitor may open, and this would theoretically cause greater interference from line disturbances. An open line filter capacitor, however, may cause entirely different effects. Owing to its position in the circuit, the receiver chassis is grounded through capacitor C-20 by the lighting mains, one side of which is grounded. The receiver installation may have no ground at all or an indifferent ground, in which case C-20 takes on a new function—that of grounding the receiver. This explains why reception (absence of hum or noise) is often improved by reversing the plug on a-c receiver installations. It also explains why a tiny spark or small shock is experienced when connecting a ground to a receiver. When C-20 is open, its grounding function is gone. The most annoying manifestation of this is known as "modulation hum"; that is, the receiver does not hum when making a hum check. The hum comes on as a station is tuned in. There will be no hum between stations. Standard procedure for modulation hum is to check the ground and capacitor C-20. Bridging capacitor C-20 with another capacitor of like value is the check for an open capacitor.

SUMMARY

Quick check for normal operation of stage.

All tubes light.
No signs of overheating.
Hum level is normal.
B+ voltage measures 200 to 300 volts.

Typical a-c power supply.

Typical a-c power supply is shown diagrammatically in the accompanying figure.

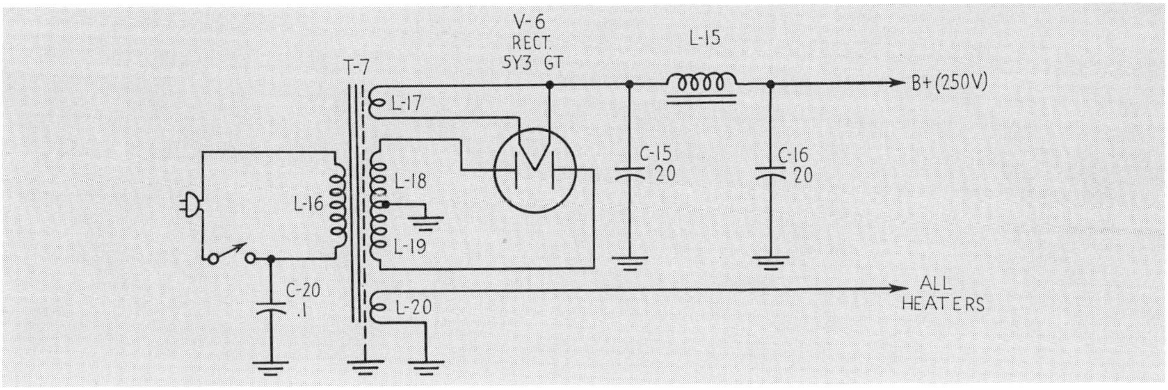

Normal resistance data.

Plug, prong to prong	2–15 ohms
Chassis to rectifier plates	35–200 ohms
Rectifier filament to *B*+, across choke *L*-15	40–400 ohms
Chassis to rectifier filament	100,000 ohms *

Normal voltage data.

Rectifier filament to filament	5 volts a-c
Across other tube heaters	6 volts a-c
Chassis to rectifier plate	200–350 volts a-c
Chassis to rectifier filament	220–385 volts d-c
Chassis to *B*+	200–360 volts d-c

* Reverse test prods and take the higher reading.

SERVICE DATA CHART

Symptom	Abnormal reading	Look for
Tubes do not light	Plug, prong to prong checks open with ohmmeter	Defective line cord and plug. Open fuse. Defective line switch *S*-1. Open power transformer *T*-7 (primary)
Rectifier-tube plates show red	Chassis-to-rectifier filament checks short circuit with ohmmeter	Shorted input filter capacitor *C*-15. Check surge voltage on replacement

Symptom	Abnormal reading	Look for
Rectifier tube overheats	*B*+ voltage checks zero. Chassis to *B*+ checks short circuit with ohmmeter	Shorted output filter capacitor *C*-16. Short circuit in *B*+ wiring
Rectifier tube overheats	*B*+ voltage low	Zero plate voltage on amplifier tubes. Short-circuited plate filter capacitor
Hum	*B*+ voltage low	Open input filter *C*-15
Hum	*B*+ voltage normal	Open output filter *C*-16 Open grid
Oscillation or motorboating	*B*+ voltage normal, or fluctuating with motorboat beats. Screen voltage normal	Open output filter *C*-16
Rectifier tube shows purplish glow		Gassy high-vacuum type of rectifier tube
Weak reception. No sign of overheating	*B*+ voltage checks low	Weak rectifier tube
No signal from speaker. No sign of overheating	*B*+ voltage checks zero (discharge filter capacitor)	Dead rectifier tube. Open filter choke *L*-15
Modulation hum		Poor ground, open line filter capacitor *C*-20, or both

QUESTIONS

1. The tubes of an a-c radio receiver do not light. List the various possible sources of trouble in the order in which you would check them.
2. An a-c receiver does not play, and the rectifier plates become red-hot. What is the most likely cause of the trouble?
3. An a-c receiver is brought in for hum. How would you check to see if the hum originates in the power-supply stage?
4. An a-c receiver does not play. A check of the receiver shows that the tubes light and that there is no sign of overheating or hum, but there is no *B* voltage. List the possible causes of the trouble, and explain how you would check for each.
5. After a shorted input filter capacitor has been replaced, what two checks should be made before checking the receiver for normal operation?
6. The power transformer of an a-c receiver overheats. The radio plays, the hum level is somewhat high, and *B* voltage is low. A voltage check of the power supply shows 180 volts a-c on one rectifier plate, 80 volts a-c on the other. What is wrong?
7. Describe the series lamp check for a short in a power transformer or its associated circuits.
8. When using the series lamp check on a receiver with an overheating power transformer, the lamp glows brightly until the amplifier filament wires are removed. Where would you look for trouble?
9. When a 5Y3-GT rectifier tube glows with a purplish glow, what is likely to be wrong?
10. Refer to the listing of replacement power transformers given in Fig. 8-6 and choose replacements for the receivers of Fig. 12-14 and Fig. 12-16.
11. The receiver of Fig. 12-14 motorboats. What component in the power supply is likely to cause this condition?
12. The hum level in a receiver is normal, but the receiver hums badly when certain stations are tuned in. What component in the power supply can cause this condition?
13. Resistor *R*-8 of Fig. 12-16 is found to be open. What should be the wattage of the replacement resistor?

A-C/D-C Transformerless Power Supply

9

The wide popularity of the a-c/d-c type of receiver is due not so much to its ability to operate on either a-c or d-c lighting mains as to the fact that it makes possible an efficient inexpensive power supply for small receivers. The percentage of a-c/d-c receivers in use is very large as compared with the percentage of homes supplied by 110-volt d-c lighting mains. The power transformer is an expensive unit, and its elimination in the a-c/d-c receiver circuit accounts for the circuit's wide prevalence.

The signal circuits of the a-c/d-c receiver are the same as in an a-c receiver, with minor differences. The signal checks and stage-gain data are approximately the same for both. The main difference between the two lies in the power supply.

Quick check for proper operation of the a-c/d-c power supply. If all the tubes in the receiver light, the hum level is normal and the *B*++ voltage measures approximately 120 volts, the a-c/d-c power supply is probably functioning normally.

Function of the a-c/d-c power supply. The function of the a-c/d-c power supply is like that of any other type: to furnish the necessary *A*, *B*, and *C* voltages to the filament, plate, and grid circuits of the rest of the receiver. In this case, the power source is the 110-volt lighting mains, to be used regardless of whether it is alternating or direct current.

The common type of a-c/d-c receiver uses a 5-tube superheterodyne circuit. The stages employed are a converter, i-f, detector and agc, first a-f, second a-f, and power supply. The detector and first a-f stage functions are combined in one tube. Printed wiring is generally used. A receiver of this type should be kept in mind while studying the power-supply stage.

The power supply can be easily subdivided into a study of the *A* or heater circuit and the *B* or plate circuit.

Heater circuits in a-c/d-c power supplies. The heater circuit in the a-c/d-c power supply is so designed as to use tubes for the receiver which draw the same heater current. The heaters are connected in series, a dropping resistor is added if necessary, and then the circuit is connected directly across the power line.

Figure 9-1 shows the heater circuit employed in widely used a-c/d-c superheterodyne receivers. Modern sets use the miniature base type of tube. Older sets, still encountered in service work, use the larger octal-base tubes in the same circuit. Equivalent tube types are shown in the diagram. The heaters will light equally well on alternating or direct current, in the same way that an ordinary lamp will.

Reference to a tube manual shows that all the tubes in this receiver are designed to operate at one value of current, 0.15 amp, a condition true for a series circuit. The voltage rating

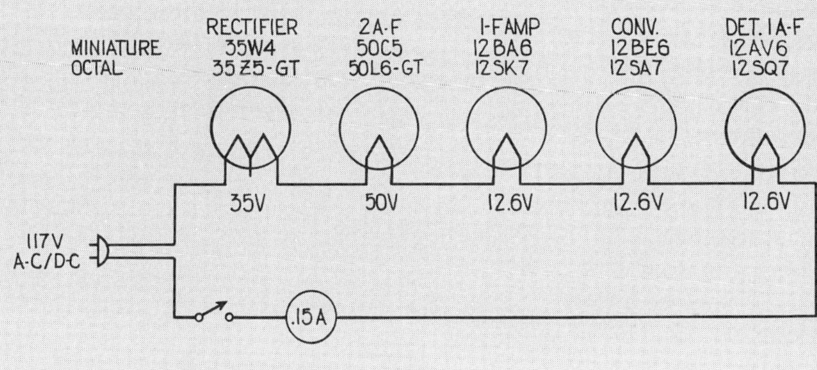

FIG. 9-1. Basic heater supply for a 5-tube a-c/d-c receiver.

for each tube is entered below the tube in the diagram. Adding these together for the series circuit to get the total voltage drop, we get 122.8 volts. Line voltage is conventionally considered to be 117 volts; but this is an average figure. Actual line voltages in different localities and at different times may vary considerably from this value. When the line voltage is lower than 122.8, the current through the heaters will be less than 0.15 ampere. At the reduced current, the voltage across each tube heater will be reduced also. Similarly, at a line voltage over 122.8, there will be a higher current through the heaters and a higher voltage across each heater than the rated value.

Differing line-voltage conditions have very little effect on the operation of the receiver. The circuit of Fig. 9-1 is considered satisfactory for operation at line voltages varying from 105 to 125 volts. Under low line voltage conditions, the main difference noted in operation on reception of local stations is a longer warmup time needed between turning on the switch and hearing the station.

In a receiver using fewer than five tubes, the tube omitted from the lineup of Fig. 9-1 is replaced by a dropping resistor. A 68-ohm/2-watt resistor is generally used to replace one of the 12-volt heaters.

The negative end of the *B* or plate power supply is connected to one side of the line, as shown in Fig. 9-2. This point is called "common negative" and corresponds to the chassis ground connection in a receiver with a transformer type power supply. It is usual practice to place the detector-first a-f tube heater at common negative or ground potential, since this is the position of minimum hum. The converter is usually placed next in line to avoid hum modulation in the oscillator stage. The sensitivity of the other tubes to hum determines the remaining order: the i-f amplifier

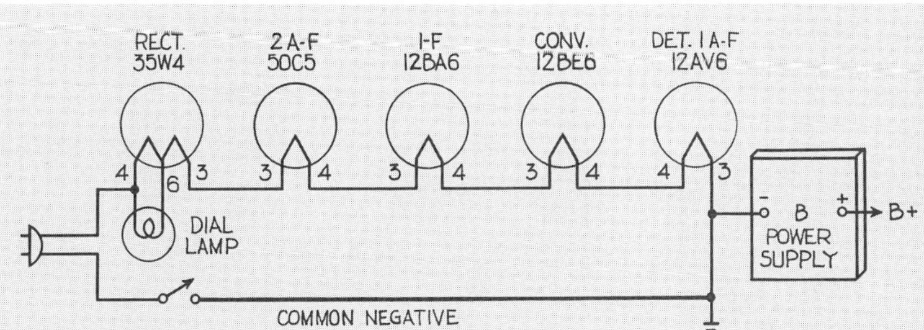

FIG. 9-2. The common negative connection in an a-c/d-c receiver. The circuit also shows provision for a dial lamp.

next, followed by the second a-f heater, and followed last by the rectifier. This order is shown in Figs. 9-1 and 9-2.

Some a-c/d-c receivers use a small lamp to light the dial, or to show that the set is turned on. The dial lamp wiring is shown in Fig. 9-2. The heater of the 35W4 rectifier tube is tapped at the number 6 pin and the dial lamp is connected across the section of the heater between the number 4 and 6 pins, as shown. The octal-base 35Z5-GT rectifier has a similar tap at pin number 3. Note that the series circuit is complete with or without the dial lamp, so that the set continues to operate if the dial lamp should open.

***B* power supplies in a-c/d-c receivers.** The *B* power supply in a-c/d-c receivers consists of a half-wave rectifier and filter circuit. A basic diagram is shown in Fig. 9-3.

Capacitor *C*-20 acts as a line filter to keep r-f disturbances like those caused by refrigerator motors, electric shavers, etc., from affecting the receiver. In addition, since one side of the line is grounded, it serves to ground the receiver. No other ground connection is used in a-c/d-c receivers. It is usually a 200-volt capacitor, whose capacitance ranges from 0.006 to 0.25 mfd.

The rectifier tube is the 35W4 miniature-base half-wave rectifier. The octal-base equivalent is the 35Z5-GT. In the half-wave rectifier circuit, current flows from cathode to plate, as indicated by the arrow in the diagram on the half cycle of a-c line current when the plate is positive with respect to common negative. When the plate is negative on the next half cycle, no current flows through the rectifier tube.

If the receiver is operated on a d-c line, the rectifier simply acts as a series resistor. However, in this case, polarity of the line plug is important, since the rectifier will act as a series open circuit if the plate is connected to the negative side of the d-c line. This fact explains the tag found near the plug on many a-c/d-c receivers which says: "If the receiver does not work on a d-c line, reverse the line plug." This connects the rectifier plate to the positive side of the d-c line.

On a-c operation, the rectifier output is 60 pulses of current per second, in the direction of the arrow in Fig. 9-3. This requires the smoothing action of a filter circuit. The *B*++ point supplies plate voltage to the second a-f or power output tube. Here the filter circuit consists only of input filter capacitor *C*-15. In order to reduce hum and maintain voltage, it must be a large capacitor. The circuit uses a 50-mfd input capacitor rated at 150 volts. The capacitance is rarely less than this value, and is often greater. Sizes as high as 100 mfd are found in some receivers. Note again that the d-c output voltage at the *B*++ point is higher

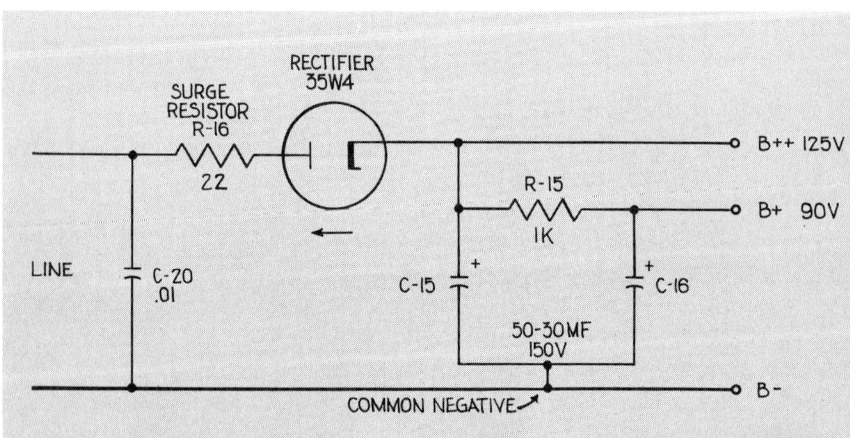

FIG. 9-3. Basic *B*-power circuit used in a-c/d-c receivers.

than the RMS input voltage. This is because the input filter capacitor tends to maintain the voltage at the peak value of the a-c input.

The initial charging current of a 50-mfd capacitor is high enough to harm the rectifier. So resistor *R*-16 is connected in series with the rectifier and input filter capacitor. It is known as a "surge" resistor. This series connection limits the charging current of the capacitor.

Small a-c/d-c receivers generally use a 4- or 5-inch speaker. These do not give a strong response at 60 cycles per second. Therefore the simple capacitor filter for the last audio stage is sufficient. However, the other stages in the receiver are followed by amplifiers, making more efficient filtering necessary. This is provided by the r-c filter consisting of filter resistor *R*-15, and input and output filter capacitors *C*-15 and *C*-16. The resistor is usually a 1-watt size ranging from 1,000 to 2,000 ohms in different receivers. The value chosen for any particular receiver is a compromise between efficient filtering and satisfactory voltage at the *B*+ point. Larger values of resistance give better filtering, but lower the voltage. Output filter capacitor *C*-16 is usually a 30-mfd or higher electrolytic type also rated at 150 volts. Both filter capacitors are usually enclosed in the same container.

Standard circuit. The heater circuit of Fig. 9-1 and the *B* power circuit of Fig. 9-3 are combined in Fig. 9-4 to show a complete power supply. This circuit appears so often that it may be taken as a standard for 5-tube a-c/d-c superheterodyne receivers.

The component marked INTERLOCK is a safety device. It is a connector in the line cord, the receptacle half of which is attached to the rear panel of the cabinet. Removing the back of the cabinet therefore automatically disconnects the line at the interlock, thereby protecting the consumer from high-voltage points within the receiver. When the technician works on such a receiver, he uses a spare line cord terminating with the receptacle half of the interlock. This special cord-connector combination is called a "cheater," and is also illustrated in Fig. 9-4.

Note that the power line is connected to the number 6 tap on the rectifier heater, so that only one section of the heater is included in the receiver series heater circuit. The other rectifier heater section brings the line voltage in a series circuit to the rectifier plate. This latter rectifier heater section thus serves a similar function to that of surge resistor *R*-16 in Fig. 9-3. Also, observe the absence of a dial light in the standard circuit. Most of these small receivers do not use one. Those sets with dial lights use the heater circuit of Fig. 9-2, and have a separate surge resistor.

Floating-chassis receiver construction. In the standardized circuit of Fig. 9-4, note the ground-symbol connection at common negative. This ground symbol indicates the metal chassis on which the receiver components are mounted. In many receivers, particularly the older ones, the metal chassis is a large metal base. In receivers with printed circuits, the chassis is reduced to a framework to hold the printed board, and many of the metal parts mounted on the board are electrically connected to the framework to take advantage of shielding effects.

The common negative connection to the chassis or to the frame presents short-circuit and shock hazards, because this connection is one side of the power line. The hazard is avoided by insulating or floating the chassis inside the cabinet so that the consumer has no contact with it. Control shafts for the volume control and tuning mechanism are recessed from the front panel and operated by means of long-shank insulating knobs which go through the front of the cabinet. Chassis holding screws do not extend to the outside of the cabinet. To get at these, the back of the cabinet must be opened. When this is done, the interlock acts as protection by disconnecting the power line.

This insulating protection is for the consumer. When the technician is working on one

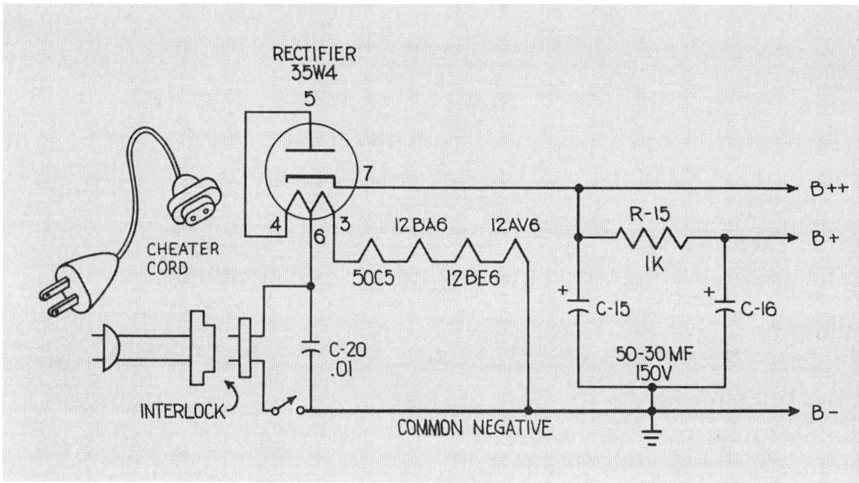

FIG. 9-4. Standard power supply circuit for an a-c/d-c receiver. The circuit also shows the interlock and the cheater cord used to replace it for service work.

of these receivers, he has no such protection and must recognize that he is working on a "hot" receiver. He must be careful to avoid grounding the receiver either directly or through himself. It is for this reason that the service bench should be of wood construction with no metal trim.

The safest way to avoid short or shock troubles is to connect the receiver to the line through an isolation transformer, which insulates the chassis from the line. A tapped isolation transformer rated at about 350 watts is a useful piece of bench equipment for this and other purposes. The 350-watt rating is recommended so that it can be used when servicing television receivers, many of which also use the same type of floating-chassis design.

NORMAL TEST DATA FOR THE A-C/D-C POWER-SUPPLY STAGE

Locating power-supply test points. Common negative and *B*++ are important test points in an a-c/d-c receiver. The technician should be able to find them quickly in any receiver that he is working on. If the set is mounted on a metal chassis, you cannot take for granted that the chassis is the common negative test point, as in a-c receivers, since other circuits are often used. Reference to Fig. 9-4 shows that common negative begins at the switch, and that the *B*++ point is at the cathode of the rectifier tube.

The volume control and switch is an easily recognized component in any receiver. The switch terminals are shown in Fig. 9-5. In the

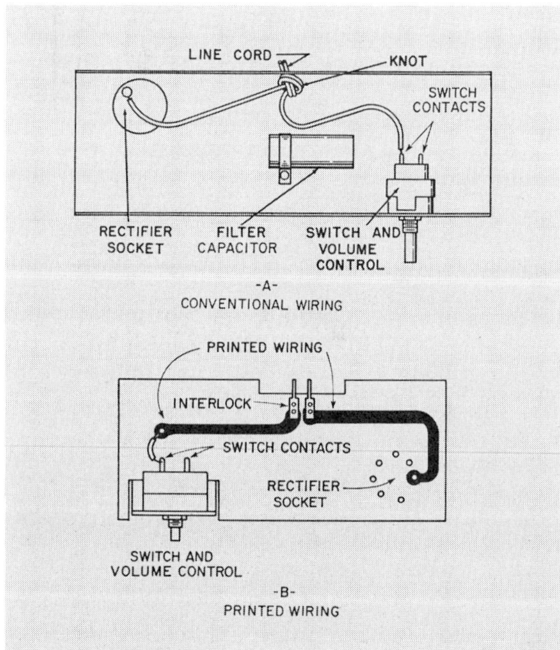

FIG. 9-5. How to identify the switch terminals and rectifier socket. Either switch terminal is common negative and the cathode pin of the rectifier is *B*++.

test procedures to be given, either switch terminal may be used for the common negative test point. If the switch terminals are not readily accessible, another easily located test point for common negative is the common negative lead on the filter capacitor.

The rectifier may be identified from a tube placement chart. When such a chart is not available and the set is one with conventional wiring, the rectifier may be found by tracing the line-cord wiring as shown in Fig. 9-5A. If the receiver uses printed wiring, the rectifier may be similarly found by tracing the printed wiring from the interlock, as shown in Fig. 9-5B. Having located the rectifier, reference to a tube manual will give the pin number for the cathode which is the *B*++ test point.

Another important test point in the power supply is *B*+. Reference to the standard diagram of Fig. 9-4 shows that *B*+ is one end of filter resistor *R*-15. This is generally a one-watt resistor with a resistance of 1,000 to 2,000 ohms, located near the rectifier tube. A voltmeter connected from common negative to one terminal of the resistor gives the voltage reading at *B*++. When the positive test prod is moved to the other terminal of the resistor, the voltage reading at *B*+ is obtained.

Quick check. The a-c/d-c power-supply stage is probably functioning properly when:

All tubes light or heat.

The hum level is normal.

There are no bad squeals or motorboating.

The *B*++ voltage measures approximately 120 volts to the common negative terminal. The *B*+ voltage measures approximately 90 volts to the common negative terminal.

This is the quick check for the stage.

The technician should familiarize himself with the normal brightness of tube heaters, since any marked variation is indicative of trouble in the heater circuit. Another aid to determining trouble in this regard is the length of time it takes for the tubes to reach their proper operating temperature. Tube brightness is not of great importance when a-c receivers are serviced, since variations in applied heater voltage are infrequent.

If the quick check indicates trouble in the power supply, disconnect the plug and discharge the filter capacitors by shorting them before proceeding to further checks. The filter capacitors may retain a charge with subsequent danger of shock or damage to test equipment.

Normal resistance data. Normal resistance data are given in the following table:

Line plug, prong to prong (switch closed)	100–200 ohms
Rectifier cathode to *B*+	1,000 ohms
Rectifier cathode to common negative	Capacitor action
B+ to common negative	Capacitor action

The reading from plug prong to prong takes in the heater circuit. There will be considerable variation in this reading depending on whether the tubes are still warm or not. Resistance of tube heaters, as of most conductors, varies with temperature. At the time of testing, the tube heaters may be anywhere from room temperature to several hundred degrees with a consequent difference in the resistance reading.

As mentioned before, readings in all a-c/d-c receivers should be taken from the common negative terminal rather than from the chassis. The readings from the common negative to *B*+ or rectifier cathode are called "capacitor action" because the electrolytic filter capacitors are connected across these points. Since there is usually no bleeder in an a-c/d-c power supply, the point at which the ohmmeter needle comes to rest will indicate the leakage of the electrolytic capacitors. Reverse the test prods and take the higher reading as the leakage resistance. This should be 100,000 ohms or greater.

Standard circuit. The standard circuit is shown in Fig. 9-6.

Normal voltage data. Normal voltage data are shown in the following table:

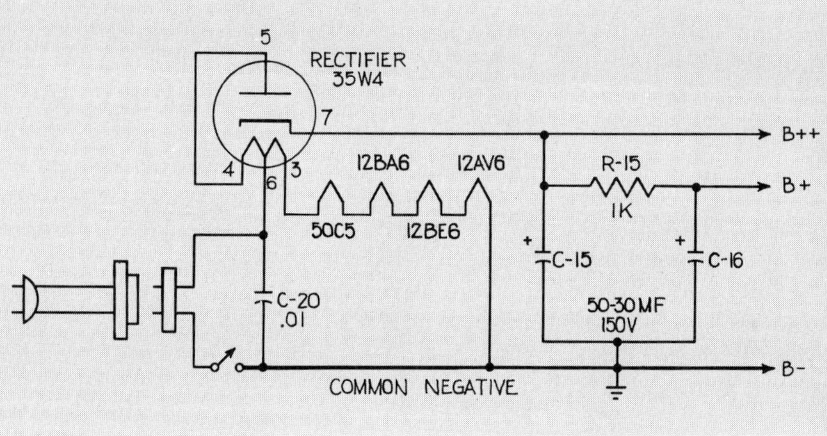

FIG. 9-6. Schematic diagram of a typical a-c/d-c power supply.

Common negative to rectifier plate	117 volts a-c (line)
Common negative to rectifier cathode	110–130 volts d-c
Common negative to *B*+	85–100 volts d-c

Heater voltages should be measured across each tube heater. The normal heater voltage for each tube is the rated voltage as found in the tube manual. An easy way of knowing the normal heater voltage is the first number in the tube designation. For example, the heater of the 50C5 is rated at 50 volts, 35W4 at 35 volts, 12AV6 at 12 volts, etc..

COMMON TROUBLES IN THE A-C/D-C POWER SUPPLY

Troubles common to the heater circuit. Since the heater circuit in an a-c/d-c receiver is a series chain, any open in any part of the circuit will cause the entire circuit to be open and the tubes will not light. The series chain includes the line cord, the interlock, the wiring, the tube heaters, and the switch. A break in any one of these can cause failure of the tubes to light, and all of them are common troubles. The technician must be able to find the break quickly and efficiently.

When the tubes in an a-c/d-c receiver do not light, a good way of determining the cause is to make a continuity check of the heater circuit with an ohmmeter. Checking across the two prongs of the line plug with the switch turned to the ON position should, of course, show an open circuit. If it shows the normal reading of 100 to 200 ohms, the receiver had been plugged into a defective or dead line outlet.

Checking the line cord, interlock, and switch. The next check should be from the common negative side of the switch to each of the line-plug prongs, as shown in Fig. 9-7. One of the line-plug prongs should show continuity (zero resistance) to the common negative wiring. If neither prong shows continuity, the trouble is in the line cord connected to the switch or in the switch itself. Which of these two is at fault is determined by further checking across the switch terminals.

If the check from common negative to the line plug shows continuity to one of the line-plug prongs, the switch and this half of the line cord are good, and the break is in the tube heaters, the wiring, or in the other wire in the line cord. It is best to make sure of the line cord first.

The plug prong that connects to the heater circuit has already been determined. Checking from this prong to the rectifier heater shows that the line cord wire is good, if a zero resistance reading is found. If these tests show

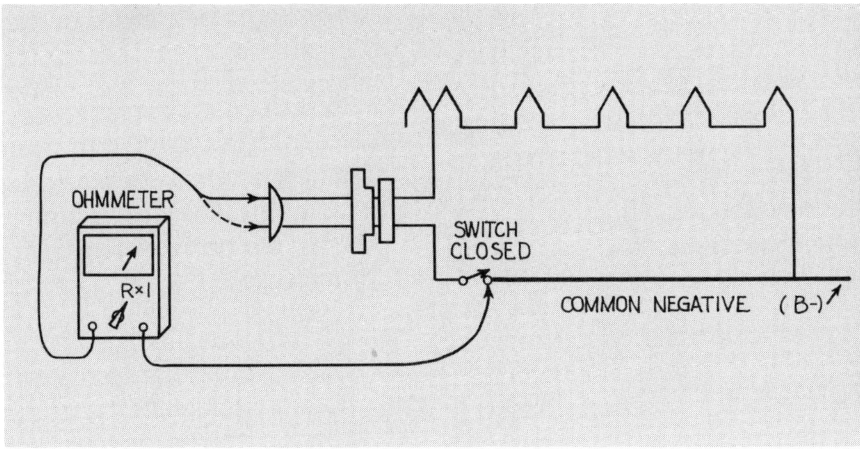

FIG. 9-7. Checking the line cord, interlock, and switch in an a-c/d-c heater circuit.

an open line cord or switch, the defective component must be replaced.

When the set uses an interlock, an open line cord or defective interlock is immediately found because the set plays when operated with a cheater cord. The repair is then made by installing a new interlock in the back panel of the radio.

Checking for an open tube heater. Having established that the line cord and switch are good, the next check is to look for an open tube heater. The tubes are checked in a tube checker, or by an ohmmeter check across each pair of heater terminals. Before replacing a tube with an open heater, one precaution should be held in mind. When any defective condition in an a-c/d-c receiver is found and repaired, the filter circuit should be checked from rectifier cathode to the common negative before plugging in. A shorted filter capacitor will ruin a rectifier tube.

Checking the heater wiring and socket contacts. If all the tubes are found to be good, we have the unusual condition where the line cord, switch, and tubes prove to be good, and yet our series circuit is open. This can only be caused by either a break in the heater wiring itself or by the failure of a tube pin to make contact in its socket.

A method for tightening or respringing socket contacts is shown in Fig. 9-8. A sharply pointed awl is inserted between the socket hole and the pin contact. This pushes the two halves of the contact together.

When the set has conventional wiring, a break in the wiring is very unusual, but is a possibility. There may also be an unsoldered or poorly soldered connection. To check for this, examine the heater wiring. Start with the line wire at the rectifier socket, and check each connection and each wire from socket to socket through the set. Grasp each wire near each connection with a pair of long nose pliers, and wiggle it gently. A broken wire will come loose. An unsoldered wire will be seen, and a poorly soldered wire will move in its soldering lug. A broken wire is replaced, and a poorly soldered connection is resoldered with a hot soldering tool.

In a receiver with printed wiring, breaks in the wiring are more common. The printed board acts as the chassis, and is subject to stresses and strains. These may cause cracks in the printed wiring. Here, it is best to check the printed wiring from terminal to terminal with an ohmmeter, because the break is often too fine to be seen by eye. The procedure is illustrated in Fig. 9-9. The test prods should be dug hard into the soldered terminals at each end of the piece of wiring being checked, to make sure of good contact through any insulating coating which may have been sprayed

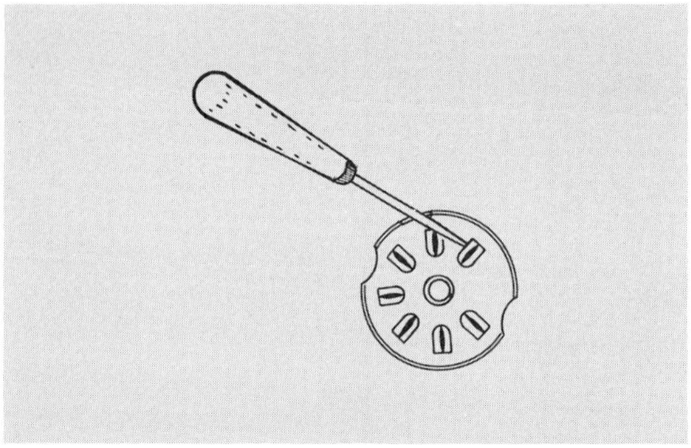

FIG. 9-8. How to tighten a socket contact.

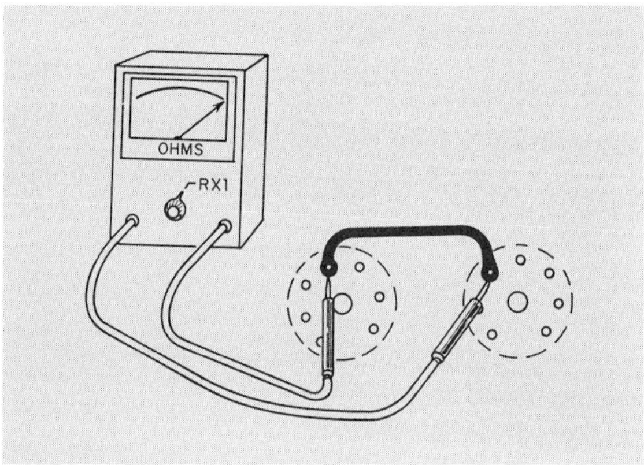

FIG. 9-9. How to find a break in printed wiring.

over the printed wiring. A continuity reading shows that the wiring is good, and a reading of "open" shows a break. The break is repaired by preparing a piece of hookup wire that is the same length and shape as the defective plating. This is placed over the plating, and the ends are soldered to the two connection points which already have well-soldered connections.

Receivers with blinking tubes. There is a type of intermittent reception during which the radio lights up and plays for a short time, and then the tubes go out and the radio dies out. After a short time, the radio lights up and plays, and the process repeats. If the radio uses a pilot lamp, it blinks on and off each time the set goes on and off. Since this is the most obvious symptom, radio men call this defective condition a "blinker."

The situation is caused by an intermittent open in one of the tube heaters. The filament expands on heating and breaks, thereby opening the series heater line. The pilot lamp goes out immediately, and all the tubes die down more slowly. In the meantime, the filament contracts, the break is temporarily reconnected, and current flows through the heater line again. The pilot light goes on immediately; the tube heaters warm up more slowly. This, of course, causes the filament to expand again, thereby repeating the process.

In servicing a complaint of this type, the problem is to find the tube which is causing the trouble. An ohmmeter check will not do for the purpose, because when the tube is cold the break is healed.

The fastest way of finding the defective tube is by using a neon test lamp. The receiver is placed in such a position that the heater wiring is readily accessible and the pilot lamp can be seen. The set is then turned on and allowed to blink on and off. The neon test lamp is now connected across each heater in turn.

When the neon test lamp is across a good tube heater, it will remain dark, regardless of whether the radio blinks on or off. When it is across the defective tube heater, it will remain dark while the set is on but will light up as soon as the pilot lamp blinks off.

Start with the rectifier and power output tubes, since these two are the most common offenders in this regard. When the cause of the blinking is found, the tube must be discarded and replaced.

When a tube with a thermal intermittent condition in its heater is present in a receiver that does not have a pilot lamp, the same symptoms of play, die down, and then play again occur. But this time, we are not guided to it by the blinking of a pilot lamp. Now, the technician must recognize that the receiver gradually loses volume, dies, and then the volume gradually builds up again. The effect is as if the line plug had been pulled out and then reinserted. The gradual nature of the fade-out and build-up should lead to suspicion of the heater circuit. Observation within the receiver then shows the tube heaters dimming as the volume dies, and the heaters brightening just before the volume starts to rise. The same servicing technique outlined before with the neon lamp shows which tube has the intermittent thermal open.

Troubles due to a short between heater and cathode. Heater-type tubes are so constructed that the heater is inside a closely fitted cathode, and shorts between the two are fairly common. In an a-c receiver, both heater and cathode are near ground potential so that a short between them will affect a bias potential or introduce a hum, confining the trouble to one stage. In an a-c/d-c receiver, a similar trouble will have more widespread effects.

Consider the circuit of Fig. 9-10 and assume a short between the cathode and heater of the 12BE6 converter tube. Since the converter cathode is connected to the common negative through the low resistance of the feedback coil, a short between its heater and cathode will short out the detector first a-f tube heater.

With glass tubes, the situation would be sufficiently obvious as trouble somewhere in the heater circuit; when metal tubes are used, the trouble would be more obscure. Signal check would show a normal second a-f stage and a dead first a-f stage, and the technician would lose time checking a perfectly good first a-f stage until he reaches the point of replacing the tube, when he would notice that it is not even warm.

Similarly, a short between heater and cathode of the 12BA6 i-f tube would short out the converter and detector first a-f tubes. In either of the above cases, the line voltage would divide itself among the remaining tubes and make them much brighter than normal.

The fastest way of recognizing troubles of this sort is for the technician to be on the alert for troubles in the heater circuit. These constitute a large proportion of all the service difficulties experienced with a-c/d-c receivers. Even if the tubes in the i-f, converter, and detector first a-f stages are metal tubes, the rectifier and second a-f tubes are usually glass. When the heaters in these tubes appear to be too bright, a heater-cathode short should be suspected and checked for.

The tubes can be tested in a tube checker which discloses cathode-heater shorts. Possibly a faster method of finding the offending tube is to remove the detector first a-f tube, fully expecting the others to remain lighted. When this happens, the cathode-heater short is confirmed and the converter tube is also removed. If the other tube heaters start to dim

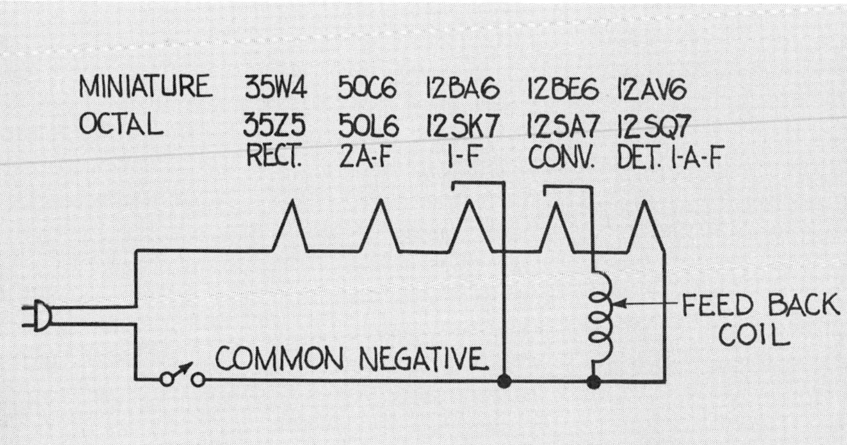

FIG. 9-10. How a cathode-to-heater short affects other tubes.

down, the short is in the converter tube; if they remain bright, the shorted tube is still in the receiver. It is probably in the i-f tube, which is then removed for confirmation.

Troubles in the *B* power supply in a-c/d-c receivers. The *B* power supply in a-c/d-c receivers is similar to the *B* power supply in a-c receivers, except for the lower voltages involved. The unit is subject to the same troubles which were described in detail in Chap. 8. To avoid repetition, the trouble and service procedures will be briefly outlined here, except for those circumstances which apply to a-c/d-c receivers only.

When the quick check discloses normal heater operation, coupled with hum, motorboating, low or no *B*+ voltage, the trouble is probably in the *B* power-supply section of the receiver.

Troubles common to the rectifier tube. Aside from trouble in the heater, the rectifier tube may become weak or entirely inactive, resulting in low or no *B*+ voltage. When this is the case, a tube checker will confirm the condition. Before replacing the tube, the technician should first check the filter circuit. This should be done because rectifier tubes used in a-c/d-c receivers have an easily fused cathode lead inside the tube. This internal cathode lead will melt on any overload, such as that caused by a shorted filter capacitor, thereby ruining the new tube.

Troubles common to the input filter capacitor. The input filter capacitor commonly opens and occasionally shorts. When the capacitor is open, the receiver hums and the *B*+ voltage is low—approximately 30 volts. The receiver may still operate weakly at the low voltage, but the quality of the reception will be badly garbled by hum.

The low voltage is due to the use of a half-wave rectifier, the output of which is half of the average of the applied voltage, when the rectifier is not followed by a capacitor input filter. This is the condition when the input filter capacitor is open. The rectifier output drops to approximately 50 volts, which is reduced to 30 volts on the *B*+ end of the filter. The best check for this condition is to substitute a test capacitor.

When the capacitor shorts, it will short the rectifier output, giving zero *B* voltage. The condition will be found on resistance check, the resistance from the rectifier cathode to *B*− checking short instead of the charge reading of the capacitor. The capacitor should be replaced with one of the same capacitance and voltage rating as the original. The rectifier tube and surge resistor, if used, will also have to be replaced since the short will have ruined them.

When replacing either filter capacitor in an a-c/d-c power supply, the technician should be careful to use the correct capacitance and voltage rating. Using capacitors with higher voltage rating is not advisable, since experience has shown that electrolytic capacitors rated at 450 volts used for replacement purposes in a-c/d-c receivers deteriorate rapidly.

Generally, there is no room to install an extra filter capacitor. Therefore, it is necessary to replace the entire unit when any capacitor in the filter unit becomes defective. A stock of 50-30 mfd/150 volt filter capacitors will be found to be satisfactory for replacement purposes for most receivers.

In printed circuit receivers, an exact replacement filter capacitor is usually to be preferred because its leads are specifically placed for mounting the part. However, there is some small leeway, in that the mounting leads may be bent somewhat to fit the holes in the printed board. When making such a replacement, remove the old capacitor by heating each contact in turn while gently pulling the part from the other side of the board. The holes are then cleared of any remaining solder. The replacement capacitor should be carefully oriented so that the leads go through the correct contact holes, and are then pushed into place. Finally, solder the leads on the underside of the board.

FIG. 9-11. Typical filter capacitor and its position in the circuit.

Troubles common to the output filter capacitor. The output filter capacitor *C*-16 also opens and rarely shorts. When it is open, the receiver operates, *B*+ voltage is normal, the receiver may hum slightly but it surely will squeal or motorboat, or both. This effect is much more apparent in a-c/d-c receivers than a similar condition in a-c receivers, since the output filter capacitor is usually the sole bypass agent for all screen and plate circuits.

The best check for this condition is to bridge the capacitor with a test capacitor.

The output filter capacitor may also short. This condition will be found by no voltage at *B*+ and almost normal voltage at the rectifier cathode. The rectifier tube may not be harmed owing to the intervening resistance of the filter resistor. The short, however, is not necessarily in the output filter capacitor. A skeleton diagram showing the *B*+ circuits of a typical a-c/d-c receiver is given in Fig. 9-12. The short may be anywhere on the *B*+ line. For example, if one of the primary leads in one of the i-f transformers were to touch the shield can, all points on the *B*+ line would show a shorted condition to common negative. A procedure for determining the location of the short is given in the next section.

The output filter capacitor may also develop internal resistance and draw heavy current. In this case, the voltage at *B*+ will measure below its normal reading of 85 to 100 volts, and the filter resistor will probably overheat. The procedure for locating shorts will also disclose a leaky output filter.

The skeleton diagram of Fig. 9-12 will help to solve another short which may be confusing. Suppose capacitor *C*-12 in the plate circuit of the second a-f tube were to short-circuit. The receiver would not play, there would be a heavy current drain through the short, and cathode resistor *R*-13 would overheat. A voltage check would give the clue. The voltage at rectifier cathode would be low—about 70 volts, indicating a heavy drain, or a weak rectifier. The overheating resistors point to the heavy drain. The voltage at *B*+ would measure about 60 volts—low, to be sure, but normally lower than the voltage at the rectifier cathode. This places the short in the second a-f stage. Continued voltage checking shows approximately 30 volts at the second a-f plate, and the same 30 volts at the second a-f cathode. An ohmmeter now confirms the short in plate capacitor *C*-12.

How to check for shorts on the *B*+ line. The diagram of Fig. 9-13 is a layout diagram of the bottom view of a conventionally wired a-c/d-c

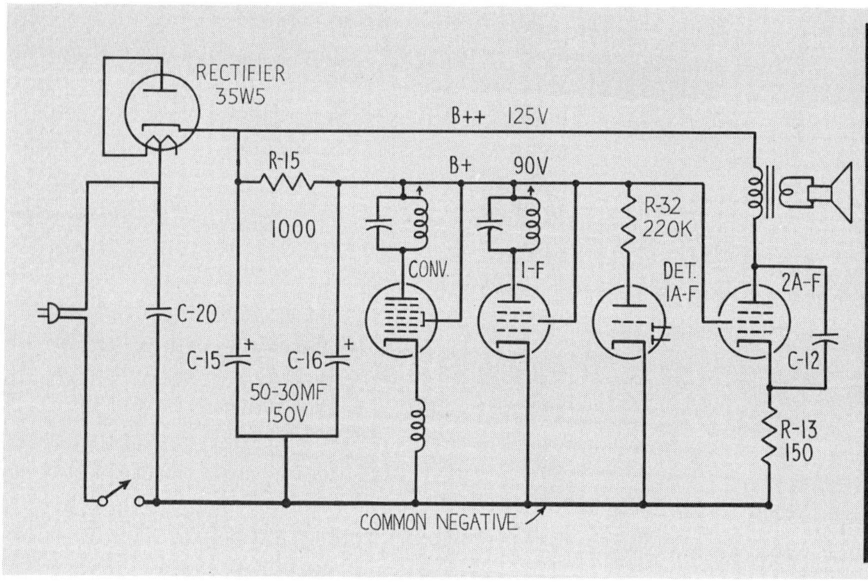

FIG. 9-12. Skeleton *B* circuit of a typical a-c/d-c receiver.

receiver, showing only the *B*+ wiring and the parts associated with it. In Fig. 9-12, note that the screen grid pins of all tubes are connected to *B*+. In Fig. 9-13, trace all *B*+ wiring to the screen grid terminals of the tube sockets.

To find the location of any short, it is necessary to open the *B*+ line. Start by removing all leads from the screen terminal of the output tube. Then check with an ohmmeter from common negative to each of the opened leads. One lead will give the "short" reading. The procedure is illustrated in Fig. 9-14. If the short is in the output filter capacitor, the latter is replaced. If, as in our previous example, the short is in an i-f can, the lead showing the "short" reading will be the one that leads to the screen grid terminal of the next tube.

All wires are then removed from this terminal, and each one is checked separately for the short. Again, only one lead will give the "short" reading. This lead is traced to the shorted component, and the short should be cleared. Replacement notes and procedures for clearing shorts in components other than the power supply are given in the chapters dealing with the stage to which the component belongs.

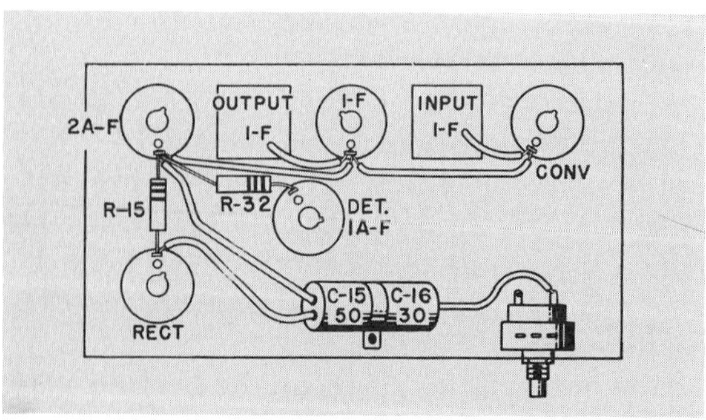

FIG. 9-13. Bottom view of a conventionally wired a-c/d-c receiver showing the *B*+ wiring.

In a printed-circuit receiver, although the schematic diagram is the same, a short on the *B*+ line requires a different servicing technique. This is because the printed wiring does not converge at the screen pins, as is the case in a receiver with conventional wiring. The drawing of Fig. 9-15 shows the *B*+ and common negative wiring of a typical printed-circuit receiver. Refer to this diagram and also to Fig. 9-12 and note how the *B*+ printed wiring takes in the screen pins of the tubes, one lead on each i-f transformer, the filter resistor, the filter capacitor, and the couplate which in-

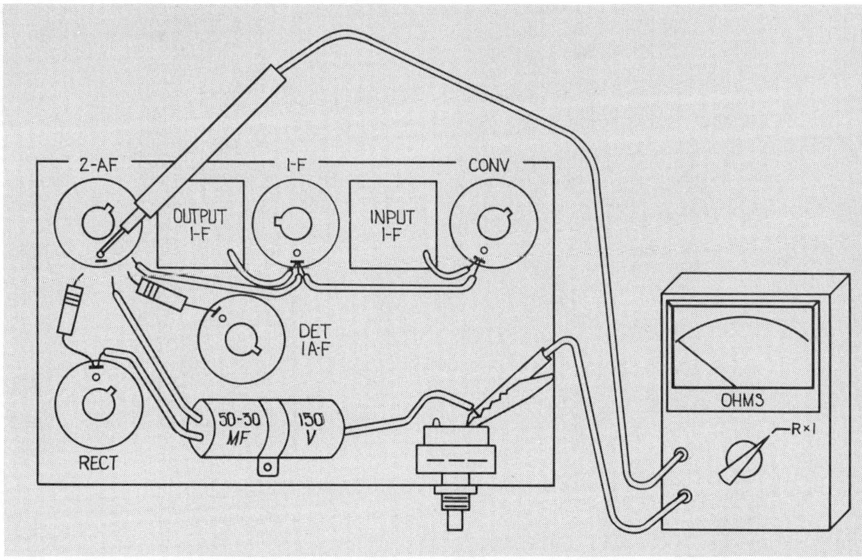

FIG. 9-14. How to check for shorts in the B+ line.

cludes resistor *R*-32 in the first a-f plate circuit. Just enough of the common negative printed wiring is shown to give some identification points. These are the common connection of the filter capacitor, the center grounding pin used on some of the tube sockets, and the shielding connection of the i-f cans.

In the printed-circuit receiver, just as in the receiver with conventional wiring, a short anywhere in the *B*+ line gives a reading of zero ohms when testing with an ohmmeter from common negative to any of the *B*+ points. Again, the *B*+ wiring must be opened to localize the short.

The procedure just given for conventionally wired receivers is adapted to the printed-circuit receiver in the following way. Choose a point near the center of the *B*+ printed wiring, and open it by making two close parallel cuts across it with a razor blade. Scrape out the foil between the cuts. The procedure is illustrated in Fig. 9-16. Now, checking with an ohmmeter shows which half of the printed wiring contains the short. A second similar cut in the middle of the shorted section further localizes the short to one or two components. The short is then found and cleared.

Finally, the cuts are repaired by bridging an

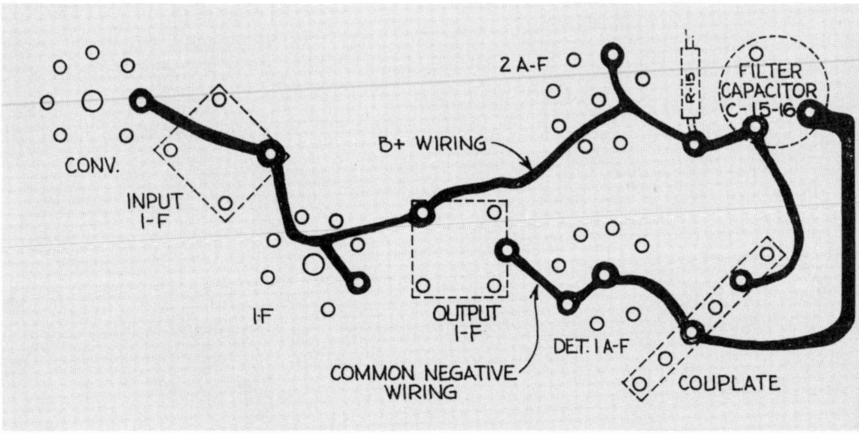

FIG. 9-15. Bottom view of a printed circuit receiver showing the B+ wiring and some of the common negative wiring.

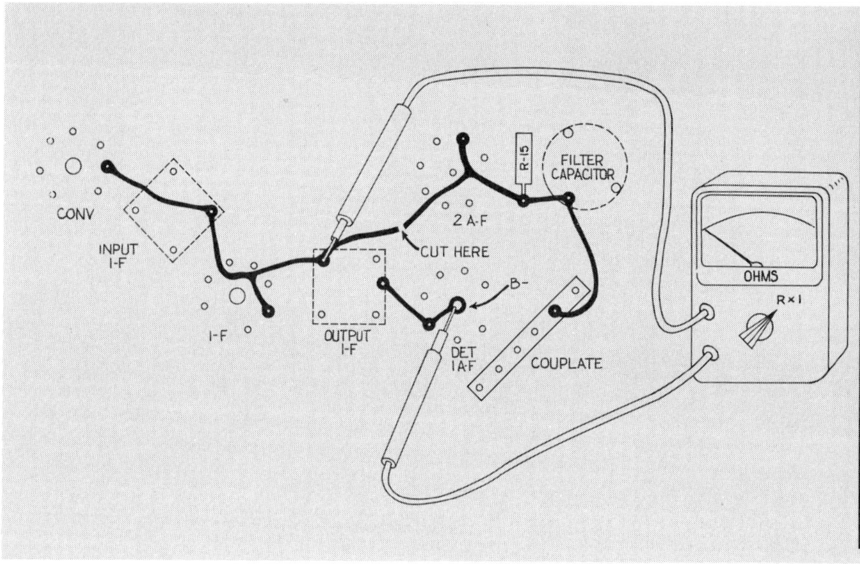

FIG. 9-16. How to check for a short on the *B*+ line of a printed circuit receiver.

entire cut section with a piece of hookup wire and soldering it securely at both ends. The cut shown in Fig. 9-16 would be repaired by preparing the hookup wire long enough to go from the *B*+ connection on the output i-f transformer to the screen connection at the number 6 pin of the second a-f tube, following the bends in the original printed wiring. After the wire is soldered in place at these two connection points, it may be held at the bends with a drop of service cement.

Troubles common to the filter resistor. As can be seen in Fig. 9-12, filter resistor *R*-15 carries the *B* current of the converter, i-f, and detector-first a-f tubes. Normally, it carries this load without difficulty. However, any short in the *B*+ line puts an overload on the resistor, causing it to overheat. This may cause it to change in value. Therefore, whenever a short in the *B*+ line is corrected, filter resistor *R*-15 should be carefully checked and replaced if there is any sign of harm caused by the overload.

Sometimes filter resistor *R*-15 burns open. In this case, voltage at *B*+ drops to zero, while the voltage at the rectifier cathode remains normal or goes up slightly. The condition is readily found by voltage and resistance checks, and the resistor is replaced. If the size of the original cannot be ascertained, a 1,000-ohm/1-watt resistor will work satisfactorily in most receivers.

Troubles common to the line filter. Capacitor *C*-20 in Fig. 9-17 is the line filter. It serves the same function as the similarly placed line filter in transformer-type receivers, and also gives the same servicing problems. Sometimes it opens; almost never does it short.

An open capacitor also produces the same symptoms—modulation hum; that is, the receiver hums on station but not between stations. The suspicion of an open line filter capacitor is confirmed when the bridging of the original with a test capacitor removes the modulation hum.

When replacing a defective line filter, you need not be too concerned about the exact capacitance. A capacitor rated at about 0.05 mfd/400v will be satisfactory for almost any receiver.

General service notes pertaining to a-c/d-c receivers. The circuit of an a-c/d-c receiver causes some problems in service procedure and techniques. For example, when a signal

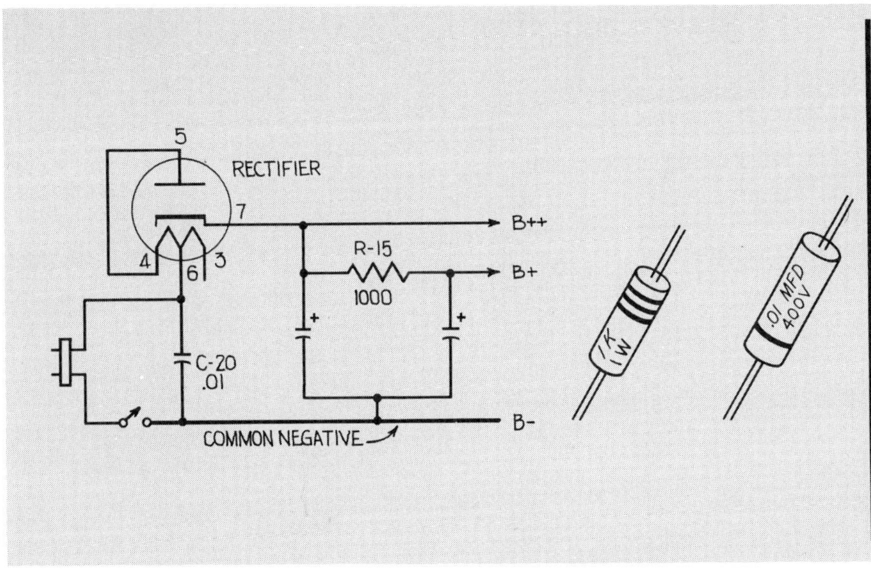

FIG. 9-17. Filter resistor *R*-15 and line filter *C*-20 and their position in the circuit.

generator is connected for signal check or alignment, a bad hum may be experienced. It can usually be avoided by connecting the shielded lead from the signal generator to the common negative rather than to the chassis. Another expedient that sometimes gives good results is to connect the shielded ground lead through a 0.1-mfd capacitor. The isolating capacitor will still be used in the "hot" lead.

Another difficulty often experienced is when the test bench is ringed with a grounded metal trim. This will cause shocks to the technician and danger of short circuits if the chassis should touch the metal trim. Another danger of short circuits exists when the signal generator test lead is connected. If this has exposed shielding, it is likely to touch any grounded object such as the plate of an electrical outlet, thereby causing a short circuit.

As previously mentioned, most of these difficulties may be overcome by the use of an isolation transformer. This is a one-to-one ratio transformer. The primary winding is connected to the line, and line voltage is available at the secondary winding because of the unity ratio. The a-c/d-c receiver is plugged into the secondary winding. Because of the structure of a transformer, the receiver is isolated from the line, one side of which is grounded.

The diagram of Fig. 9-18 shows the setup for using an isolation transformer when connecting a signal generator to an a-c/d-c receiver. The use of the isolation transformer is recommended for all service work on a-c/d-c receivers.

FIG. 9-18. Use of an isolation transformer with an a-c/d-c receiver.

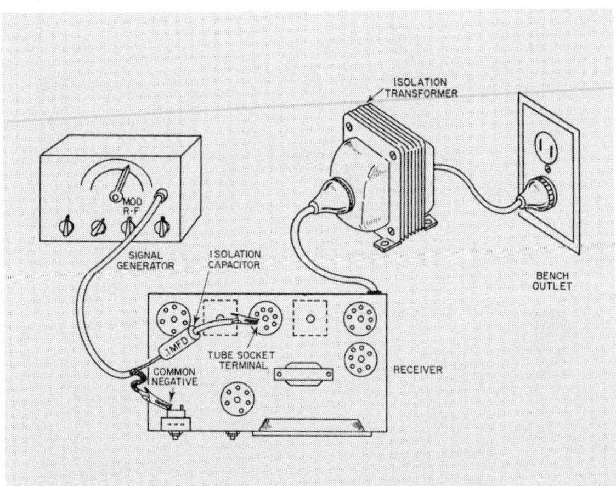

SERVICE DATA CHART

SYMPTOM	ABNORMAL READING	LOOK FOR
Tubes do not light	Plug prong to prong checks open	Open line cord or interlock. Open switch. Open heater in one of the tubes. Break in wiring.
Some tubes are overly bright, others do not light or warm up		Cathode-heater short circuit in one of the tubes
Tubes light—no reception	No *B*+ voltage	Dead rectifier tube, short-circuited input filter capacitor, or both
Tubes light—no reception	No *B*+ voltage. Low voltage from common negative to rectifier cathode	Short-circuited output filter capacitor. Short in the *B*+ wiring
Tubes light—no reception	No *B*+ voltage. Voltage from common negative to rectifier cathode measures 150 volts	Open filter resistor. (Discharge filter capacitor before checking)
Tubes light—no reception	*B*++ low. *B*+ low. Second a-f cathode voltage high	Shorted second a-f plate by-pass capacitor
Tubes light—bad hum	*B*+ voltage measures 30 volts	Open input filter capacitor
Tubes light and receiver motorboats, squeals, or both	*B*+ voltage normal but jumps with the motorboat	Open output filter capacitor
Modulation hum	All tests normal	Open line filter capacitor C-20

SUMMARY

Quick check for normal operation of the A-c/D-c power supply.

All tubes light at normal brightness.
The hum level is normal.
There are no bad squeals or motorboating.
The *B*++ voltage measures about 120 volts to common negative.
The *B*+ voltage measures approximately 90 volts to the common negative terminal.

Standard circuit. This circuit is shown in the accompanying figure.

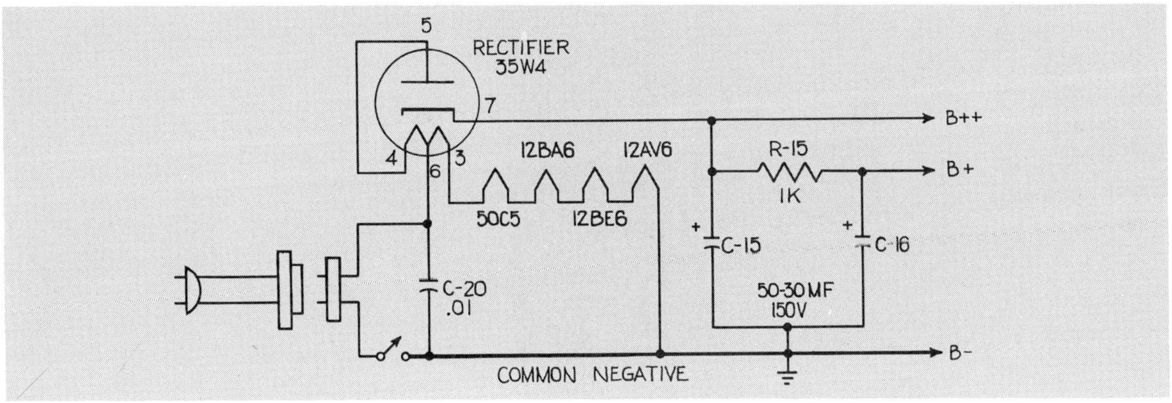

Normal resistance data.

Plug, prong to prong (switch on)	100–200 ohms
Rectifier cathode to *B+*	1,000 ohms
Rectifier cathode to common negative	Capacitor action
B+ to common negative	Capacitor action

Normal voltage data.

Common negative to rectifier plate	117 volts A-C (line)
Common negative to rectifier cathode	110–130 volts D-C
Common negative to *B+*	85–100 volts D-C

QUESTIONS

1. The tubes in an a-c/d-c receiver do not light. List the possible causes and explain how you would check for each.
2. A dead a-c/d-c receiver is brought in for repair. All the tubes light, there is no hum or squeal, and the *B+* voltage measures zero. List the likely causes of trouble and outline how you would check for each.
3. How would you check for the most probable cause when an a-c/d-c receiver hums?
4. What precautions should be taken before replacing a dead rectifier tube in an a-c/d-c receiver?
5. In an a-c/d-c receiver, the 35Z5-GT and the 50L6-GT tubes light up very brightly. The receiver does not play. How would you check for the likely cause of the trouble?
6. What is the most probable cause of a tunable hum (modulation hum) in an a-c/d-c receiver?
7. When an a-c/d-c receiver squeals badly, which unit in the power supply can cause this condition? How would you check to determine whether or not this unit is at fault?
8. In the *B* circuit breakdown of Fig. 9-12, assume that an ohmmeter check from common negative to *B+* gives a reading of 15 ohms. List some of the likely locations of the short. Outline a procedure that could be used in tracking down the short.
9. List the bench provisions and techniques that should be observed when working with a-c/d-c receivers relative to the use of grounds, accidental grounding, and connections to a signal generator.
10. What is the most probable cause of a blinking pilot light in an a-c/d-c receiver? Outline a test procedure that will determine the exact cause.

Power-Supply Variations

10

The power-supply circuits described in the previous chapters are standard or basic in that they incorporate the most widely used practices. Power-supply circuits found in individual receivers, however, differ somewhat, and may require adaptations of the servicing procedure that has been outlined for the standard circuits. In this chapter, the power supplies in some representative receivers will be described, showing the variations, together with applicable servicing techniques where required.

Power supplies that use two-section filters. The filter system in the standard transformer-type power supply described in Chap. 8 uses a choke and two capacitors. The filter system in the standard transformerless power supply of Chap. 9 uses a resistor and two capacitors. Although the single-section filters are generally adequate in reducing hum in receivers, some manufacturers employ a second filter section, and sometimes even a third, for better hum elimination.

An a-c/d-c power supply using a two-section filter is shown in Fig. 10-1. Resistor *R*-15 and capacitors *C*-15*A* and *C*-15*B* make up the first section of the filter which supplies the maximum *B*+ voltage for the power output stage. Note the low ohmic value of resistor *R*-15, which is in series with *B*+ Max. This will always be a low value so that the voltages available at the high *B*+ point will be as high as possible, consistent with good hum elimination. Resistor *R*-16 and capacitors *C*-15*B* and *C*-15*C* make up the second section of the filter. These provide a lower value of *B*+ voltage with very good hum elimination for the rest of the receiver.

The filter sections shown in Fig. 10-1 are known as R-C filters, since they are made up of resistors and capacitors. When a choke is used in place of a filter resistor, the circuit is known as an L-C filter.

The two-section filter circuit of Fig. 10-1 calls for three capacitors, *C*-15*A*, *C*-15*B*, and *C*-15*C*. These are generally contained in one filter capacitor unit. A typical unit of this type is included in Fig. 10-1. Note the semicircle over capacitor *C*-15*B* in the diagram, and the triangle over capacitor *C*-15*C*. These symbols are used to identify the different capacitors in the filter unit. Note also the absence of a symbol over capacitor *C*-15*A*. Now note that the accompanying drawing shows similar symbols and markings. The semicircle or triangle symbol is cut into the insulation near each soldering terminal of the capacitor unit. When a filter unit contains four capacitors, as would be the case in a three-section filter, a square symbol is added for the fourth capacitor.

Servicing a set with a two-section filter. The

complete schematic diagram of a Motorola clock radio which uses a two-section filter is given in Fig. 10-2 as a typical example of a receiver of this type. The diagram is reproduced from the Motorola service manual so that you can note differences in drafting practices and become familiar with all types of service diagrams.

Examine the lower part of the diagram, which shows the power-supply circuit. Note the indication at the plug that the receiver is for use on alternating current only. The set itself uses a standard a-c/d-c circuit and will work on either type of current, but the clock will burn out on direct current; hence the notation. The clock motor indicated by the coil is across the line at all times. The on-off switch for the radio is included in the clock unit so that it can be preset to go on and off at specified times.

Tracing the hot lead from the clock, we find a heater line and rectifier circuit exactly like our standard circuit. Note the marking of 125v at the rectifier cathode. This is the normal voltage to be expected at this point as measured from common negative. It is identified both on the circuit diagram and on the printed-circuit board by three adjacent squares. Normal voltage readings are found at the important socket terminals throughout the diagram. Also note the black half-circles on the heater line wiring. Similar half circles are included on the printed board in the receiver for easy identification of the heater circuit. In the same way, a single square identifies the *B*+ line in the diagram and on the printed wiring, while two adjacent squares denote the *B*++ wiring.

Tracing the lead from the rectifier cathode, we come to the two-section filter, consisting of resistors *R*-10 and *R*-11 and the three capacitors marked *C*-6*A*, *C*-6*B*, and *C*-6*C*. The three capacitors are in a single can similar to the unit shown in Fig. 10-1. Note that capacitor *C*-6*A* is uncoded, *C*-6*B* has a triangle, and *C*-6*C* has a square. These codes identify the individual capacitors in the can. Similar markings identify the capacitors on the insulation around the terminals and on the label of the can.

When servicing a receiver of this type, the same general rules and checks already studied will apply. We still want to know whether the tubes light, whether the set hums or squeals, and whether the voltage at the rectifier cathode measures between 110 and 125 volts.

If any of these checks shows that there is trouble in the power supply, you would then apply the same tests and service procedures you learned in connection with the standardized power-supply circuit. The heater power circuit and the rectifier are standard. In the filter circuit, the main difference is that you have three filter capacitors to check instead

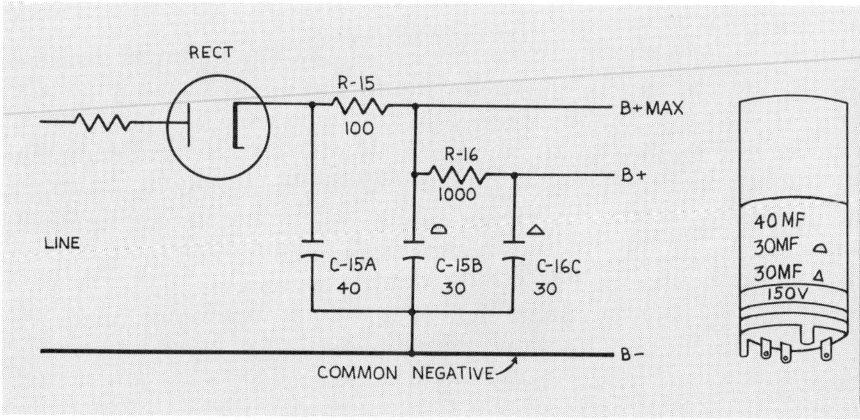

FIG. 10-1. Two-section R-C filter and typical filter capacitor used in this circuit.

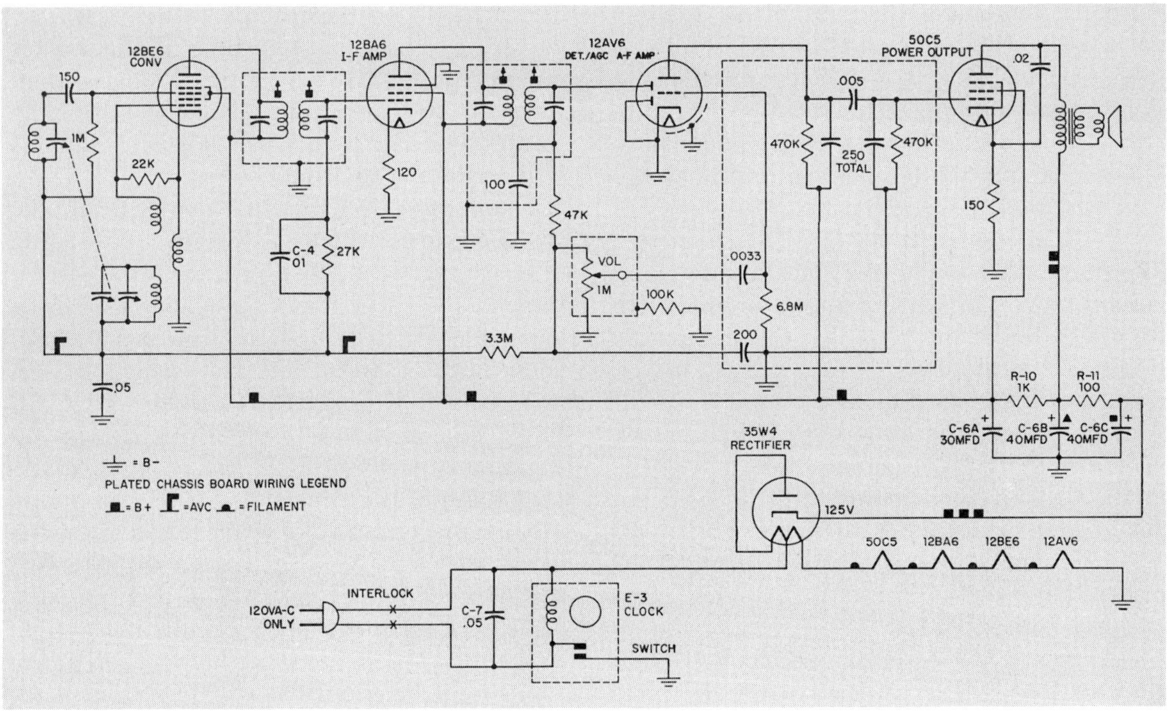

FIG. 10-2. Schematic diagram of Motorola clock radio which uses a two-section filter circuit.

of only two. Filter capacitors are subject to the same ills regardless of whether there are two or three: they open or lose capacitance, develop large leakage currents, and sometimes short.

If the input filter capacitor opens, the voltage at the rectifier cathode will drop to approximately 50 volts. The hum level will go up, although it will not be as high as is the case with a single-section filter because the two remaining capacitors now function as a single-section filter.

If the intermediate capacitor with the triangle marking should open, the voltage at the rectifier plate would be normal, but the hum level would increase because of insufficient filtering for the second a-f plate-supply voltage.

If the output filter capacitor with the square marking were to open, the result would be the same as when the output capacitor in a single-section filter circuit opens. That is, there will be increased hum, loss in volume, and probably squeals and motorboating.

The check for an open filter capacitor is the same as has been given before. Connect a test capacitor across each suspected unit and see whether it clears up the difficulty. In this case, there would be three test points instead of two.

Similarly, excess leakage in any of the filter capacitors would cause a decrease in voltage output with a consequent lack of volume. There would also be excess heating in the filter resistors affected. A completely shorted capacitor would make the output *B+* voltage drop to zero, and the set would not play at all. These defects would be located by the same procedures that were studied before except that this time the procedure would be applied to three capacitors instead of two.

If the filter capacitor unit is found to be

defective, it should be replaced. A stock 50-30-mfd replacement filter with an additional 20-mfd/150-volt capacitor could be used, but you might find difficulty squeezing the extra capacitor into the chassis. There is the problem of space requirements in these small receivers. It would be much better, therefore, to obtain an exact replacement, even though it may take more servicing time to get the part.

Use of output transformer in the filter circuit. In some receivers, the primary of the speaker output transformer is tapped, providing a section which is included in the filter circuit. Figure 10-3*A* shows the circuit used in a transformer-type power supply. The corresponding circuit in an a-c/d-c power supply is shown in Fig. 10-3*B*. In both circuits, transformer *T*-6 is the output transformer that couples the signal energy from the audio output tube to the loudspeaker. The primary winding *L*-12 is tapped, and the lower section is in series with filter resistor *R*-15. The tap is near the *B*+ end of winding *L*-12, not in the center. The purpose of the tapped connection is to reduce the hum level of the speaker. The lower section of coil *L*-12 is known as the hum-bucking winding.

The complete schematic diagram of an RCA a-c/d-c receiver is given in Fig. 10-4 as a typical example of a receiver which uses a tapped output transformer in the filter circuit.

Study the power-supply section. Note that the heater line is like the standard circuit of Chap. 9, except that provision is made for the use of a dial lamp, by connecting it between pins 4 and 6 of the rectifier heater. Resistor *R*-6 is connected in series with the rectifier plate to take over the surge-limiting function of this portion of the rectifier heater in the standard circuit.

The power-supply section of the diagram shows only filter capacitor *C*-10*B*. The rest of the filter circuit appears in the audio section. Trace the lead from rectifier cathode to the tap on the output transformer, through the hum-bucking winding to filter resistor *R*-12. The other end of this resistor is the *B*+ lead

FIG. 10-3. Filter circuit which uses a tapped section of the output transformer.

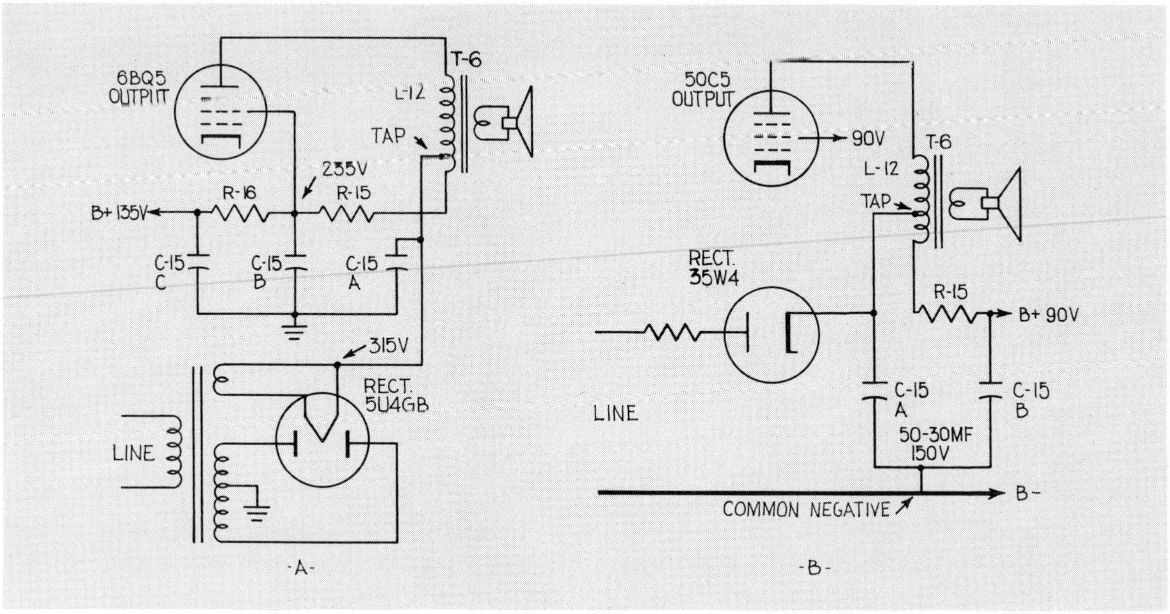

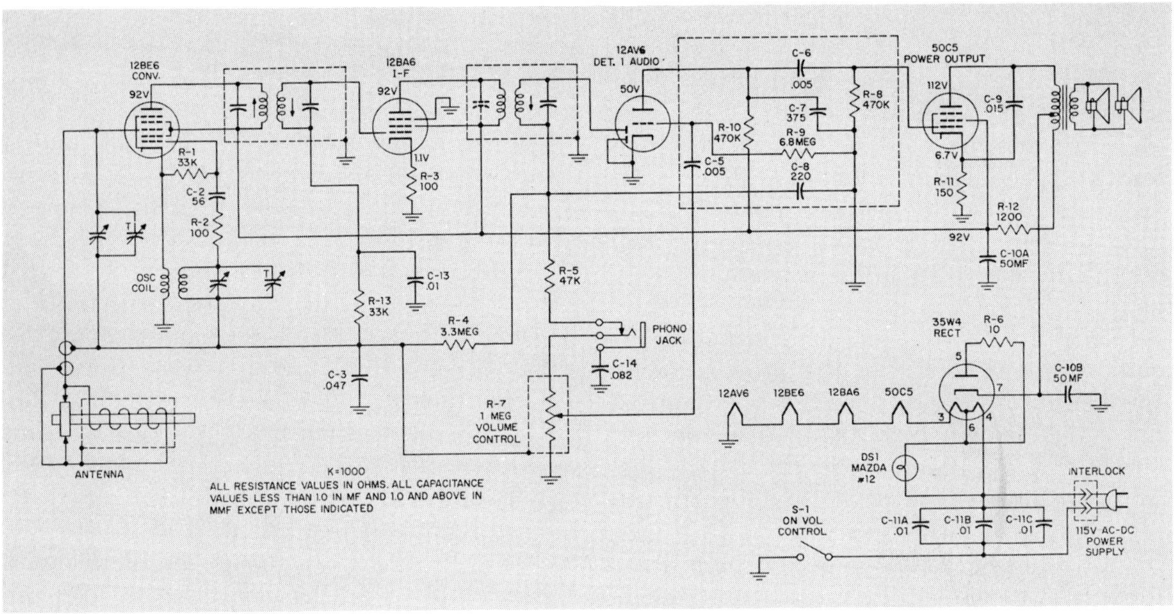

FIG. 10-4. Schematic diagram of an RCA a-c/d-c receiver incorporating a tapped section of the output transformer in the filter circuit.

to the rest of the receiver. It is marked 92 volts in the diagram, and is the connection point for output filter capacitor *C*-10*A*.

From the servicing point of view, all checks are the same as for a conventional receiver. The only problem that is encountered is that an exact replacement must be obtained, if the output transformer should become defective. In replacing the transformer, the leads must be connected properly, since any reversal would cause increased hum, weak output, or both.

If an exact replacement transformer cannot be obtained and the receiver is large enough to include an extra filter capacitor, a repair can be effected by using a standard matching transformer, omitting the hum-bucking winding. If the hum level is then excessive, an extra filter circuit is added, as shown in Fig. 10-5. The original filter circuit is shown by *R*-15 and capacitors *C*-15*A* and *C*-15*B*. The added filter, consisting of resistor *R*-15*D* and capacitor *C*-15*D*, will make up for the hum-bucking effect of the original transformer.

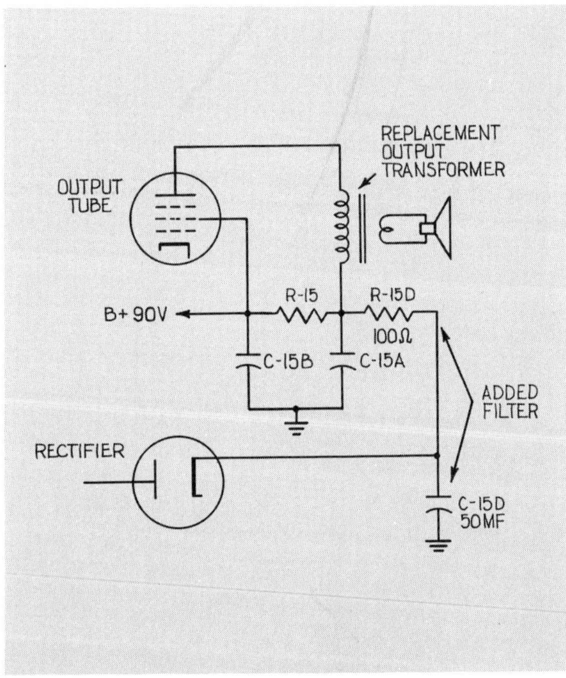

FIG. 10-5. Replacing a tapped output transformer with a standard replacement transformer, and adding a filter, if needed.

Compact a-c/d-c receivers using the 100 milliampere heater string. The a-c/d-c receiver described in Chap. 9 consumes about 30 watts. Although these sets can be made quite small, the cabinets are generally somewhat larger than necessary because of the heating factor of the 30 watts that is dissipated. To appreciate this, imagine a 30-watt lamp enclosed in a cigar box, turned on for 4 hours or more. The entire box would become quite hot. Even with the oversize cabinet, many a-c/d-c receivers become uncomfortably warm, especially if they are pushed close against a wall, thereby cutting down on air circulation which is usually provided by vents in the back cover. This, incidentally, gives a cure for a radio that becomes too warm. Pull it away from the wall.

The heaters in the tubes of our standard circuit draw a current of 0.15 ampere. Tubes with heaters designed to operate at 0.10 ampere have been developed; these give a reduction of 6 watts in heater circuit consumption, and a more compact a-c/d-c receiver can be made. The schematic diagram of such a receiver is shown in Fig. 10-6.

The power supply is at the bottom of the diagram. Although it is drawn somewhat differently than our standard circuit, trace the leads and you will see that it is exactly the same. First check the heater line. Add up the first two numbers of each tube type, and you will have a total of 122, which is just about right for operation from normal line voltage. Note that the 18FY6, the last in the series chain, is the detector-first a-f tube. Check the other tubes, and you will see that they are in the same order as our standard

FIG. 10-6. Schematic diagram of an Arvin compact a-c/d-c receiver.

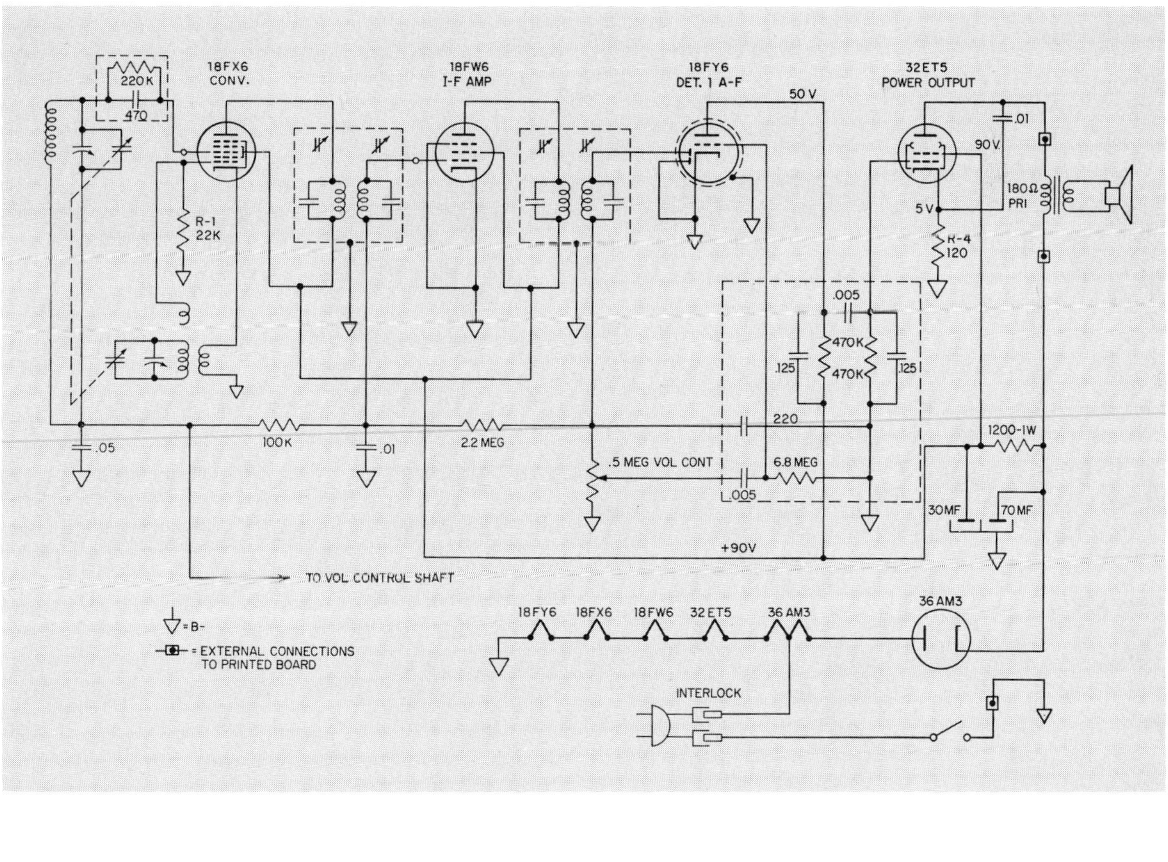

circuit. Note the different symbols used for the interlock, common negative, and the filter capacitor.

From the servicing point of view, this receiver is checked for any trouble exactly like the standard circuit. To make matters easier, these tubes use the same pin numbers as the equivalent miniature-base 0.15-ampere tubes of the standard circuit. The only precaution needed is to use the correct type numbers for replacement tubes.

Six-tube a-c/d-c receivers. Some a-c/d-c receivers have six tubes in the filament line. The sixth tube is usually found in sets designed for greater sensitivity, where an r-f amplifier stage is added to the basic five-tube circuit.

The standard five-tube a-c/d-c receiver has a carefully worked out series heater circuit where the heater potentials add up to 120 volts, so that it can be fed directly from the line. If another tube were added, the line voltage would be insufficient, and all tubes would be operated below their normal brilliancy.

The series-heater problem was solved by the development of special tubes. The diagram of Fig. 10-7 shows the heater circuit of a six-tube receiver, where the added tube is a 12BA6 r-f amplifier. The 50C5 power-output tube is replaced by a 35C5 which has very similar characteristics, except for a 35-volt/0.15-ampere heater. The voltage saved by this change is used to light the 12BA6 heater. The heater potentials now add up to 118 volts, which is correct for the average lighting line.

Servicing the power supply in one of these six-tube receivers is no more difficult than servicing the standard five-tube circuit. The only complication that has been added is that there is one more tube heater to check in case the tubes do not light.

Power supply for an a-m/f-m receiver. The schematic diagram of Fig. 10-8 shows the power supply of a large Westinghouse a-m/f-m phono receiver. The diagram was chosen as a typical example of receivers of the type where one power supply furnishes current for the various sections of a large receiver. The receiver is built on two chassis. One contains the tuner and i-f amplifier; the other contains the a-f amplifier and power supply. Plugs and cables interconnect the two chassis.

Capacitor *C*-20 and resistor *R*-20 in the primary circuit of the power transformer form the line filter. These rarely give any service difficulties. Note that the ohmic resistance values given near each winding of the power transformer conform to the low readings to be expected when checking a large transformer, as shown in the resistance chart of Chap. 8. The high-voltage and rectifier circuit are the same as in the basic circuit.

Note that there are two 6-volt heater windings. The one for the a-f amplifier heater is center-tapped, with the tap grounded. This arrangement is to minimize hum. The 6-volt winding for the i-f amplifier heaters is grounded, as in the standard circuit. Note r-f choke *L*-1 and capacitor *C*-1 in the feed line to the tuner heaters. These components form an r-f filter in this line. The tuner tubes operate at very high frequencies (88–108 mc) and the filter is used to prevent feedback effects in the heater line which may cause interaction at these high frequencies. The filter shown is present in the power supply. There is generally another connected at the socket in the tuner assembly. The r-f filter in the heater line is operating in a low-voltage circuit, and almost never gives servicing difficulty.

The rectifier output feeds the *B*+ maximum point through a filter circuit consisting of resistor *R*-15 and capacitors *C*-15*A* and *C*-15*B*. The filter circuit differs from the basic circuit in that the filter choke is replaced by a resistor, forming what is known as an R-C filter. The choke in the L-C filter of the basic circuit is more effective in reducing hum, but the overall effectiveness of the filter is not impaired, since the capacitors in an R-C filter circuit are generally much larger. Note the

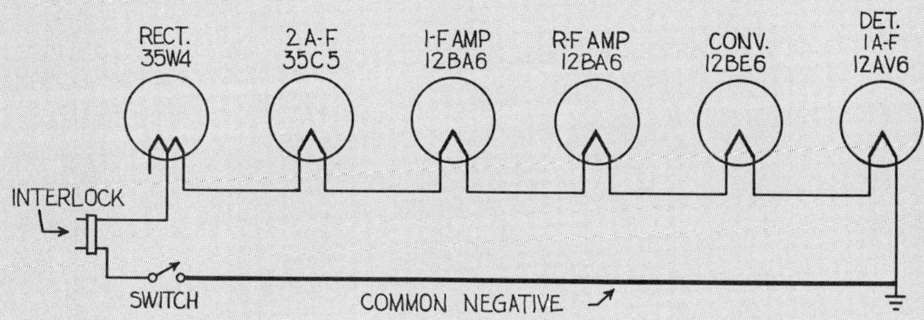

FIG. 10-7. Heater circuit of a six-tube receiver using an r-f amplifier stage and a 35C5 in the second a-f socket.

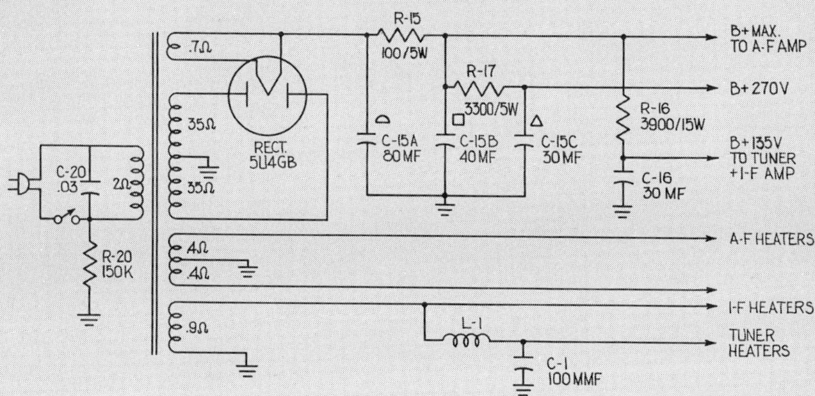

FIG. 10-8. Power supply in a Westinghouse a-m/f-m/phono receiver with stereo amplifiers.

80-40 mfd capacitors used in the circuit of Fig. 10-8 as compared to the 20-20 mfd units of the basic circuit.

From the servicing point of view, the R-C filter presents few new problems. An open filter resistor is found by the same procedure that shows an open filter choke. However, a filter resistor is more subject to harm from temporary overloads than a choke. A shorted condition further down on the *B*+ line would drive heavy current through the filter. The resultant overheating is less likely to do permanent damage to a choke than to a resistor. It is wise, therefore, when correcting such a condition to check filter resistors carefully and to replace any that have changed in value from the overload. In making the replacement, the technician should be careful to use the same ohmic value and wattage rating as the original.

Another interesting feature of the circuit of Fig. 10-8 is the voltage distribution system. The *B* circuit has been redrawn in Fig. 10-9 to show how *B* voltage is supplied to the various stages. The maximum *B*+ voltage is fed directly to the stereo power-amplifier stages. The plate circuits of the a-f amplifier stages are fed through resistor *R*-17. Resistor *R*-17 and capacitor *C*-15*C* serve a dual function. The resistor reduces the voltage to the a-f amplifiers, and the resistor-capacitor combination provides a second-section filter for further reduction of hum. Similarly, resistor *R*-16 and capacitor *C*-16 provide a lower voltage and a second section filter for the r-f and i-f amplifiers. Note the 15 watt size of resistor *R*-16. This is because of the heavy load of the r-f and i-f amplifier tubes. The receiver has five of these, although the diagram indicates only two because of lack of space.

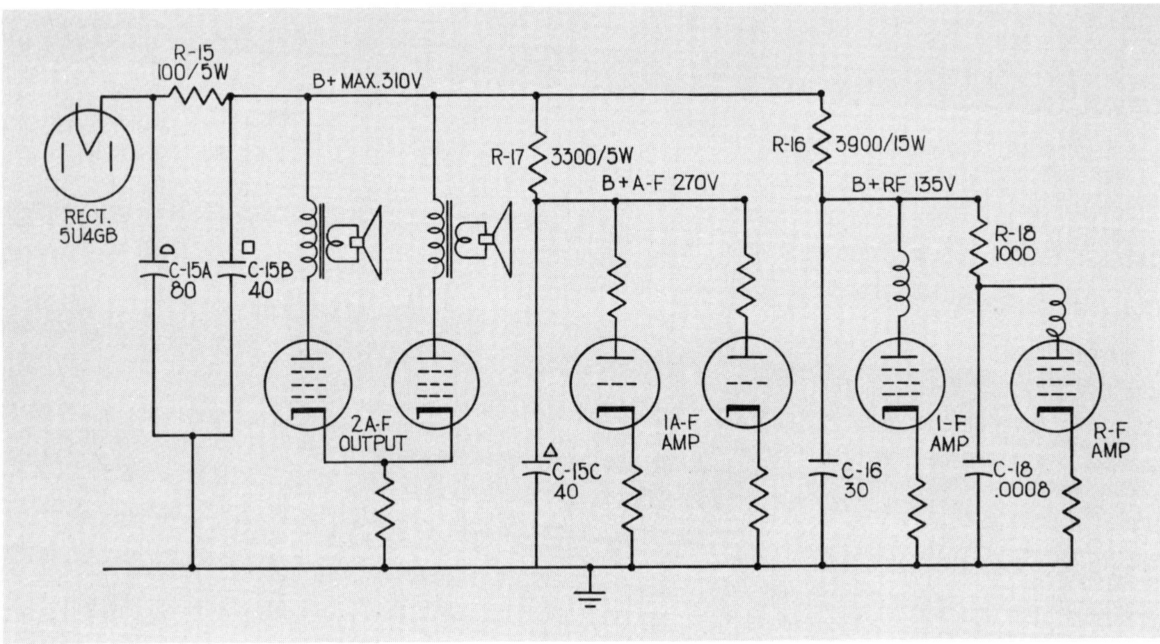

Fig. 10-9. Skeleton *B* circuit of the Westinghouse a-m/f-m/phono receiver.

The plate circuits of the tuner and i-f stages are often supplied through individual R-C filters represented by resistor *R*-18 and capacitor *C*-18 in the feed circuit to the last tube in the diagram. These act as decoupling filters to prevent interaction between the high-frequency stages, along the common *B*+ lead. They are located in the stage, and so the decoupling filters do not appear in the power-supply diagram of Fig. 10-8. However, they are in the power-supply feed lines, and a shorted condition in capacitor *C*-18, for example, would affect the entire power supply. The receiver would not play on a-m or f-m but would operate on phono, although possibly at reduced efficiency. The overload caused by the shorted capacitor would show below normal readings at *B*+ 310 and *B*+ 270, and a very low reading at *B*+ 135. Checking along this line, the technician would come to the overheating resistor *R*-18, and zero plate voltage at the associated r-f or i-f plate. Capacitor *C*-18 is now suspected of the short, and the condition is confirmed by an ohmmeter check.

The shorted capacitor is now replaced, taking care to position the replacement capacitor and its leads exactly as the original. This is done so that the lead dress is not disturbed, a factor which may be important in high-frequency decoupling circuits. It is also considered good practice in preventive maintenance to replace resistor *R*-18, since it may have been harmed by the overheating.

A-c power supplies with 6-volt heater-cathode type rectifier tubes. Many transformer-type power supplies make use of a 6CA4, or the older 6AX5 or 6X4 cathode-heater types of rectifier tubes. A basic circuit of this type is shown in Fig. 10-10. The circuit is generally found in smaller receivers.

The heater-cathode type of construction permits the rectifier tube to be operated from the same heater line with the other amplifier heaters as shown in the diagram. This avoids the need for a separate rectifier filament winding in the power transformer. Examination of

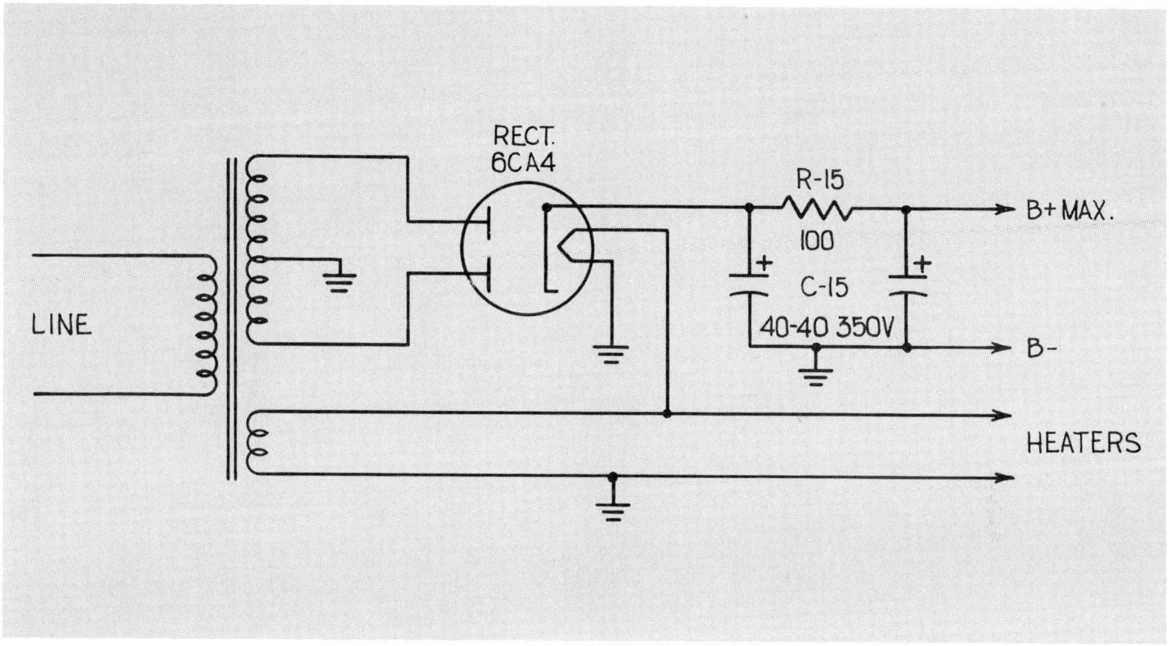

FIG. 10-10. A-c power supply with 6-volt heater-cathode rectifier.

the circuit shows that the heater is at ground potential, and the cathode is at the maximum *B+* voltage. There must therefore be good insulation between the heater and cathode. The tube manual rates the heater-cathode insulation of the 6CA4 as being good for 500 volts maximum. This is sufficient for most receivers because *B+* maximum rarely exceeds 350 volts.

In service work, however, the heater-cathode rectifier quite often turns out to be the cause of power-supply failure. Reference to Fig. 10-10 shows that leakage or a short between heater and cathode in the rectifier shorts out the power supply, just like a short in input capacitor *C*-15. Voltage at *B+* and rectifier cathode will check zero, and the rectifier will overheat. Both the rectifier tube and capacitor *C*-15 should now be suspected as possible causes. An ohmmeter check, taken from rectifier cathode to *B−* may not show the short, because this is taken with the set turned off. The heater which in its hot expanded condition was touching the cathode, may now, when cold and contracted, have cleared the short. The best procedure, therefore, is to make the ohmmeter capacitor check. If the result is a normal capacitor charge reading, the tube is probably at fault, and operation is tried with a new tube. If the result is a low-ohmic reading, the short may be in the tube or a filter capacitor. The tube is now removed from its socket, and an ohmmeter reading taken again. If the result is the normal capacitor charge reading, the tube is defective. If the low ohmic reading persists, the tube is good, and the filter circuit is checked for the short in the usual way.

The schematic diagram of the power supply used in a Magnavox f-m receiver is given in Fig. 10-11 as a typical example of the use of a cathode-heater rectifier. This is a receiver of intermediate size, as can be confirmed by noting the ohmic values given for the primary and high-voltage windings of the power transformer.

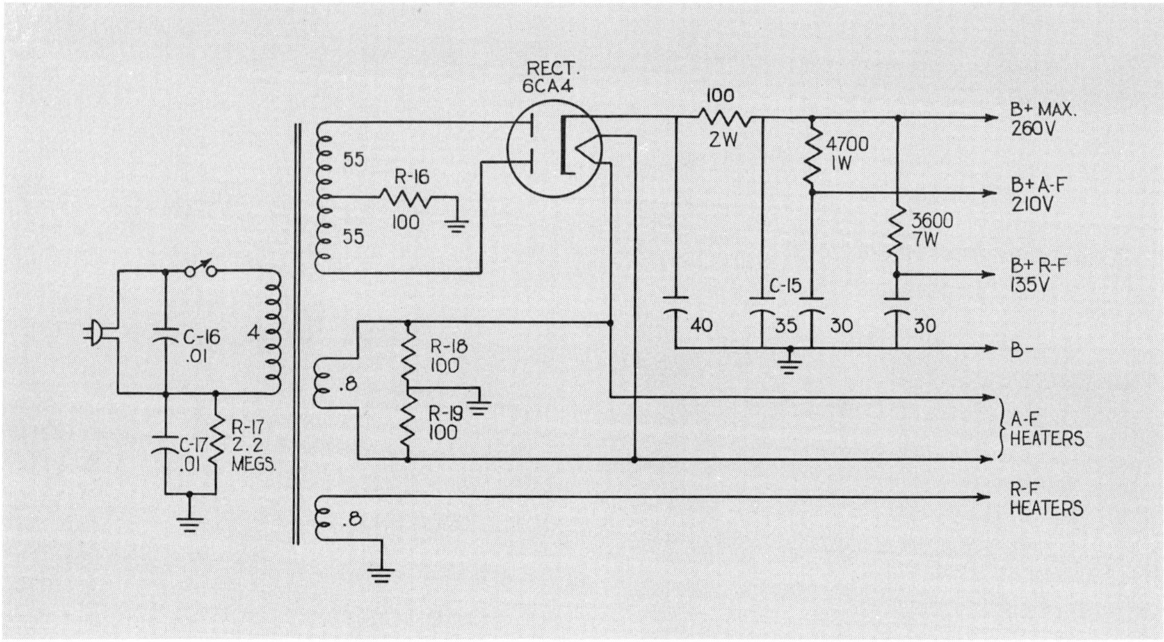

FIG. 10-11. Power supply of a Magnavox f-m receiver.

The circuit shows two features of power-supply design not previously described. One is the arrangement used to ground the six-volt heater winding for the a-f tubes. Two 100-ohm resistors, *R*-18 and *R*-19, are connected across the winding in series, and the junction of the resistors is grounded. This provides an electrical center tap for grounding the heater winding. In some receivers, the two resistors are replaced by a potentiometer, the arm of which is grounded. The potentiometer has a slotted shaft which acts as a hum adjustment. The use of the resistors or the potentiometer rarely causes any servicing difficulty.

The other feature is the presence of resistor *R*-16 connected in series with the center tap-to-ground connection of the high-voltage winding. This reduces the initial charge current of the input filter capacitor so as not to exceed the plate current-carrying capacity of the rectifier plates. The resistor serves the same function as the surge resistor in a-c/d-c power supplies. Refer to resistor *R*-16 in Fig. 9-3. The series connection of the resistor in the center tap of the high-voltage lead in Fig. 10-11 permits the use of a single surge resistor in a full-wave rectifier circuit.

In service work, surge resistors often open. In a receiver like that of Fig. 10-11, there may be some difficulty in that the presence of the surge resistor may not be expected. The condition would be found by the following sequence. If the surge resistor is open, the tubes would light, but the set would not play. Routine checks would show no voltage at *B*+, and no a-c input voltage at either rectifier plate. The high-voltage winding of the power transformer is now under suspicion, and a resistance check would show both plates open. A resistance check from plate to plate would show a normal reading for the entire high-voltage winding, and would therefore point to an open circuit in the center tap-to-ground lead. The tracing of this lead, or reference to the schematic diagram, would reveal the presence of the resistor, which is then checked for the open circuit.

Use of semiconductor diode rectifiers. In many receivers, silicon or selenium semiconductor rectifiers, sometimes known as dry rectifiers, replace the high-vacuum tube rectifiers of our basic power-supply circuits. They are generally found in a-c/d-c receivers using six or more tubes. The power-supply circuit described in Chap. 9 was developed around a five-tube a-m superheterodyne receiver. When f-m or stereo functions are added to this circuit, more tubes are needed, and the simple five-tube series heater string must be redesigned. A semiconductor rectifier does not have a heater. Its use to replace the 35W4 tube rectifier gives an easy solution in that 35 volts is thus made available to heat extra tubes in the series heater string.

Drawings of silicon and selenium rectifiers, and their symbols are shown in Fig. 10-12. Note the plus sign near one terminal of each. The connection marked with the plus sign corresponds to the cathode of a vacuum-tube diode rectifier. The other connection, generally unmarked, corresponds to the plate.

The schematic diagram of a Granco a-m/f-m receiver is shown in Fig. 10-13 as a typical example of a receiver using a silicon rectifier. The power supply is redrawn in Fig. 10-14. The heater line shows a series string of six tubes connected directly across the a-c line. The heaters of the r-f and i-f tubes are bypassed to ground, as is usual in high-frequency circuits. The feed line to the f-m tuner tube, the last in the chain, is drawn right through its bypass capacitor. This symbol indicates that the capacitor is incorporated in the feed-through connection into the shielding of the tuner sub-chassis.

FIG. 10-12. The symbol for a semiconductor diode, and silicon and selenium types used in receivers.

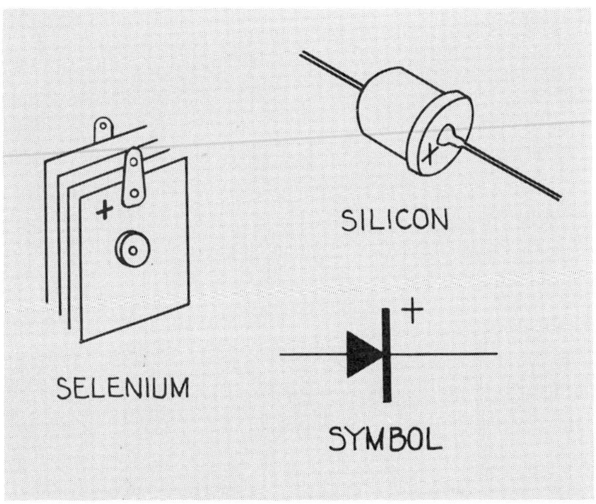

In the *B*-supply circuit, the silicon rectifier replaces the vacuum-tube diode. Note the plus sign near the terminal which corresponds to the cathode of the tube rectifier. Because of the extra tubes in the receiver, the current requirements in the *B* circuit are greater than in the case of the five-tube superheterodyne receiver. As a result, larger filter capacitors are needed to maintain voltage. The receiver of Fig. 10-14 uses a capacitor unit of 80-50 mfd, as compared with the 50-30 mfd unit of the basic circuit.

Note also that the voltage at *B*++ is marked at 134 volts, somewhat higher than the average of 120 volts for the five-tube superheterodyne set. With a half-wave rectifier, the maximum rectifier output voltage is the peak value of the a-c input and is 168 volts for a 120-volt line. The actual d-c voltage present will be less than 168 volts, depending on the size of the load, the size of the filter capacitors, and the voltage drop across the rectifier. Silicon rectifiers cause a smaller drop than the corresponding tube rectifiers, and so the output voltage is generally higher in the former case. Some manufacturers use filter capacitors rated at 200 volts, rather than the 150-volt rating common in a-c/d-c receivers.

Resistor *R*-16, in series with the rectifier, is the surge resistor whose purpose is to protect the rectifier from the large initial charging current of the filter capacitors. It also has a second function. In case of a short or overload on the *B*+ line, it passes a heavy current and

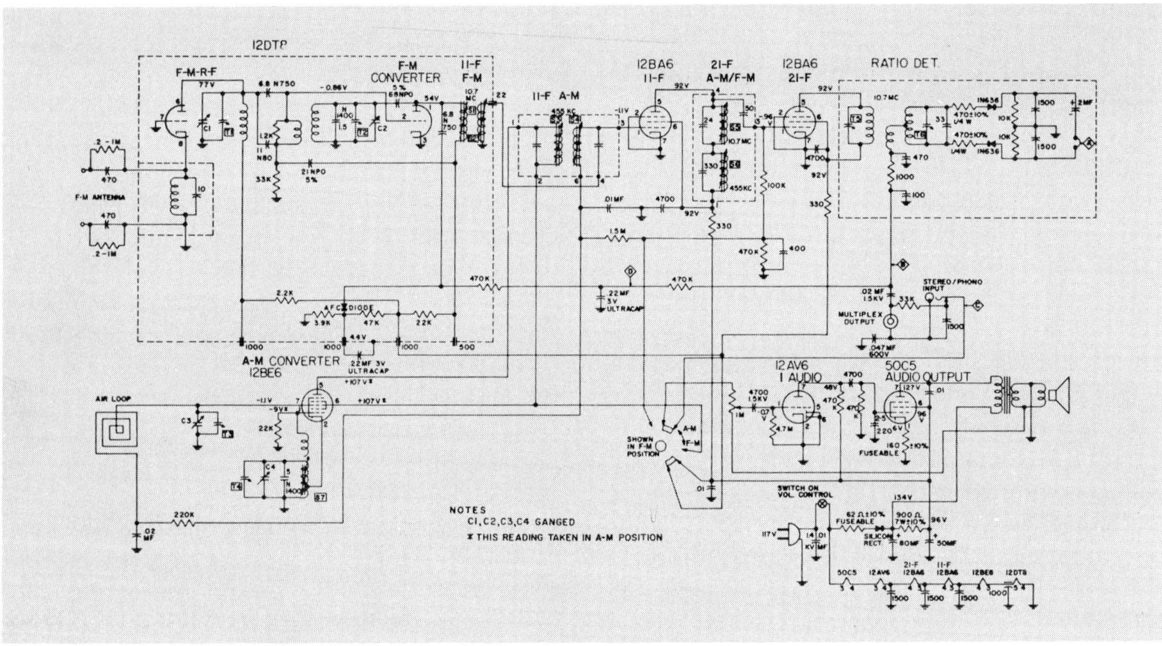

FIG. 10-13. Schematic diagram of a Granco a-m/f-m receiver.

FIG. 10-14. Power supply in the a-m/f-m receiver of Fig. 10-13.

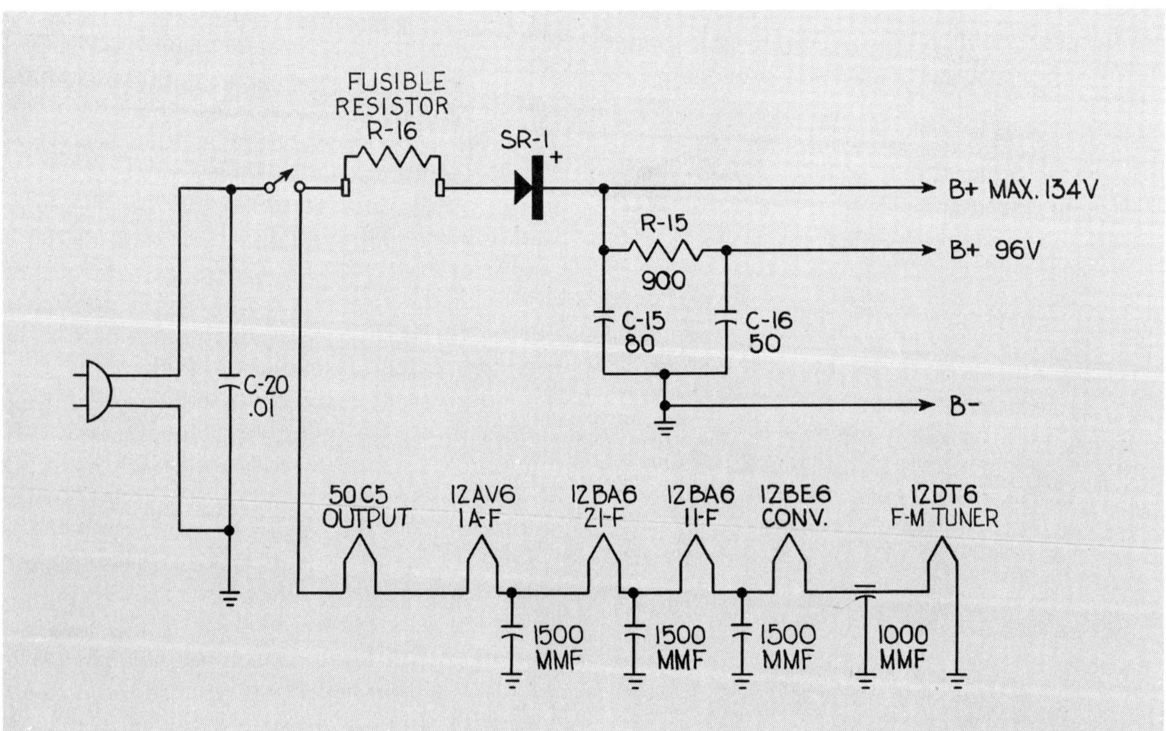

burns out before the rectifier is harmed. For this reason, it is known as a fusible resistor. It is usually a wire-wound resistor equipped with pin leads for easy replacement.

Servicing receivers with silicon or selenium rectifiers. The servicing procedures for an a-c/d-c receiver with a semiconductor rectifier is the same as the one described in Chap. 9, for the standard a-c/d-c receiver. The only difference is that the rectifier is not as easily changed for test purposes. To illustrate, let us suppose that the receiver of Fig. 10-13 does not play and is brought in for repair.

Preliminary examination shows no sound from the speaker, all the tubes light, and the voltage at *B*++ is very low or zero. A study of the power supply in Fig. 10-14 indicates that the condition could be caused by an open in fusible resistor *R*-16, a dead rectifier, a short on the *B*+ line, or some combination of the three conditions.

The first step should be to examine fusible resistor *R*-16 for signs of being burned open. If these signs are evident, there is likely to be a short on the *B*+ line. This short must be found and cleared before the resistor is replaced. If the resistor looks normal, it should be checked with an ohmmeter. If the reading is open, the *B*+ line should also be checked with the ohmmeter to make sure it is not shorted, before replacing the resistor. This resistor may have opened due to such a short without showing signs of burning.

If the surge resistor checks good and the *B*+ line gives its normal capacitor charge reading, the trouble is most likely in the rectifier. This is now checked in the manner shown in Fig. 10-15. Disconnect the original rectifier from the circuit by opening one lead. If the surge resistor *R*-16 is equipped with pin leads, this makes a convenient way of opening the rectifier circuit, as shown in the diagram. Provide yourself with a replacement rectifier equipped with clips for test purposes. The diagram shows a selenium rectifier; either selenium or silicon could be used as a test rectifier, regardless of the type of the original. A rectifier rated at 130 volts and 500 ma is recommended for test purposes, since this size is heavy enough to work in most radio or television receivers. Clip in the test rectifier as shown in the diagram, being careful to observe proper polarity. Connect the voltmeter from common negative to *B*++ and turn on the receiver. If the output voltage is restored to normal, the original rectifier is defective and must be replaced.

In choosing a replacement rectifier, an exact duplicate is desirable, but is not absolutely essential. Check your service notes for the specifications of the original rectifier. It is generally rated at 130 volts. The current rating may be as low as 50 ma and, depending on the receiver, may go as high as 500 ma. Choose a rectifier with a current rating at least as high as the original one. In general, use the same type of rectifier; that is, replace selenium with selenium and silicon with silicon. However, if a selenium rectifier of the correct rating cannot be found, a silicon unit with the proper rating may be substituted. The reverse is rarely possible, since a set which originally used a silicon rectifier is not likely to have the space for the much larger selenium unit. When wiring in the new rectifier, be careful to observe the correct polarity. An improperly polarized rectifier may harm the filter capacitors.

Power supplies where the heater string is fed from the d-c *B*-supply. The diagram of Fig. 10-16 shows the power supply of an 8-tube Westinghouse a-m/f-m receiver. This circuit is interesting because the series 8-tube heater string, which requires more voltage than is obtainable from the a-c line, is connected to the d-c output of the rectifier. As was mentioned before, the combined action of the rectifier and input filter capacitor, particularly when low-loss rectifiers and large filter capacitors are used, provides a voltage which may be close to the peak value of the line at this point. Note the unusually large values of the capacitors in filter unit *C*-45. These very large values are needed to maintain the recti-

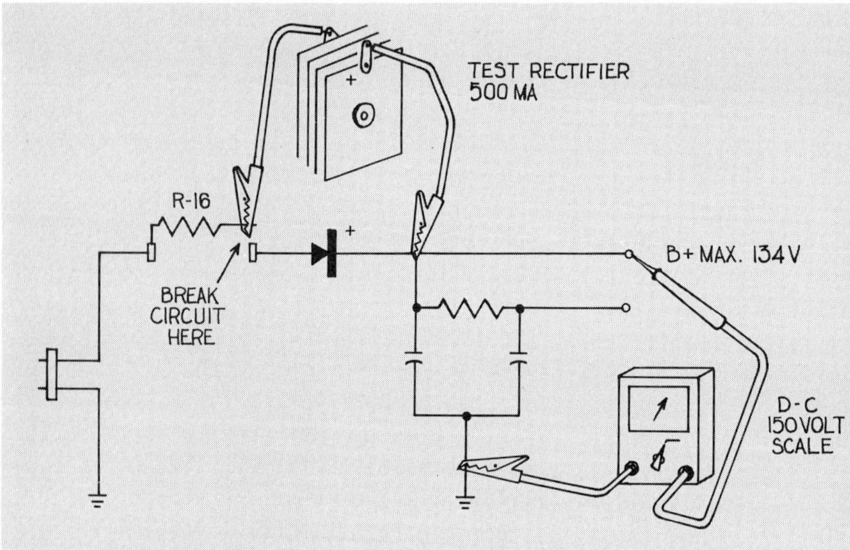

FIG. 10-15. How to check a semiconductor rectifier by substitution.

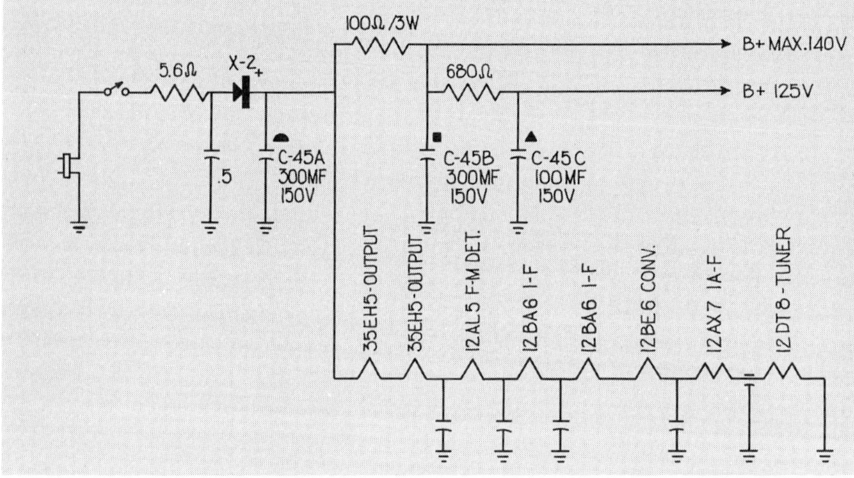

FIG. 10-16. Power supply in a Westinghouse a-m/f-m receiver.

fied voltage, since the load on the rectifier and filter system now includes the heater current as well as the plate current requirements of the receiver. Rectifier *X*-2 is rated at 375 ma in this receiver.

Service work in a power supply of this type may cause some difficulty, unless the technician is aware of the differences caused by the feeding of the heater line from the d-c *B*-supply. For example, if one of the heaters should open, the entire heater string would go out, just as in any a-c/d-c set. If the heater line is then checked with an a-c voltmeter, the d-c voltage which is present will give erroneous readings. If it is checked with an ohmmeter, and the filter capacitors remain charged, the ohmmeter may be harmed when the capacitors discharge through the instrument as it bridges the open heater. Also, failure of the tubes to light may not necessarily be caused by an open heater, open line cord or switch. It can also be caused by a defective rectifier or filter. Finally, as the rectifier and filter capacitors age, the symptoms in the conventional a-c/d-c receiver are weak reception and hum. In this receiver, the first sign of reduced output from the rectifier and filter system is an increase in the warmup time of the receiver. In either case, a voltage check at the *B*++ point indicates the cause of the trouble.

Power supplies with voltage-doubler circuits. The voltage at *B*+ max in a transformerless receiver is limited to approximately 150 volts. This is adequate for efficient operation of all receiver circuits. However, the usual audio output stage delivers approximately 2 watts of audio power to the loudspeaker when the voltage supply is 120 to 150 volts. Where larger outputs and larger speakers are required, common practice is to make use of a power transformer where the high-voltage winding produces *B*+ outputs of 250 volts or more. The power transformer is an expensive component, and so the a-c/d-c circuit is generally found in small inexpensive receivers. In this latter case, by use of a circuit known as a voltage doubler, it is possible to obtain a *B*+ voltage of approximately 250 volts, without the use of a power transformer. Voltage doubler circuits are found in many radio and TV receivers. We will now study the operation of the most commonly used voltage doubler, so as to gain a background for servicing that type of set.

How the voltage doubler works. Refer to Fig. 10-17. When the a-c line has the polarity shown in Fig. 10-17*A*, rectifier *SR*-1, like any half-wave rectifier, conducts and charges capacitor *C*-1 to line voltage or higher with the polarity shown. Now examine the circuit of Fig. 10-17*B*. This shows the condition on the reversal of line polarity. Now rectifier *SR*-1 does not conduct and is an open circuit. For this reason, it is omitted from the circuit diagram. But the line voltage across the capacitor *C*-1 is now in series with the line voltage, and with adding polarity. This double voltage is applied through conducting rectifier *SR*-2 to give a positive voltage of about twice line voltage at the load. The next alternation of line voltage only serves to charge up capacitor *C*-1 once again. This circuit is called a half-wave voltage doubler because the double voltage appears across the load only on half the input cycle.

The circuits of Fig. 10-17*A* and *B* are combined in Fig. 10-17*C* to show the complete voltage doubler. The series-connected heaters are shown in their usual position across the a-c line. Surge resistor *R*-16 serves its protective function in both rectifier circuits. It is usually a fusible type. Filter capacitor *C*-2 maintains the double voltage across the load.

If there were no load connected, capacitor *C*-1 would be charged to peak value of the line voltage, approximately 160 volts, and capacitor *C*-2 would be charged to twice this value or 320 volts. When the load is applied, the voltage drops to an extent depending on the size of the load and the capacitance of the filters. For the usual values in receivers, the double voltage available is 240 to 280 volts, as marked for *B*+ max on the diagram. Note that capacitor *C*-1 is rated at 150 volts and capacitor *C*-2, which is subjected to the double voltage, is rated at 300 volts.

Rectifiers *SR*-1 and *SR*-2 have the same rating. They are sometimes contained in one unit.

A typical voltage doubler. The schematic diagram of the power supply of a Sylvania 3-tube stereo phonograph is shown in Fig. 10-18 as a typical example of a voltage doubler. Connector *S*-1 is the interlock through which the amplifier chassis is connected to the line supply. There is no switch in this power supply, since this function is taken over by an automatic control on the record changer. The heater line includes series resistor *R*-1, since the three tubes require 112 volts, which is a little less than normal line voltage. The voltage-doubler circuit is exactly like the one shown in Fig. 10-17*C*, with a two-section R-C filter added for hum elimination. Note that the main 80-40-20 mfd/300v filter is in one unit, as shown by the semicircle, the square, and the triangle over each capacitor symbol. Capacitor *C*-1, in series with the line, is in a separate container.

Servicing voltage doublers. The voltage

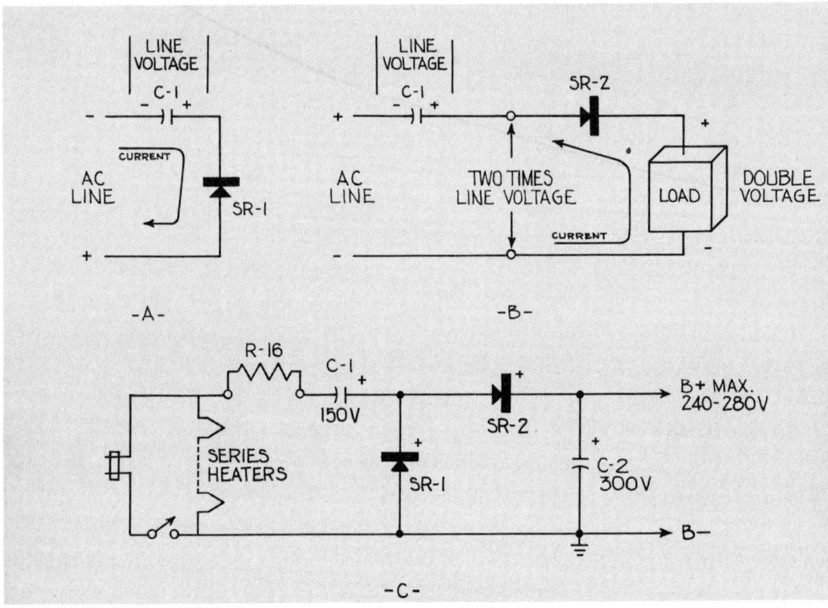

FIG. 10-17. How the half-wave voltage doubler works.

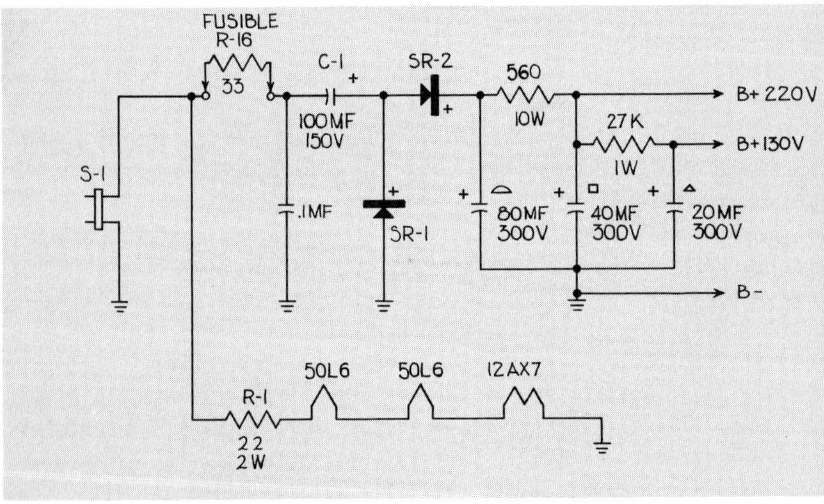

FIG. 10-18. Schematic diagram of the power supply of a Sylvania stereo phonograph as a typical example of a voltage-doubler circuit.

doubler is subject to the same defects that happen to any *B*-power supply. The series heater string may open. When it does, the tubes do not light and the set is dead. This condition is serviced like any a-c/d-c heater circuit.

In the *B*-power supply, the rectifiers weaken with age, the electrolytic filter capacitors dry up and lose capacitance, or an overload may develop on the *B*+ line. Any of these conditions causes the voltage at *B*+ max to drop below its normal reading of 240 to 280 volts. Also, the fusible resistor may open, in which case the a-c input to the rectifiers is open and the voltage at *B*+ measures zero.

If the tubes light and the voltage at *B*+ measures zero, the most likely cause is the open fusible resistor (*R*-16 in Fig. 10-18). This is confirmed by checking directly across the resistor with an ohmmeter. If the resistor checks open, look for a short on the *B*+ line before replacing it. Such a short, if present, caused the original resistor to open and will damage the replacement resistor if the short is not removed. The *B*+ line is checked with an ohmmeter, measuring from common negative to *B*+. The reading at these points represents the leakage resistance of all the capacitors on the *B*+ line, as well as the resistance across the two rectifiers. When making the check, reverse the test prods and take the higher of the two readings. If this reading is 10,000 ohms or more, you may assume no shorts and you may replace the resistor. If the meter gives a low ohmic reading. the short must be found and cleared before the resistor is replaced.

If the voltage at *B*+ measures considerably below normal, the trouble is probably a weak rectifier or filter capacitor, or both. The *B*-power supply of the receiver of Fig. 10-18 is repeated in Fig. 10-19 to illustrate the servicing procedure.

It is easy to check filter capacitors, so this is done first. When working on a voltage doubler for an open capacitor, do not make the mistake of connecting the negative end of a test capacitor to common negative and probing around the various *B*+ points with the positive lead of the capacitor. First examine the schematic diagram of the receiver being tested and the receiver itself, to locate the rectifiers and filter capacitors which are equivalent to *SR*-1, *SR*-2, *C*-1, and *C*-2 in Fig. 10-19. Schematic diagrams always indicate the positive terminal of rectifiers and electrolytic filter capacitors. Note that capacitor *C*-1 is not connected to ground nor to a *B*+ point. It will be in a separate container from capacitor *C*-2, which may be combined with other capacitors in the main filter unit.

Set up your multimeter as a d-c voltmeter on the 300-volt scale and leave it connected to common negative and *B*+, to observe the results of your tests on the output voltage. Observing proper polarity, connect a test capacitor directly across first one and then the other filter capacitor. If the output voltage comes up to normal when either capacitor is checked, that capacitor is open and must be replaced. If there is no change or little change when the capacitors are checked, they are probably good, and the rectifiers are defective.

To check a rectifier, unsolder one terminal, as shown in Fig. 10-19. This disconnects the original. Now, observing proper polarity, clip in a test rectifier and observe the output voltage. If the reading comes up to normal, the original rectifier is defective and must be replaced. If there is no change, the second recti-

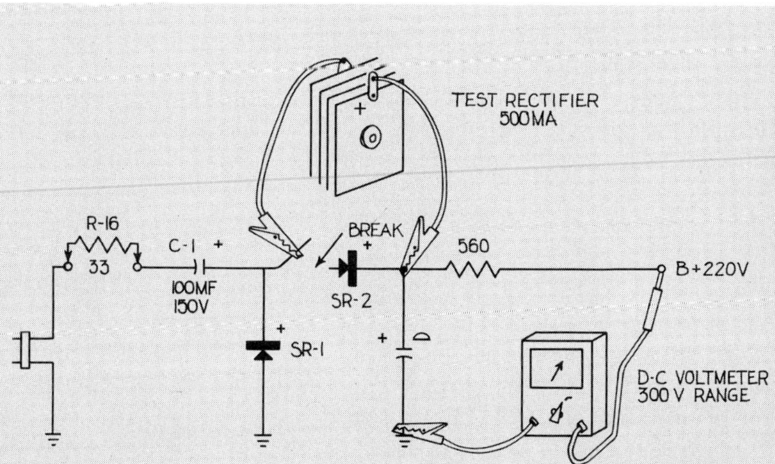

FIG. 10-19. How to check a rectifier in a voltage-doubler circuit.

fier in the power supply is checked in the same manner.

If either rectifier proves defective, it is usually considered good practice to replace both. If an exact replacement is not available, check your service notes for the rating of the original rectifiers and choose replacements of a similar type with similar ratings. Be careful to observe proper polarity when wiring in the replacement.

Small transformer-type power supplies. It is generally considered that the transformerless type of power supply is used in small receivers, and the transformer type for large ones. This is not entirely true, since the transformerless type of power supply with the very efficient silicon rectifiers and a voltage doubler circuit can be adapted for use in very large radio and TV receivers. A better reason for use of one or the other type of power supply involves the cost and quality characteristics.

A power transformer is an expensive component and will not be used in an economy line of receivers, either large or small. On the other hand, the primary winding of a power transformer isolates the receiver circuits from the power line. This permits the receiver to be grounded, reduces short and shock hazards, and introduces less r-f disturbances from the line into the receiver. The transformer type of power supply is therefore used in all receivers, large or small, where these advantages are considered important enough to warrant the cost of the component.

Small power supplies are found in small receivers and in self-powered tuners. A tuner is a radio receiver with all stages through the i-f and detector stages, but omitting the audio amplifiers. There is a trend in receiver design whereby a manufacturer can achieve a wide selection of styles and models by combining two or more self-powered units in one cabinet or console. The manufacturer makes two or three sizes of self-powered audio amplifiers. He also makes various self-powered tuners. The units are so designed that interconnecting any tuner and any amplifier produces a complete receiver. The system is used extensively in the a-m/f-m, hi-fi, phono, stereo combination field.

The power supply for a tuner is small, since the *B*+ requirement is generally 90 to 135 volts at a current drain of 30 to 50 ma. Some self-powered tuners use a transformer full-wave rectifier circuit, like the basic power supply of Chap. 8. These are serviced by the procedures described in that chapter. Just remember that components will be small, transformer-winding resistance readings high, and voltages low. Other self-powered tuners use a semiconductor rectifier in a transformer-type half-wave circuit.

The schematic diagram of a typical power supply of this type is shown in Fig. 10-20. Tuners generally contain most of the controls of the combination. When the on-off switch is located in the tuner, it becomes a complicated device because it must make provision for operation of the other units in the combination. Study the circuit around switch *S*-1. The switch is shown in the OFF position. Here the feed line to the tuner power supply is open at the top portion of the switch. The second section feeds current to the phono receptacle so that the switch on the changer can operate the phono motor and audio amplifier. The ON position of switch *S*-1 connects the tuner and amplifier to the line and opens the feed to the phonograph.

Transformer *T*-1 is the power transformer for the tuner. Note the comparatively high resistance given for the primary winding. The resistance reading of 20 ohms given for the secondary winding cannot be considered low because the winding produces only 100 volts, and so cannot be compared with the usual high-voltage winding. Rectifier *S*-2 is a 50-ma selenium unit. The rectifier output is smoothed by a three-section R-C filter.

From the servicing point of view, the power supply develops the same defects as any other. Rectifiers, whether tubes or semiconductors,

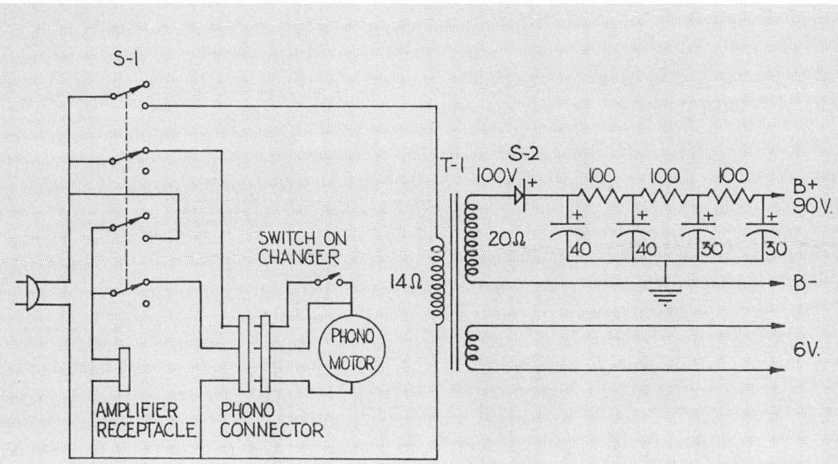

FIG. 10-20. Power supply in a tuner, showing typical switching arrangement.

weaken with age, and filter capacitors dry up and lose capacitance. In either case, the voltage at *B*+ will be below normal, and squeals may develop. The capacitors are checked by bridging them with a test filter capacitor; the rectifier by substituting a test rectifier. A resistance check from chassis to the *B*+ line is not very valuable in this circuit because semiconductor *S*-2 conducts on resistance check, giving a low reading which masks the normally high leakage resistance of the filter capacitors.

Low-voltage circuits rarely develop shorts. As a result, the power transformer and filter resistors rarely give any service difficulty. Similar low-voltage, low-current transformer-type power supplies are used in separate chassis multiplex units.

Unusual power supply. When the individual chassis in a combination unit are not self-powered, one chassis must furnish power to the other. Sometimes, the power supply is split up between the two chassis. The circuit of Fig. 10-21 is an example of such a circuit. Power transformer *T*-1 and the power supply for the audio amplifier are mounted on the amplifier chassis. Power transformer *T*-1 also furnishes 110 volts a-c and 6 volts a-c to supply the tuner. The rest of the tuner power supply is mounted on the tuner.

Note that the amplifier power supply is a standard a-c/d-c circuit, even to the series heater string. The main difference is that it is fed by winding *L*-2 of the power transformer, rather than from the line. Another difference is the absence of a surge resistor. The surge resistor function is taken over by the ohmic resistance of winding *L*-2.

Service work in this power supply would be as follows: If the tubes light, transformer *T*-1 is functioning and the trouble is in the *B*-power supply. This is then checked as outlined in Chap. 8. If the tuner tubes light, but the amplifier tubes do not, one of the amplifier tube heaters is open. These are then checked in the usual way. If no tubes light, trouble is in the line cord, switch, or transformer *T*-1. If transformer *T*-1 proves to be defective, an exact replacement is necessary.

Trouble in the tuner is indicated when the set gives normal operation on the phono, but not on the radio positions of the function switch. The trouble is traced to the tuner power supply when the voltage at *B*+ in the tuner is below normal. It is then traced to a defective rectifier or filter capacitor by the usual methods.

Power supply furnishing negative voltage for *C*-bias. In some of the large audio ampli-

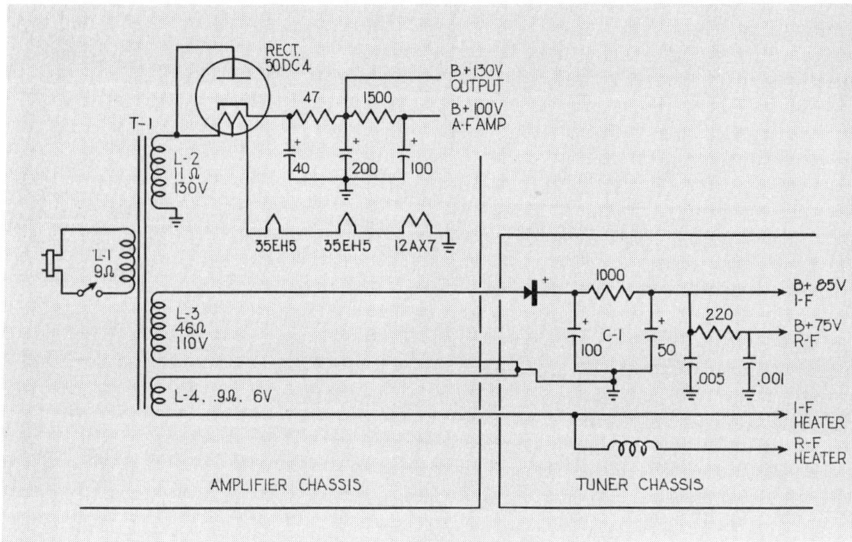

FIG. 10-21. Power supply in a General Electric a-m/f-m tuner and stereo amplifier.

fiers used in high-power stereo installations, the power supply furnishes a fixed negative voltage (with respect to ground), which is used to bias the output audio tubes. Such a circuit is shown in Fig. 10-22.

The high-voltage winding is tapped near the center, producing about 12 volts a-c at the tap, as measured from ground. The center goes to ground in a standard full-wave rectifier circuit. The 12-volt a-c supply is rectified by a selenium or germanium diode *S*-1. The rectified output is smoothed by the R-C filter circuit

FIG. 10-22. Power supply furnishing *B*+ and fixed bias for output tubes.

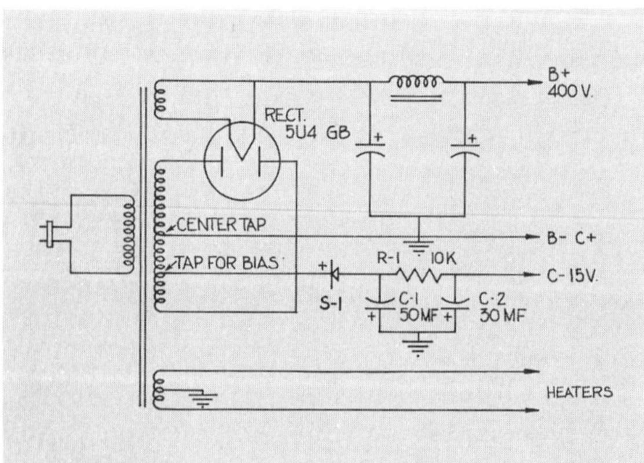

consisting of *R*-1, *C*-1, and *C*-2. The filter capacitors are rated at 50 volts. The grounded center-tap of the high-voltage winding becomes the positive end of the bias supply (*C*+), as well as the negative end of the *B* supply (*B*—). A 12-volt tap on the transformer produces a negative potential of about 15 volts which is used for biasing the output 7189 tubes. Note that the positive terminals of *C*-1 and *C*-2 go to ground. In some circuits, a separate winding on the power transformer is used for the bias supply.

The bias supply is a low-voltage, low-current circuit, and the components in it rarely give any servicing difficulty. Failure of any of the components would cause poor tone or hum or both. The standard servicing procedure of voltage check, followed by bridging filter capacitors and substituting rectifiers would quickly locate the cause. When replacing components in the bias supply, a wide latitude in replacement parts is permissible, since none of the values is critical. But this does not hold true for power transformer *T*-1. If the power transformer should prove defective, an exact replacement is necessary. The usual precautions relating to the polarity of rectifier *S*-1 and filter capacitors *C*-1 and *C*-2 must be observed.

Loudspeakers 11

Quick check. To determine whether a loudspeaker is functioning, turn on the receiver and momentarily unseat the second a-f tube. A loud click should be heard. Where the output stage is of the push-pull type, removing either tube will produce the same result.

Function of the loudspeaker. The loudspeaker is a device that takes electrical energy or power at audio frequencies from the second a-f output stage and converts it into sound energy. Its fidelity of reproduction depends on its ability to convert into sound all the component frequencies at the second a-f output.

Types of loudspeakers. Many varieties of loudspeakers have paraded across the stage throughout the period of radio evolution. The type of loudspeaker most widely used at the present time is the moving-coil dynamic loudspeaker. Therefore, the balance of the description that follows will concern itself primarily with that type.

Theory of operation of the dynamic loudspeaker. The theory of operation of a dynamic speaker is quite simple. In these speakers, the a-f signal from the second a-f stage is impressed across a small, free-floating coil of wire (called the "voice" coil), which is suspended in a strong stationary magnetic field. The a-f current causes a varying magnetic field around this coil. This varying field reacts with the stationary field and causes motion of the voice coil. The latter is cemented to a paper cone which vibrates with the voice coil and produces the audible sound waves.

Two main varieties of moving-coil dynamic loudspeakers have been developed. The difference between the two lies in the manner in which the stationary magnetic field is produced. The two types are the electromagnetic dynamic speakers and the permanent-magnet (P-M) dynamic speakers, the latter being almost universally used.

In the P-M dynamic speaker, a strong stationary magnetic field is created by attaching a powerful alnico permanent magnet to the pot of the speaker. The front end of the alnico magnet is shaped into a small cylinder and is known as the pole piece. Suspended by means of its attached paper cone, the voice coil rides freely back and forth over the pole piece. The basic structure of the P-M speaker is shown in Fig. 11-1A.

The power-output tube of the receiver feeds a-f currents to the voice coil through the output transformer. (The output transformer is often mounted on the speaker unit itself.) The reaction of the varying magnetic fields, created by the a-f currents in the voice coil, with the stationary magnetic field of the pole piece produces the back-and-forth vibratory motion of the voice coil and its attached paper cone.

The outer edge of the paper cone is attached either directly or by means of soft leather or plastic to the basket of the speaker, so that the voice coil may float freely over the pole piece. A flexible membrane, called a spider, is cemented to the voice-coil form and at its outer edge, to the pot, The spider guides the motion

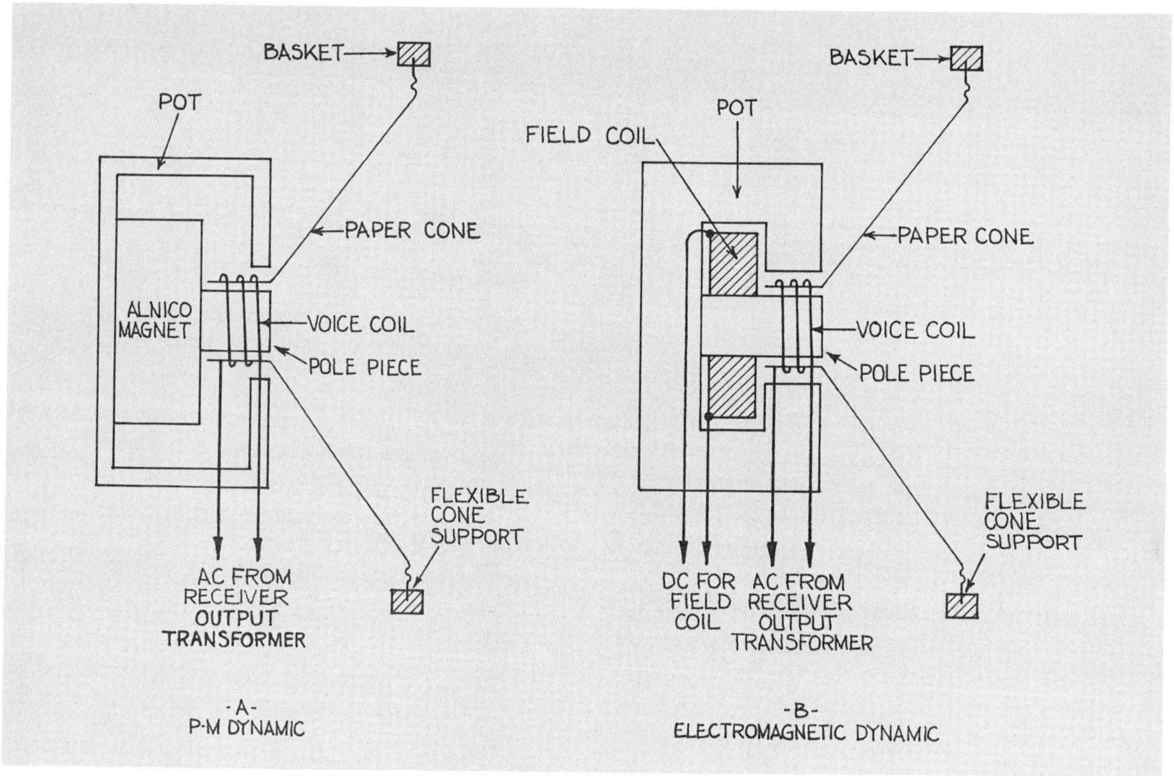

FIG. 11-1. Typical moving-coil dynamic speakers.

of the voice coil over the pole piece. Sometimes, the outer edge of the spider is screwed to the pot, rather than cemented, to provide for replacement of a defective voice coil and paper cone. A dust cap, usually made of felt, is cemented at the front end of the voice-coil form to prevent dust or other grit from getting between the voice coil and the pole piece.

The electromagnetic dynamic speaker is similar to the P-M dynamic speaker, except that an electromagnet, energized by direct current, is wound over an iron pole piece to produce the stationary magnetic field. Its basic structure is shown in Fig. 11-1*B*. Note that it has four leads, whereas the P-M speaker has only two.

CHECKS FOR LOUDSPEAKER OPERATION

When the quick check indicates trouble in the speaker or if the servicing complaint is rattles or poor tone quality, the speaker should be carefully tested. The following section describes the quick check in detail and discusses other tests that may be applied to the loudspeaker.

Quick check for speaker operation. In the quick check, the second a-f tube is unseated. When this is done, a click should be heard in the speaker. Unseating of the tube causes the *B*+ voltage to the plate pin of the tube to rise to maximum, with a consequent surge through the primary of the output transformer. This surge, induced in the secondary of the transformer, momentarily energizes the voice coil and produces the click.

This quick check does not tell us how well the speaker is functioning, but merely whether the voice coil is open or short-circuited. It is equally applicable to both types of dynamic speakers.

To determine if the field coil of an electromagnetic dynamic speaker is open, a blunt piece of iron, like a socket wrench, should be held near the center pole piece. A perfect field coil will cause the tool to be attracted strongly. An open field coil will give either no attraction or a slight attraction due to residual magnetism. Unfortunately, the dust cover may in some cases make this test somewhat unreliable. Of course, this latter test is not necessary for a P-M dynamic speaker.

Checking the voice coil for opens and shorts. When the quick check produces no clicks (and the field coil in the case of the electromagnetic dynamic speaker checks good), the defective condition is probably either an open or short-circuited voice coil. To check its condition, remove the receiver and the speaker from the cabinet, with all connecting leads in place.

Examine the speaker and note two leads from the voice coil that are cemented to the paper cone. From the paper cone, two thin flexible wires bring the voice-coil leads to two lugs on a fibre strip mounted on the speaker frame. The secondary winding of the output transformer is connected to the voice coil via these two lugs.

Disconnect either one of the transformer leads to its lug. The voice coil may now be checked with an ohmmeter, as shown in Fig. 11-2. Its resistance should have the value indicated by the receiver manufacturer. If this information is not available, assume the general value of about 3 to 15 ohms, the higher values being found in larger speakers. However, if the ohmmeter reading is infinite ohms, the voice coil is open. A short in the voice coil is indicated if the reading is zero ohms.

Checking the speaker as a source of distortion. The speaker with an open or shorted voice coil will make the receiver dead. However, a defective voice coil that is neither open nor shorted may produce a distorted output. To check for this condition, open the same lead as directed in the previous check. Connect a test bench P-M speaker to the leads from the secondary winding of the output transformer, as shown in Fig. 11-3. If the output from the receiver now sounds well, the original speaker is defective. Of course, this check must be made with the set turned on.

TROUBLES COMMON TO THE LOUDSPEAKER

From the servicing point of view, the loudspeaker may be responsible for many receiver defects. Consider the P-M dynamic speaker first. The receiver may be dead because the voice coil is either open or shorted. Receiver reception may be distorted because the voice coil is off-center or warped, or the paper cone is warped. There may also be strange rattles from the set because of loose parts in the speaker mounting, a torn paper cone, dirt between the voice-coil form and the pole piece, vibrations of other parts within the receiver, or loose voice-coil windings.

All these defects could also exist with electromagnetic dynamic speakers. But in addition, the set could be dead or weak because of an open or shorted field coil. Each defect will be described from the point of view of its source with instructions for repair.

Troubles common to the voice coil. The most frequent speaker defect is an open voice coil. When this condition exists, the set is dead. The open is almost always in the leads to the voice coil rather than in the coil itself. The points to inspect carefully are those where the voice-coil wires are connected to the flexible leads on the paper cone and where the flexible leads are connected to the lugs on the insulated strip. These broken connections may be repaired by resoldering.

One of the leads may be broken, rather than a broken connection. These leads are made of very thin wire and must be handled with the greatest of care. Holding the broken lead with a tweezer so as to avoid tugging, gently scrape the ends of the broken lead with the blade of a knife to remove the enamel insulation. Then resolder the break, using rosin-core solder. To avoid too much tension on the repaired lead, it may be necessary to solder a jumper across the break. If so, use a single

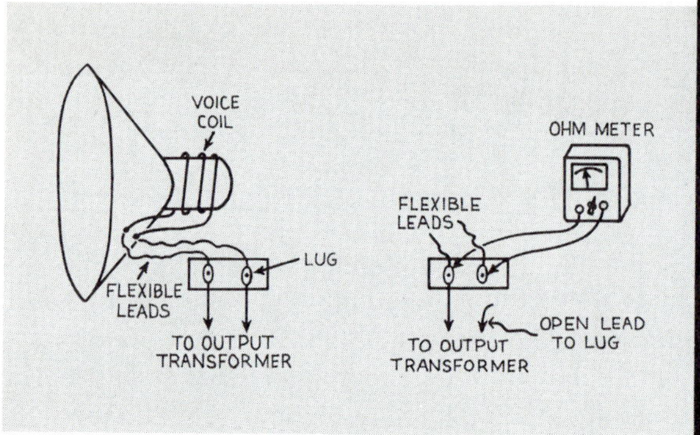

FIG. 11-2. Checking the voice coil for an open or short.

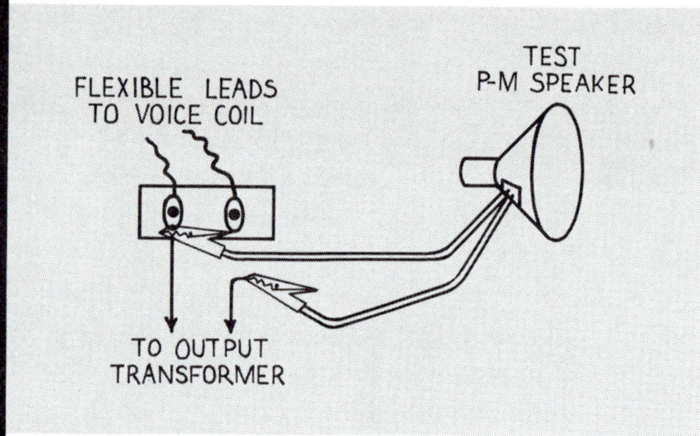

FIG. 11-3. Checking a speaker with a test bench speaker.

strand of fine copper wire whose ends have been pretinned. Finally, paint the new wire and its connections with speaker cement.

A short-circuited voice coil is most likely to be the result of the connection points shorting either to each other or to the speaker frame. Here, too, the set is dead. Carefully unsolder one of the flexible leads to its lug. Check with an ohmmeter to determine whether the short has been removed. If it has not, inspect the connection points on the paper cone to see whether conductive dirt has been deposited there. With a pointed knife, carefully and gently scrape the space between the two points on the cone. However, if the first check cleared the short, the defect was on the lug. Improvise a new insulated tie point on the speaker frame and solder the flexible lead and its output transformer lead to this new tie point.

The remaining voice-coil troubles manifest themselves by distortion of the output or by buzzing sounds accompanying the output. These may be caused by a warped voice coil that rubs against the pole piece, by loose voice-coil windings, by dirt between the voice-coil form and the pole piece, or by voice-coil leads that have broken loose from the cement on the cone. The last-named condition is readily repaired by applying speaker cement to the leads and gently pressing them against the cone until the cement is dry.

The other conditions are not so easily repaired. If the speaker is not a very expensive one, the average technician had best replace the entire speaker with a new one. This is because there are too many cemented areas that must be opened to make the repair. In Fig. 11-4 all the points that are usually cemented are indicated by the letter C. Cement holds the speaker edge to the basket, the voice coil to the paper cone, the dust cap to the paper cone, the spider to the speaker frame or basket and to the voice-coil form, the voice-coil leads to the paper cone, and the voice-coil turns to each other. When an expensive speaker is involved and repair of these defects is warranted, it would be best to send the speaker back to the manufacturer or to a service shop that specializes in such repairs.

Troubles common to the paper cone. The paper cone may cause distortion because it has dried out and warped. Or it may cause strange buzzes and rattles because the cement at its outer edge or at its connection to the voice-coil form or at the dust cap has loosened. A hole or tear in the paper cone may also be the cause of rattles.

The last named defect may be repaired, if the tear is not too large, by applying special speaker cement to the edges of the hole or tear and pressing the edges together until the cement is dry. Do not try to fix the break with masking tape or similar adhesive; they will not hold permanently.

The loose dust cap may be repaired by applying a very small amount of cement to the point where the cap appears to be loose and pressing the cap to the cone. As stated previously, the other defects would normally best be fixed by replacement of the complete speaker assembly or by sending the speaker out for specialist repair.

Replacing voice-coil and cone assembly. When the expense of the speaker warrants it, the cone and voice-coil assembly may be removed and replaced. To do this, the cement at the various holding points must be loosened with a cement solvent, usually acetone: at the dust cap connection to the cone, where the edge of the cone is held to the basket, and where the outer edge of the spider is attached to the basket. Sometimes the job is easier because the latter two joints are not cemented, but are held by screws. Extreme care must be taken when removing the dust cap that the cement binding the cone to the voice-coil form is not loosened, particularly if the same latter assembly is to be used again. Now the cone and voice-coil assembly, with the attached spider, may be drawn forward and out.

We may now proceed with the repair. To remove dirt from the space between the pole piece and the pot, blow into the area. Then gently run a pipe cleaner through the space while holding the pot facing downward. Blow out the area once again. If this is the only defect, the cone and voice-coil assembly is then replaced in the manner described below.

If the voice-coil wires have become loose, apply a small quantity of cement to the wires and gently wipe away the excess. Be sure that no blob of cement stands up, to rub in the pole piece space. Allow it to dry. Similarly, if the cone cement has broken away from the voice-coil form, apply a little cement and allow to dry thoroughly. Then replace the coil and cone assembly.

FIG. 11-4. Loudspeaker showing all cemented joints. Cemented points are labelled C.

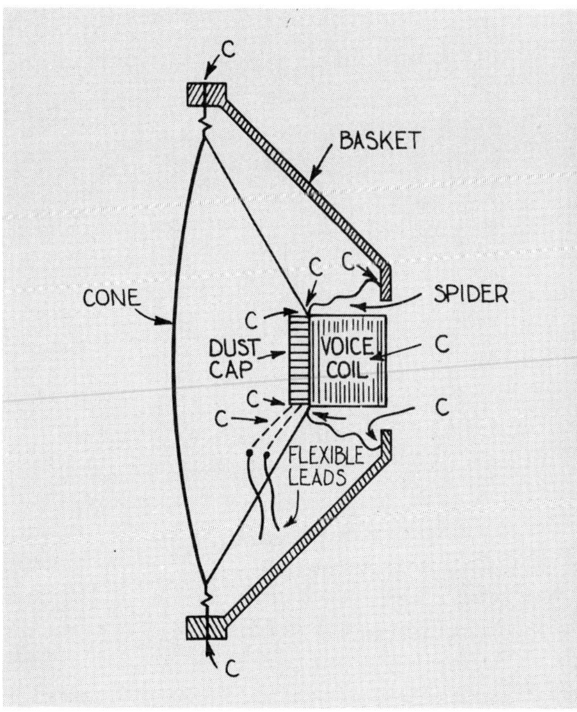

FIG. 11-5. Using shims to replace and center a coil and cone assembly.

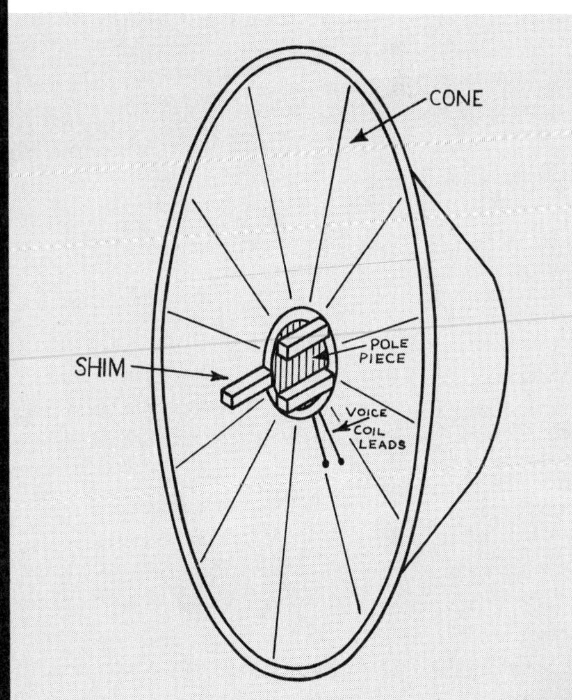

If the voice coil is warped, do not try to straighten it. Obtain an exact duplicate voice-coil and cone assembly and use that as the replacement.

The reinsertion of the voice coil and cone is not a simple task. The voice coil must be properly centered in the space between the pole piece and the pot so that it can move without rubbing. Of course, where the defect was only an off-center rubbing voice coil, this recentering procedure is necessary.

First apply cement with care to the outer edge of the spider and to the edge of the pot to which it is to be attached. Of course, if this outer edge is held by screws, this will not be necessary. Now insert the voice coil with its attached cone over the pole piece, but do not press or screw the spider into position. It is first necessary to center the voice coil for free movement. This is done by inserting three or four equal-thickness nonmetallic shims, spaced symmetrically, between the pole piece and the inner side of the voice-coil form, as shown in Fig. 11-5. These shims should fit snugly but should not be forced into position, lest it warp the voice-coil form. Now press the outer edge of the spider to the pot and allow to dry. If the spider is held by screws, tighten them in place.

With the shims still in place, apply cement to the outer edge of the cone and to the rim of the speaker basket. Press the two edges together and permit the bond to dry thoroughly.

In many receivers, there is a felt or cardboard ring that is fastened around the rim of the basket over the edge of the cone. Its purpose is to prevent acoustic continuity between the speaker and the baffle to which it is attached. This ring should now be either cemented or bolted back into position.

When all cemented joints are completely dry, remove the shims and test that the voice coil is properly centered. Place your fingers close to the center of the cone and press the voice coil back. Then allow it to slowly come forward. Repeat this several times and feel for any rubbing. When satisfied that the coil moves freely, apply some cement to a new dust cap and press it into position on the cone. Be careful that no excess cement seeps into the gap in which the voice coil floats. The cone and voice-coil replacement is now complete.

Sometimes, the spider is of a different type from the one just described. It may be a plastic disc with perforations in it, and it is cemented to the cone in front, rather than behind. It is then bolted to the pole piece at the center of the disc. Such a spider, shown in Fig. 11-6, does not require cement when replacing the coil and cone assembly. As seen in the figure, the shims are placed in position through the perforations in the disc and the center bolt is tightened. Otherwise, all steps given previously may be followed.

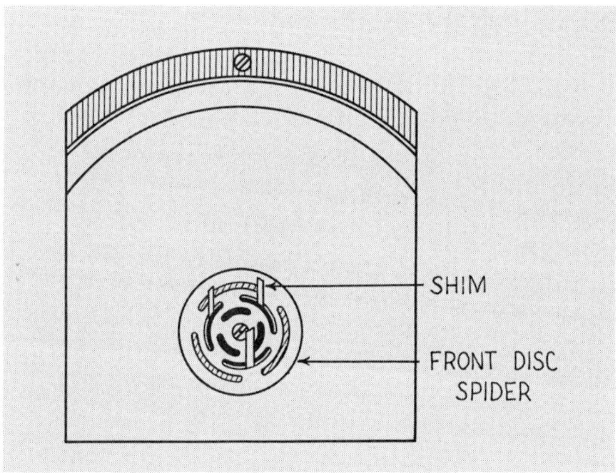

FIG. 11-6. Speaker with front spider showing shims inserted.

Troubles common to speaker field coils. The electromagnetic dynamic speaker is found only in older receivers. All that has been said about P-M dynamic speakers applies equally to the older type of speaker and is handled similarly. The only difference lies in the troubles common to the speaker field coil.

The function of the field coil is to develop a stationary magnetic field; therefore it must be fed with direct current. The most usual procedure is to get this direct current supply by employing the speaker field as a filter choke in the power supply, as shown in Fig. 11-7.

A defective speaker field will normally be located in a power supply check. It will be noticed from Fig. 11-7 that an open coil will cut off the *B+* supply, so that all stages will be inoperative. The receiver will be brought in dead, even though all tubes light. A check of the power supply will show no *B* (plate) voltage.

Disconnect the power plug and discharge the filter capacitors. An ohmmeter check for continuity will confirm the open field. The break may be in the field leads or in the connection between the field wire itself and the lead. These should be inspected and, if found at fault, resoldered. If the break is not readily found, you had best change the complete speaker, in the manner explained later.

Another field-coil defect is that of shorts between the field winding and the pole piece or the pot. If the speaker is mounted on the chassis, the short will cause partial or complete loss of *B++* voltage and possible damage to the power supply. This condition will be found in a check of the power supply. The power-supply check would seem to indicate a shorted filter capacitor. The actual defect would be found when removal of the suspected filter capacitor does not remove the short from the circuit. If the speaker is not mounted on the chassis, the speaker case will become "hot" with high voltage, but the receiver operation may not be affected. Once again, complete speaker replacement is recommended.

Troubles common to speaker assembly and mounting. After continuous operation, various parts within the receiver may have a tendency to become loose. Various screws in the speaker or associated with its mounting may also loosen, because of continuous operation. Where such is the case, rattles and buzzes may mar the speaker reproduction. The technician may verify this condition by connecting in a substitute test speaker from his test bench and observing if improvement results. If the rattles and buzzes disappear, a careful hunt must be made for any loose part prone to vibrate. This is done by holding various parts with the fingers while the receiver is operating, and observing if the vibrations are damped. No part should be beyond suspicion. Even the cabinet must be inspected for loose or cracked parts.

Replacing an electrodynamic with a P-M speaker. Sometimes an electrodynamic speaker has to be replaced, and a similar speaker is unobtainable. In such a case, a P-M dynamic speaker of proper voice coil wattage and size may be used with some provision to replace the field coil in the circuit of the receiver.

Connect the voice-coil wires of the new speaker to the secondary winding of the output transformer, just as with the old speaker.

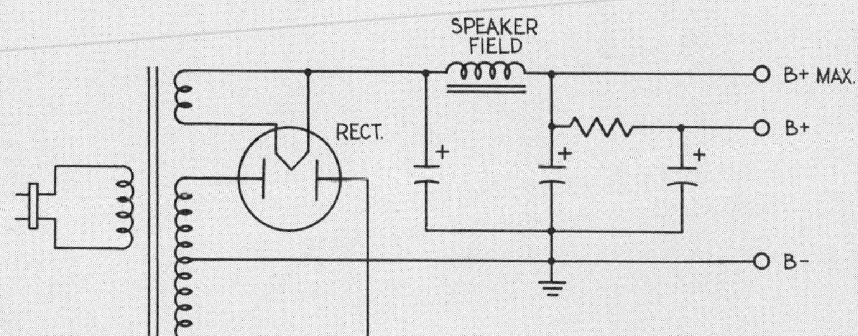

FIG. 11-7. Speaker field used as part of the filter in a transformer-type a-c power supply.

Where the old field coil was wired into the filter circuit of the power supply, solder a 10-watt resistor with the same resistance as that of the speaker field coil in its place. This change gives us an R-C filter which is not as efficient as the original L-C filter. To obtain a comparable filtering action, it becomes necessary to approximately double the capacitance of the input filter capacitor. Be sure to observe proper polarity when replacing this electrolytic capacitor.

Replacing a complete loudspeaker. Many speaker defects require the complete replacement of the entire loudspeaker assembly. Where such is the requirement, an exact replacement is most desirable. This may not always be possible, and a speaker that resembles the original as closely as possible must be used.

Several factors must be kept in mind by the technician. Is there sufficient space within the cabinet? Can the new speaker be mounted with sufficient ease? Is the resistance of the field coil, if an electrodynamic speaker is used, similar to that of the original? Is the current-carrying capability of the field coil of the replacement speaker sufficient for the receiver? Is the impedance of the new voice coil the same as that of the old speaker? And finally, is the power-handling capability (wattage) of the voice coil of the new speaker sufficient for the receiver?

Replacement speakers are usually listed according to the following factors:

1. Diameter of the basket in inches.
2. Voice-coil impedance in ohms.
3. Voice-coil wattage.
4. Field-coil resistance in ohms.

Although the current-carrying capability of the field coil is not listed, it is an important factor that must not be overlooked. In this consideration, the size of the pot is an indication of the current-carrying capability of the field coil. The pot of the replacement speaker should be no smaller in size than that of the old speaker.

If the wattage output of the receiver is not indicated in the manufacturer's schematic, the proper wattage for the voice coil may be determined in another manner. The tube or tubes used in the second a-f or output stage are determined by inspection. Then reference to a tube manual will give the undistorted power output for that tube. This wattage may be considered as the voice-coil wattage.

Often in making a speaker replacement, the old output transformer, mounted on the old speaker, may be removed and used with the new replacement speaker. The transformer primary will thus match the second a-f stage. Care must then be taken that the transformer secondary impedance matches that of the voice coil of the new replacement speaker.

If the chosen replacement speaker has a voice-coil impedance differing considerably from the original, this will necessitate changing the output transformer for proper impedance match. Replacement notes on output transformers are found in Chap. 12.

E.I.A. color code for loudspeakers. The various terminal wires of a loudspeaker may often be identified for servicing by means of the E.I.A. color code, tabulated below:

Voice coil	
1. Green	Finish
2. Black	Start
Field coil (if any)	
1. Black and red	Start
2. Yellow and red	Finish
3. Slate and red	Tap (if any)

Dual speaker systems. Some receivers are built with two speakers within the same cabinet. Where a condition develops in such a receiver that one of these speakers must be replaced or requires a voice-coil and cone replacement, the procedure is similar to that described for single-speaker replacements.

However, a new consideration develops. If one voice coil moves in while the other moves out, interference effects develop and reduce

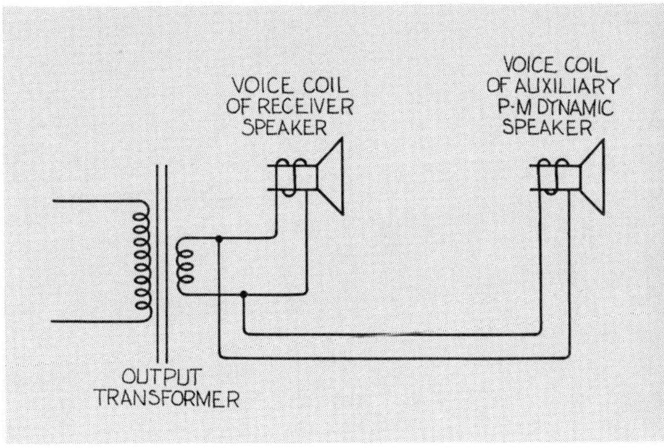

FIG. 11-8. Adding an auxiliary P-M dynamic speaker to a receiver.

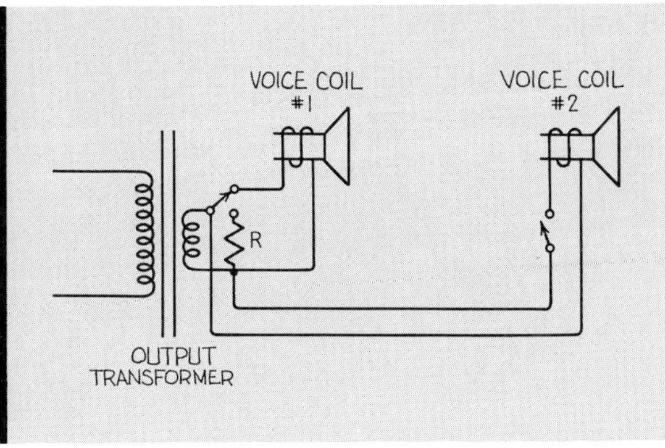

FIG. 11-9. Switching the receiver speaker and the auxiliary speaker, individual control.

the volume of total output. This condition is undesirable and may be remedied by reversing the voice-coil or field-coil leads to one of the speakers. The voice coils will then move in and out together, and the speakers are said to be in phase.

To determine if the speakers are properly phased after replacement, turn on the receiver and tune to a non-station position. Place your hands on the cones of the two speakers. Then apply the voltage of a dry cell across the output transformer secondary. The movement of the cones will be felt and seen, and proper phase may be found.

Adding a speaker to a receiver. In some cases, a customer may desire a second speaker connected to his receiver and installed in another room. Since the speakers are remote from each other, phasing is not important.

A simple procedure in this requirement is to obtain a P-M dynamic speaker and connect its voice coil in parallel with the voice coil of the receiver speaker. Of course, this will cause mismatch with the output transformer secondary, but the effect will not be too poor. Besides, the larger the impedance of the voice coil of the P-M dynamic speaker, the less will be the total mismatch, although less power will be fed to the auxiliary speaker. This may be of advantage, since it is generally desirable to operate the auxiliary speaker at a reduced volume. The combination is shown in Fig. 11-8.

In the setup described above, both speakers will operate simultaneously. If it is desired to shut off the receiver speaker while the auxiliary speaker functions, it is necessary to use a single-pole double-throw switch to cut out the first voice coil. In addition, a resistor, of comparable impedance to that of the voice coil just cut out, should be connected across the secondary of the output transformer. A second switch is connected at the auxiliary speaker to cut it out when not in use. This setup is shown in Fig. 11-9.

It is now possible to control the volume of the receiver and auxiliary speakers only by means of the receiver volume control. If it is desired to vary the volume of the auxiliary speaker at the speaker itself, a standard *L* pad control may be inserted across the voice coil of the auxiliary speaker. The ohmic rating of the pad should match the impedance of the auxiliary voice coil. The complete setup is now shown in Fig. 11-10. No switch is required at the second auxiliary speaker, since the *L* pad can replace its function. The minimum position of the *L* pad will cut out the auxiliary speaker.

Adding headphones to a receiver. A custo-

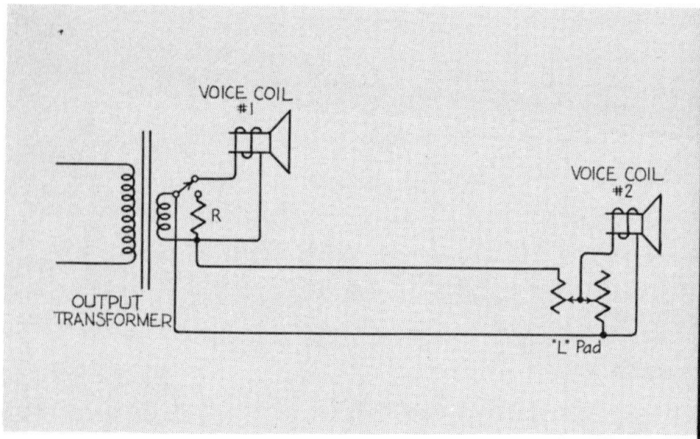

FIG. 11-10. Volume control for an auxiliary speaker.

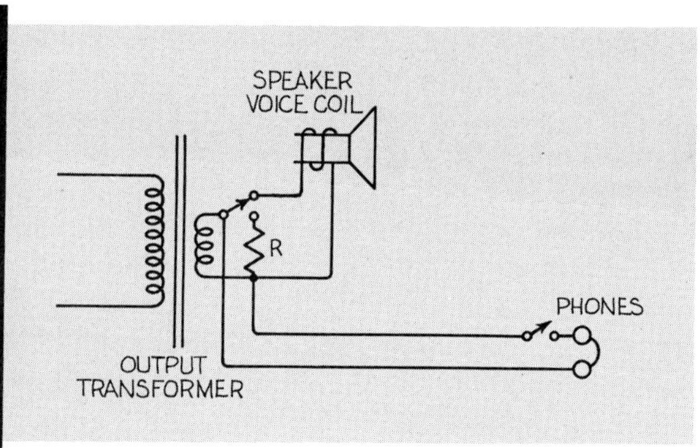

FIG. 11-11. Circuit for connecting headphones to a receiver.

mer may request that headphones be installed on his receiver, so that he may turn off the loudspeaker and still listen to the radio late at night. The simplest procedure is to connect the phones across the voice coil. The high impedance of the phones will cause great mismatch and keep power fed to the phones low, giving low volume. A switch may be installed to cut out the speaker voice coil and to cut in an equivalent impedance resistor. A second switch may be used to cut out the phones. The setup is shown in Fig. 11-11.

The same effect may be achieved by the installation of a circuit-switching phone jack, as shown in Fig. 11-12. Pushing the phone plug only part way in will allow simultaneous operation of the phones and speaker. Pushing the phone plug all the way in will cut out the speaker voice coil and allow operation of the phones alone.

FIG. 11-12. Circuit for adding headphones to a receiver, using jack and plug.

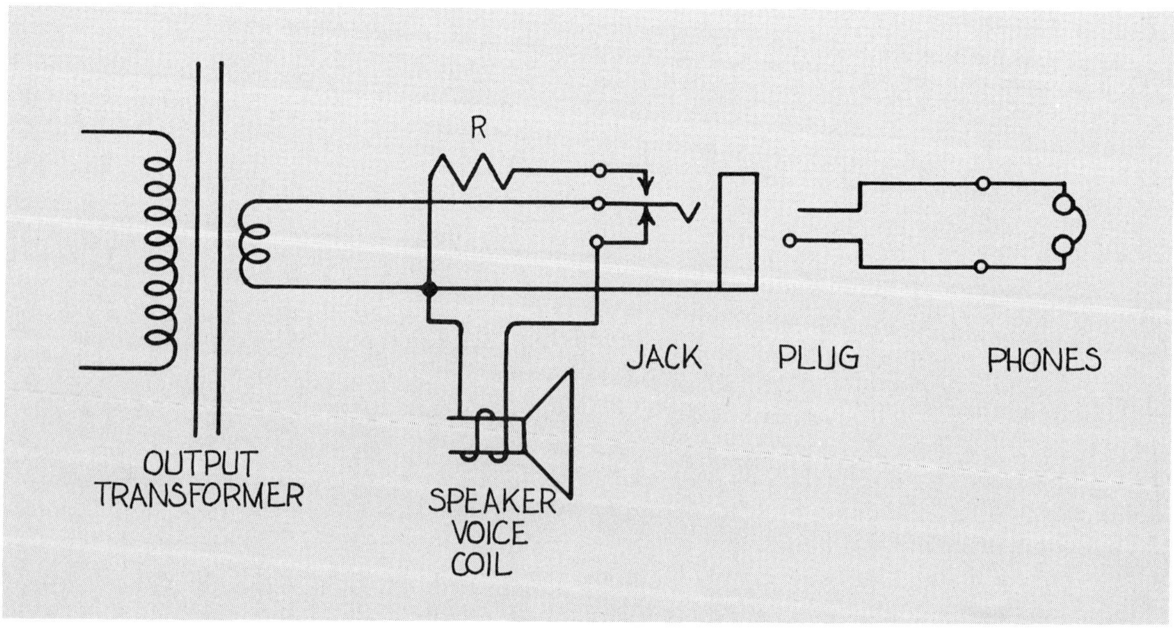

SERVICE DATA CHART

Symptom	Abnormal reading	Look for
No reception	No $B+$ voltage	Open field
	No click on quick check	Open voice coil. Open voice-coil leads
Weak reception	Weak click on quick check	Deenergized field (open field or low excitation voltage). Jammed voice coil
Distortion		Rubbing voice coil. Low field excitation voltage. Warped cone. Factors within the receiver (eliminate the speaker by substitution test)
Rattles		Rubbing voice coil. Loose voice coil. Torn paper cone. Cone loose from basket rim. Loose spider. Grit and dirt in field gap. Loose dust cap. Loose screws on speaker. Loose parts in radio

QUESTIONS

1. A receiver is brought in as dead. No plate voltage appears to be present. If you suspect that the speaker is defective, what part would you suspect? How would you test for it?
2. A receiver requires a new speaker. An exact replacement is not obtainable. What considerations must be made in selecting a new speaker?
3. A rattling, rasping speaker is reported by a customer. Examination shows an off-center voice coil. What remedial measures would you make?
4. A receiver has a dual speaker system. One speaker requires replacement of a voice coil and cone. List the steps in order by which you would make this replacement.
5. A customer has a receiver in his living room. He wants to add an auxiliary speaker to operate in the cellar. He wants to be able to operate either or both speakers and also to control the volume of the cellar speaker in the cellar. Design a circuit for these requirements.
6. A customer wants to use headphones with his receiver at night, so that he can cut off the loudspeaker. Design a circuit for him.
7. Explain how you would replace an electromagnetic speaker in an old set with a P-M dynamic speaker.

Second, or Power, Audio-Amplifier Stage

12

The second a-f amplifier stage is also called the "power-amplifier" stage, or "output" stage.

Quick check. If a plugged-in soldering-iron tip or finger is placed on the control grid of the second a-f amplifier tube and causes a low growl to be heard in the speaker, the second a-f stage is probably functioning properly, and the trouble shooter moves on to the first a-f stage.

Standard circuit. Figure 12-1 represents our standard second a-f stage.

Function of second a-f stage. The control-grid circuit is the signal input of the stage; the plate circuit is the signal output. The signal fed into the stage is an a-f voltage, the magnitude of which would be about sufficient to operate headphones. It is the function of the second a-f stage to amplify this signal to an amount sufficient to operate a loudspeaker. To get an idea of the magnitude of the signal voltages handled by the second a-f stage, a 6V6-G tube (most commonly used) gives an output of approximately 5 watts with an input grid signal voltage of 13 volts peak. A smaller signal-input voltage would give a smaller output power; 13 volts is the maximum the tube will handle without undesirable distortion. The input signal is fed from the preceding first a-f stage. The plate or output circuit of the stage feeds the amplified signal to the speaker.

Regardless of whether the receiver is a-c, a-c/d-c, or battery-operated, the function and operation of the second a-f stage is the same. Indeed, this is true for all stages but the power-supply stage.

FUNCTIONS AND VALUES OF COMPONENT PARTS

Grid-load resistor *R*-12. Resistor *R*-12 is the grid-load resistor, and the input signal is impressed across it. Its value usually is 470,000 ohms. When a different ohmage is used, a lower value would result in lower gain and better frequency response, while a higher value would give slightly higher gain at a sacrifice of tone quality.

Self-bias. Since grid-bias voltage affects tone quality and amplification and is a valuable indication of trouble to the technician, the theory underlying self-bias circuits should be thoroughly understood.

Let us first remember that, in order to maintain a grid-bias voltage, the grid must be made negative with respect to its cathode. Assume no signal input voltage, and examine the amplifier circuit redrawn as in Fig. 12-2, with components unnecessary to the self-bias circuit eliminated. Observe that resistor *R*-13 and the tube *V*-5 are in series across the *B* power supply. Tracing the screen circuit, current flows from *B*− through *R*-13, through the tube to the screen and *B*+. Tracing the plate circuit, current flows from *B*− through *R*-13, through the tube to the plate, and finally through *L*-12, the output-transformer primary, to the maximum *B*+ point. It is seen that both screen and plate currents flow through *R*-13 and, as a result, a voltage drop is developed across it. Note also that the cathode is made positive with respect to *B*− by this voltage drop. Then, when the grid is returned to *B*− through *R*-12, the grid is negative with respect to the cathode. A negative grid will not attract electrons and, as a result, there is no current in the grid circuit through *R*-12 and no voltage drop across it. The full voltage developed across *R*-13, therefore, is applied to the grid as the bias voltage.

This system of obtaining grid-bias voltage is known as "self-bias," since the tube's own screen and plate currents cause the voltage drop, which is used for biasing the grid. The self-bias circuit can be used with a triode type of tube also, in which case the voltage drop is caused by the plate current alone.

The ohmic value of *R*-13 will depend on the tube used and its operating potentials. Several common values follow:

6V6-G	330 ohms
6BQ5	150 ohms
6EH5	68 ohms
35C5	150 ohms
50C5	150 ohms

Self-bias bypass capacitor *C*-13. The input circuit of the tube is between grid and cathode. This involves grid-load resistor *R*-12 and self-bias resistor *R*-13. The input-signal voltage divides itself between them, most of the signal being across the larger grid-load resistor *R*-12. The output circuit of the tube is between plate and cathode. This includes the output-transformer primary *L*-12, the *B* power supply, and self-bias resistor *R*-13. The output signal divides itself among these three, most of it being across *L*-12, which has the highest impedance to the output signal. Resistor *R*-13 is common to both the input and the output circuits, and some of the input signal and some of the output signal will mix in *R*-13. This is coupling. Since a tube's output signal is 180 deg out of phase with its input signal, cancellation takes place where the two signals are coupled, as across *R*-13. This effect of coupling, where cancellation takes place, is known as "degeneration" and results in a decrease in the gain of the tube.

The degenerative action can be minimized by bridging *R*-13 with a capacitor. The current through *R*-13 has components made up of the d-c screen and plate currents of the tube, the a-c input signal, and the a-c output signal. When *R*-13 is bridged by a capacitor, the impedance of the parallel combination to the signal current is reduced, while its oppo-

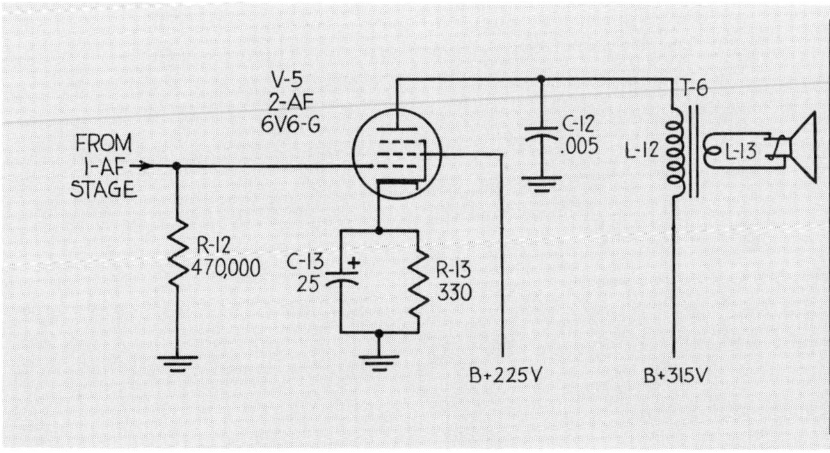

FIG. 12-1. Standard circuit for a typical second a-f stage.

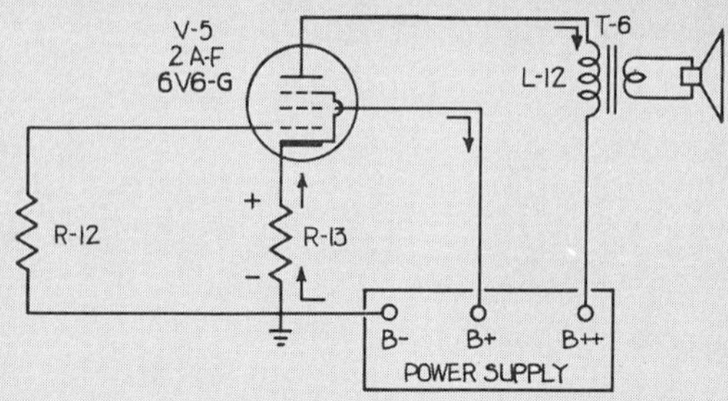

FIG. 12-2. Circuit illustrating self-bias.

sition to direct current remains the ohmic value of *R*-13. The voltages across the parallel combination therefore are reduced as regards signal voltage, while the d-c bias voltage remains the same. The reduced input- and output-signal voltages across the parallel combination decrease the degenerative effect.

The action of the parallel capacitor has been called "bypass," since, from one point of view, the signal current is taken out of resistor *R*-13 and passed around it through the capacitor. The bypass action depends on the impedance of the capacitor to the signal frequencies. To be effective, it should be lower than the ohmic value of the resistor being bypassed.

Since the signal in the second a-f stage is at audio frequency, a high-capacitance capacitor will be necessary for adequate bypass action. Capacitor *C*-13 is usually a low-voltage electrolytic type. Capacitances from 10 to 50 mfd will be found in various receivers. The higher capacitances will provide better bypass action, with a consequent improvement of the response, especially at the low audio frequencies.

Self-bias circuits similar to *R*-13 and *C*-13 are used to obtain bias voltages for the r-f and i-f tubes. The action in these tubes is similar, except that the cathode bypass capacitors need be only 0.1 mfd for adequate signal bypass, owing to the higher signal frequencies at radio frequency and intermediate frequency.

Output capacitor *C*-12. Capacitor *C*-12 across the signal-output circuit bypasses high audio frequencies to ground. The pentode and beam-power tubes introduce a considerable amount of harmonics, which will be most noticeable in the high a-f range. Placing *C*-12 across the signal output circuit bypasses some of the signal away from the output transformer. This effect will be greatest at the high audio frequencies, since the impedance of a capacitor decreases as the frequency increases. Therefore, the harmonic content will be reduced by the action of this capacitor.

An average value for capacitor *C*-12 is 0.005 mfd. In individual receivers, this value may vary from 0.001 to 0.02 mfd. Receivers using the higher capacitance values have been designed to favor the bass register, since the higher capacitance bypasses more of the high frequencies out of the output transformer and speaker, making the response deeper by comparison.

Output transformer *T*-6. Transformer *T*-6, called the "output" transformer, is often mounted on the speaker. Its function is to couple the output circuit of the tube to the speaker. The average beam-power tube requires a load of 5,000 ohms, and the average speaker voice coil has an impedance of 8 ohms at audio frequencies. The coupling transformer is designed, therefore, to have a primary impedance of 5,000 ohms and a secondary of 8 ohms. Obviously, if the output transformer should become defective, the original manufacturer's part should be obtained for best results. However, where this is not possible, a universal-type transformer may be used satisfactorily. The replacement notes on output transformers explain this procedure more fully.

Vacuum tube *V*-5. Vacuum tube *V*-5 is called the "power" tube, sometimes the "output" tube, as well as the second audio tube. The tube most commonly found in this stage is the 6V6-G beam-power amplifier. Smaller receivers, where the *B* supply voltage and the power output are lower, use the 6EH5 power-amplifier pentode. A-c/d-c receivers use the 35C5 or 50C5 tubes. Some a-c/d-c phonographs use the 25L6 or 50L6 tubes.

All these tubes are characterized by high power sensitivity; that is, a low signal-input voltage causes a high power output. For example, in the case of a 6V6-G, an input signal of 13 volts gives an output power of 5 watts. By way of comparison, the very much older 45 power-amplifier triode requires an input signal of 50 volts to give an output power of 1.6 watts.

NORMAL TEST DATA FOR THE SECOND A-F STAGE

Check for normal stage operation. The signal check for the second a-f stage is shown in Fig. 12-3. The signal generator is adjusted to give an a-f signal and the attenuator is set for maximum output. The signal generator ground lead is connected to the receiver chassis, and the "hot" lead is connected through a 0.1-mfd/600-volt capacitor to the plate terminal of the second a-f tube (pin No. 3 for a 6V6-G tube). The purpose of the capacitor is to prevent the d-c voltage, present at the plate of the second a-f tube, from affecting the signal generator circuits. Normally, the full a-f output of the signal generator is just sufficient to cause an audible note in the speaker. The hot lead of the signal generator is then shifted to the grid terminal (pin No. 5 for a 6V6-G tube) of the second a-f tube. The signal-generator note should be heard in the speaker at a much greater volume. The gain in volume is an indication of the gain of the tube.

Experienced technicians rarely go to the trouble to use this method. A much faster

FIG. 12-3. Signal check of the second a-f stage.

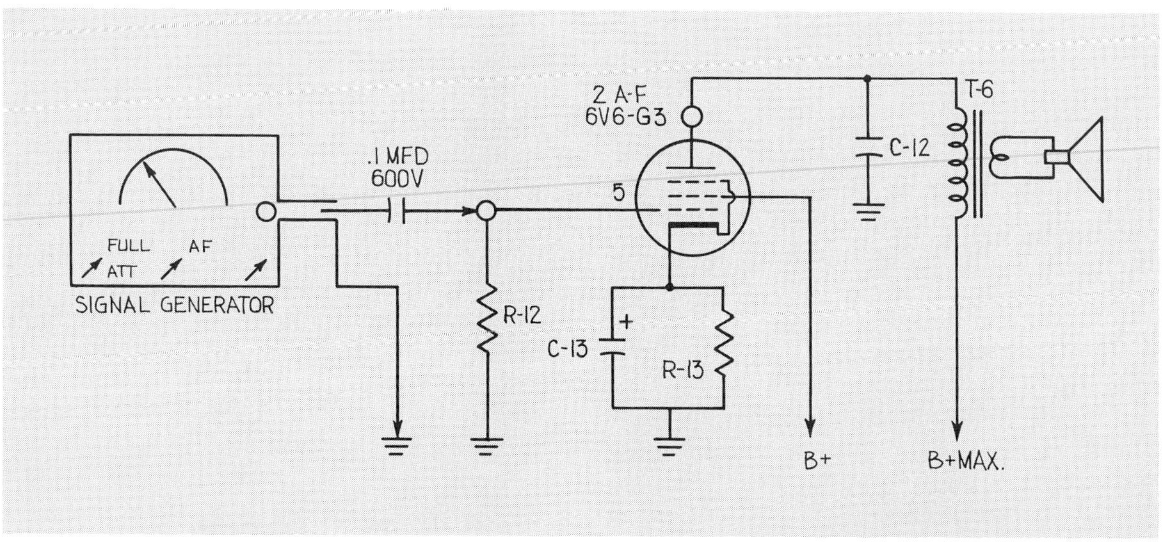

check, given as the quick check at the beginning of this chapter, is to touch the grid terminal with a finger or the tip of a plugged-in soldering tool. In either case, a low growl will be heard from the speaker, indicating that the stage is functioning.

Normal second a-f voltage data. Voltages are measured from chassis or common negative to tube terminal indicated. In some a-c/d-c receivers, where the circuit insulates *B*− from the chassis, the negative terminal of the voltmeter is connected to the common negative. This is most easily found at the line switch. See Chap. 9 on a-c/d-c power supply.

Tube terminal	25L6 and 6V6-G pin No.	A-c receivers, volts	A-c/d-c receivers, volts	6EH5 and 50C5 pin No.
Plate	3	200–325	110–120	7
Screen	4	175–215	90–100	6
Grid	5	0	0	2,5
Cathode	8	10–13	5–7	1

The voltages given for a-c transformer-type receivers vary as shown in the table, depending on the d-c output of the power supply. In a-c/d-c receivers, there is very little variation.

A positive voltage reading at the grid is indicative of breakdown of the coupling capacitor *C*-32 in the preceding stage. Service notes on this fault will be found in the chapter describing the first a-f stage.

Normal second a-f stage resistance data. These data are given in the following table:

Chassis to cathode	330 ohms
Chassis to control grid	470,000 ohms
Plate to *B*+ max	200–600 ohms

The 330 ohms of resistance from chassis to cathode is the ohmic value of self-bias resistor *R*-13. When a tube other than the 6V6-G tube is used, a different value will be found. Refer to the diagram of the receiver being tested, or to the table described under self-bias. The plate to *B*+ max reading measures the resistance of *L*-12, the primary of the output tranformer.

COMMON TROUBLES IN THE SECOND A-F STAGE

Troubles common to the grid-load resistor. Resistor *R*-12 rarely causes trouble. Occasionally it may open, thereby opening the grid circuit and causing lack of grid-bias voltage. This will result in bad distortion. At other times, signal voltage at the grid may build up and discharge periodically through dirt at the socket terminals, acting as a resistance parallel to *R*-12 and allowing a surge of current with each discharge. These surges may be heard in the speaker as a put-put known as "motorboating." If the surges come more rapidly, they will take the form of a low-pitched growl. The latter is sometimes mistaken for hum, which is also a low-pitched growl. Standard procedure in trouble shooting for hum is first to check the filter capacitors in the power supply and then to look for an open grid-load resistor in this or any other stage.

Open *R*-12 would be found in a voltage check of the stage, since an open grid causes a much heavier plate current. This condition makes for a greater than normal voltage drop across *L*-12 and, as a result, the plate voltage is lower. Since the screen voltage remains near its normal value, there will be a greater than normal difference between plate and screen voltages. Since conditions other than an open grid-load resistor will cause heavy plate current, confirmation must be obtained. This can be done with an ohmmeter.

In replacing resistor *R*-12, nothing in particular need be stressed. An exact duplicate of the original is desirable although not necessary. An ohmic value differing by as much as 20 percent either way will cause no noticeable difference, and the wattage rating is unimportant. However, the soldering must be

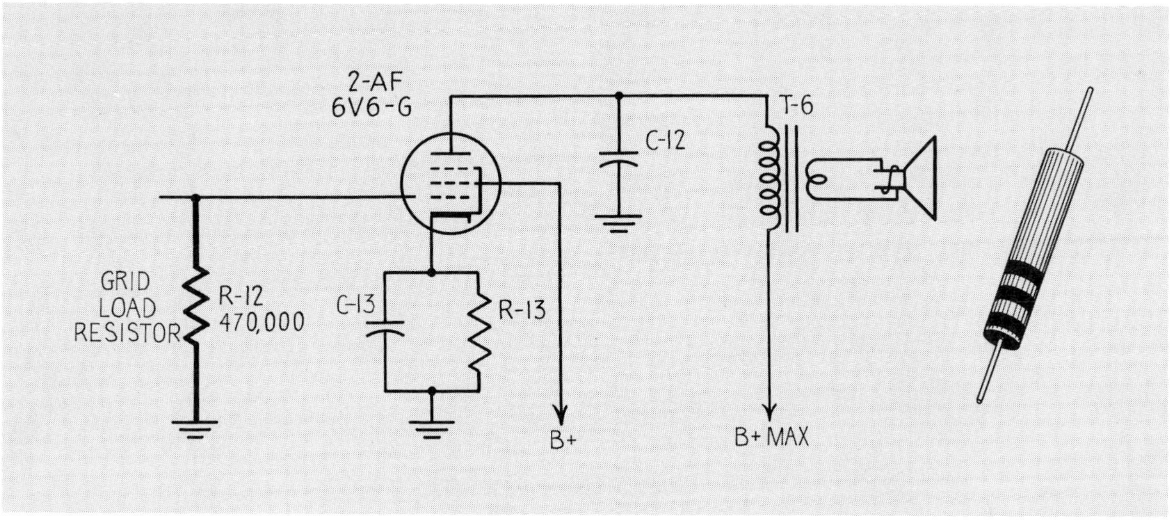

FIG. 12-4. A typical grid-load resistor and its position in the second a-f circuit.

carefully done, and the socket must be cleaned of dirt and excess rosin.

Troubles common to the cathode bypass capacitor. The cathode bypass capacitor *C*-13 often causes trouble. Like all electrolytic capacitors, it is likely to dry out and lose capacitance. As *C*-13 loses capacitance, approaching an open capacitor, the stage would give low gain and poor low-frequency response. This condition would be found in checking for low gain or for poor tone by bridging the capacitor with a test capacitor. A 20 MF/450 volt test capacitor, used for checking filter circuits in the power supply, may be used for test purposes here also.

Less frequently, *C*-13 shorts or leaks badly, acting as a partial or complete short across

FIG. 12-5. Typical audio cathode bypass capacitor.

R-13. This will result in poor tone quality due to lowered bias and would be found by a voltage check. Since plate current increases at lowered bias, a greater than normal voltage drop across the output-transformer primary would be produced. This results in a lowered plate voltage. A shorted *C*-13 may have been caused by an open cathode resistor *R*-13. Check this condition before replacing.

In replacing capacitor *C*-13, the technician watches for proper polarity, the positive side being connected to the cathode. The defective capacitor should be removed. A capacitance larger than the original may be used since, if anything, this will improve the low-frequency response. However, a capacitor of capacitance lower than the original will adversely affect the low-frequency response. Low-voltage electrolytic capacitors are usually rated at 25 or 50 volts. Either will do for *C*-13, since the voltage across the capacitor is approximately 12.5 volts for a-c receivers and 6 volts for a-c/d-c receivers. This is the voltage developed across *R*-13 for self-bias.

It is important to emphasize again that a shorted *C*-13 may have been caused by an open bias resistor *R*-13. Therefore, when replacing a shorted C-13, the bias resistor should be checked immediately after the shorted capacitor has been removed.

Troubles common to the self-bias resistor. Self-bias resistor R-13 is a likely source of trouble. It carries considerable current and is subject to heating. Sometimes it changes in ohmic value, and sometimes it opens. A change in ohmic value affects the bias voltage and, therefore, the tone quality. When R-13 is open, the cathode circuit is completed by the leakage resistance of parallel capacitor C-13. Since this leakage resistance is comparatively high, the voltage drop across it will be high, making for abnormally high bias voltage. The condition would be found in a voltage check. The plate voltage would be high, since at the high bias voltage, plate current would be low, the voltage drop across the output-transformer primary would be low, and plate voltage would be high. The open would result in a high cathode bias voltage which might damage parallel capacitor C-13.

Any change in ohmic value of R-13 is found in a voltage check of the stage. If it becomes low in ohmage, cathode voltage will be low, resulting in high plate current and a large voltage drop across the output transformer, increasing the voltage difference between screen and plate.

When replacing bias resistor R-13, use the same ohmic value as the original. As for the wattage rating, choose at least the same size, or the next higher size.

Troubles common to the a-f bypass capacitor. Capacitor C-12, the high a-f bypass, often comes up as the cause of a dead radio. Its position in the receiver is not only at a high d-c potential but also where the a-c signal potential (audio variation) is at its highest. This high voltage causes frequent breakdown of insulation, resulting in a shorted capacitor, which shorts out the audio signal from the primary of T-6 and also the power supply at this point. This condition is quickly found in a voltage check. Plate voltage equals zero, and screen or $B+$ voltage is low, since the power-supply voltage drops with the heavy load.

Capacitor C-12 may also open. However, a radio will rarely come in for this defect alone, since an open C-12 will merely increase the high-frequency response, and the customer may overlook this. In some radios, an open C-12 may cause a high-frequency oscillation. If this is the case, bridging C-12 with a similar capacitor or with a higher capacitance capacitor is the standard check procedure.

When capacitor C-12 is replaced, a good quality of capacitor should be used. Regardless of the original value, the voltage rating of the replacement capacitor should be at least 600 volts. The outside foil lead or ground lead should be connected to the chassis. Capacitor C-12 sometimes is connected from plate to $B+$. In that case, the outside foil lead is connected to $B+$. The replacement capacitor should have the same capacitance as the original. If the capacitance of the replacement capacitor is changed for any reason, it should be borne in mind that a higher capacitance will cut more highs out of the signal delivered to the speaker, while a lower capacitance will increase the high-frequency response.

Troubles common to the output transformer. Output transformer T-6 is also a common source of trouble. In addition to carrying the audio signal, the primary winding also carries the normal d-c plate current of the tube. An open primary often results. When the plate circuit opens, the positive screen attracts the total cathode emission. It was not intended to carry so heavy a current, and the screen mesh becomes red-hot. This can be seen in the case of a glass pentode and is one of the things the experienced technician looks for when making a visual inspection of the receiver. In the case of a metal tube, the condition cannot be

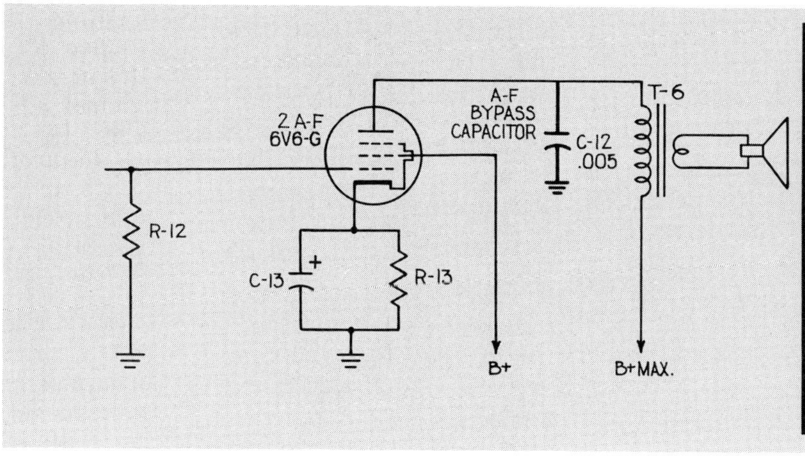

FIG. 12-6. The a-f plate bypass capacitor and its position in the second a-f circuit.

seen and will be found by voltage analysis, since the open plate circuit will cause zero plate voltage.

As explained before, the output transformer should be replaced with an exact duplicate where obtainable. When this is not possible, a universal output transformer may be substituted. These usually come with an instruction sheet, but technicians sometimes find the sheet confusing and connect the transformers improperly. This results in poor tone quality. A bit of theory might help to clear up this matter.

The output transformer, as an impedance matching device, works on the principle of reflected load, a term the average technician shies away from. Let us first try to explain it.

Assume a power transformer that is being used to light lamps. For simple arithmetical figures, let us also assume a 100-volt line, rather than the usual 110 or 120 volts, and lamps requiring 10 volts at 1 amp each. For further simplification, assume 100 percent efficiency in the transformer; that is, watts input equals watts output. The transformer has a 10 to 1 step-down ratio to furnish the 10 volts needed for the lamps. Each lamp has a resistance of 10 ohms ($R = E/I = 10/1 = 10$ ohms).

When one lamp is connected, as in Fig. 12-7, 1 amp flows through the lamp. Wattage dissipated is 10 watts ($W = E \times I = 10 \times 1 = 10$ watts). To satisfy watts input equals watts output, the primary current will be 0.1 amp ($I = W/E = 10/100 = 0.1$ amp). To the 100-volt line, the transformer primary looks like a 1,000-ohm impedance or resistance load, since it will drive only 0.1 amp into it ($Z = E/I = 100/0.1 = 1{,}000$ ohms). Now let us light two lamps from the same transformer, as in Fig. 12-8. Two 10-ohm lamps in parallel have a combined resistance of 5 ohms and will draw 2 amp ($I = E/R = 10/5 = 2$ amp) from the secondary, which remains at 10 volts. The actual impedance therefore is 5 ohms. Watts con-

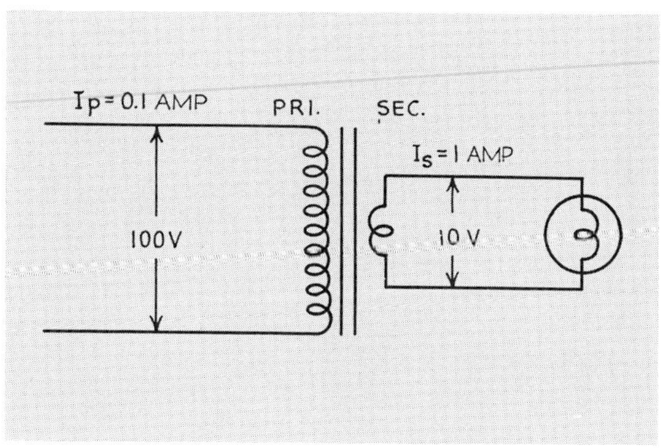

FIG. 12-7. Transformer lighting one lamp.

sumed is 20 watts ($W = E \times I = 10 \times 2 = 20$ watts). Once again to make watts input equal watts output, the primary current must now increase to 0.2 amp ($I = W/E = 20/100 = 0.2$ amp). The 100-volt line now looks at the transformer primary as though it were a 500-ohm impedance, since it must furnish 0.2 amp to it ($Z = E/I = 100/0.2 = 500$ ohms). This is called "reflected load." Under the above conditions, a 10-ohm actual load reflects back to the primary a 1,000-ohm load, while a 5-ohm actual load gives a 500-ohm reflected load in the primary. Note also the ratio of reflected to actual load, 100 to 1, which is the square of the turns ratio 10 to 1; that is, a transformer with a 10 to 1 turns ratio would make the reflected load in the primary 100 times (10^2) as great as the actual load in the secondary.

Now let us apply this bit of transformer theory to the output transformer. Assume a 10-ohm voice coil connected to the same 10 to 1 transformer that was used before to light lamps. The connections are shown in Fig. 12-9. The primary would look like 1,000 ohms to any line feeding it. Obviously, this transformer would not do to couple the 10-ohm voice coil to a 6V6-G tube, which requires a 5,000-ohm load resistance for optimum results. A turns ratio of 20 to 1 would make a much better match, since the reflected load in the primary of a 10-ohm voice coil in the secondary would be $(20)^2$, or 400 times as great (4,000 ohms). A turns ratio of 22.4 to 1 would be exactly right.

A universal output transformer is one supplying many possible combinations of turns ratio, so that almost any voice coil may be matched to almost any tube or combination of tubes. A typical universal output transformer is shown in Fig. 12-10. The primary is center-tapped for use in push-pull circuits. In second a-f stages using a single tube, the center tap should be taped up and disregarded. Then, either end of the primary winding is connected to the plate, and the other to *B*+. The secondary usually has six taps, numbered 1 to 6, and a great number of turns ratio combinations is possible.

In using a universal output transformer as a replacement, the first requisite is to use the proper size. They are rated by wattage. Physical size of the transformer is a rough indication of the wattage. Make sure that the replacement is as large as the original. Confirmation may be obtained by comparing the wattage size used with the tube-manual rating for the output tube or tubes in the receiver. The tube manual will also give the recommended load impedance.

The next step is to determine the voice-coil impedance. To do this, determine its resistance on the low-ohm scale of your ohmmeter. Then multiply the reading by 1.25. (This rule of thumb is close enough for general service work.) Then check with the in-

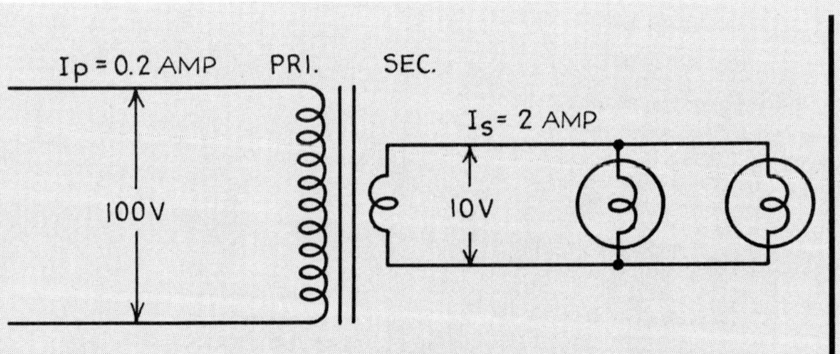

FIG. 12-8. Transformer lighting two lamps.

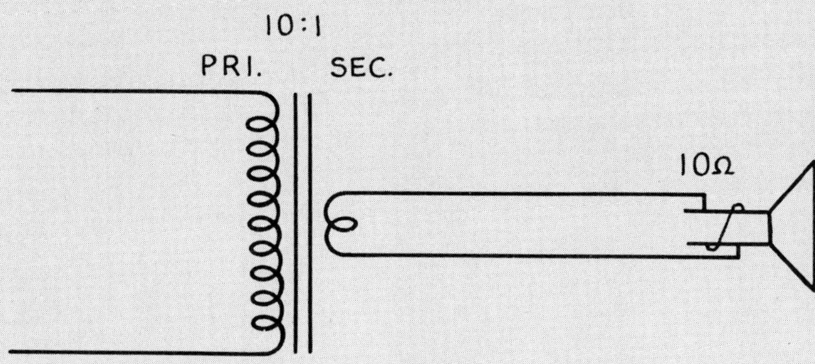

FIG. 12-9. Transformer feeding a 10-ohm voice coil.

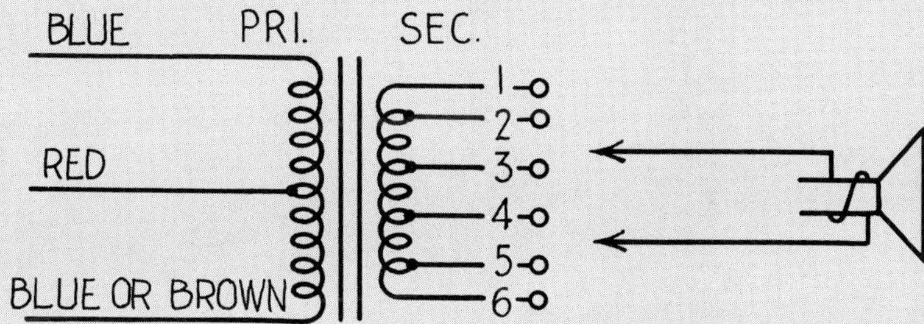

FIG. 12-10. A typical universal replacement output transformer.

struction sheet, which comes with the universal output transformer for the proper taps to use. Figure 12-11 is a sample instruction sheet. As an example of how the sheet is to be used, let us find the proper taps for the standard receiver. The voice coil is measured on the low-range ohmmeter and found to be 5 ohms. Multiplying by 1.25, its approximate impedance is found to be about 6 ohms. The single 6V6-G output tube requires a load impedance of about 7,000 ohms. The output transformer must therefore match about 7,000 ohms to 6 ohms. Look for the single 6V6-G, which is found in the column headed by 7,000 as the primary load impedance. Run down this column into the voice-coil impedances looking for 6 ohms. Then, read across horizontally to find secondary taps 1 and 4, which are to be used.

There is sometimes an inverse feedback lead connected from the secondary of the output transformer back to a previous point in the audio amplifier circuit. In this case, when the output transformer is replaced, the voltage fed back may be in the wrong phase and cause an audio oscillation or squeal, which will be present with or without any signal being fed into the amplifier. When this happens, reversal of either the primary or the secondary leads will clear up the difficulty. More will be said regarding this matter in the section on circuit variations dealing with inverse feedback.

Troubles common to the second a-f tube. The tube itself may be the cause of poor operation of the stage. Low emission will cause low gain and poor power-handling capacity. A tube checker usually shows this condition, or it will show up on voltage analysis. Low emission results in low plate current, consequently low self-bias voltage, and a too small difference between plate and *B*+ max voltages.

SIMPLIFIED CHART SHOWING PROPER USE OF SECONDARY TAPS

Primary load impedance	18,000	14,000	10,000	8,000	7,000	4,000	2,000
Single			3V4	89 pentode-1G5G-6K6G-6A4/LA	6V6 6B5-6F6-6N6G-GA-	45-50-71A-6L6-6BQ5	2A3-6A3-6Y6G-35C5 50C5 25L6-6EH5 35L6GT
Push-pull		41-47-	42-2A5-6AC5G-6B5-6F6-6N6G 6V6	43-6BQ5	45-6L6 Class AB_1 6EH5	2A3-6A3-6L6-25L6	
Secondary tap	Voice-coil impedance						
2–3	0.97	0.75	0.54	0.43	0.38	0.22	0.11
3–4	1.2	0.90	0.04	0.51	0.45	0.26	0.13
4–5	1.8	1.4	1.0	0.80	0.70	0.40	0.20
1–2	3.2	2.5	1.8	1.4	1.2	0.71	0.36
2–4	4.2	3.3	2.4	1.9	1.6	0.94	0.47
5–6	4.8	3.7	2.7	2.1	1.9	1.1	0.53
3–5	5.9	4.6	3.3	2.6	2.3	1.3	0.65
1–3	7.7	6.0	4.3	3.4	3.0	1.7	0.85
2–5	11.6	9.0	6.4	5.1	4.5	2.6	1.3
4–6	12.5	9.7	6.9	5.6	4.9	2.8	1.4
1–4	14.8	11.5	8.2	6.6	5.8	3.3	1.6
3–6	21.3	16.5	11.8	9.5	8.3	4.7	2.4
1–5	27.0	21.0	15.0	12.0	10.5	6.0	3.0
2–6	31.5	24.3	17.4	14.0	12.2	7.0	3.5
1–6	54.5	42.4	30.2	24.2	21.2	12.1	6.1
Primary load impedance	**18,000**	**14,000**	**10,000**	**8,000**	**7,000**	**4,000**	**2,000**

FIG. 12-11. Universal output transformer instruction sheet.

Type No. 2774	4-watt size
Type No. 2776	6-watt size
Type No. 2780	10-watt size
Type No. 2782	12-watt size
Type No. 2788	18-watt size

FIG. 12-11. Continued

The tube also might be noisy (possible loose elements) or cause hum (cathode-to-filament leakage). The best check for these conditions is to substitute a similar type of tube, known to be good.

A fairly common trouble, particularly in the case of 25L6, 35C5, and 50C5 tubes, is known as "grid emission." The complaint here is that the radio starts playing normally, but after 5 min or so begins to distort badly. A voltmeter connected from chassis to grid will begin to show positive at the grid as the distortion begins.

CIRCUIT VARIATIONS OF THE SECOND A-F STAGE

Tone control in the second a-f stage. There are many tone-control circuits. The diagram of Fig. 12-12 shows how the tone control circuit is included in the output stage. Capacitor *C*-112 and variable resistor *R*-112 are in parallel with capacitor *C*-12. Like capacitor *C*-12, capacitor *C*-112 bypasses high audio frequencies out of the speaker circuit. Capacitor *C*-112 has a comparatively high capacitance, 0.05 mfd being usual. By itself, it would remove most of the high audio frequencies from the signal and make the low notes seem more prevalent by comparison. Variable resistor *R*-112, which by its setting allows more or less of the bypassing of high frequencies through *C*-112 to take place, constitutes a tone control. The usual value of *R*-112 is 50,000 ohms.

All tests for the standard second a-f stage are equally applicable to this variation, and all notes applying to capacitor *C*-12 may also be used for tone capacitor *C*-112. If this capacitor should short, however, the path for the high *B*+ voltage to ground would be through the tone-control variable resistor *R*-112. This would give a variable shunt path depending on the tone-control setting. At the maximum bass position, there would be a very low resistance from plate to ground through the shorted capacitor, the *B* voltage would be low, and the receiver would not operate. At the minimum bass position there would simply be a 50,000-ohm shunt path for the *B* supply, and the receiver would operate. The erratic action of the tone control would, of course, focus the technician's attention to this circuit, and the defect would be found by ohmmeter check. When a shorted tone capacitor is replaced, the heavy current through *R*-112 may have damaged the tone control. It would therefore be wise to replace the tone control also. Replacement notes on volume controls given in Chap. 13 may be applied to the tone control.

A trend in the use of tone controls is to replace the variable resistor *R*-112 with a

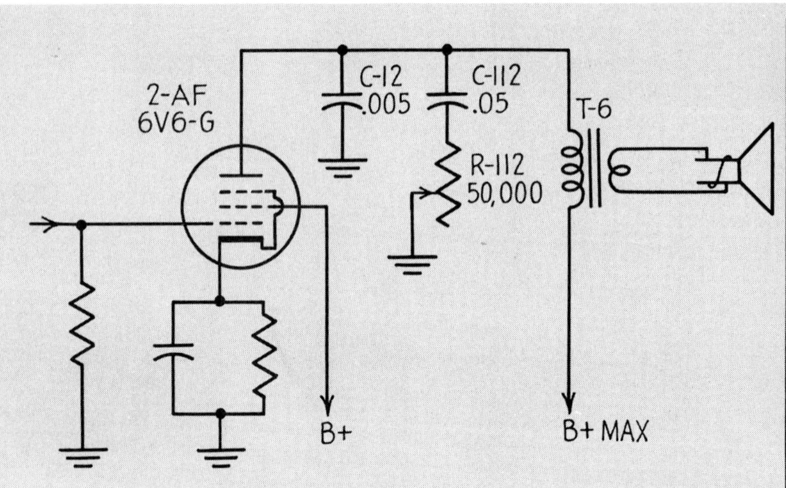

FIG. 12-12. Tone-control circuit in the second a-f stage.

switch, thereby making a 2-point tone control. An example of this is shown in Fig. 12-13. Note the tone capacitor *C*-23 and its associated switch in the plate circuit of the 50L6 tube. When the switch is open, there is no shunting action, and this is the treble position. In the bass position, where the switch is closed, *C*-23, which is 0.04 mfd, shunts some of the high audio frequencies out of the speaker.

Figure 12-14 shows a similar 2-point tone control. In this case, the shunting action of *C*-21 and its associated switch is in the input circuit of the second a-f tube.

Inverse feedback in the second a-f stage. Inverse feedback is a form of desirable degeneration often used in the audio amplifiers of radio receivers. There are many types of inverse feedback circuits in common use. In all of them, part of the output signal is fed back in an out-of-phase relationship (hence the name "degeneration") to some point in the input signal circuit, to provide improved over-all audio fidelity. Inverse feedback is always accompanied by a loss in gain, but the amplifier is designed for higher than normal gain to compensate for this loss.

In Fig. 12-13, the cathode bypass capacitor has been omitted to provide degeneration through self-bias resistor *R*-3, which is common to both the input and output circuits of the 50L6 tube. Since the input and output circuits of a tube are 180 deg out of phase, degeneration is automatic. Capacitor *C*-18, the high-frequency bypass capacitor, is returned directly to cathode rather than to ground, so that the degenerative effect is greater at the higher frequencies (especially the high harmonic frequencies), thereby making for more uniform response for the stage.

In Fig. 12-14 the inverse feedback circuit is shown by the heavy lines. The feedback voltage is taken from the plate of the 6V6-G output tube and fed through resistor *R*-14 back to the plate of the first a-f section of the 6SQ7 tube.

As a general rule, inverse feedback circuits do not cause many complications to the technician. All tests and service notes pertaining to the standard amplifier circuit may be applied. Resistor *R*-14 in the feedback circuit, represented in Fig. 12-14, will rarely cause any service difficulty.

A servicing problem pertaining to inverse feedback circuits occurs when the feedback voltage is taken from the output-transformer secondary, as shown in Fig. 12-15. In this case, the feedback voltage is reintroduced into the cathode circuit of the first a-f tube, which has no bypass capacitor. Another

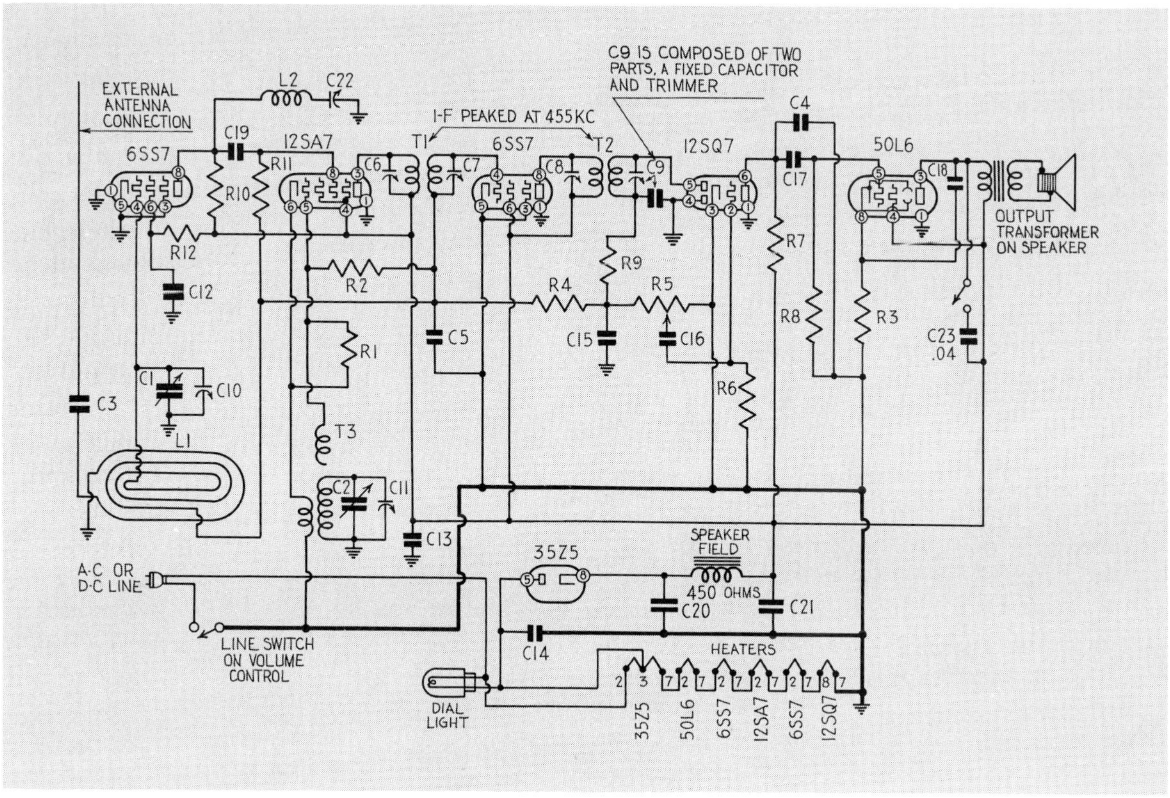

FIG. 12-13. Schematic diagram for Emerson receiver.

FIG. 12-14. Schematic diagram for a Stromberg-Carlson receiver.

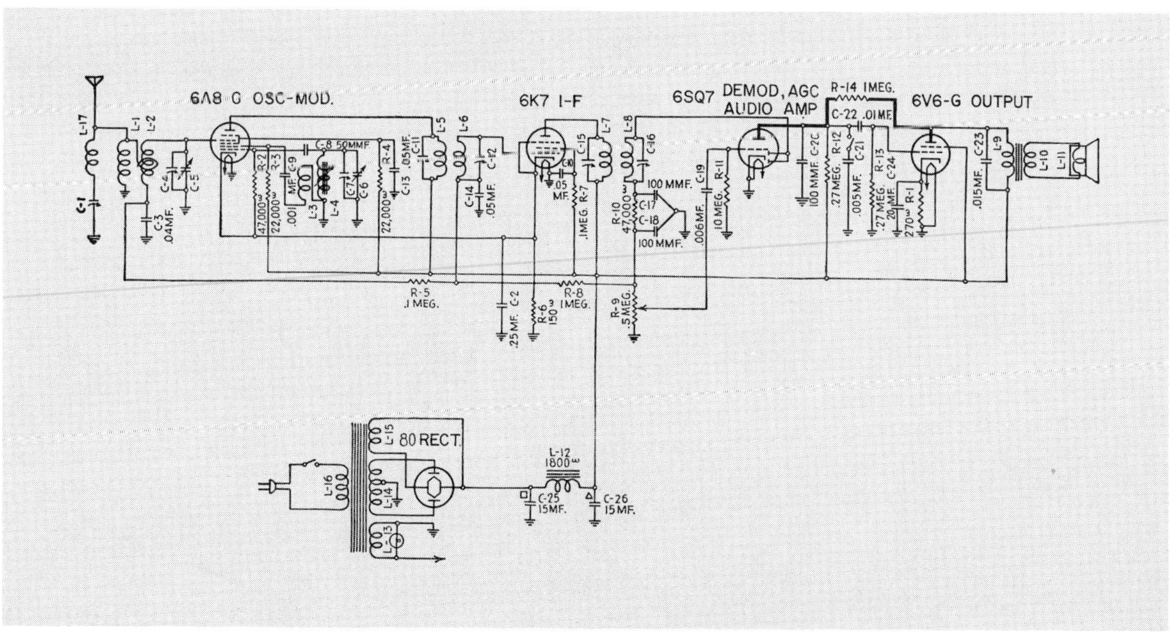

variation of this same circuit introduces the feedback voltage into a tap in the grid load of the first a-f tube. In either case, if the output transformer leads should become reversed, as may easily happen when the output transformer is replaced, the feedback voltage will be in phase with the signal voltage rather than out of phase. This will produce regeneration rather than degeneration, and the audio amplifier becomes an audio oscillator. The oscillation appears in the speaker, usually as a high-pitched squeal, and will, of course, be unaffected by tuning the receiver. The technician must be aware of this possibility when replacing the output transformer, since the usual service procedure for oscillation will not disclose it. Reversing the primary or secondary leads, whichever is simpler, will clear up the difficulty.

Voltage distribution in smaller receivers. The power supply and audio section of an Airline a-m/f-m receiver is shown in Fig. 12-16, as a typical example of voltage distribution in a smaller receiver. Note that the circuit is very similar to our basic standard circuit. A smaller power transformer is used, and so the rectifier output or *B*+ max is 245 volts, rather than over 300 volts as indicated for the standard circuit. The power supply filter circuit consists of a two-section R-C filter made up of resistors *R*-8 and *R*-9 and capacitors *C*-8, *C*-9, and *C*-10. All the latter have a capacitance of 40 mfd.

The plate circuit of the 6V6 power output tube is fed directly from input filter capacitor *C*-10, making for a normal potential of 230 volts at the plate of the tube, as indicated in the diagram. The filtering at this point is insufficient for the screen of the output tube, so this is fed from the output of the first R-C filter section, making a potential of 180 volts normal for the screen. Even more filtering is required for the rest of the receiver, which is therefore fed from the output of the second R-C filter section. The potential at the *B*+ point for most of this receiver is only 90 volts.

In the second audio-amplifier stage, note that self-bias resistor *R*-7 is 270 ohms—very close to the 330 ohms of the standard circuit. But the lowered plate and screen voltage causes the tube to operate at lower plate and screen currents. As a result, the voltage across the bias resistor is given as 9.4 volts rather than the 13 volts of the standard circuit. The tube will therefore handle input signals up to 9.4 volts, producing a speaker output of something less than 4 watts. Note the tone control circuit consisting of capacitor *C*-6 and resistor *R*-6 in the input circuit of the 6V6 tube.

When servicing a set of this type, its normal voltage distribution must be kept in mind. In all other respects, the signal tests and service notes pertaining to the standard amplifier circuit are equally applicable.

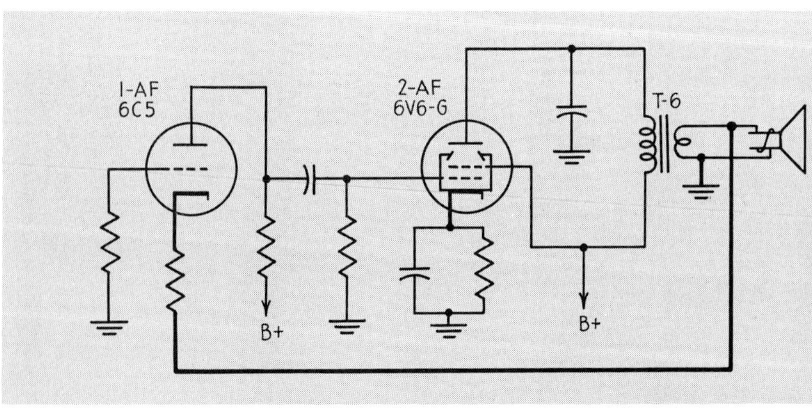

FIG. 12-15. Inverse feedback circuit where the feedback is taken from the secondary of the output transformer.

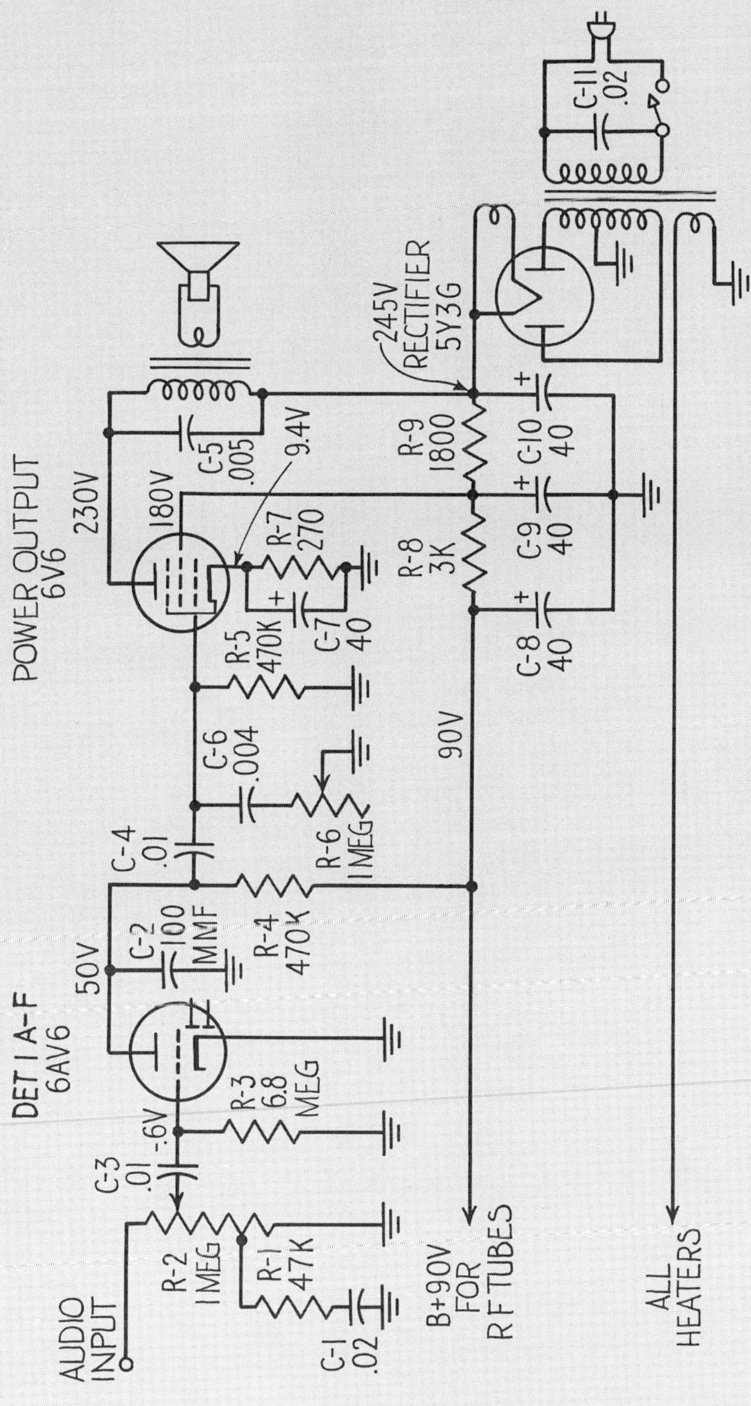

FIG. 12-16. Audio section and power supply of an Airline a-m/f-m receiver.

SERVICE DATA CHART FOR THE SECOND A-F STAGE

Symptom	Abnormal reading	Look for
No signal from the speaker	Plate voltage = 0. Screen voltage = 0	Trouble in the power supply. See Chaps. 8, 9, 10
	Plate voltage = 0. Screen voltage low	Short-circuited high a-f by-pass capacitor *C*-12
	Plate voltage = 0. Screen voltage normal or high. (Screen of a glass second a-f tube glows)	Open primary winding of output transformer *T*-6
	Plate and screen voltages normal or high. Cathode voltage low	Weak or dead second a-f tube
	Plate and screen voltages normal or high. Cathode voltage high	Open self-bias resistor *R*-13
Poor tone quality	Plate voltage low. Screen voltage normal	Defective second a-f tube. Short-circuited cathode by-pass capacitor *C*-13. Open grid-load resistor *R*-12. Shorted or leaky coupling capacitor *C*-32 (see Chap. 13)
	Voltages normal	Open cathode bypass capacitor *C*-13. Mismatched replacement output transformer
Squeal or oscillation	Voltages normal	Open output filter capacitor *C*-16. Open high a-f bypass capacitor *C*-12. Degenerative feedback connection from replacement output transformer incorrectly phased

Symptom	Abnormal reading	Look for
Motorboating		Open output filter capacitor *C*-16. Open grid-load resistor *R*-12

SUMMARY

Test for normal operation of the second a-f stage.

The tip of a plugged-in soldering tool applied to the grid of the tube causes a growl to be heard in the speaker.

Diagram of a typical second a-f stage.

The accompanying figure shows the typical second a-f stage.

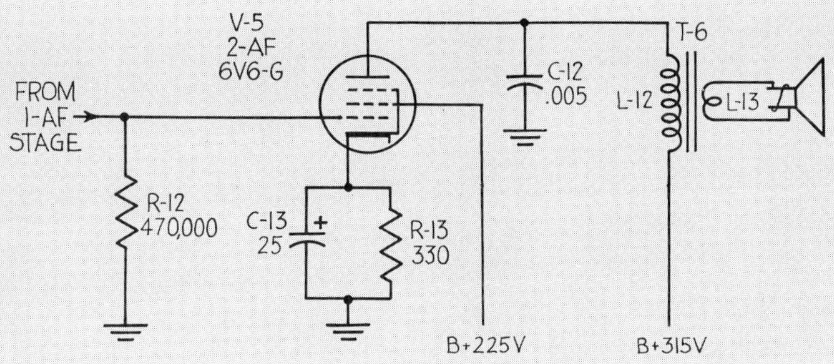

Normal voltage data.

Voltage is measured from the chassis or common negative lead. Voltage data are given in the accompanying table.

Tube terminal	25L6 and 6V6-G pin No.	A-c receiver, volts	A-c/d-c receiver, volts	6EH5 50C5 pin No.
Plate	3	200–235	110–120	7
Screen	4	175–215	90–100	6
Grid	5	0	0	2,5
Cathode	8	10–13	5–7	1

Normal second a-f stage resistance data.

Chassis to cathode	330 ohms
Chassis to control grid	470,000 ohms
Plate to *B*+	200–600 ohms

The 330 ohms of resistance from chassis to cathode is the ohmic value of self-bias resistor *R*-13. When a tube other than the 6V6-G is used a different value will be found. Refer to the diagram of the receiver being tested or to the table on page 105.

The plate to *B*+ reading measures the resistance of *L*-12, the primary of the output transformer.

QUESTIONS

1. A receiver is brought in for repairs, the complaint being "no reception." Visual inspection shows a red-hot screen grid in the type 6F6-G power tube. What is likely to be wrong? Indicate the tests that should be made to confirm your assumption.
2. In a dead receiver, the power supply is found to be operating normally. A voltage check of the second a-f stage shows the following:

Plate	300 volts
Screen	300 volts

What are the likely causes of the trouble? Indicate the tests that should be made to confirm the actual cause of the trouble.

3. An a-c receiver, using a 6V6-G tube in the second a-f stage, gives a high-pitched squeal regardless of the setting of the volume control or tuning dial. What are the possible causes of the trouble? How would you check for each?
4. The receiver of Fig. 13-25 has an open output transformer. If an original replacement is not obtainable, use the universal output-transformer chart of Fig. 12-11 for reference and choose (1) the type of transformer that should be used, and (2) the secondary taps that should be used.
5. The receiver of Fig. 12-14 has low volume and sounds tinny. A voltage check shows normal voltage readings. Substitution of the bench test speaker causes no improvement. What should the next check be?
6. The receiver of Fig. 12-14 motorboats. Bridging the output filter capacitor *C*-26 with a 20-mfd/450-volt capacitor causes no improvement. What should the next check be?
7. The receiver of Fig. 12-13 begins to distort after it has been playing for 15 min. What would you suspect is wrong? How would you confirm your suspicion?
8. A distorting receiver gives the following voltage check for the 6V6-G tube in the second a-f stage:

Plate	180 volts
Screen	200 volts
Grid	0 volts
Cathode	2 volts

What is likely to be the cause of the distortion? How would you confirm your assumption?

9. The receiver of Fig. 12-13 is brought in as dead and gives the following voltage readings for the second a-f stage:

Plate	110 volts
Screen	110 volts
Cathode	30 volts

What is likely to be the cause of the trouble? How would you confirm your assumption?

10. What precautions should be observed in replacing a shorted high-a-f bypass capacitor?

First Audio-Amplifier Stage

13

Quick check. If a wet finger or a plugged-in soldering iron is applied to the input of the first a-f stage and a very strong growl comes out of the speaker, the stage is probably functioning properly, and the technician moves on to the next stage.

Function of first a-f stage. The control grid circuit is the stage input and is coupled to the detector output circuit. The plate circuit is the stage output, which is in turn coupled to the grid or input circuit of the second a-f stage. The detector has an output of roughly 1 volt of a-f signal. The second a-f stage, if it contains a 6V6-G beam-power amplifier, requires an input signal of 13 volts to drive the speaker to full volume. It is therefore the function of the first a-f stage to build up the detector output signal voltage (1 volt) to the level necessary to drive the second a-f stage (13 volts).

Theory of operation, functions, and values of component parts. From the function of the stage, to amplify 1 volt of signal to 13 volts, it would seem that a voltage amplification of 13 for the stage would be sufficient. However, the detector output may be less than 1 volt, in which case there would be insufficient volume. The first a-f stage, therefore, is usually designed for high voltage gain, 50 or higher, so that low input signals can be amplified to the required level to operate the second a-f stage. Then, should the input be excessive, the detector signal level feeding the first a-f stage is reduced through a potentiometer, which is the manually operated volume control of the receiver.

The first a-f stage is called a "voltage amplifier," while the second a-f stage is called a "power amplifier." The reason for these descriptions lies in their functions. The second a-f stage drives the speaker and must furnish power to vibrate the speaker cone and the surrounding air. Electric power is measured in watts, which incorporates both voltage and amperage. The second a-f tube, the output transformer, and the speaker are all rated in watts. The second a-f stage, therefore, is a power amplifier developing enough power to drive the speaker. The first a-f stage, on the other hand, furnishes the grid excitation for the second a-f tube. The grid of the second a-f tube is always kept at a negative potential by the bias voltage supply, and the signal voltage does not normally exceed the bias voltage. As a result, the grid circuit does not draw current from the previous stage, and the signal grid excitation therefore requires voltage but not current. For this reason, the first a-f stage, which furnishes the grid excitation for the second a-f stage, is called a "voltage amplifier." If the signal voltage at the second a-f grid should exceed the bias voltage and grid current result, the first a-f stage would also be furnishing power. Likewise, if the first a-f stage were used to drive a pair of headphones, it would be operating as a power amplifier.

The tube used as the first audio amplifier is usually a high-mu triode. Most often, it is the triode section of a dual-purpose diode and high-mu triode, like the 6SQ7, which will be used in our standard circuit. The diode section is used as the detector and will be described in Chap. 14.

Standard circuit. Potentiometer *R*-27 is the manual volume control for the receiver. Its usual value is 500,000 ohms. The detector signal output is connected across *R*-27, and the position of the potentiometer arm determines how much of the detector signal output voltage is fed to the audio amplifier. For example, if the arm is near the grounded end, little of the detector output voltage developed across *R*-27 gets amplified, and this is the low-volume position. If the arm is nearer the ungrounded end, more of the available signal voltage gets amplified, and this is the high-volume position.

Capacitor *C*-31 is the coupling capacitor. It feeds the audio signal voltage from the volume control to the grid or input circuit of the tube and is usually 0.005 mfd. It may vary in different receivers from 0.001 to 0.02 mfd.

Resistor *R*-31 is the grid load. It returns the grid directly to the cathode in a circuit known as "contact bias." As will be explained, the grid-load resistor in a contact bias circuit usually is high: 2 to 15 megohms. The average size for the standard circuit is 10 megohms.

Operation of contact bias. When the schematic diagram is studied, it would seem at first glance that there is no grid-bias voltage on the triode section of *V*-4, since the grid goes to ground through *R*-31 and the cathode is also at ground potential. To understand how a bias voltage is developed between grid and cathode, first assume a condition of no signal input. In the tube, the cathode is emitting electrons which are attracted by the positive plate, as shown in Fig. 13-2. Some of these electrons impinge on the grid located between cathode and plate, as shown in Fig. 13-3. These will flow through the grid load *R*-31 back to cathode. Since *R*-31 usually has a high resistance, it will not require very much grid current flow to develop a voltage across it. By applying Ohm's law, $E = I \times R$, a current of only 0.1 microampere (0.0000001 amp) will develop 1 volt across 10 megohms, the usual size of *R*-31. Note the arrow showing direction of electron flow through *R*-31 in Fig. 13-3. Since electrons flow from negative to positive, the grid end of *R*-31 is negative, with respect to the ground or cathode end, by this voltage drop. Therefore, a small negative bias is established on the grid. This negative bias remains constant for a particular circuit because, as fast as electrons leak off the grid across *R*-31, new electrons impinge on it, and therefore a condition of equilibrium is set up whereby a slight negative bias is maintained on the grid. Capacitor *C*-31 prevents electrons from leaking across *R*-27 to ground.

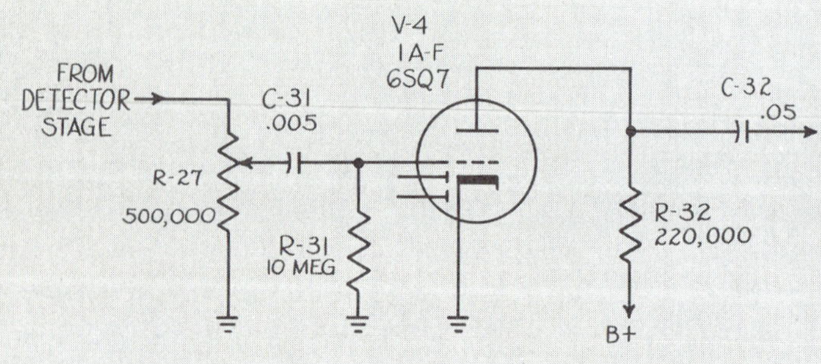

FIG. 13-1. Typical first a-f amplifier stage.

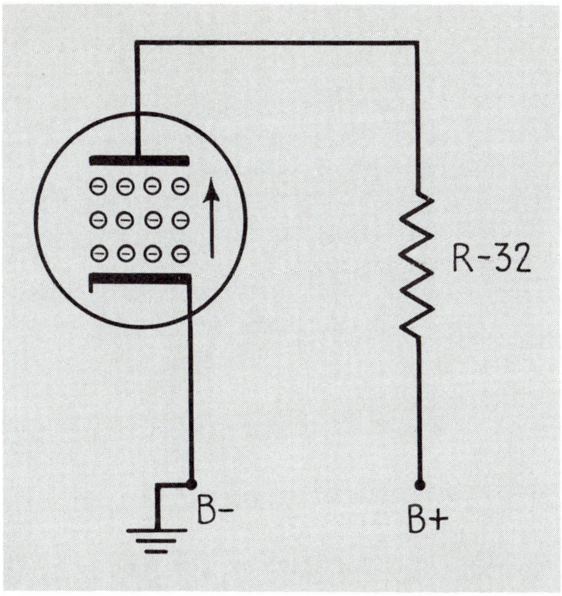

FIG. 13-2. Electrons being attracted from the cathode to the positive plate.

In amplifiers used in radio receivers, grids are maintained at all times at a negative potential. When the signal voltage is placed on the grid, it drives the grid more negative or less negative with each alternation. If the signal voltage should be larger than the steady negative grid-bias voltage, the grid will be driven positive on the positive half of the signal cycle, resulting in serious distortion. For this reason, the signal voltage must always be lower than the grid-bias potential. In the case of contact bias, the grid-bias potential is low, and as a result the signal handling capacity is low. Contact bias, therefore, is used only in the first audio stage where the input signal is at a low level of potential.

Tubes used in the first a-f stage. Vacuum tube *V*-4 is the voltage amplifier tube. The one most often used in the first a-f stage is the high-mu triode section of the type 6Q7 or 6SQ7 tube. Receivers with miniature tubes use the similar 6AV6 tube. When lower gain for the stage is desired or the stage is to be followed by transformer coupling, the type 6R7 tube is employed. Where a separate diode is used for the detector stage, the tube employed for the first a-f stage is a 6F5 or 6SF5; these have the same characteristics as the triode section of the 6SQ7. Even in the latter case, the 6SQ7 is often used with the diode plates grounded. Receivers of the a-c/d-c type use the 12SQ7 octal base or the 12AV6 miniature-base tubes in a similar circuit.

Coupling circuit to the second a-f stage. Resistor *R*-32 is the plate load of the first a-f tube. It may go as high as 0.5 megohm and as low as 0.1 megohm. Higher values would give somewhat greater gain; lower values would result in reduced gain. The average size chosen for the standard circuit is 220,000 ohms. When the first a-f tube is a low-mu triode like the 6SR7, resistor *R*-32 is lower in value, 47,000 to 100,000 ohms being usual. In all cases, wattage dissipation is relatively unimportant. The resistors generally in use are the ½-watt size.

Capacitor *C*-32 is the audio coupling capacitor. This capacitor, plate-load resistor *R*-32, and grid-load resistor *R*-12 of the following stage make up a resistance coupling circuit between the two stages, as shown in Fig. 13-4. Its function is twofold: It conducts the a-f signal from the plate circuit of the first a-f tube to the grid of the second a-f tube;

FIG. 13-3. Electrons impinging on the grid of a tube in developing contact bias.

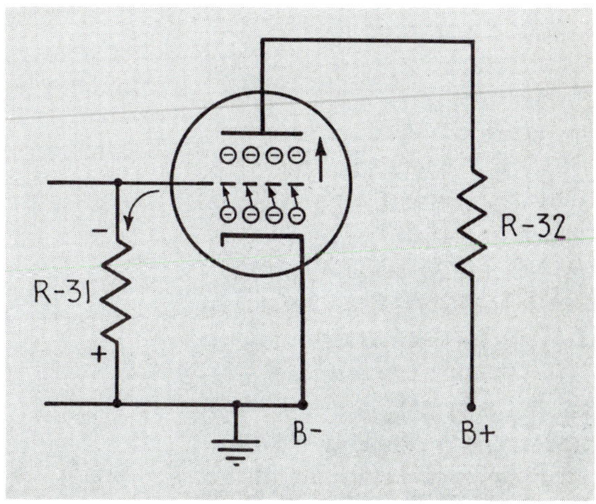

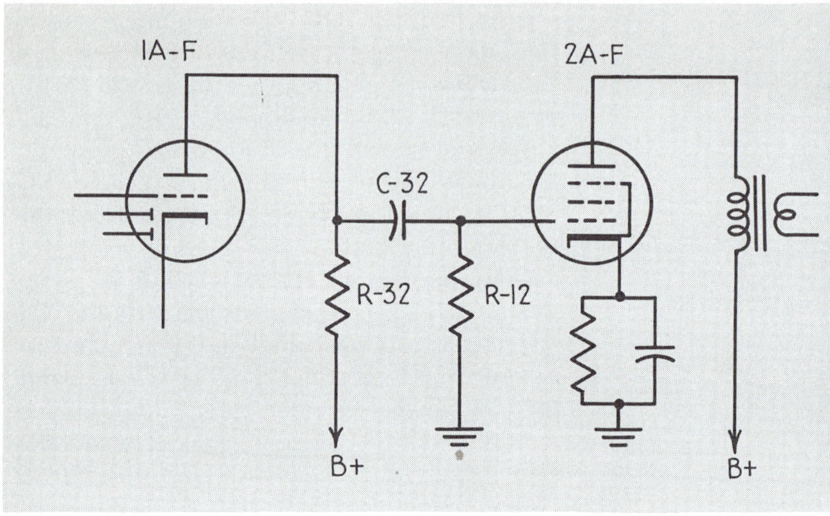

FIG. 13-4. Resistance coupling between the first and second a-f stages.

at the same time, it keeps the positive plate potential of the first a-f tube from affecting the grid of the second a-f tube.

The capacitance of coupling capacitor *C*-32 varies considerably with different receivers. Capacitances ranging from 0.01 to 0.1 mfd are common. The standard receiver uses 0.05 mfd. The larger capacitances give better bass frequency response. Some receivers purposely use a small-capacitance capacitor at *C*-32 and are generally designed to give a poor response to low audio frequencies so as to minimize the hum frequency (120 cycles for a full-wave and 60 cycles for a half-wave rectifier).

The insulation of capacitor *C*-32 must be good, since any leakage would put a positive bias on the second a-f grid from the first a-f plate. Paper tubular capacitors are usually used with a voltage rating of 400 or 600 volts DC.

NORMAL TEST DATA FOR THE FIRST A-F STAGE

Signal check. In the signal-substitution method of service procedure, only the final audio stage is measured as a single unit. Thereafter, as each stage is added, the test is over-all. In the case of the first a-f stage, the test signal is applied to the first a-f stage input circuit while the output indication is taken from the speaker.

Most signal generators provide a pair of terminals, where a 400-cycle current is available for the testing of a-f circuits. When this test signal is applied to the input of an a-f amplifier, a 400-cycle note is heard in the speaker.

When the audio output from a signal generator is not readily available, a good substitute is found on every service bench. The tip of the soldering tool, either iron or gun, which is energized by 60-cycle current, is usable as a source of signal input voltage for a-f amplifiers. The test frequency is low, 60 cycles, which accounts for the note heard in the speaker being described as a growl. Also, the human body seems to pick up some 60-cycle voltage, and many practical technicians use a moist finger as their signal source. This last procedure is not recommended for beginners, who might accidentally touch a plate lead at 300 volts instead of a grid lead at zero volts.

Quick check for the first a-f stage. If a wet finger or a plugged-in soldering iron tip is applied to the ungrounded (called the "hot") end of the volume control with the control in the full ON position, a very strong growl should be heard in the speaker. If it is not heard or if it is not considerably stronger than the growl heard when the second a-f stage was checked, the trouble is in the first a-f stage.

The quick signal check can also be used for further narrowing down the location of the trouble. Assume normal response from the second a-f grid (a low growl) and no response from the ungrounded (hot) end of the volume control, as in Fig. 13-5.

Then, if the test signal is applied to the plate of the first a-f tube, normal response (a low growl in the speaker) indicates that coupling capacitor *C*-32 is functioning and the trouble is before the first a-f plate. No response at this point indicates an open coupling capacitor, or a first a-f plate-to-ground short.

If there is normal response from the first a-f plate, the test signal is shifted to the first a-f grid. Normal response (a strong growl) from this point indicates trouble in the volume control or coupling capacitor *C*-31. No response means that the trouble is between the first a-f grid and the plate. The likely causes are:

1. An inoperative first a-f tube. Confirm by substituting a good tube.
2. A grounded grid lead. Confirm with an ohmmeter. (The ground is probably caused by defective shielding.)
3. An open plate-load resistor *R*-32. Confirm by voltage and resistance checks.

Use of output meter. The ear, judging differences in sound intensity, can make only a rough estimate. Except at very low sound levels, the judgment of the ear is not very reliable. A more quantitative check for all receiver testing is to measure the actual signal power that is put into the speaker.

Radiomen usually work to a definite level of output from any receiver and then make comparisons of input signal necessary to attain that output. This reference level is called "standard output" and is defined as 50 mw (0.05 watt) of signal power into the speaker. Note that the 50 mw is well below the output capabilities of any radio receiver and, therefore, the test signal level at any point in the receiver, necessary to attain standard output, will not overload any tube.

The output power may be determined by measuring the signal voltage across the speaker voice coil with an a-c voltmeter. For example, if we have a 5-ohm voice coil, 0.5 volt will correspond to standard output.

$$W = \frac{E^2}{R} = \frac{0.5 \times 0.5}{5} = \frac{0.25}{5} = 0.05 \text{ watt}$$

A reading of 0.5 volt may be read on a meter which has a low a-c voltage range of 0–5 volts or less. A diagram showing how such a meter may be used to measure standard output is shown in Fig. 13-6A.

Some multimeters, however, are not equipped with so low a range. In this case, a more easily read output indication is obtainable at the primary of the output transformer

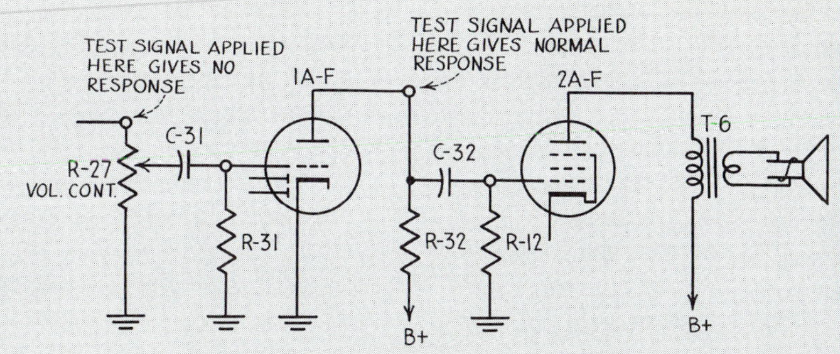

FIG. 13-5. Trouble shooting an inoperative first a-f stage by signal check.

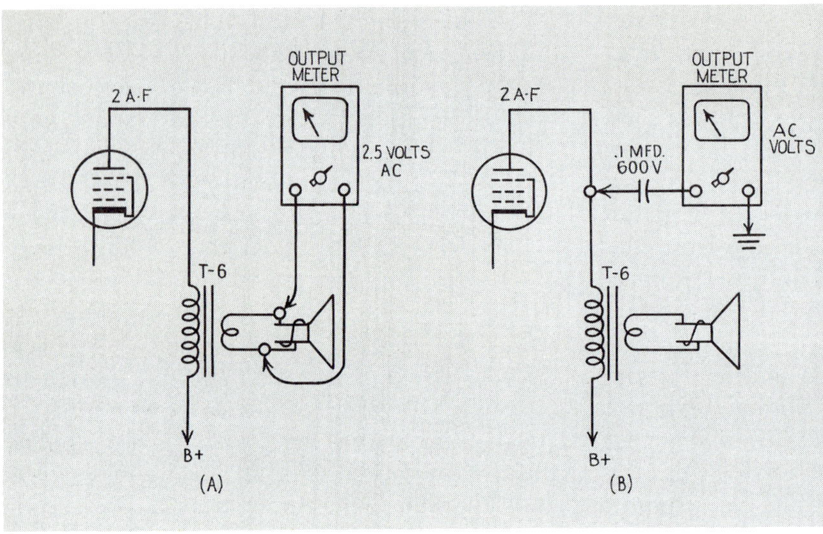

FIG. 13-6. Connection of an a-c voltmeter as an output meter.

where, owing to the turns ratio of the transformer, standard output will correspond to approximately 16 volts. The primary of the output transformer, however, is in a circuit where direct current, the plate current of the second a-f tube, is flowing. The signal itself is a pulsation of this current. To keep the direct current of the plate circuit from affecting the a-c meter, a capacitor must be inserted in series, so that the meter will read only the a-c signal component. This is shown in Fig. 13-6*B*, which indicates the connections for an output meter. A convenient size for this series capacitor is 0.1 mfd/600 volts. Some multimeters have the output capacitor built in, in which case there will be test jacks on the instrument labeled OUTPUT METER, and the 0.1-mfd capacitor need not be connected externally. The meter should be used on a suitable a-c range where 16 volts will give a good indication. (About half scale is best.) It might be advisable for the technician to work to a reading of 15 or 20 volts as his reference level, to take advantage of a convenient marker on the meter scale. Then as far as his test bench is concerned, 15 or 20 volts, as the case may

FIG. 13-7. Test leads for the output meter.

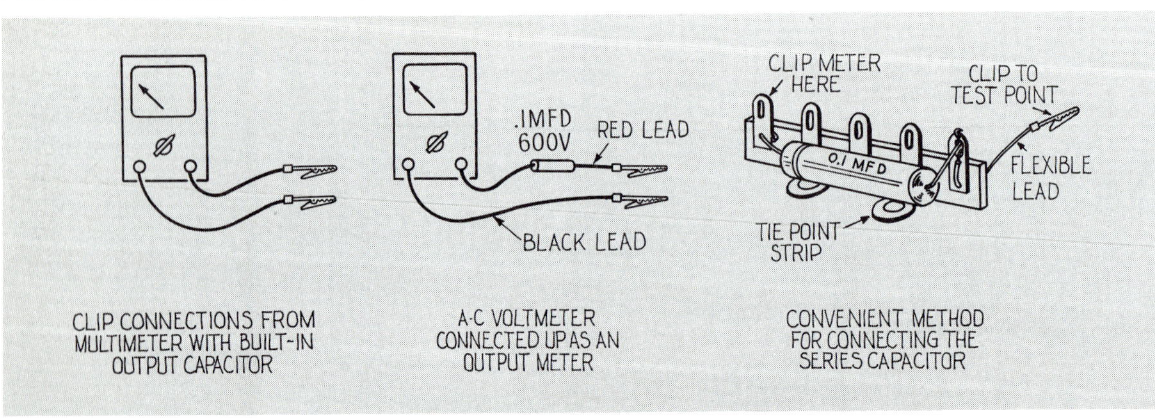

be, is standard output, and he will work at this level except where service notes issued by the manufacturer of the receiver concerned specify differently. The voltage chosen to represent standard output will not vary too much from 50 mw and is sufficiently accurate for any service work.

The technician would do well to provide himself with some special test leads for convenience in checking the output voltage. If the multimeter has a built-in output capacitor, a pair of test leads terminating in alligator clips will be all that is needed. If the output capacitor is not built in, a 0.1 mfd/600v capacitor, mounted on a tie-point strip, as shown in Fig. 13-7, is a convenient method of making a series connection. The same setup may be used as a dummy antenna in connecting a signal generator, and as a test capacitor, for bridging suspected open capacitors of approximately the same capacitance.

Stage-gain measurements. Now, having established standard output, let us make some gain checks on a receiver known to be perfect to determine how this information may be used in later servicing. Figure 13-8 shows the audio amplifier of the standard receiver. The output meter and the a-f output of the signal generator are connected to make gain checks. The capacitor in the hot lead of the signal generator (which may be connected internally) serves to keep d-c plate potentials out of the signal generator circuit when the hot lead is connected to a plate terminal in the radio. The receiver volume control is set for maximum output (full ON) and the tone

FIG. 13-8. Audio stage-gain measurements.

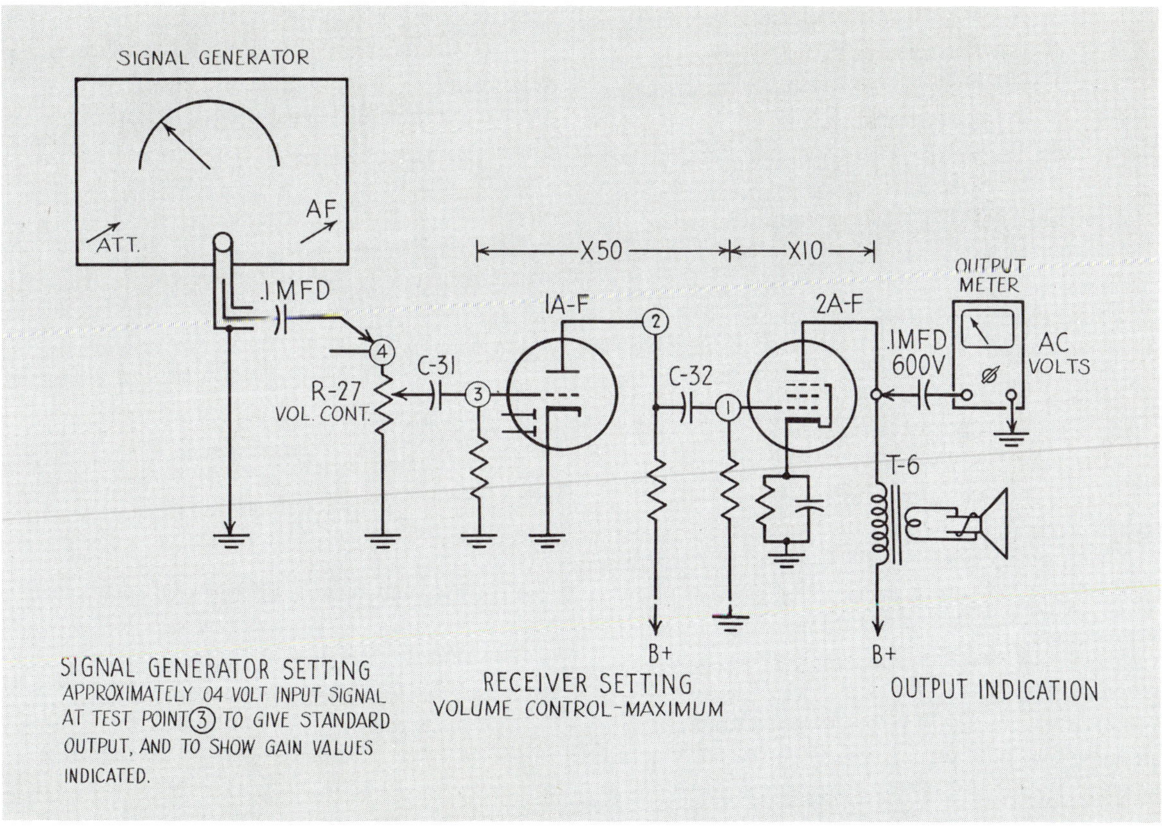

control, if any, is set for the minimum bass position.

The gain per stage is approximately 50 for the first a-f stage and 10 for the second a-f stage, as is indicated in Fig. 13-8. Now let us assume our test bench works to a reference level of 20 volts at the second a-f plate as standard output indication. Then when the hot lead of the signal generator is connected to point ①, the second a-f grid, a 2-volt signal will be needed to give standard output from this point, since 2 volts input times 10, the amplification of the stage, equals 20 volts output. It is not necessary to measure the input signal voltage. Accurate stage-gain measurements would call for expensive test equipment and, although this would be of advantage in design engineering, service work to find a poorly operating stage does not require anything more than comparative data. For an idea of 2 volts input, simply note the position of the attenuator on the a-f signal generator to obtain standard output on this perfect receiver.

When the test signal is connected to point ②, the first a-f plate, the signal-generator attenuator will have to be advanced slightly to maintain 20 volts on the output meter, to compensate for the loss caused by coupling capacitor C-32.

When the test signal is connected to point ③, which is the grid of the first a-f tube, only 0.04 volt will be needed to give standard output, since

$$\begin{array}{ccccccc}\text{Input volts} & \times & \text{gain of first a-f stage} & \times & \text{gain of second a-f stage} & = & \text{output volts} \\ 0.04 & \times & 50 & \times & 10 & = & 20\end{array}$$

The signal-generator attenuator position is again noted for the 0.04-volt position.

Moving the test signal to point ④, the hot end of the volume control, will again require a slight increase in signal input voltage to compensate for the loss caused by coupling capacitor C-31.

Having established comparative reference points on his signal generator and output meter, by trying the above procedure on a number of perfect receivers, the technician is in a position to determine the normal gain to be expected from any audio stage of a receiver brought in for servicing.

Normal first a-f voltage data. Voltages are measured from chassis or common negative to tube terminal indicated. In some a-c/d-c receivers where the circuit insulates $B-$ from the chassis, the negative terminal of the voltmeter is connected to the common negative. This is most easily found at the line switch. Normal data are given in the accompanying table.

Tube terminal	12SQ7 and 6SQ7 pin No.	A-c receivers, volts	A-c/d-c receivers, volts	12AV6 pin No.
Plate	6	100–170	40–60	7
Grid	2	−1	−1	1
Cathode	3	0	0	2

Voltages vary with different receivers and also with the ohms-per-volt rating of the multimeter. Since the plate-load resistor R-32 has an average value of 220,000 ohms, the plate circuit is a high-resistance circuit, and the plate voltage as read by a meter will depend on the extent to which the meter loads the circuit.

Normal first a-f resistance data. These data are given in the following table:

Chassis to cathode	0 ohms
Chassis to grid	10 megohms
$B+$ to plate	220,000 ohms

COMMON TROUBLES IN THE FIRST A-F STAGE

Troubles common to the volume control. Volume controls sometimes open. Since a signal check may give normal response even with an open volume control, this difficulty may not be found until the detector stage is checked, the volume control being also an important component of the detector stage.

More often, volume controls are noisy in operation, usually because of dirt between

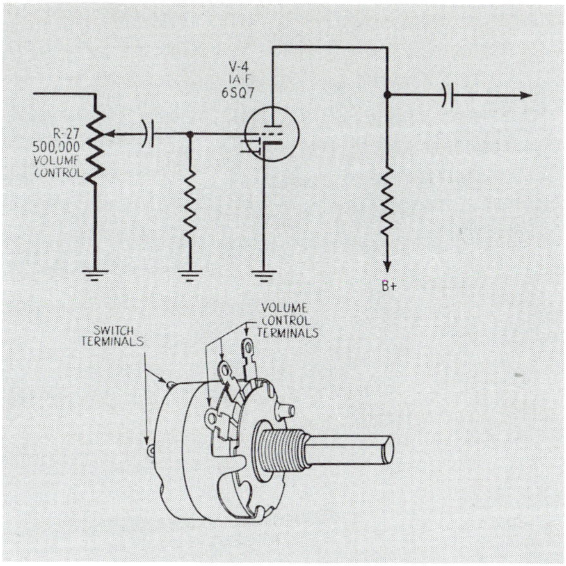

FIG. 13-9. Typical volume control and its position in the first a-f circuit.

the sliding arm and its contact ring. Although a temporary repair is often possible by a cleaning, such procedure is questionable, since a noisy control is also a possible cause of intermittent operation or fading. Debit the control to normal wear and tear of a moving part and replace it with a new one.

In replacing the volume control for electric and mechanical defects, it is best to obtain an exact replacement. When this is not possible, a replacement control as similar to the original as possible must be selected. When choosing the replacement control, the technician must keep several factors in mind:

1. *Space requirement.* The replacement must not be physically larger than the original unless there is room for it.

FIG. 13-10. A volume-control extension shaft.

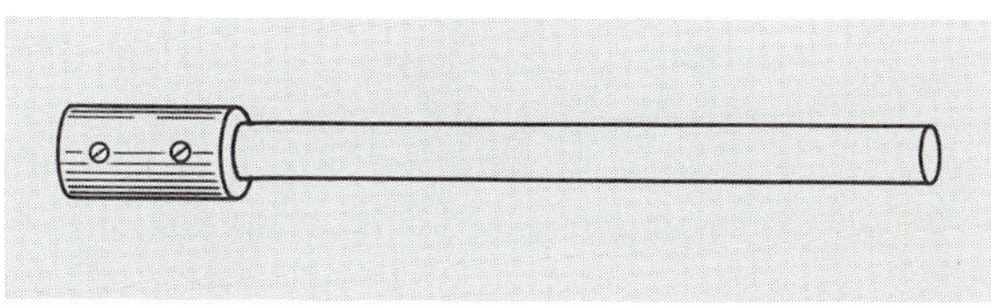

2. *Shape of shaft.* The shaft of the replacement control may be longer but not shorter than the original. The excess can be cut off. If the original has an unusually long shaft, an extension shaft (see Fig. 13-10) may be used.
3. *Flat side of shaft.* The volume-control knob should be examined. If it fastens with a setscrew, any shape of shaft may be used. If it is a spring push-on type of knob, the knob must fit the shaft snugly with spring tension. Too small a shaft will not do, since the knob will be loose.
4. *Resistance and taper.* The total resistance and taper of the replacement control should be the same as the original. The wrong taper will cause the control to bunch all its action in a small segment of the control rotation, while the rest of the turn has very little effect. The technician need not concern himself too much about the taper, however, since the replacement-control manufacturers have gone into the matter thoroughly and specify the proper taper to use in accordance with the circuit arrangement of the control.
5. *Switch.* Volume (or tone) controls are usually combined with the line ON-OFF switch in one unit. When this is the case, if the volume control is defective, the switch is replaced at the same time. Similarly, if the switch is defective, the volume control is replaced at the same time.

How to replace a volume control

1. Choose a proper replacement control as described above.
2. Do not remove the wiring from the old control. Loosen the mounting nut, slip the shaft through the hole, and let the old control dangle from its leads.
3. If necessary to cut the shaft of the re-

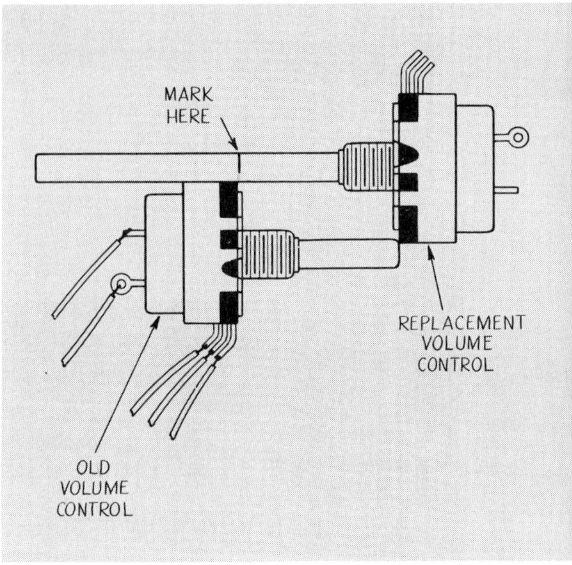

FIG. 13-11. Measuring the replacement volume control for length.

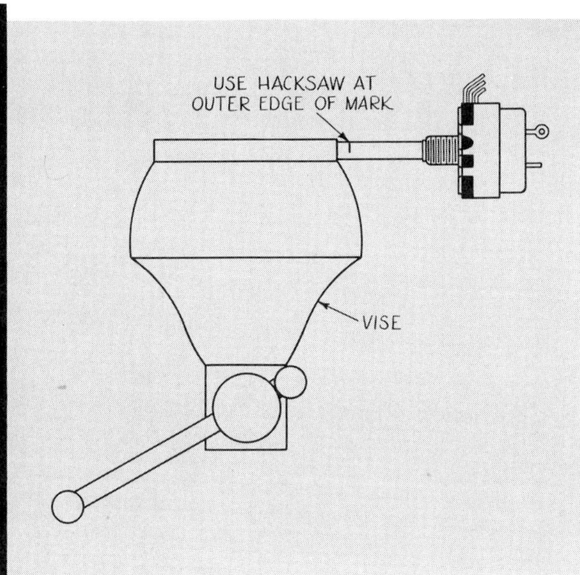

FIG. 13-12. Cutting the volume-control shaft to size.

placement control, proceed as follows:

a. Measure it against the original as shown in Fig. 13-11 and mark the proper length.

b. Clamp excess portion in vise with mark showing and cut to the mark with a hack saw, as shown in Fig. 13-12.

c. Remove saw burr with a file.

4. If shaft is to be filed for a push-on type of knob, proceed as follows:

a. Measure against the original, as shown in Fig. 13-13, and indicate with a mark the amount of shaft to be removed.

FIG. 13-13. Marking the volume-control shaft for a push-on knob.

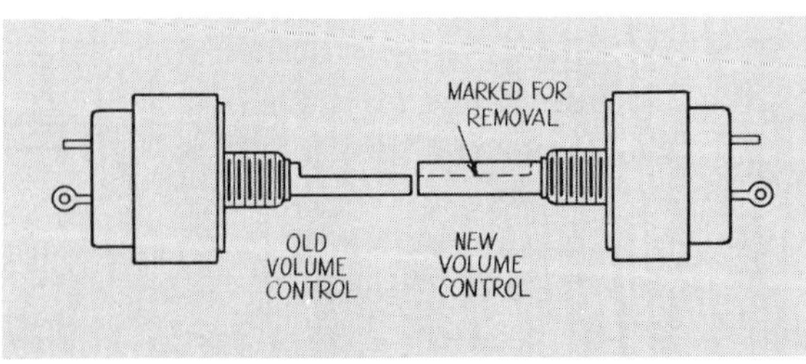

b. Clamp in vise with mark showing. Cut vertically at *A* with a hack saw, as shown in Fig. 13-14. Stop cutting before reaching the horizontal line. File the material away, almost down to the line with the file held horizontally.

c. Try the push-on knob. If too tight, one or two file strokes will bring the shaft down to the line where the knob spring should fit just right.

5. Slip the replacement control through the chassis hole, using a lock washer or locating pin, as shown in Fig. 13-15. If there is no hole for the locating pin, bend it down if it is metal or snap it off if it is bakelite. If this is not done, the locating pin will force the control at an angle when the nut is tightened up, either damaging it or giving the control erratic action. Tighten the mounting nut with an open-end wrench. An open-end wrench marked ½ in. on one end and $\frac{9}{16}$ in. on the other will handle most volume-control mounting nuts. A hollow-shaft socket

wrench of the proper size will be a more convenient tool for the purpose.

6. Remove the wires from the old control, one at a time. Each wire is to be soldered to the corresponding terminal lug on the replacement control.

If the wiring has been disturbed before the new control is in place, it will be necessary to trace the leads before soldering them into place. First the switch leads are traced, one to the line cord and the other to the power-transformer primary. Next the wire to the first a-f control grid through capacitor *C*-31 is found and soldered to the center terminal of the potentiometer. The last two leads go to ground and the detector circuit, and the technician must be careful not to reverse them or the control will work backward. The easiest way to be sure is to turn the control to the full ON position and imagine the position of the arm inside the control. At the full ON position the arm is stopped at the detector circuit end of the control, and the detector lead is soldered to the lug that stopped the arm. The final soldering lug connects to the chassis. These connections are illustrated in Fig. 13-16.

To check the volume-control action, tune the receiver to a strong local station. Turn the volume control to the position just before the switch shuts off power. The sound from the speaker should be just a whisper or completely off. As the volume control is rotated in a clockwise direction, the volume should gradually increase. At the halfway point, the volume should be just about right to fill the average

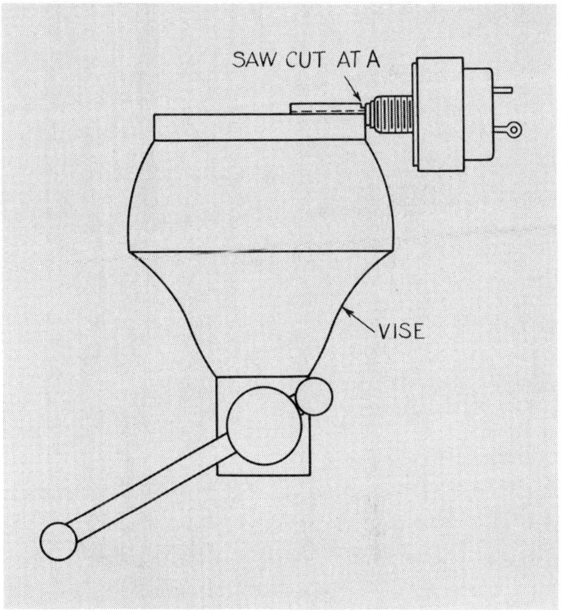

FIG. 13-14. Cutting the volume-control shaft for a push-on knob.

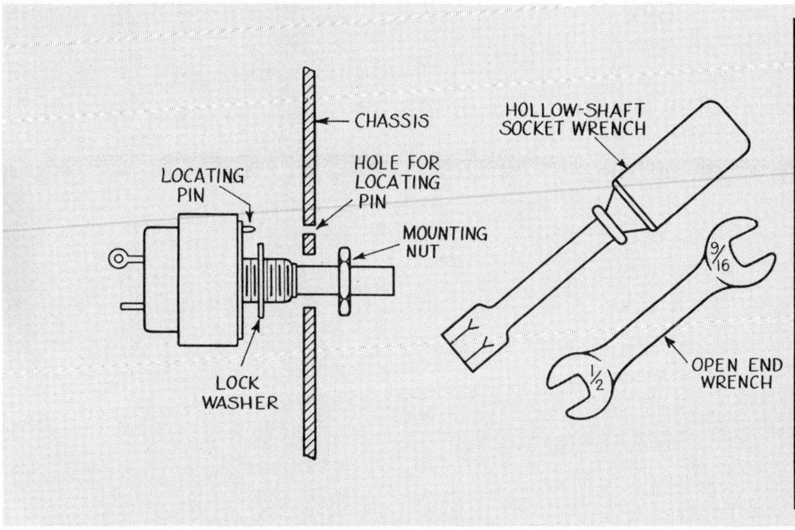

FIG. 13-15. Mounting the volume control.

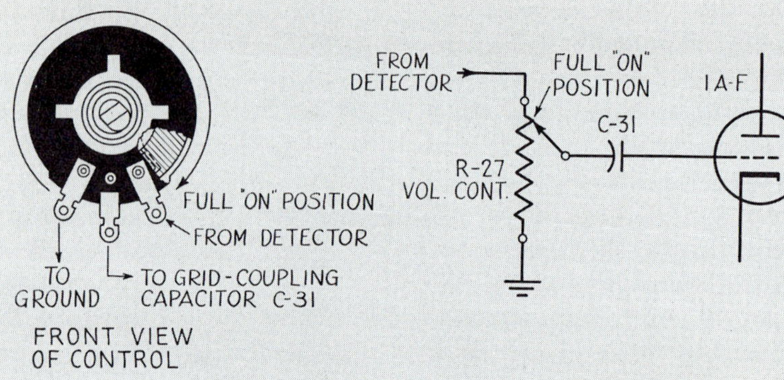

FIG. 13-16. Volume-control connections.

home living room. As the rotation is continued, the volume should increase. Beyond the three-quarter point there will probably be distortion, rattling of the speaker, and microphonics.

To check the volume control for noisy action, the r-f section of the receiver is made inoperative by removing the i-f tube. Then rotate the volume control while listening for noise. In case of an a-c/d-c receiver, where a tube cannot be removed without stopping all operation, the r-f section of the receiver may be made inoperative by grounding the i-f grid or oscillator capacitor stator. Grounding the oscillator capacitor stator is a standard servicing procedure. A description of how the oscillator section of the gang tuning capacitor may be easily recognized is found in Chap. 14 in the discussion of the signal check for the detector and agc stage.

Troubles common to the input coupling capacitor. Coupling capacitor *C*-31 rarely causes any service difficulties. It may open, in which case the condition would be found by a signal check: normal response from the grid of the first a-f tube and no response from the arm of the volume control.

When replacing the capacitor, be sure to use one with the same capacitance as the original. Place the capacitor in the same position as the original and dress the leads in the same manner. The positioning of the capacitor and leads is stressed because any hum picked up at this point is amplified by the entire audio amplifier that follows. Also follow the original for the placement of the outside foil lead, although this procedure may be unimportant, since either end of the capacitor is equally "hot."

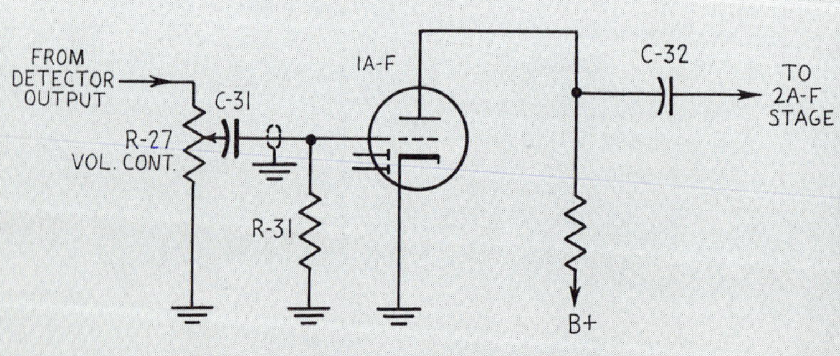

FIG. 13-17. The input coupling capacitor and its position in the input circuit of the first a-f stage.

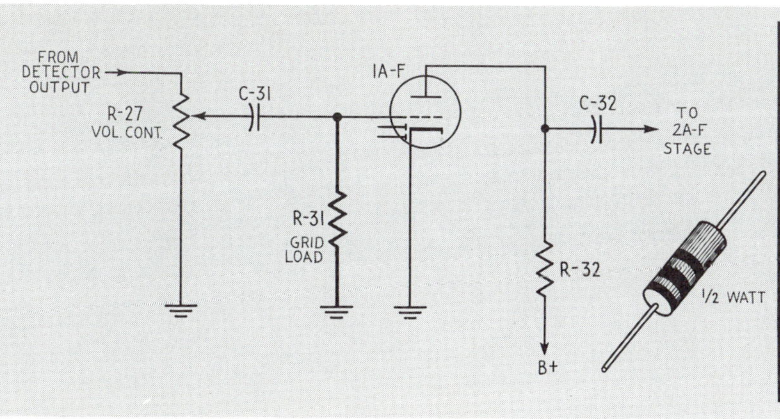

FIG. 13-18. The first a-f grid-load resistor and its position in the circuit.

Troubles common to the grid-load resistor. Grid-load resistor *R*-31 may open, resulting in motorboating as described for *R*-12, the grid-load resistor of the second a-f stage, in Chap. 12. This would be found by the standard check for motorboating, which is to check the filter capacitors in the power supply and then to look for an open grid circuit.

When replacing *R*-31, be sure to use the same ohmic value as is called for in the receiver diagram. A wrong value here would change the contact bias (see operation of contact bias in Chap. 13) and result in poor tone.

Troubles common to the first a-f amplifier tube. The first a-f amplifier tube *V*-4 is the most likely source of trouble in the stage. Hum, no reception, weak reception, noisy reception, and intermittent reception might all be due to the tube. The best check is to substitute a similar tube known to be good. When the signal check shows normal response from the first a-f plate and weak or no response from the first a-f grid, the tube is a likely suspect.

Troubles common to the plate-load resistor. Plate-load resistor *R*-32 sometimes opens. The signal check would show normal response from the first a-f plate and no response from the first a-f grid. A voltage check would then show no voltage at the first a-f plate.

Troubles common to the output coupling capacitor. Coupling capacitor *C*-32 is subject to many ills that impair performance of the receiver. It opens, shorts, becomes leaky, and opens intermittently.

An open capacitor would result in a dead receiver and is found by a signal check. There would be a normal response from the second a-f grid and no response from the first a-f plate. Such a response could also be caused by a plate-to-ground short, which should be checked. This last possibility would be eliminated by a normal plate-voltage reading. The open capacitor would then be confirmed by substituting a test capacitor.

If *C*-32 is shorted or has low leakage resistance, the tone quality would be badly affected. Positive voltage from the first a-f plate would leak over the defective coupling capacitor to the second a-f grid, disturbing the bias voltage on the second a-f tube, with distortion as a result. The condition would be found in a voltage check of the second a-f stage. Insufficient or positive bias on the second a-f tube grid would cause heavier than normal plate current and result in an abnormally large potential across the output-transformer primary and an unusually large potential difference between plate and *B*+ voltages. This check is more reliable than a positive indication on the second a-f grid, which may be small and therefore missed in

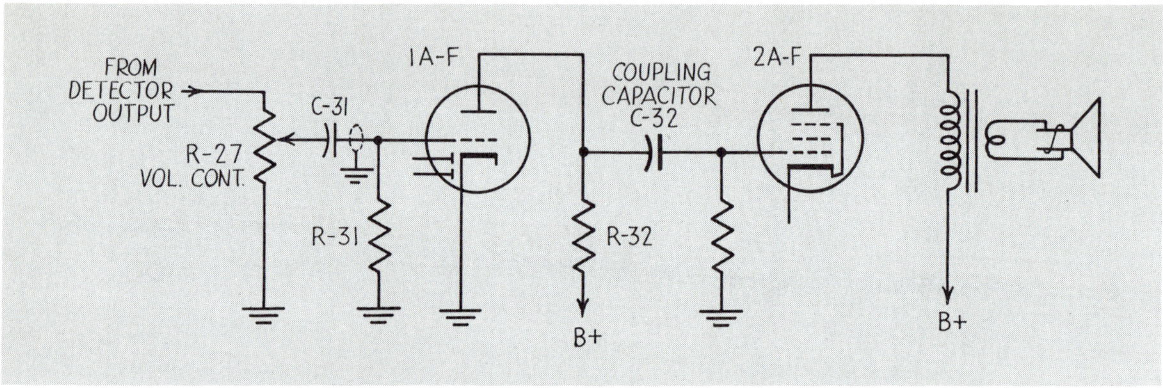

FIG. 13-19. Coupling capacitor *C*-32 between the first a-f and second a-f stages.

the case of high leakage resistance. In the latter case, even though a small positive voltage leaks across the coupling capacitor, it will still decrease the applied bias voltage, with consequent increased plate current and reduced signal handling capabilities. Since a leakage resistance of several megohms would be hard to measure on the average ohmmeter but would still cause distortion, a good confirmation check would be the following: Open the coupling capacitor from its grid connection, and check for voltage to ground, as shown in Fig. 13-20. With a good capacitor, the voltmeter needle will swing up as the capacitor charges and return to the zero position, when the capacitor is fully charged. Leakage resistance in the capacitor will cause the voltmeter needle to remain at some position higher than zero. Owing to the high activating voltage at the first a-f plate, a leakage resistance of several megohms will cause a readable deflection on the voltmeter.

FIG. 13-20. How to check an audio coupling capacitor for leakage.

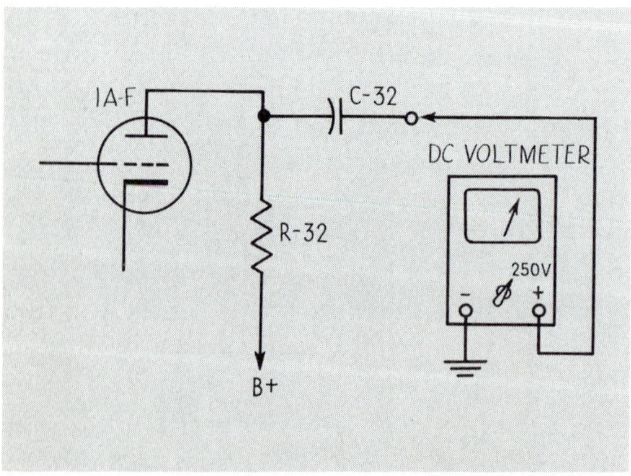

If the coupling capacitor is intermittently open, fading will result; the receiver will not operate when it is open and will resume operation when it is closed. This condition is due to a poor contact between one of the capacitor leads and the tin foil of the capacitor plates. Usually the condition can be confirmed by yanking gently on the capacitor leads, thereby starting and stopping reception. Parenthetically, it may be added that, when a receiver is serviced for fading, all coupling capacitors should be replaced as a matter of course.

When coupling capacitor *C*-32 is replaced, a good-quality capacitor should be used. The capacitor should have a rating of 600 volts. Although a 400-volt capacitor is sufficient for the voltages normally found in this circuit, the thicker dielectric of the 600-volt size makes for less likelihood of leakage. The capacitance used should be the one called for in the receiver diagram. If a different capacitance is used, the technician should remember that a higher capacitance will give a better low-frequency audio response.

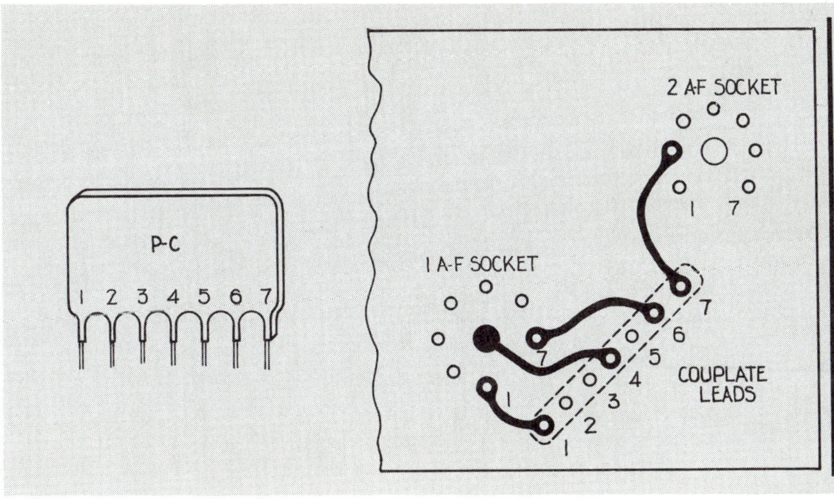

FIG. 13-21. A couplate and some of its connections in a receiver with printed wiring.

CIRCUIT VARIATIONS IN THE FIRST A-F STAGE

Printed coupling circuits. In many receivers, particularly those using printed wiring, the audio coupling capacitors and resistors are incorporated in a single unit called a couplate. A unit of this type, together with some of its wiring, is illustrated in Fig. 13-21. The schematic diagram of an RCA a-c/d-c receiver is given in Fig. 13-22 as a typical example of a receiver using this type of construction.

In this diagram, note the dotted line labeled PC 1 which encloses all of the coupling in the audio stages. Observe the circled numbers,

FIG. 13-22. Schematic diagram of an RCA receiver using a couplate.

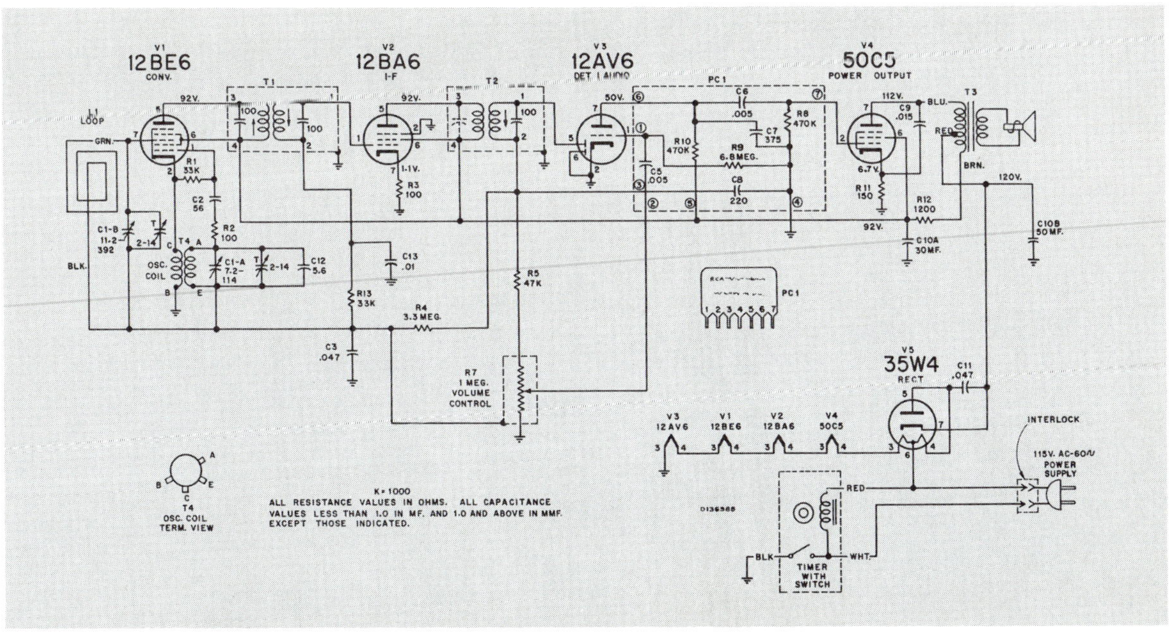

1 through 7, which are placed near each wire entering the dotted enclosure. These are the leads, likewise labeled 1 through 7, which appear on the drawing of the couplate directly beneath the dotted enclosure. The couplate contains all the resistors and capacitors indicated. Note further that the circuit is the same as our standard circuit, with the addition of bypass capacitor *C*-7 connected from the first a-f plate to the common negative or ground.

In servicing a receiver of this type, all of the test procedures given previously may be applied. For example, assume that the receiver does not work and an audio signal applied to the second a-f grid gives a normal response. When the test signal is shifted to the first a-f plate, there is no response. As explained previously, this condition may be caused by an open in coupling capacitor *C*-6, or by a plate-to-ground short in the first a-f tube. A voltmeter establishes a normal 50-volt reading at the first a-f plate, eliminating the possibility of the short. A test capacitor connected from the first a-f plate to the second a-f grid restores reception, thereby confirming the open capacitor.

Having proved one unit defective, the difference in servicing procedure is that the entire couplate must be replaced with an exact duplicate. Fortunately, two or three types serve as replacements for a large variety of receivers that use couplates. When replacing the couplate, use the technique previously described for replacing components on a printed board. First make a note as to the correct orientation of the leads. Hold the soldering iron to each connection in turn while wiggling the couplate from the other side of the board. When it is completely free, remove any excess bits of solder and clear all the holes. Then thread the leads of the replacement unit, correctly oriented, through the holes. Clip off the excess length and solder each connection carefully.

Bass compensation circuit. It is characteristic of the human ear to be less sensitive to low audio frequencies than to high ones at reduced volume levels. To compensate for this deficiency, the circuit of Fig. 13-23 is found in many receivers.

Potentiometer *R*-27 is a tapped volume control with the tap located in the low-volume area. When the arm is in the high-volume position near the ungrounded end of the volume control, *C*-127 has little effect. As the volume is reduced and the arm approaches the tap, *C*-127 bypasses some of the high a-f signal from the amplifier, thereby making the low audio frequencies seem stronger. The effect is greatest at the tap which will be at the low-volume position for the particular receiver. Resistor *R*-127, which may be omitted from some circuits, is to keep the

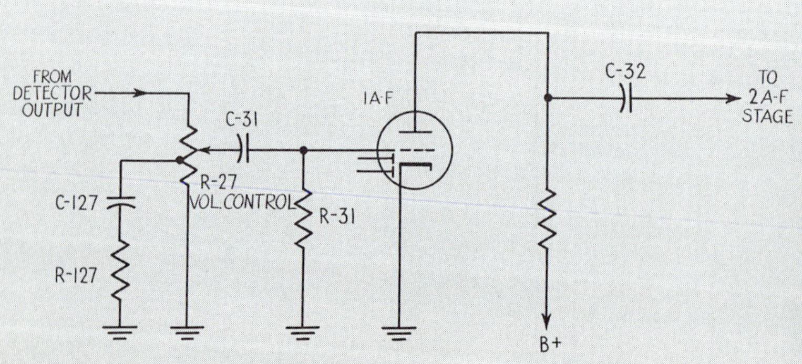

FIG. 13-23. Bass compensation at the first a-f stage.

bypassing effect from being too pronounced at the tap and to broaden the region around the tap where the bass compensation circuit is effective.

Some receivers carry out the tone compensation at two points on the volume control, as shown in Fig. 13-24. The volume control is tapped at two points. Again, at the high-volume position the high-frequency bypass circuits have little effect. As the volume is reduced, a slight amount of bass compensation is attained through *C*-127 and *R*-127. As the volume is reduced further, more bass compensation is attained through capacitor *C*-131. All checks and operations are the same as for the standard circuit.

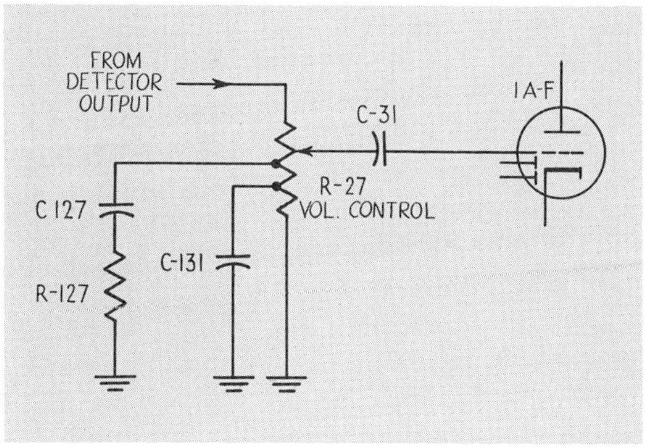

FIG. 13-24. Bass compensation at two points in the volume range.

Capacitors *C*-127, *C*-131, and resistor *R*-127 rarely if ever give any service difficulty. Volume control *R*-27, however, is subject to all the ills of volume controls. In replacing *R*-27, the technician must find an exact replacement for proper operation of the bass compensation circuit.

The schematic diagram of a Motorola a-c/d-c receiver is given in Fig. 13-25, as a typical example of the use of bass compensation. The volume control is the 1-meg potentiometer *R*-9, which is tapped at 300,000 ohms for bass compensation. The remainder of the compensation circuit consists of resistor *R*-8 and capacitor *C*-6. Note that the mounting bracket which will make contact with the metal shell and shaft of the volume control is floated from common negative by capacitor *C*-7. Other metal parts with which the consumer

FIG. 13-25. Motorola receiver showing bass compensation circuit.

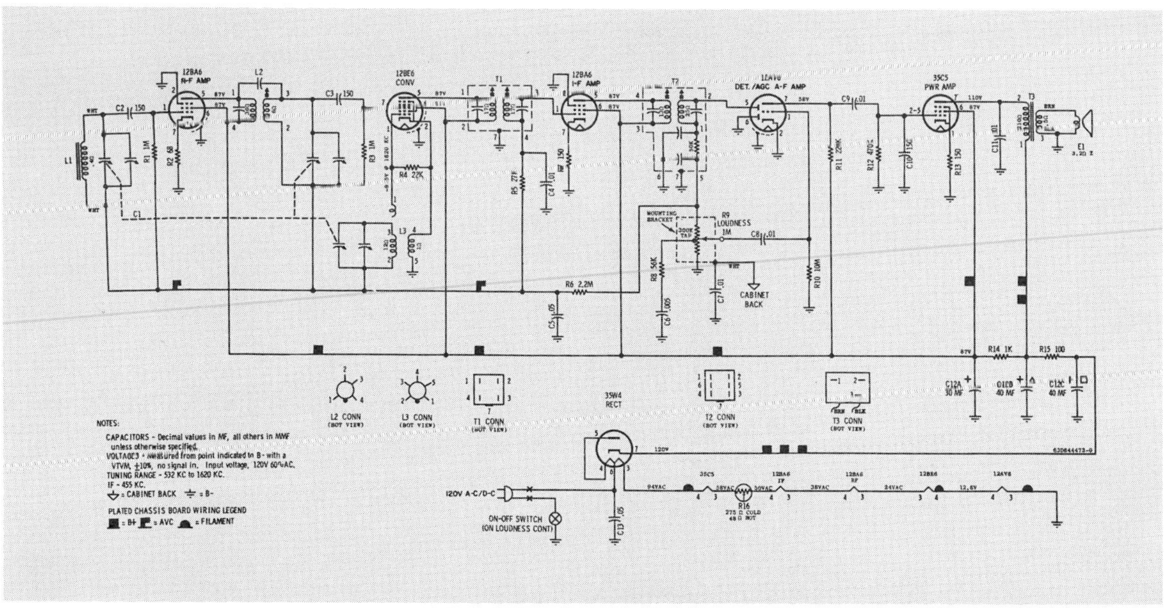

may come in contact, like the back of the cabinet, are also connected here for similar protection against the shorting and shock hazards of the a-c/d-c circuit, where common negative is one side of the power line. Note also that potentiometer *R*-9 is labeled LOUDNESS, a term which may supplant the older name of volume control.

The rest of the first a-f stage follows the standard circuit. In the second a-f or power amplifier stage, self-bias resistor *R*-13 is unbypassed, and the high audio-frequency response is limited by capacitor *C*-10 in the grid circuit, as well as by capacitor *C*-11 in the plate circuit.

In the power supply, note the use of thermistor *R*-16 to limit the current surge to the tube heaters. Also note the use of the symbols which serve to identify the printed wiring on the plated chassis board.

SUMMARY

Quick check for normal operation of the first a-f stage.

A wet finger or a plugged-in soldering-iron tip applied to the ungrounded end of the volume control causes a very strong growl to be heard in the speaker.

Standard first a-f diagram.

The accompanying figure shows the standard first a-f diagram.

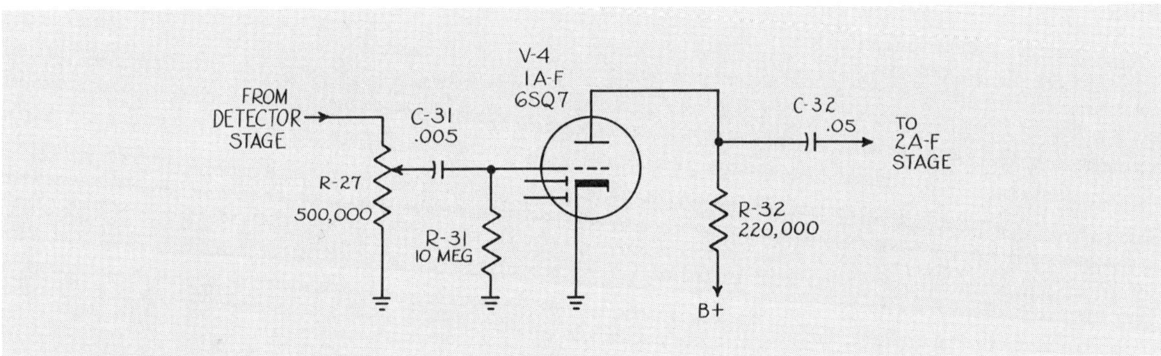

Normal first a-f voltage data.

Voltage is measured from chassis or the common negative in an a-c/d-c receiver. Data given in the accompanying table.

Tube terminal	12SQ7 and 6SQ7 pin No.	A-c receivers, volts	A-c/d-c receivers, volts	12AV6 pin No.
Plate	6	100–170	40–60	7
Grid	2	−1	−1	1
Cathode	3	0	0	2

Normal first a-f resistance data.

Chassis or common negative to cathode	0 ohms
Chassis or common negative to grid	10 megohms
Plate to *B*+	220,000 ohms

SERVICE DATA CHART FOR AN INOPERATIVE FIRST A-F STAGE

Assume an inoperative first a-f stage as shown by normal response when an a-f test signal is applied to the second a-f grid, and no response when the test signal is applied to the ungrounded end of the volume control. The following service procedure is recommended.

Step	Signal check	Response	Trouble
1	Apply a-f test signal to first a-f plate	None or weak	Look for open coupling capacitor *C*-32 or first a-f plate short-circuiting to chassis
		Normal	Proceed to step 2
2	Apply a-f test signal to first a-f grid	None or weak	Look for plate voltage on first a-f plate (open *R*-32). Substitute a good first a-f tube. Look for a shorted grid lead (shielding)
		Normal	Proceed to step 3
3	Apply a-f test signal to volume control arm	None or weak	Look for open coupling capacitor *C*-31. Look for grounded volume-control arm (shielding)
		Normal	Open volume control. Grounded "hot" end of volume control

SERVICE DATA CHART FOR OTHER SYMPTOMS

Symptom	Abnormal reading	Look for
Poor tone quality	First a-f plate voltage low	Short-circuited or leaking coupling capacitor *C*-32
	Voltages normal	Short-circuited or leaking coupling capacitor *C*-31. Incorrect value of grid load *R*-31
Motorboating		Open grid load *R*-31
Hum	Voltages normal	Defective first a-f tube. Incorrectly dressed grid leads. Positioning of coupling capacitor
Intermittent reception (fading)		Coupling capacitors *C*-31 and *C*-32 may open intermittently. Defective first a-f tube. Defective volume control

QUESTIONS

1. A receiver is being serviced for weak reception. A signal check shows no gain for the first a-f stage. Outline a test procedure for determining the cause of the trouble.
2. The receiver of Fig. 13-25 has poor tone quality. A voltage check shows 20 volts on the first a-f plate. What is likely to be wrong and how would you confirm your assumption?
3. A receiver like the one in Fig. 13-25 motorboats. How would you check to find the cause in the power supply? In the second a-f stage? In the first a-f stage?
4. An a-c receiver hums excessively. When the first a-f tube is removed from its socket, the hum level drops to normal. How would you check the various possibilities for hum in the first a-f stage?
5. What are the possible causes of intermittent reception in the first a-f stage? How would you check for each?
6. A receiver gives normal response when an a-f test signal is applied to the first a-f grid and a very weak response when the test signal is shifted to the hot end of the

volume control. What are the possible causes of the defect and how would you check for each?

7. A receiver gives normal response when an a-f test signal is applied to the second a-f grid and no response when the test signal is shifted to the first a-f grid. What are the possible causes of the trouble and how would you check for each?

8. What is a good test for high leakage resistance in a coupling capacitor between a first a-f plate and a second a-f grid?

9. The receiver of Fig. 12-14 has been completely overhauled and reconditioned. As part of the servicing procedure the first a-f grid-load resistor *R*-11 had been found to be open and replaced. However, it had been replaced with a 1-megohm resistor in error. The customer later complains that his radio does not sound so clear as before. Could the incorrect first a-f grid-load resistor be the cause of this condition? Explain your answer to this question.

10. The receiver of Fig. 12-16 does not play. A 400-cycle test signal applied to the second a-f grid produces a normal note in the loudspeaker. When the test signal is shifted to the first a-f grid, no response is heard in the speaker. A voltage check of the first a-f stage shows 0.5 volt negative at the grid pin, and zero at the plate pin. What are the likely causes of the trouble, and how would you check for each?

Detector Stage—agc 14

Quick check for operation of the detector stage. The signal generator is adjusted for modulated output at the receiver intermediate frequency and its output is applied to the grid of the i-f tube. The frequency control of the generator is then wobbled back and forth through the intermediate frequency. When the stage is functioning properly, the modulation note will be heard in the speaker, at or near the intermediate frequency of the receiver, and the technician moves forward to the next stage.

Since the agc (automatic gain control) action is dependent on the operation of the r-f converter and i-f stages, there is no quick check for the automatic gain function at this time.

Function of the detector and agc stage. In the standard superheterodyne receiver, detection and automatic gain control are accomplished in one circuit and, although they are two separate functions, must be treated together.

The input signal, normally fed to the detector stage, is an alternating voltage at the intermediate frequency of the receiver and modulated by the audio component of the original signal picked up by the antenna. The signal that appears across the output of the detector stage is the audio component only. One function of the detector and agc stage, therefore, is to demodulate the signal; that is, to remove the audio component and pass it on to the audio amplifier.

The detector stage or tube is sometimes called the "demodulator," the reason for which is obvious from its function. It is also sometimes called the "second detector" to distinguish it from the mixer tube, an old name for which was "first detector."

Agc action can be described as follows: A strong local station delivers a strong signal to a receiver. A station at some distance away will deliver a much weaker signal to it. Yet it is desirable for each of these stations to produce approximately the same volume from the speaker. This effect could be performed manually by means of a volume control, but it is far superior if this effect is performed automatically. That is the function of the agc system. It is also sometimes called automatic volume control (avc). In this book, we shall use the more descriptive term, automatic gain control (agc).

The upper limit of sensitivity of a receiver is set by the design characteristics of the receiver itself. However, the agc circuit reduces the sensitivity of the receiver more or less below the upper limit—more for a strong signal and less for a weaker signal. This effect is produced by the use of remote-cutoff (variable-mu) tubes in the r-f and i-f stages of the receiver. The gain of these tubes changes with different control grid-bias voltages: at greater negative bias, the gain is lower; at lower negative grid bias, the gain is greater. In an agc circuit, the station signal itself develops negative bias voltage for the control grids of the remote-cutoff tubes. A strong signal develops a large negative bias voltage which reduces the gain of the controlled tubes. A weak signal develops a smaller negative bias voltage which does not reduce the gain of the controlled tubes so much. As a result, a fairly constant volume is obtained from the speaker, regardless of the original strength of the receiver signal within the limits of the sensitivity of the receiver.

Theory of operation. The detector and agc stage in modern receivers performs its functions in a circuit arrangement very similar to that of a power supply; that is, it also employs a diode rectifier and filter circuit. Since power-supply circuits are generally understood, a parallel will be drawn to explain the operation of the detector and agc stage.

Consider the half-wave rectifier circuit shown in Fig. 14-1, common in a-c/d-c receivers. The input is 110 volts a-c. Only when the positive phase of the input voltage is impressed on the plate will current flow through the tube. The circuit is completed through load resistor *DL*. Capacitors C_x and C_y and choke *L* make up the smoothing filter. The wave forms of Fig. 14-1 show the complete action of the circuit. Note the polarity of the voltage across load resistor *DL* and the hum ripple that is present. If it is desired to eliminate the hum ripple, a second section L-C filter would be added, as in Fig. 14-2.

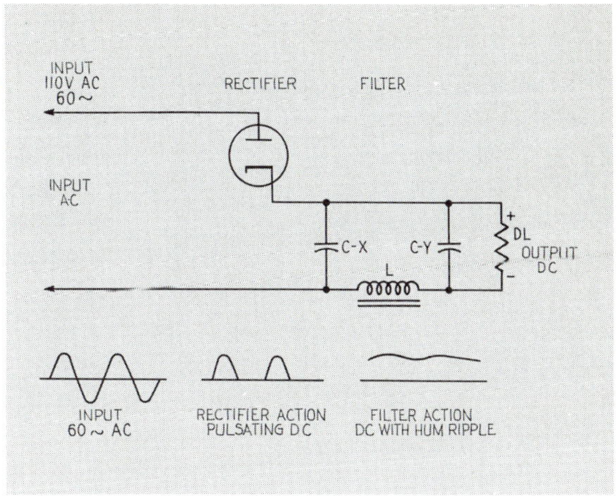

FIG. 14-1. Half-wave *B* power supply.

In the detector stage, to draw a parallel, the input voltage is across the tuned secondary of the i-f transformer *T*-5, as shown in Fig. 14-3. The graph below *T*-5 represents the input voltage at the intermediate frequency and modulated by its audio component. Similar to the action in the power supply, the

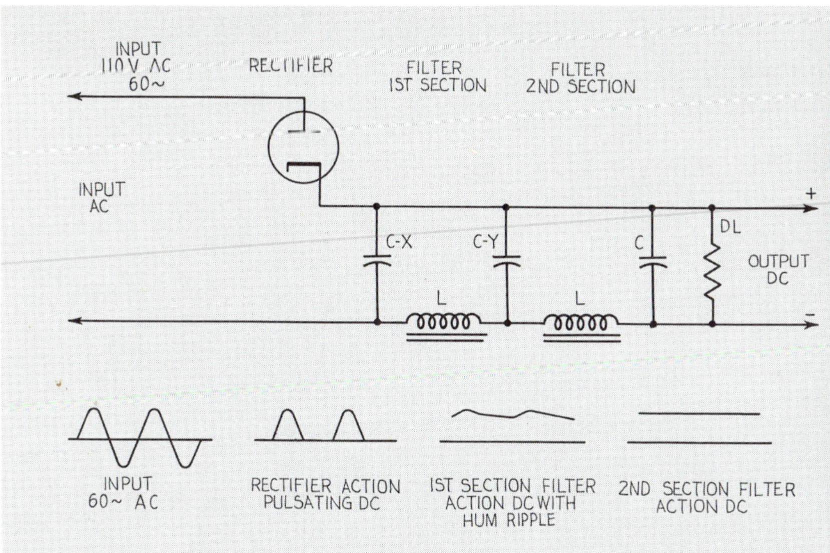

FIG. 14-2. Eliminating hum ripple by means of a second section filter.

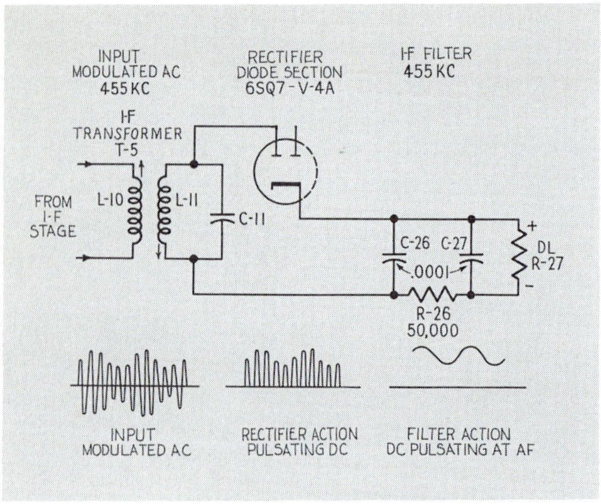

FIG. 14-3. Diode detector operation—developing the audio output signal from the modulated i-f input signal.

rectifier chops off the negative half of the input voltage, as represented in the graph under the rectifier tube V-4A. Now let us examine the filter circuit. A filtering resistor *R*-26 has been substituted for the choke. It serves a similar function. In the power supply, the filter capacitors are usually 20 mfd each. In the detector filter circuit, *C*-26 and *C*-27 are usually 0.0001 mfd apiece. This filter circuit will not give unvarying direct current at its output but will make an effective filter at the intermediate frequency (455 kc). The output at this point will be the audio component of the signal which is impressed across the resistor *DL*, since the audio signal cannot be bypassed across the low-capacity capacitors *C*-26 and *C*-27. Resistor *DL* is called the "diode load" and is usually the manual volume control of the receiver. With the audio signal across the volume control, the position of its arm determines the strength of the signal fed to the audio amplifier.

The audio signal, owing to its strong pulsations, is not suitable for use as an automatic bias voltage, since any bias voltage should be pure direct current. Therefore a second section filter, *R*-28 and *C*-28, is added after the audio circuit to smooth it to direct current, as shown in Fig. 14-4. The capacitance of *C*-28 is 0.05 mfd to make it effective at audio frequency.

Now note again the polarity of the voltage across the diode load. If the diode cathode is grounded, the voltage at *R*-28 will be negative with respect to ground, and therefore suitable for use as bias voltage. The amount

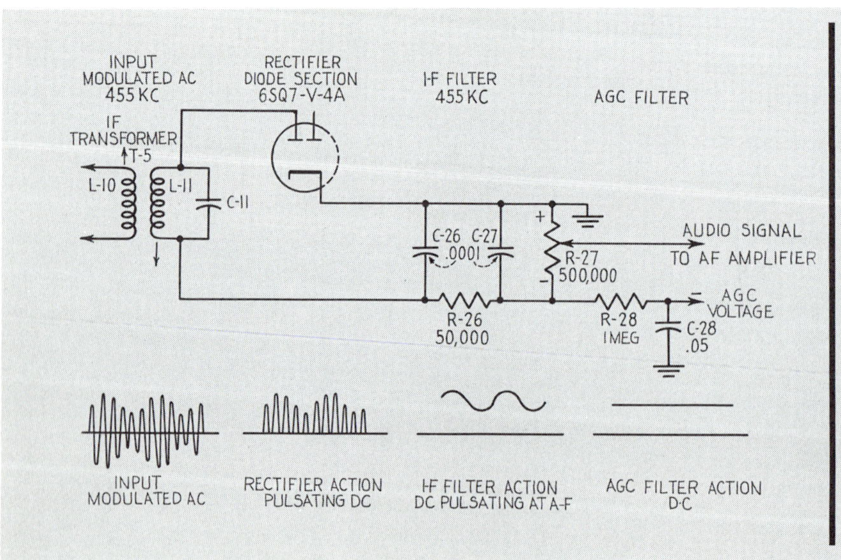

FIG. 14-4. Developing audio signal and agc voltage from the modulated i-f signal.

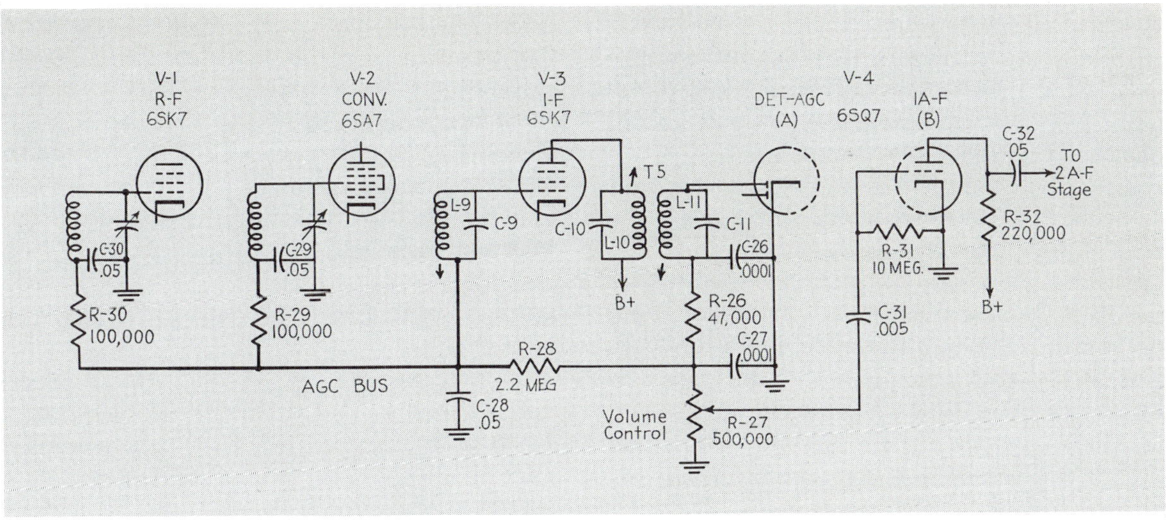

FIG. 14-5. Typical detector and agc circuit.

of voltage available at *R*-28 will depend on the voltage of the signal impressed across the secondary of the i-f transformer *T*-5, since it is the rectified and filtered output of the signal voltage. For strong signals, the signal voltage across *T*-5 is high, the agc bias voltage is high, and the amplification of the controlled r-f and i-f tubes is reduced. For weak signals, the signal voltage across *T*-5 is low, the agc bias voltage is low, and the amplification of the controlled tubes is greater.

Figure 14-5 shows the detector and agc system, including the control-grid circuits of the controlled tubes, and the coupling to the first a-f stage. The wire that feeds the agc voltage to the controlled tubes is known as the "agc bus."

Resistor *R*-30 and its associated capacitor *C*-30 in the r-f grid return lead isolate the r-f stage from the other stages. This is called a "decoupling" filter, which will be described in a later section. Resistor *R*-29 and capacitor *C*-29 serve a similar function for the converter stage.

Functions and values of component parts. Potentiometer *R*-27 is the manually operated volume control for the receiver. In the detector stage, it acts as the diode load resistor, and the audio component of the signal voltage is developed across it. A portion of this voltage is taken off at the volume-control arm and is amplified as was described in Chap. 13. The ohmic value of *R*-27 is usually 500,000 ohms, although higher values are sometimes found in circuits where, at the increased load resistance, a higher value of audio output voltage is possible.

Capacitors *C*-26 and *C*-27 and resistor *R*-26 make up the i-f filter circuit. In this circuit, the i-f pulsations are removed, leaving the audio envelope. Resistor *R*-26 is usually 47,000 ohms, and capacitors *C*-26 and *C*-27 are usually 0.0001 mfd for an intermediate frequency of 450 to 480 kc. Sometimes these capacitances are a little higher, not so much for more efficient filtering as for attenuation of high audio frequencies with resultant improvement of the apparent low a-f response.

Resistor *R*-28 and capacitor *C*-28 form the additional filter for the agc voltage. In this circuit, audio pulsations are removed. Since the controlled grid circuits do not require current, *R*-28 can have a high value of resistance for efficient filtering of the a-f pulsations, and *C*-28 bypasses the remainder to ground. In receivers containing an r-f stage and decoupling filters, *R*-28 is usually 2.2 megohms. In receivers that do not employ an r-f stage, *R*-28 is usually higher, 3.3 megohms being the average size. *C*-28 is almost always 0.05 mfd.

The diode employed in the detector and agc stage is the duo-diode section of the 6SQ7 tube. The other diode plate is generally grounded. The triode section functions as the first a-f amplifier tube. Receivers that use the loctal type of tubes employ the 7C6, and receivers with miniature tubes use the 6AV6 or 12AV6. When a separate tube is used for the detector, it is the 6H6 or 6AL5 twin diode. Occasionally, a receiver is found that uses a germanium diode like the 1N60 for the detector and agc function.

Transformer assembly *T*-5 couples the output of the i-f stage to the detector. The transformer assembly includes capacitors *C*-10 and *C*-11, so that both primary and secondary windings are in resonance at the intermediate frequency of the receiver. For alignment purposes, resonance adjustments are provided so that the inductance of the windings may be changed by altering the position of powdered-iron slugs in the coils. In older receivers, the inductances of the windings are fixed, and capacitors *C*-10 and *C*-11 are small variable capacitors called trimmers, whose capacitances may be varied by a screwdriver adjustment. Typical slug-tuned and trimmer-tuned transformer assemblies are illustrated in Figs. 14-8 and 14-9.

NORMAL TEST DATA FOR THE DETECTOR AND AGC STAGE

Signal check. The input of the detector stage is the i-f transformer *T*-5. However, when the stage is checked, the signal generator is connected to the grid of the i-f tube, as shown in Fig. 14-6. There are several reasons for this connection:

1. Since the detector input is the last step in the r-f chain, the signal voltage at this point is high, higher than the r-f output of most signal generators, and the amplification of the i-f tube may be needed to make the signal more easily heard in the speaker.

FIG. 14-6. Signal check connections for the detector and agc stage.

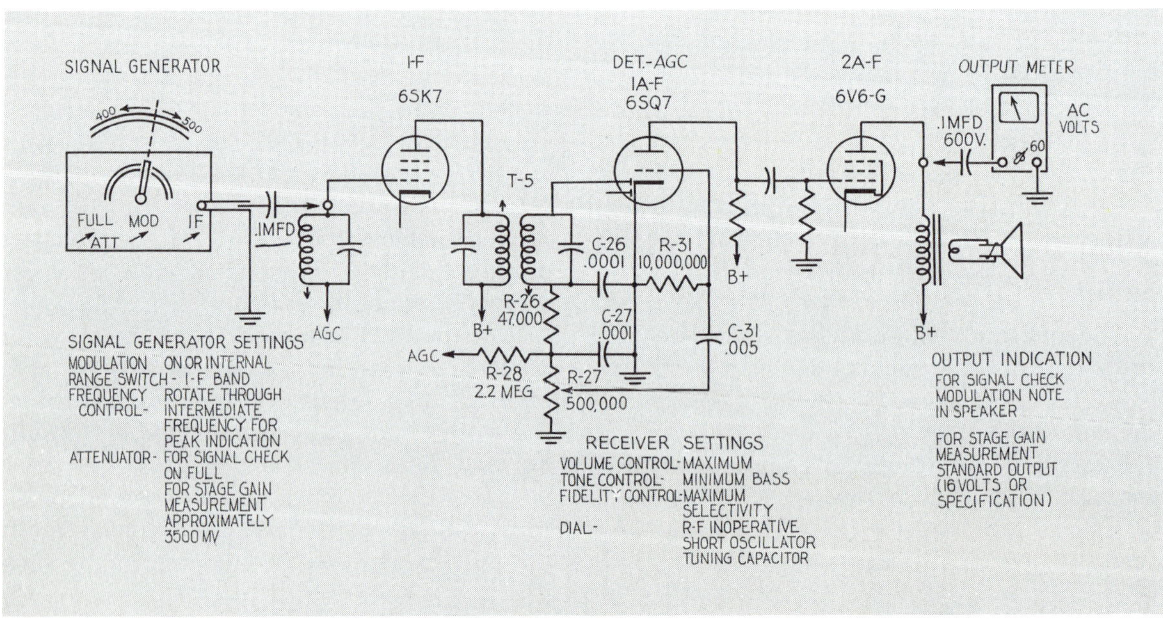

2. If the added capacitance of the signal-generator leads were connected to the i-f plate, the normal input of the detector stage, it would seriously detune the primary circuit of *T*-5, making the response broad and possibly at an off-frequency setting.

The 0.1-mfd capacitor in the hot lead of the signal generator acts to isolate the signal generator from d-c receiver potentials in case the signal input is connected to a plate lead. It is also the standard dummy antenna capacitance (coupling device) between the signal generator and the receiver for i-f measurements. The output indication is the signal-generator modulation note in the speaker. This can be measured by connecting the output meter (35- to 60-volt a-c range of the multimeter with a 0.1-mfd/600-volt capacitor in series) from the second a-f plate to ground, as described for stage-gain measurements for the first a-f stage.

When the signal check is made, it is also wise to check the intermediate frequency of the receiver, which is always listed in the manufacturer's service notes. In modern receivers it is usually 455 kc. Some automobile radios use an intermediate frequency near 260 kc. In very old receivers intermediate frequencies of 260, 175, and 130 kc have been used. In checking the alignment and operation of the stage, the previous stages of the receiver should be made inoperative. This is done by shorting the oscillator section of the tuning capacitor. To determine which of the sections of the gang tuning capacitor is the oscillator, the technician should trace the circuit; or it is sometimes possible to locate the oscillator section by faster methods. In some receivers, the oscillator rotor plates are smaller than the other rotor plates in the gang capacitor. Another method that can be used when the receiver is operating on a station is to touch only the stator plates of the various sections. When the r-f and converter sections are touched, there will be little difference observed. When the oscillator section is touched, the added capacitance of the body will cause the station to disappear. A short piece of flexible wire with a clip at each end will serve as the short. One end is clipped to either stator terminal lug; the other is clipped to the capacitor frame.

To check operation and alignment, the signal generator dial is rotated slowly back and forth about 50 kc each side of the specified intermediate frequency. This rotation procedure is called wobbling. The modulation note at or near the intermediate frequency indicates that the stage is operating. Failure to hear the note indicates an inoperative stage or i-f tube. The latter is checked by tube substitution and voltage analysis, as described in Chap. 15. The presence of two peaks, broad tuning, too low an output, or the peak at a considerable distance from the specified frequency—all indicate misalignment.

Sensitivity measurements for the detector stage. When making sensitivity and stage-gain measurements, since it is unlikely that the test oscillator has a calibrated output, the technician should run checks on several receivers in perfect condition, as was done for the audio amplifier (see Chap. 13), until he has a basis of comparative data for normal gain to be expected from the stage.

The signal generator, receiver, and output meter are hooked up, as shown in Fig. 14-6. The receiver is adjusted for maximum output as follows: The volume control is set at maximum; the tone control is set at the minimum bass position; the fidelity control (if present) is set for maximum selectivity. The r-f portion of the receiver is made inoperative by shorting the oscillator section of the gang variable capacitor.

The signal generator is adjusted to give a modulated signal at the intermediate frequency of the receiver. The signal-generator dial is rotated back and forth through the intermediate frequency while the output meter is being watched, and is carefully adjusted for peak deflection. If the output meter deflection goes off scale, the signal input is reduced by adjusting the signal-generator attenuator. After the peak deflection has been obtained,

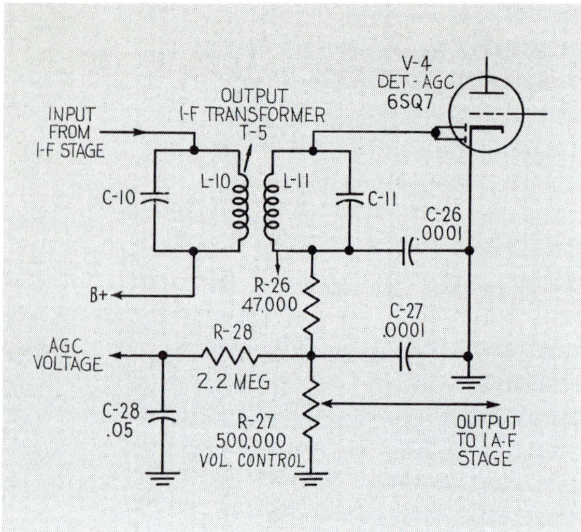

FIG. 14-7. Typical detector and agc stage.

the signal-generator attenuator is further adjusted to give the standard output of 50 mw in the speaker. Standard output corresponds to an output meter reading of 16 volts (see the discussion of the use of the output meter in testing the first a-f stage, Chap. 13).

The average signal input at the i-f grid necessary to give standard output is 3,500 microvolts. The attenuator setting just obtained, therefore, corresponds to 3,500 microvolts. After several perfect receivers have been checked by the above procedure, a reference point corresponding to 3,500 microvolts has been duly established. It would be more important for the technician to remember this average attenuator setting for his signal generator rather than the corresponding 3,500 microvolts. For example, if his average attenuator setting turns out to be 50 × 100, or 5,000, he knows that a setting of approximately 5,000 on the attenuator of his signal generator should produce standard output when connected to the i-f grid of any receiver. Any substantial variation from his average or normal attenuator setting indicates trouble in the stage.

Normal voltage data for the detector stage. The voltages normally present in the detector and agc stage are the signal voltage and the developed agc voltage. Normal-voltage data are usually given as an aid in determining the cause of defective operation. Since measurements of these voltages would require expensive equipment and are therefore not easily obtained, normal voltages will not be given, and defects for this stage will be localized by means of resistance measurements.

Standard circuit for the detector and agc stage. A typical detector and agc stage is illustrated in Fig. 14-7.

Normal resistance data. See the table at the bottom of this page.

COMMON TROUBLES IN THE DETECTOR AND AGC STAGE

Troubles common to the output i-f transformer assembly. I-f transformers are enclosed in a shield can which also contains the associated capacitors. Two commonly

Primary (*L*-10) of output i-f transformer (*T*-5)	Iron core Air core	5–15 ohms 30–50 ohms
Secondary (*L*-11) of output i-f transformer (*T*-5)	Iron core Air core	5–15 ohms 30–50 ohms
Chassis to diode plate Across entire volume control Chassis to agc bus		547,000 ohms 500,000 ohms 2,700,000 ohms

used slug or permeability-tuned types are illustrated in Fig. 14-8. A trimmer-tuned i-f transformer, together with its symbol in the detector stage is shown in Fig. 14-9. The trimmer capacitors rarely cause any service difficulty, except in relation to alignment. The fixed capacitors in the slug-tuned type likewise cause little difficulty.

In operation, i-f transformers of both types open and cause noisy reception. Should either winding of the transformer open, the receiver would become inoperative. A signal check would locate the stage; a resistance check would show the open winding. Noisy reception, when it is caused by the transformer, is due to corrosion of the fine wire in the windings. A resistance check discloses this condition also, since the resistance of a corroded winding is several hundred ohms instead of the 5 to 50 ohms that the winding should read.

There is a rather wide divergence in the design of individual i-f transformers, and the technician should make every effort to secure an original replacement. Where this

FIG. 14-8. Permeability or slug-tuned i-f transformer.

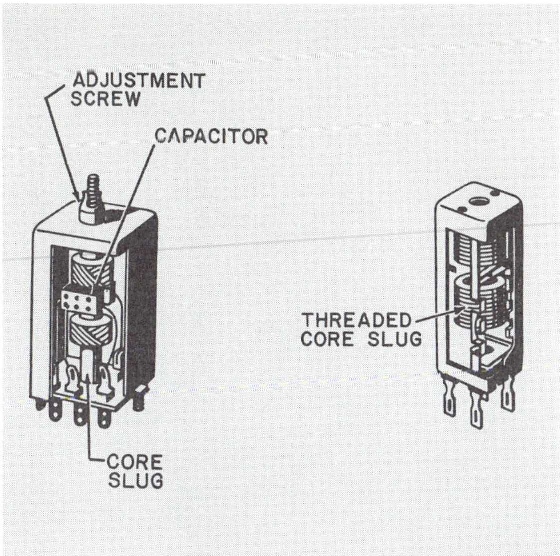

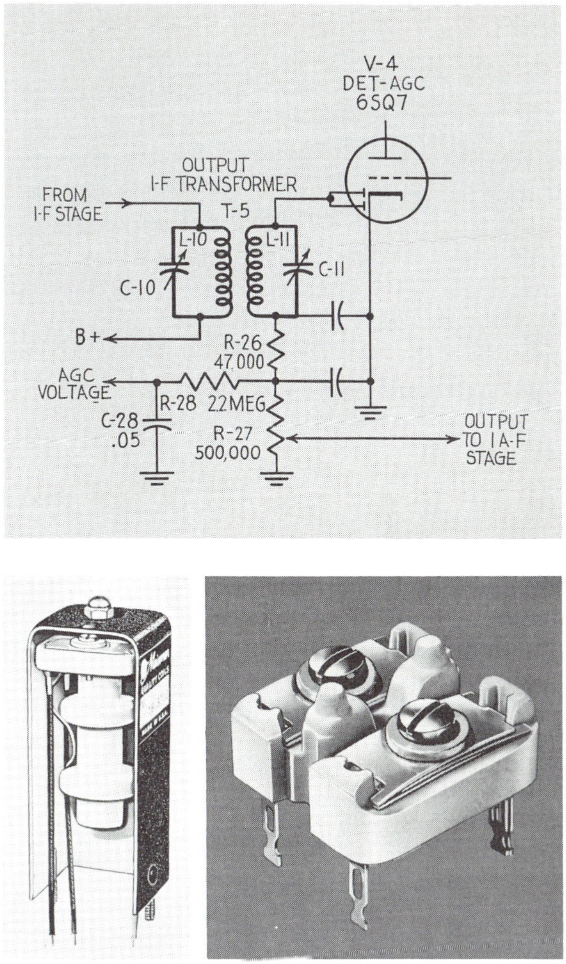

FIG. 14-9. A trimmer-tuned i-f transformer and its symbol in the detector stage. Enlarged view of trimmers is shown at the lower right.

is impossible, coil manufacturers offer a rather large variety of universal replacement i-f transformers. These are listed by the following factors:

---

1. Size of the shield can.
2. Type of core (air or iron).
3. Type of aligning adjustment (trimmer or permeability tuning).
4. I-f range of the transformer (scope of trimmer).

5. Type of transformer (input, interstage, output).
6. Adaptability for use on printed-board receivers.

Sometimes the i-f filter circuit composed of *R*-26, *C*-26, and *C*-27, or part of it, is mounted with the transformer and trimmer assembly inside the shield can. When this is the case and an exact replacement is unobtainable, provision should be made to reinsert the filter circuit which was discarded with the defective transformer.

Replacement i-f output transformers are usually color-coded in accordance with the E.I.A. specifications as follows:

Blue	Plate lead
Red	*B*+ lead
Green	Diode plate lead
Black	Diode return

Before removing an i-f output transformer for replacement, the technician should study the wire dress of the leads, since oscillation can result from incorrectly dressed wiring. If the leads have already been disturbed, the following general notes should be observed. The leads are usually well separated as they come out of the shield can. In the case of a square shield can, the leads come out of the four corners. Before the replacement transformer is mounted, it should be so turned that the blue plate lead points toward the i-f tube socket and the green diode plate lead points toward the detector-tube socket. These are the "hot" leads. They should not cross, and they should be dressed close to the chassis and routed directly to their connection terminals. Replacing an i-f transformer on a printed-board receiver involves choosing a transformer with correct lead placement.

When the transformer has been replaced, the trimmers should be aligned in accordance with the receiver manufacturer's service notes or the general alignment instructions given in Chap. 18.

Troubles common to the i-f filter circuit. The voltages and currents encountered in this circuit are so small that there is no danger of burned-out resistors and capacitors. The capacitors sometimes develop leakage resistance. Check for this condition with an ohmmeter.

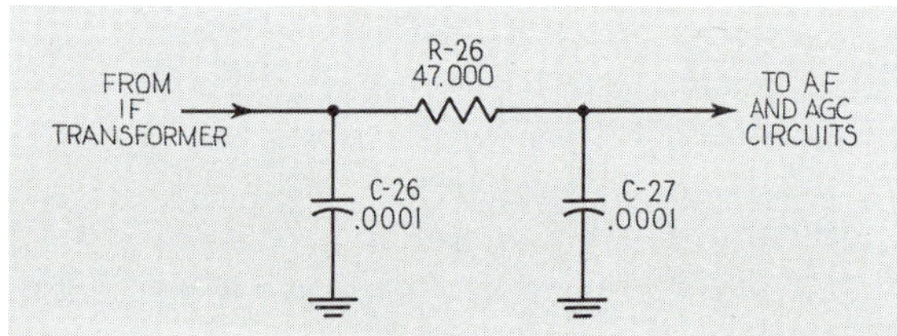

FIG. 14-10. The i-f filter circuit.

Troubles common to the volume control. The volume control sometimes opens. When this occurs, the receiver becomes inoperative. A signal check will show that the audio amplifier is working but the detector stage is not. A resistance check of the components in the detector stage will confirm the open control.

The volume control is also part of the audio amplifier. Replacement notes on the volume control are found in Chap. 13.

Troubles common to the detector tube. The tube is the most likely source of trouble in the stage. A defective tube may cause hum, no signal, weak signal, or distortion. When checking for these symptoms, substitute a similar tube known to be good. In the case of a multi-unit tube like the 6SQ7, which combines both the detector and the first a-f functions, there is a possibility that the a-f portion operates normally but that the detector does not. The technician should not assume that the tube is good because it shows normal operation as an audio amplifier.

Troubles common to the agc filter and decouplers. Figure 14-11 illustrates the agc circuit and shows it connected to the r-f, converter, and i-f stages. Resistor *R*-28 and capacitor *C*-28 make up the agc filter described previously.

Strictly speaking, the purpose of the agc

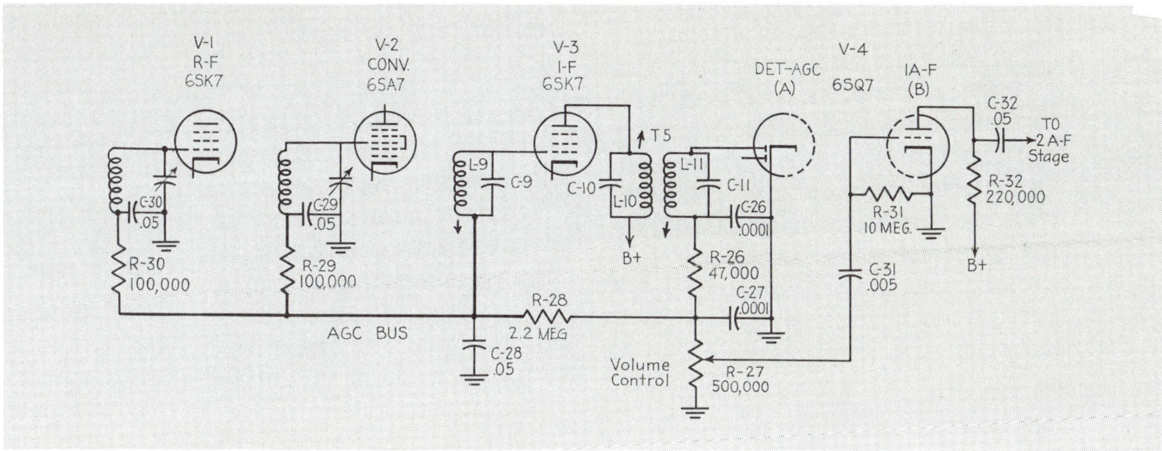

FIG. 14-11. The agc circuit.

circuit is to develop a biasing voltage, and it would seem best to test it by means of a voltage test. However, such a measurement would require a vacuum-tube voltmeter, since the instrument would be across a low-voltage high-impedance circuit. It would also require an accurately calibrated attenuator on the signal generator, and too often it is not accurate. Therefore, analysis of troubles in the agc circuit will be made from the symptoms encountered.

Resistor *R*-28, being one of high resistance, may have a tendency to open. If it does so, the receiver will become inoperative and may develop hum because the grid returns to ground of the associated tubes will be open. Replace the resistor with one of similar value.

The agc filter capacitor *C*-28 may open or become leaky. If it opens, the signal will become weak and oscillation may result. This condition would be found in a signal check of the i-f stage. The gain of this stage would be abnormally low. Also, the i-f tuning would be very broad and possibly off true frequency, with adjustments of trimmer capacitor *C*-9 ineffective.

If capacitor *C*-28 becomes leaky, the agc voltage would drop to an extent dependent on the resistance of the leak. This would result in insufficient bias to handle a strong signal. As a result, the receiver would overload and distort on strong local stations. Reducing the setting of the manual volume control would have little effect on this distortion. Whether the capacitor is open or leaky, confirmation of the condition would be obtained by substituting a similar capacitor that is known to be good. If the trouble disappears, the capacitor was defective. In replacing *C*-28, the technician should be careful to use the same capacitance value as the original capacitor. Even though the voltage across it is quite low, it is advisable to use a capacitor of high voltage rating so that the leakage resistance will be quite large.

Associated with the agc circuit are the decouplers, *C*-29 and *R*-29 for the converter and *C*-30 and *R*-30 for the r-f stage. As a rule, the resistors cause little trouble and are therefore of little consequence to the service technician. However, capacitors *C*-29 and *C*-30 can cause trouble. If either one opens, reception would be very weak. This condition would be confirmed with a signal check when their respective trimmer capacitors would not produce a peak. Capacitor *C*-30 is a particularly odd one. When it opens, the tuning circuit in the r-f stage becomes inoperative, with a resulting drop in signal output. At the same time the loss of signal in the r-f stage causes the agc voltage to drop, resulting in high sensitivity so that the noise level goes up. The receiver sounds exceptionally lively even though strong local stations come in as weak ones do when the receiver is normal.

Capacitors *C*-29 and *C*-30 may become leaky. When this is the condition, the developed agc voltage will be low and the receiver will overload and distort. If the external antenna (when used) is disconnected and the sound of the receiver improves, the technician should hunt for leaky capacitors.

VARIATIONS IN DETECTOR AND AGC STAGE

Use of electron-ray tuning indicator. Unless the superheterodyne receiver is tuned exactly to a station, serious distortion due to side-band cutting may result. Many receivers use some form of tuning indicator as an aid in tuning correctly, so as to avoid this distortion. The tuning indicator in most general use in modern receivers is an electron-ray (often called a "magic eye") tube like the 6U5/6G5. This is a cathode-ray tube which shows a wide deflection when a low voltage is applied to its grid. The deflection narrows as the applied grid voltage is increased. The magic-eye grid is connected to the agc bus as shown in Fig. 14-12. At no signal, the agc voltage is zero and the deflection is wide; as a signal is tuned in, the agc voltage increases

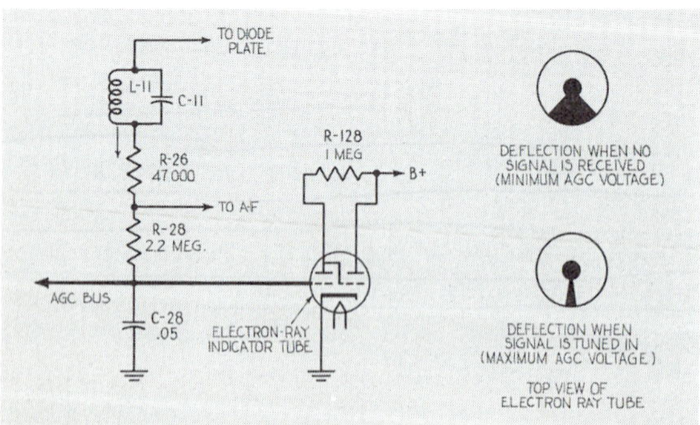

FIG. 14-12. Electron-ray tube connected to the agc bus as a tuning indicator.

and the deflection narrows. When the signal is tuned accurately, the agc voltage is at a maximum and the deflection is at its narrowest. To tune any station accurately, simply tune the receiver for the narrowest deflection of the magic-eye tube.

Since this tube must be located on the front panel of the receiver, its socket is not on the chassis. The tube is usually supported in position by a clamp, with a cable of connecting leads running down to the chassis.

Resistor *R*-128 is a 1-megohm/¼-watt resistor. In an a-c/d-c receiver, it is a ½-megohm/¼-watt resistor. In either case, it is usually located inside of the tube socket.

From a service point of view, the magic-eye tube adds few complications. All checks and tests are the same as for the standard receiver. If the tube does not glow, a new tube is needed; if the tube glows but the deflection does not change as stations are tuned in, and if the receiver operation is normal in all other respects, *R*-128 is probably open. To change *R*-128, the tube socket must be opened.

A receiver equipped with an electron-ray tube has a virtual vacuum-tube voltmeter already connected to the output of the r-f and i-f stages. It can be used as an indication of the agc voltage and as an output meter for alignment purposes.

Use of a crystal as a detector. Some receivers use a semiconductor germanium crystal as the detector, instead of a diode tube. As we have previously seen in the study of power supplies, the semiconductor crystal diode operates as a rectifier in much the same manner as a vacuum-tube diode. No further theory need be given on this point since the detector circuit, whether tube diode or crystal diode, uses a rectifier for demodulation and for the establishment of the agc voltage.

Crystal diodes are found most often as the a-m detector in a-m/f-m receivers. The schematic diagram of the a-m tuner and detector of a Silvertone a-m/f-m receiver is given in Fig. 14-13 as a typical example of the use

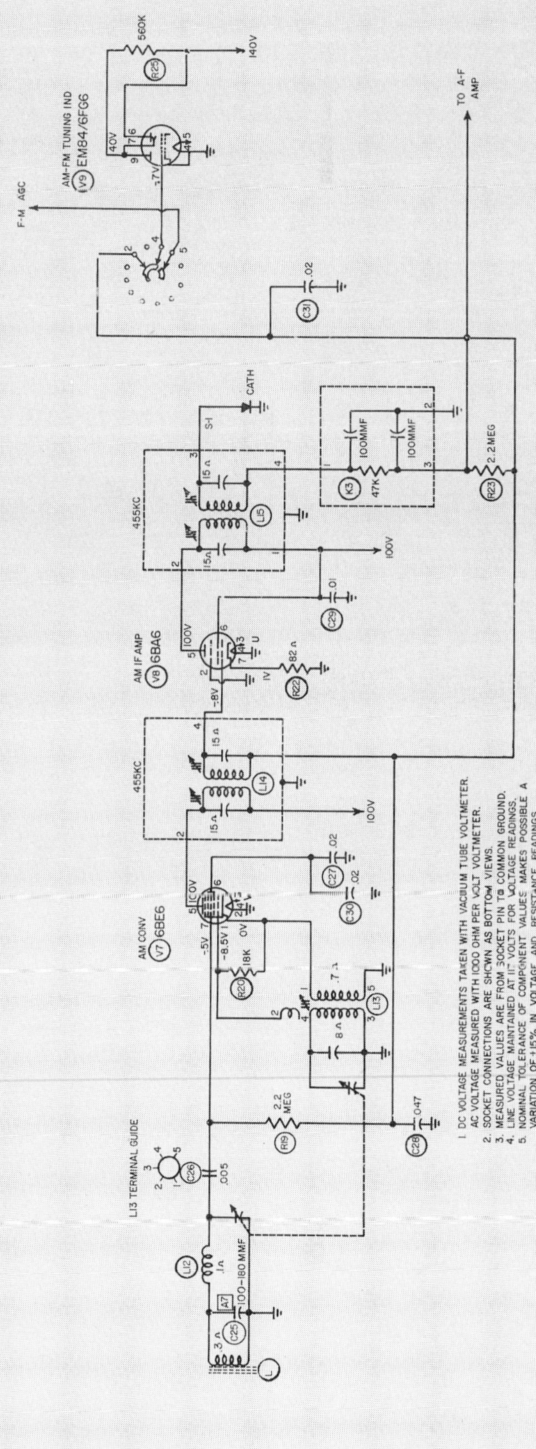

FIG. 14-13. Silvertone a-m tuner with crystal diode detector, used in an a-m/f-m receiver.

of a crystal diode as a detector. Rectifier S-1 is the detector diode. Note that the detector-agc circuit is exactly like our standard circuit with the diode crystal substituted for the diode section of the 6SQ7 tube. Also note the dotted line around the i-f filter. This indicates that the 470,000-ohm resistor and the two 100-mmf capacitors are contained in one combination unit or couplate.

The signal check and the servicing procedure is exactly the same as for a tube diode. If it becomes necessary to try operation with a new diode, the substitution procedure to be followed is shown in Fig. 14-14. A replacement crystal 1N60 or 1N64 is equipped with clips, one connection to the original is opened, and the replacement is clipped in, being careful to observe correct polarity. All crystal diodes indicate the lead which is the equivalent of the cathode of a tube diode by the letter K, or by notches or a dot, or by the symbol for a crystal diode, as shown in Fig. 14-14, in which the lead nearest to the line of the symbol is the cathode. Incorrect polarity will put a positive, instead of a negative, bias on the agc bus, thereby overloading a controlled r-f or i-f tube.

If a replacement crystal is needed, an exact replacement is preferred but is not necessary. A G.E. type 1N64 or 1N541 or a Sylvania type 1N60 will work satisfactorily in most receivers.

Germanium crystals require considerable care in handling. The leads may be easily pulled out of the body of the crystal, and the crystal will deteriorate if subjected to excessive heat. To avoid failure due to these causes, always leave some slack in the connecting leads. This relieves the tension which might cause the lead to pull out of the crystal. When soldering, grasp the lead with long-nose pliers as shown in Fig. 14-14, between the crystal and the end of the lead to be soldered; then solder. The metal mass of the tool absorbs the heat of soldering so that the portion of the lead going into the crystal is relatively cool.

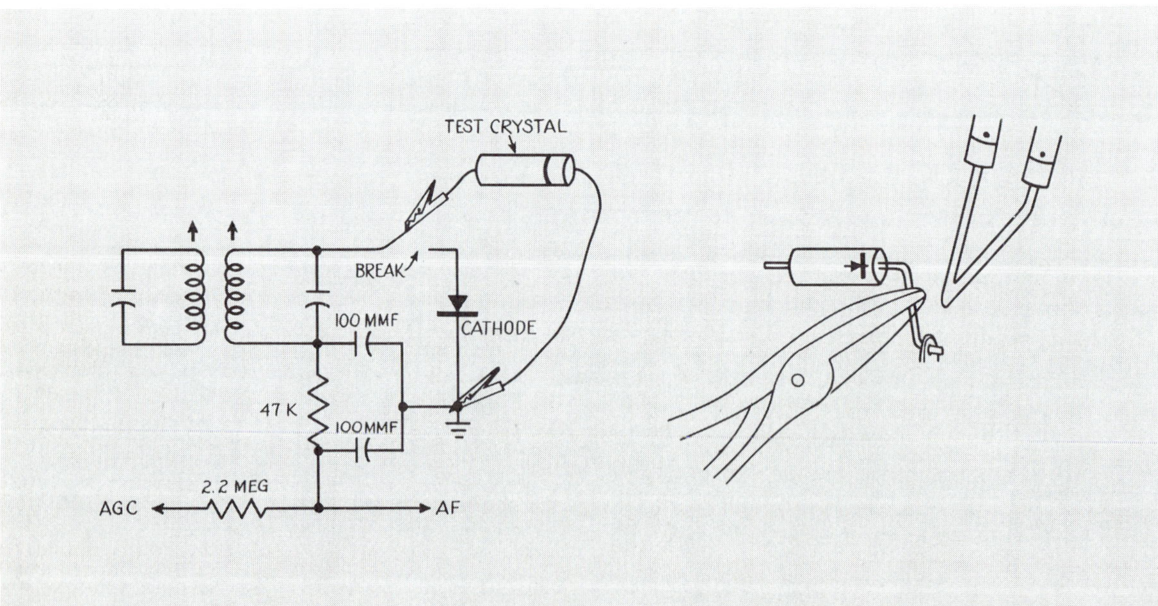

FIG. 14-14. Testing and replacing germanium diodes.

SUMMARY

Quick check for normal operation of the detector and agc stage.

The signal generator is adjusted for modulated output at the receiver intermediate frequency and its output is applied to the grid of the i-f tube. The frequency control of the generator is then wobbled back and forth through the intermediate frequency. When the stage is functioning properly, the modulation note will be heard in the speaker, at or near the intermediate frequency of the receiver.

Diagram of standard detector and agc stage.

A diagram of standard detector and agc stages is given in the accompanying figure.

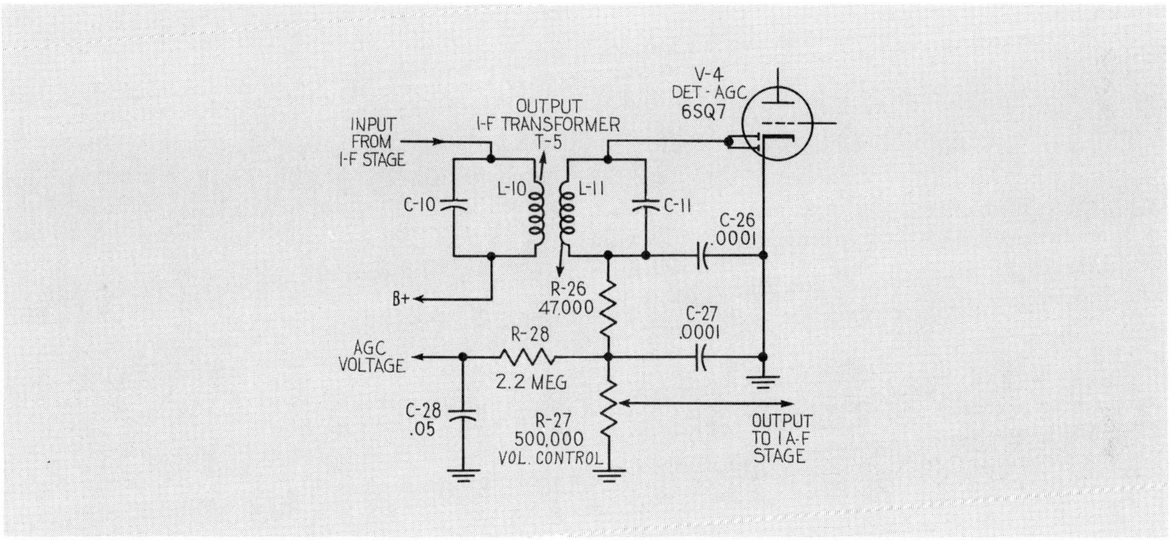

Resistance data.

These data are given in the accompanying table.

Primary of output i-f transformer	Iron core	5–15 ohms
	Air core	30–50 ohms
Secondary of output i-f transformer	Iron core	5–15 ohms
	Air core	30–50 ohms
Chassis to diode plates		547,000 ohms
Across entire volume control		500,000 ohms
Chassis to agc bus		2,700,000 ohms

SERVICE DATA CHART

Assume an inoperative detector stage, as shown by normal response when an a-f test signal is applied to the ungrounded end of the volume control, and no response when a modulated test signal at the intermediate frequency is applied to the i-f grid.

Step	Check	Response	Trouble
1	Advance the signal-generator attenuator and rotate the dial through the intermediate frequency	The modulation note is heard at an off-frequency setting	The i-f transformer is out of alignment
		The modulation note is not heard	Proceed to step 2
2	Apply the i-f test signal to the i-f plate. Rotate the signal-generator dial and advance the attenuator to full output	The modulation note is heard in the speaker	The trouble is in the i-f tube or its supply voltages. See Chap. 15 on the i-f stage
		The modulation note is not heard	Defective tube. Substitute a detector tube known to be good. The trouble may be an open i-f transformer winding, a shorted trimmer capacitor, etc. Make a resistance check of all components in the stage

SERVICE DATA CHART FOR OTHER SYMPTOMS

Symptom	Abnormal reading	Look for
Hum		Defective detector tube. Substitute a good one. Poorly dressed leads in the diode plate and plate return circuits.
Weak reception and oscillation	If equipped with an electron-ray tuning-indicator tube, the eye will not close fully	Incorrect alignment. Open agc bypass capacitors *C*-28, *C*-29, and *C*-30

Symptom	Abnormal reading	Look for
Distortion on strong signals		Leaky agc bypass capacitors *C*-28, *C*-29, and *C*-30

QUESTIONS

1. A dead a-c receiver gives a normal response when checking the a-f stages but gives no response when a test signal at the proper intermediate frequency is applied to the i-f grid. Outline a service procedure to be followed in finding the cause of the trouble.
2. List the likely sources of trouble that will cause a receiver to give no response when an i-f test signal is applied to the i-f plate and normal response when an a-f test signal is applied to the "hot" end of the volume control.
3. A radio-phonograph combination has a distorted output when it is tuned to local stations. The tone quality is normal when it plays phonograph recordings. Would you check the audio stages for the trouble? Why? What is likely to be wrong? How would you prove it?
4. When a receiver with weak reception is checked, it is noted that the trimmer across the input i-f secondary has no effect on the output. What circumstances can cause this condition? How would you check for each?
5. Which components in the detector stage may cause hum? How would you check each?
6. What precautions in regard to lead dress should be taken when replacing an output i-f transformer? What conditions might result from improper lead dress?
7. A receiver equipped with an electron-ray tuning-indicator tube operates normally, but the magic-eye tube deflection does not change as stations are tuned in. What can be wrong and how can it be checked?
8. A radio is brought in with a complaint that reception is weak. The technician also notices that the noise level is high. What is likely to be wrong? How can this condition be checked?
9. You suspect an open crystal detector in the receiver of Fig. 14-13. How would you confirm the suspicion? List the precautions to be observed when replacing the crystal.

I-f Amplifier Stage 15

Quick check. A modulated signal is applied to the converter signal grid and tuned back and forth through the intermediate frequency of the receiver. If the modulation note is heard at or near the intermediate frequency, the i-f stage is working, and the technician proceeds to check the next stage.

Function of the i-f stage. The input i-f transformer couples the i-f stage to the previous converter. The output i-f transformer couples the i-f stage to the succeeding detector stage. The signal at the input of the stage contains components at the oscillator frequency, the received signal frequency with its modulation, and sum and difference values of these two frequencies with the same modulation as the signal. The signal at the output of the stage should be at the difference or intermediate frequency and will also contain the modulation component of the original signal. The function of the i-f stage, therefore, is to tune and amplify at the intermediate frequency.

THEORY OF OPERATION

Standard circuit. This is shown in Fig. 15-1.

Functions and values of component parts. The tuning function of the i-f amplifier is accomplished by the action of the four tuned circuits of the input and output transformers: *L*-8 and *C*-8, *L*-9 and *C*-9, *L*-10 and *C*-10, and *L*-11 and *C*-11. All are tuned sharply to the intermediate frequency, and the four tuned circuits make possible the well-known selectivity of the superheterodyne receiver.

The amplification function of the i-f amplifier is dependent on two factors: the design of the transformers *T*-4 and *T*-5, and the amplification of the tube. In a circuit of this type, the transformers are usually designed for high gain; the possible gain of the 6SK7 tube is also high. So the overall gain of the stage is generally 100 or better. A discussion of stage-gain measurements will be given in the section on the signal check.

Input i-f transformer. Input i-f transformer assembly *T*-4 includes the primary coil *L*-8 with its associated capacitor *C*-8, and the secondary coil *L*-9 with its capacitor *C*-9. This transformer is the coupling device between the converter and the i-f stages. It is very similar to the output i-f transformer *T*-5 and, although there may be some design differences between them, the input and output transformers are usually a matched pair. Both transformers are tuned to the intermediate frequency of the receiver. The transformers are permeability-tuned. The tuning device is a screw arrangement which varies the position of powdered-iron core plugs in the coils. In older receivers, the coils are fixed and the associated capacitors are made variable. Transformers of this type are known as trimmer-tuned i-f transformers.

We might at this time refer to the intermediate frequencies in common use. Older receivers operate at an intermediate frequency of 130 or 175 kc. Later receivers operate at some frequency between 450 and 480 kc, to minimize image-frequency interference (see Chap. 17 on the r-f stage). An intermediate frequency often encountered is 260 kc. The trend in modern receivers is to standardize at 455 kc. In almost all cases, the intermediate frequency used in a particular receiver is indicated on the schematic wiring diagram of that receiver.

I-f tube. The tube employed in the i-f stage is usually the type 6SK7 remote-cutoff pentode. Other remote-cutoff pentodes commonly used as i-f amplifiers include the miniature 6BA6, and the lock-in 7A7, all of which have similar characteristics. A-c/d-c receivers use the 12SK7, 12BA6, or 14A7 tubes.

Minimum-bias circuit. Resistor *R*-23 in the cathode circuit of the i-f tube functions in a self-bias circuit similar to self-bias resistor *R*-13 in the second a-f stage (see Fig. 12-1). To see the similarity more clearly, assume for the moment that the grid return goes to ground instead of the agc bus, as in Fig. 15-2. Plate and screen currents flow through the cathode resistor *R*-23, making the cathode 1 volt positive with respect to ground. Since the grid is at ground potential, it is 1 volt negative with respect to cathode. This is the grid-bias voltage.

When the grid is returned to the agc bus, as in Fig. 15-3, there is no agc voltage when no signal is present, and the grid is therefore at zero or ground potential. This makes the condition similar to that of a grounded grid return; that is, the grid is at ground potential, the cathode is at a potential of plus 1 volt due to the self-bias resistor *R*-23, and the grid is therefore 1 volt more negative than the cathode. When a signal is tuned in, it develops an agc voltage, which is negative with respect to chassis, thereby making the grid negative with respect to chassis by an amount equal to the agc voltage. The cathode is still positive with respect to chassis because of self-bias, and therefore the actual bias on the grid of the tube is the sum of the agc and the cathode voltages. The weaker the signal, the lower the agc voltage will be, and the less it will add to the minimum grid bias. However the grid-bias voltage cannot fall below the self-bias voltage, even when no signal is received. Since the self-bias circuit of *R*-23 sets a minimum limit to the grid-bias voltage, it is called the "minimum-bias" circuit.

Cathode resistor *R*-23 is usually a ½-watt resistor, and its ohmic value is usually 68 or 100 ohms. A higher value would mean a higher

FIG. 15-1. Typical circuit of an i-f amplifier stage.

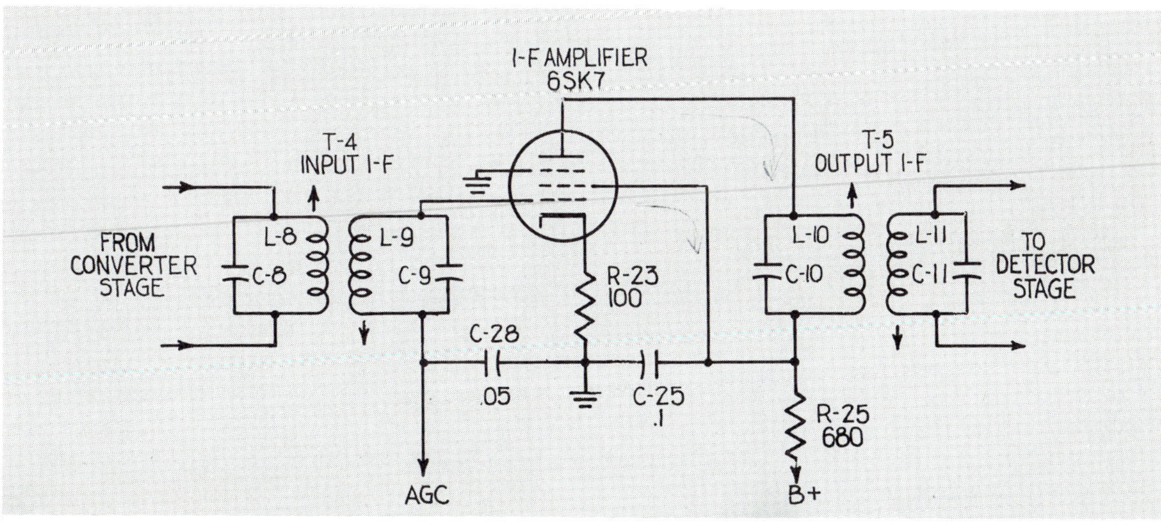

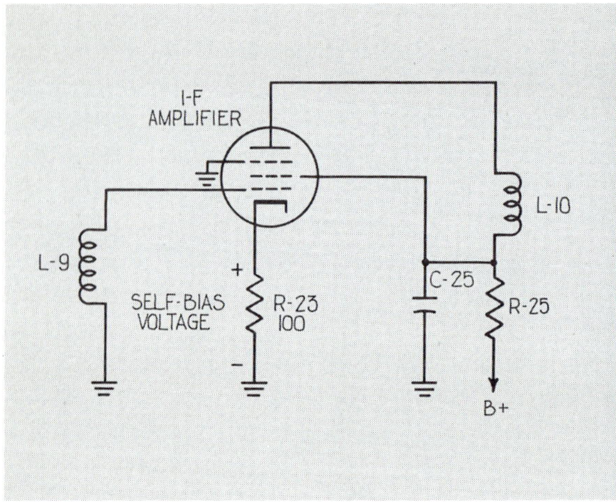

FIG. 15-2. Self-bias in an i-f amplifier without agc.

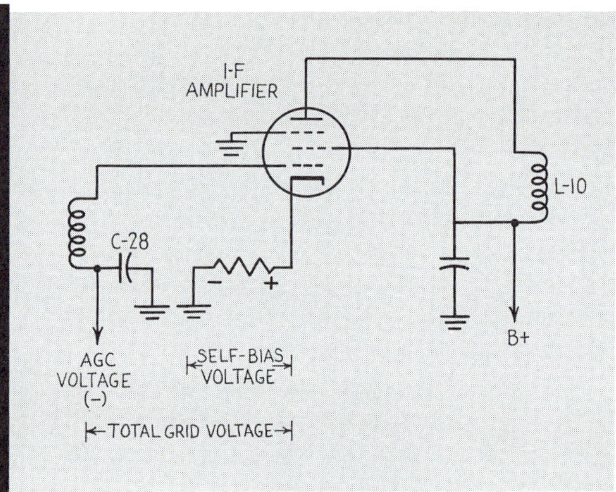

FIG. 15-3. The grid bias applied to the i-f tube is the sum of the self-bias and agc voltages.

minimum-bias voltage and less possible amplification for the stage. It is usually unbypassed, to provide a little degeneration which reduces the gain at zero agc voltage, thus adding to the stability of the circuit. In those receivers where a bypass capacitor is used, the size chosen is generally 0.1 mfd, rather than the 25 mfd found in the similar audio self-bias circuit. Since this amplifier is operating at 455 kc, a 0.1-mfd capacitor provides sufficient bypass action. In some receivers, cathode resistor *R*-23 is omitted, with the cathode going directly to ground.

Output i-f transformer *T*-5. Output i-f transformer *T*-5 couples the output of the i-f stage to the detector stage. Replacement notes for *T*-5 are found in Chap. 14, which describes the detector and agc stage.

Decoupling filters. Whenever two or more stages are operated from the same voltage supply, there is a possibility of coupling between the stages through the common power supply. This is illustrated in Fig. 15-4.

If we consider the signal voltage in the plate circuit as being from plate to cathode, the signal voltage of tube *V*-1 is across *L*-4, *C*-16 in the power supply, and *R*-1. The signal voltage of *V*-2 is across *L*-8 and *C*-16 in the power supply. The signal voltage of *V*-3 is across *L*-10, *C*-16 in the power supply, and *R*-23.

Let us consider the plate circuit of tube *V*-1. The greater part of the signal voltage will be where it is wanted—across the high impedance of *L*-4, where it will be transferred to *L*-5 and the grid circuit of the following tube. There will also be some signal voltage drop across the low impedance of *C*-16 in the power supply and *R*-1 in the cathode circuit.

Now let us consider the plate circuit of *V*-2. Again, the signal voltage will be mainly across *L*-8, but there will be some across *C*-16. Note that the signal voltages of tubes *V*-1 and *V*-2 have a common circuit in *C*-16 in the power supply.

When we consider the plate circuit of *V*-3, again, most of the signal is across *L*-10, but a small part will be across *C*-16, which is common to all three plate circuits.

If the signals from any of the tubes are in phase, oscillation may result owing to regenerative feedback through the common coupling, *C*-16.

The coupling through the common power supply is usually avoided by the addition of a resistor and capacitor known as a "decoupling filter" or isolation circuit, as shown in Fig. 15-5.

The decoupling filter consists of *R*-25 and *C*-25. Capacitor *C*-25 offers a low opposition path to ground for the signal, and *R*-25 offers a high opposition path to the signal. The net result is to keep the signal voltage of *V*-3 out of the power supply, so that it cannot mix with the signal from any other tube. An r-f choke is sometimes used instead of *R*-25. This also offers high opposition to the signal.

The decoupling filter may be applied in the plate circuit of tube *V*-1 instead of tube *V*-3. The result would be the same, since in this case the signal voltage of *V*-1 would be kept out of the power supply and therefore would not react with the signal from any other tube.

In different receivers, there is considerable variation as to the placement of the decoupling filter. Sometimes it is in the plate circuit of *V*-1, sometimes in the plate circuit of *V*-3, sometimes in both. Also, the plate circuit of *V*-2 may be tied to either that of *V*-1 or *V*-3, or have its own filter. Since there is no standardization in the placement of the decoupling filters, a decision as to the placement in the standard receiver circuit (Fig. 1-1), which attempts to show the most commonly used practices, has to be reached. In the standard receiver circuit, a decoupling filter is placed

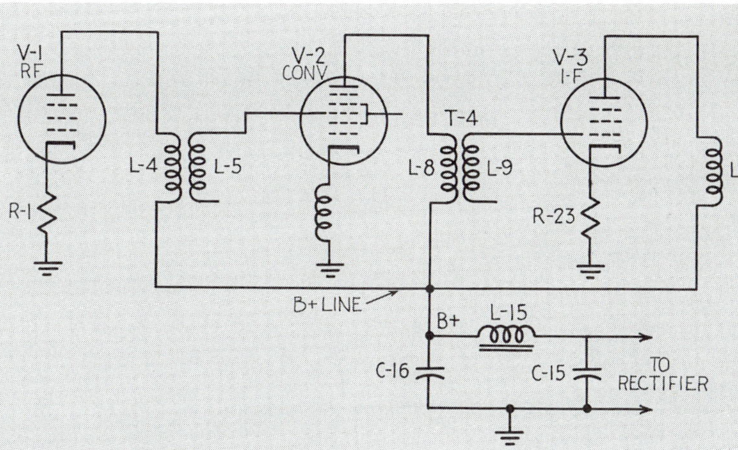

FIG. 15-4. Coupling in the plate circuit due to a common *B*-power supply component.

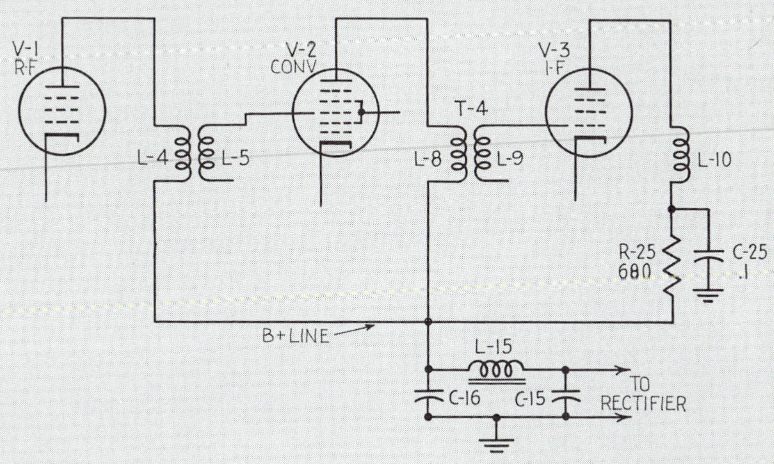

FIG. 15-5. Decoupling filter in the plate circuit of tube V-3.

in the plate and screen circuit of each tube, as shown in Fig. 15-6, and servicing procedures are dealt with so as to include the filter. From the above discussion, it is to be hoped that the technician will expect an individual receiver to differ somewhat from the standard in that one or more decoupling filters may be omitted.

By a similar line of reasoning, there could be undesirable regenerative coupling, if the cathodes of three stages were connected together and fed from a common cathode to ground resistor for equal self-bias voltages. The same thing could happen with the screen-voltage supply, or the grid returns through the agc bus. Where we have three stages operating at similar frequencies through a common coupling, decoupling filters will be found in at least one of these circuits.

In the standard circuit, the cathodes of the r-f, converter, and i-f tubes have individual self-bias resistors to avoid coupling. The screen of each tube is tied to its plate supply, as shown in Fig. 15-6. The plate circuit decoupling filters therefore also decouple the screens.

Decoupling filters in the grid returns of the r-f and converter tubes are rarely omitted. In this case, the standard circuit is indeed standard. In Fig. 15-7, resistor *R*-30 and capacitor *C*-30 make up such a decoupling filter for tube *V*-1, while *R*-29 and *C*-29 make up a similar filter for tube *V*-2.

A great many receivers do not use an r-f stage. In this case, since there are fewer stages with a common coupling component, the probability of regenerative feedback is lessened, and there is little necessity for decoupling filters.

To get back to the i-f stage, the plate decoupling filter consists of *R*-25 and *C*-25, as

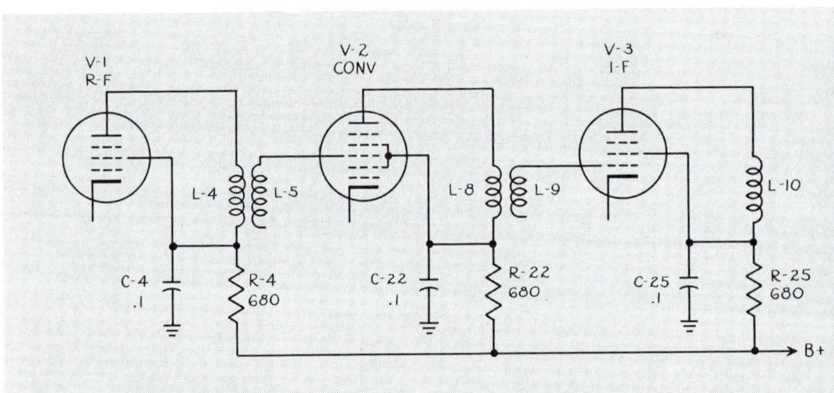

FIG. 15-6. Decoupling filters in the plate and screen circuits of the r-f, converter, and i-f tubes in the standard circuit.

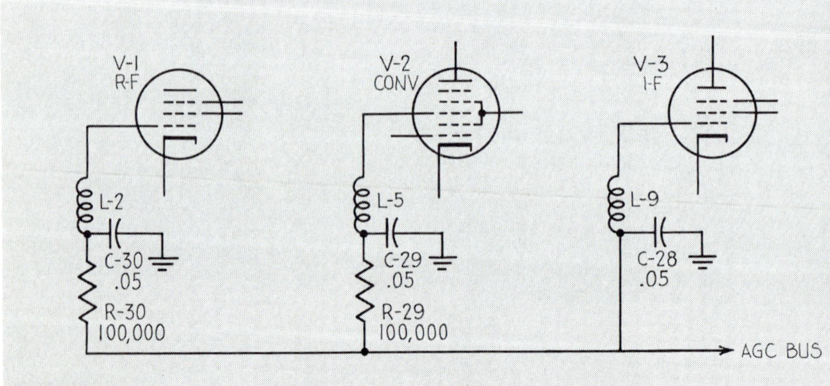

FIG. 15-7. Decoupling filters in the grid-return circuit to avoid coupling in the common agc voltage supply.

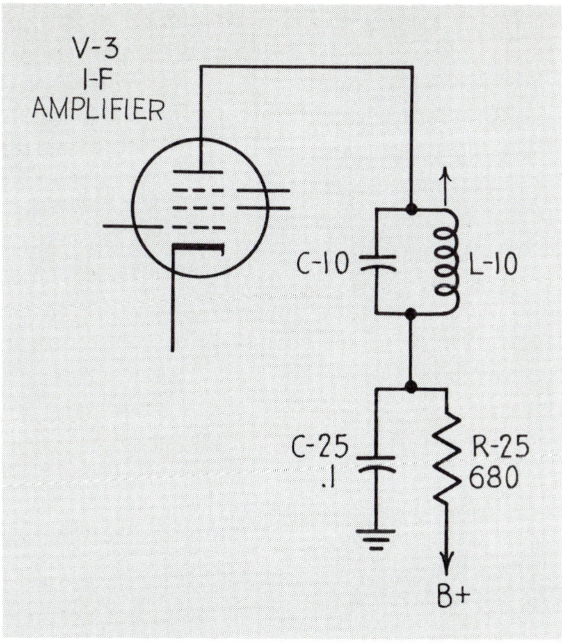

FIG. 15-8. Plate-circuit decoupling filter in the i-f amplifier stage.

shown in Fig. 15-8. Resistor *R*-25 varies from 400 to 1,000 ohms in different receivers, and *C*-25 varies from 0.05 to 0.25 mfd. These values are not critical.

NORMAL TEST DATA FOR THE I-F STAGE

Signal check. The test point for the signal check of the i-f stage is the converter signal grid, as shown in Fig. 15-9. As was the case in the signal check for the detector stage, the input signal is applied to a previous tube, to avoid the detuning effect of the capacitance in the signal generator. The converter-grid test point is readily available at the stator-plates terminal of the converter tuning capacitor, *C*-5. The output indication is the modulation note of the signal generator in the speaker, or its amplitude, as shown by the output meter.

The output meter should be adjusted for a high-voltage range at the start of the signal check, since, from this test point, the amplification of the receiver is considerable. Until the signal generator's attenuator is adjusted, the output signal may be high enough to harm the meter, if it is at its usual 25- to 60-volt range for output measurements. The r-f portion of the receiver is made inoperative by shorting the oscillator section of the gang tuning capacitor, as explained in Chap. 14 in the discussion of the signal check for the detector and agc stage.

FIG. 15-9. Connections for the signal check of the i-f stage.

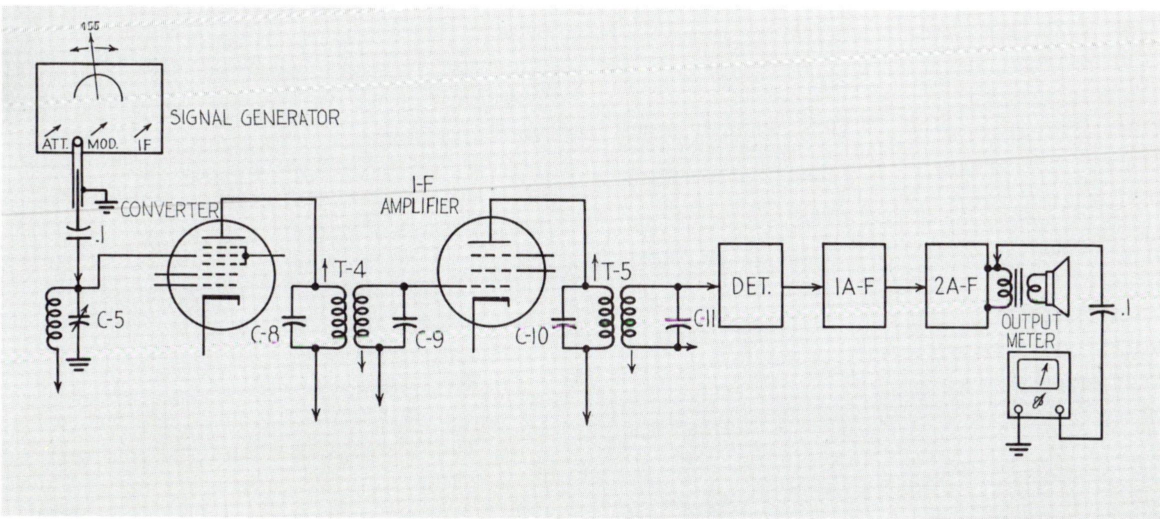

The signal check consists in rotating the frequency control of the signal generator (wobbling) through the receiver's intermediate frequency, while listening for its modulation note in the speaker or observing the output meter reading. Unless the response is considerably stronger than that heard from the i-f grid (quick check for the detector stage), the i-f stage bears investigation for trouble. This is the quick check for the i-f stage.

At the same time, the signal check may be used to check alignment, operation at the proper intermediate frequency, and the presence of oscillation. The presence of two peaks close together is not necessarily an indication of misalignment. This may be the normal response from an overcoupled i-f transformer. This is explained in the variations section dealing with broad-band i-f amplifiers.

When the modulated i-f signal is applied to the converter grid and there is no response or abnormally low response, the trouble may be in the converter tube or its operating potentials. This can be checked by shifting the signal generator test lead to the converter plate. In this case, normal response is a somewhat stronger signal from the converter plate than was obtained from the i-f grid (quick check of the detector stage). Trouble in the converter tube or its operating potentials is handled in Chap. 16, which deals with the converter. If there is no signal response from the converter plate, the trouble is definitely in the i-f stage.

Normal voltage data. Readings are taken from the chassis or common negative terminal to tube elements. The data are given in the accompanying table.

Tube elements	A-c receivers, volts	6SK7 or 12SK7 pin No.	A-c/d-c receivers, volts	6BA6 or 12BA6 pin No.
Plate	90–135	8	90	5
Screen	90–135	6	90	6
Cathode	1	5	0–1	7

Normal resistance data. These data are presented below.

	Resistance, ohms	
	Air core	Iron core
Across *L*-8, primary of *T*-4	30–50	5–15
Across *L*-9, secondary of *T*-4	30–50	5–15
Across *L*-10, primary of *T*-5	30–50	5–15
Across *L*-11, secondary of *T*-5	30–50	5–15
Cathode to chassis	0–100	
Control grid to chassis	2.7 meg.	

A wide divergence is given for the coils *L*-8, *L*-9, *L*-10, and *L*-11, to allow for differences between receivers. In any one receiver, however, owing to the common use of matched transformers, these coils should all check very close to the same value within the limits given.

Sensitivity check of the i-f stage. As was done for the previous stages, the technician should run some checks on receivers known to be good, so as to have a basis of comparative data as to the operation of his test equipment and the normal gain to be expected from the i-f stage.

The receiver test oscillator and output meter are connected, as shown in Fig. 15-9. The receiver's r-f section is made inoperative by shorting the oscillator section of the gang tuning capacitor. The receiver is set to the full volume and minimum bass positions. A selectivity control, if any, is set for the maximum selectivity position. The test oscillator is adjusted for modulated output on the i-f band. The output meter is set at a high a-c voltage range for safety's sake, although the range will be reduced for the final check of the standard output voltage.

The signal generator is connected to the converter grid, and the frequency-control dial is rotated carefully through the receiver's intermediate frequency for peak deflection on the output meter. At peak, the attenuator of the signal generator is adjusted to give the standard output of 50 mw in the speaker. Standard output corresponds to an output meter reading of 16 volts (see the discussion of the use of the output meter in testing the first a-f stage, Chap. 13).

The average i-f signal input at the converter grid, necessary to give standard output, is 50 microvolts for a modern high-gain receiver. The attenuator setting just obtained, therefore, corresponds to 50 microvolts. After several good receivers have been checked by the above procedure and the results have been compared, a reference point, corresponding to 50 microvolts, has thus been established on the signal-generator attenuator dial.

COMMON TROUBLES IN THE I-F STAGE

Troubles common to the input i-f transformer. Replacement notes and troubles of the input i-f transformer *T*-4 will be outlined briefly here. For a more detailed discussion, the replacement notes on the similar output i-f transformer are equally applicable (see Chap. 14).

The i-f transformers sometimes open. When this is the case, the receiver will not operate, and a signal check will indicate the defective stage. An ohmmeter check then shows the open transformer.

The i-f transformers also cause noise. This condition is usually due to corrosion of the windings. It will be found by an ohmmeter check, since a corroded winding will check several hundred ohms instead of its normal value of 15 to 50 ohms.

If an exact replacement transformer is not available, the suggestions for replacing the output i-f transformer assembly given in Chap. 14 should be helpful.

Input i-f transformer color code. The E.I.A. color code given below will help to identify the leads.

Blue	Plate lead
Red	*B*+ lead
Green	Grid lead
Black	Grid return

When a new transformer is installed, grid and plate leads should be short and direct and away from each other and all other wiring.

Troubles common to the agc bypass capacitor. Replacement notes on the agc bypass capacitor *C*-28 are found in the discussion of the agc filter and decoupler in Chap. 14.

Troubles common to the minimum-bias resistor. The voltages and currents encountered in the cathode circuit of the i-f tube are such that there is no overload on minimum-bias resistor *R*-23, and the resistor rarely gives trouble. If it should open, the stage will not operate and the condition would be found in a voltage check. The cathode-to-ground voltage would check abnormally high, since the test voltmeter, with its high resistance, would bridge the open resistor in the circuit.

The original should be duplicated as to ohmage and wattage. If the exact ohmage value is not available, a considerable tolerance may be allowed, since the value is not critical and will cause little effect on the overall performance of the receiver.

Troubles common to the plate decoupling filter. If present, the decoupling filter may be a source of trouble. Capacitor *C*-25 may short, with the result that there will be no plate voltage, the receiver will be inoperative, and resistor *R*-25 will probably burn.

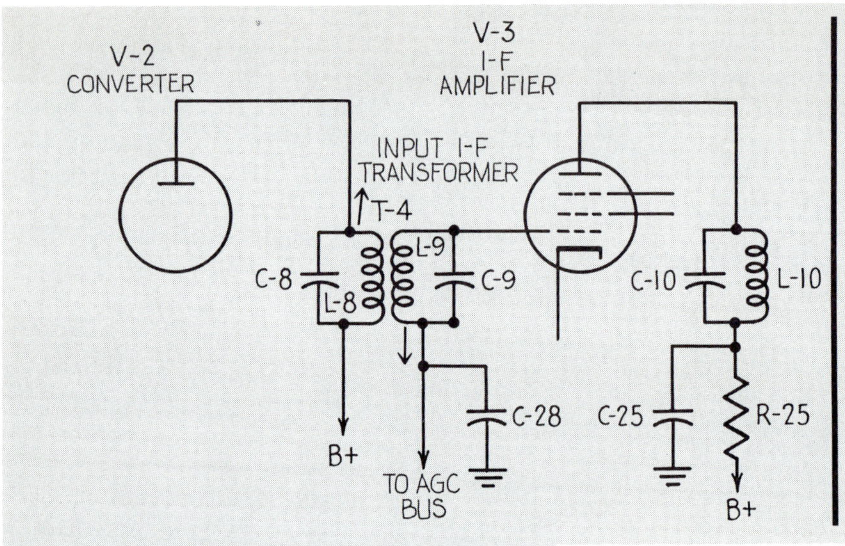

FIG. 15-10. The input i-f transformer and its position in the circuit.

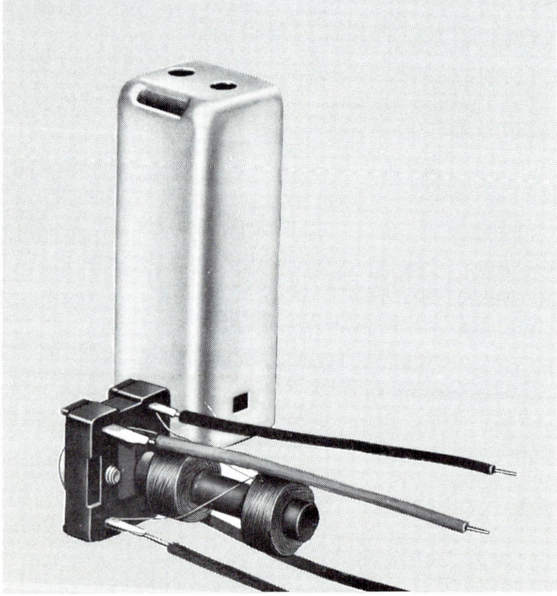

This condition would be found very early in the trouble-shooting procedure, since the *B*+ voltage will be very low. To find the short, however, might be more difficult, since there are several circuits in parallel with the capacitor. Refer to Fig. 15-11. An overheating *R*-25 would be one indication. Another helpful device is to make a resistance check from all plates to ground. If capacitor *C*-25 were shorted, the i-f plate would check approximately 40 ohms to ground (the resistance of *L*-10), while all other plates would check their normal plate load plus the resistance of their decoupling filter, if any, plus the resistance of *R*-25.

It is not unusual to find only a bypass capacitor connected at *B*+ of an r-f or i-f tube, even though no other form of decoupling filter is used. This capacitor therefore is connected from *B*+ to chassis and is in parallel with the power-supply filter capacitor *C*-16. When this is the case, the ohmmeter check from each plate to ground would give no definite clue, since, with no decoupling resistors, all plate-to-ground readings would show their normal plate load. It would be necessary then to open the *B*+ wiring, one circuit at a time, to find the short as described in Chap. 8.

A decoupling filter capacitor may also open. In this case, all voltages would show normal readings, but the receiver would have a tendency toward oscillation. Since it is common practice, in trouble shooting for oscillation, to bridge all bypass capacitors with a good capacitor, the open decoupling capacitor would be found in this manner.

When replacing capacitor *C*-25, voltage rating is important. A 600-volt rating is recommended for all replacements. The capacitance is not critical, so that a wide tolerance may be allowed. If a shorted *C*-25 is being replaced, the resistor *R*-25 in the decoupling filter should also be replaced, since it has been damaged by feeding heavy current to the short. Unless *C*-25 has been shorted, *R*-25 will, of itself, cause no service trouble.

Troubles common to the i-f amplifier tube. The amplifier tube is the most common cause of trouble in the stage. The best check, of course, is to compare operation with a similar tube known to be good.

Broad-band i-f amplifiers. The i-f transformers of receivers, like that of the standard, are designed for great selectivity and gain. Figure 15-12 shows the frequency-response curve for such transformers. However, such a circuit has a defect in that it is too selective and attenuates the high-frequency audio signals. This defect is known as "sideband cutting." In high-fidelity reproduction, where the high-frequency audio notes are desired, it is necessary to broaden the response curve of the i-f transformers to that shown in Fig. 15-13. This is commonly done in f-m and a-m/f-m receivers.

In high-fidelity receivers, where the response curve of the i-f amplifier is broadened, the amplification of the stage is reduced. Usually, a second broadly tuned stage is therefore added to make up for this loss. The over-all gain of the two-stage i-f amplifier is somewhat greater than the gain of a single-stage amplifier, and the over-all selectivity is equally good owing to the extra tuning circuits of the added stage. Figure 15-14 is a graphic representation of the response curve of each stage of a two-stage i-f amplifier and the over-all response of the amplifier.

Several methods are in common use to obtain the desired broad-band response. One method is known as "overcoupled" transformers. In any i-f transformer, the relative position of the primary and secondary windings to each other is called the "coupling." When the two windings are far apart, the energy transfer from primary to secondary is small, and the transformer will give low gain and good selectivity. As the two windings are brought closer together, the gain of the transformer increases and the selectivity becomes somewhat broader up to a critical

FIG. 15-11. The i-f amplifier plate-decoupling filter.

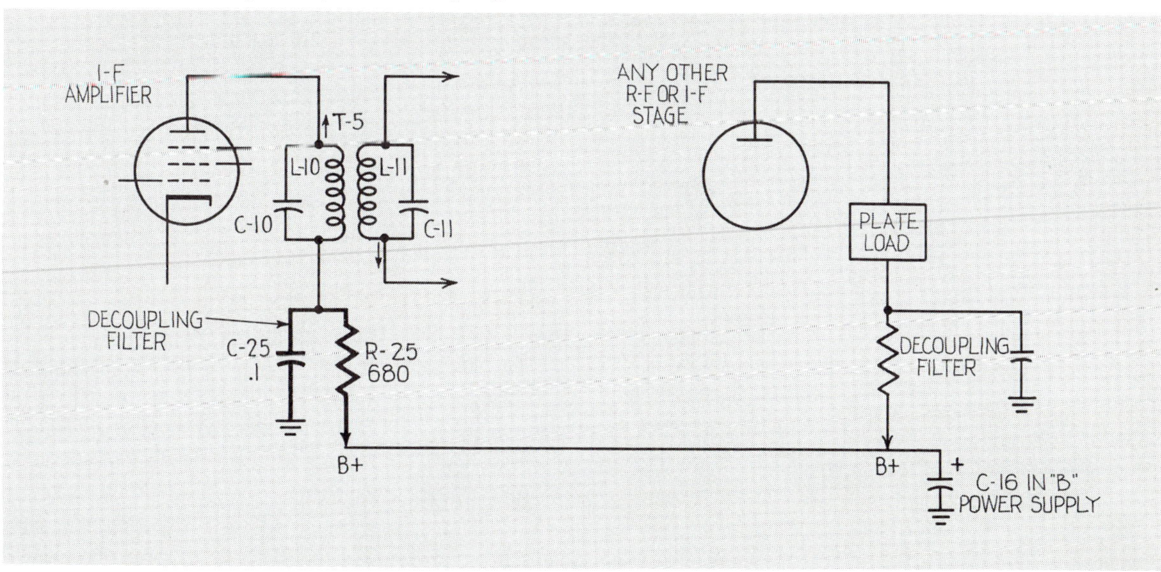

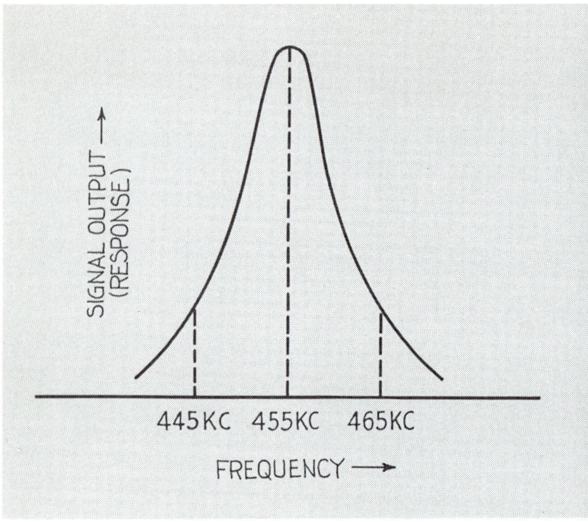

FIG. 15-12. Frequency-response curve of the usual i-f transformer.

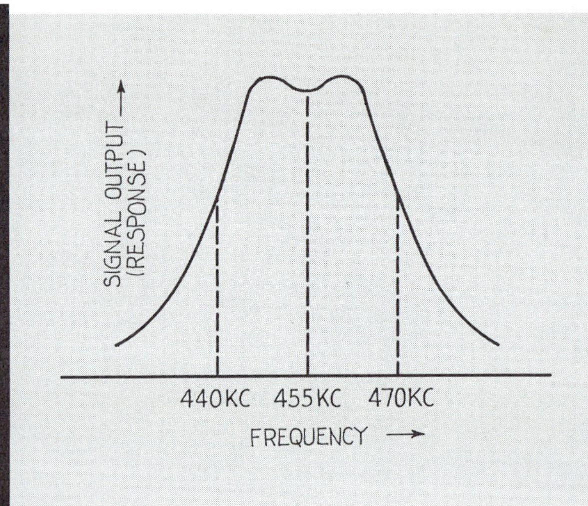

FIG. 15-13. Frequency-response curve of a high-fidelity i-f transformer.

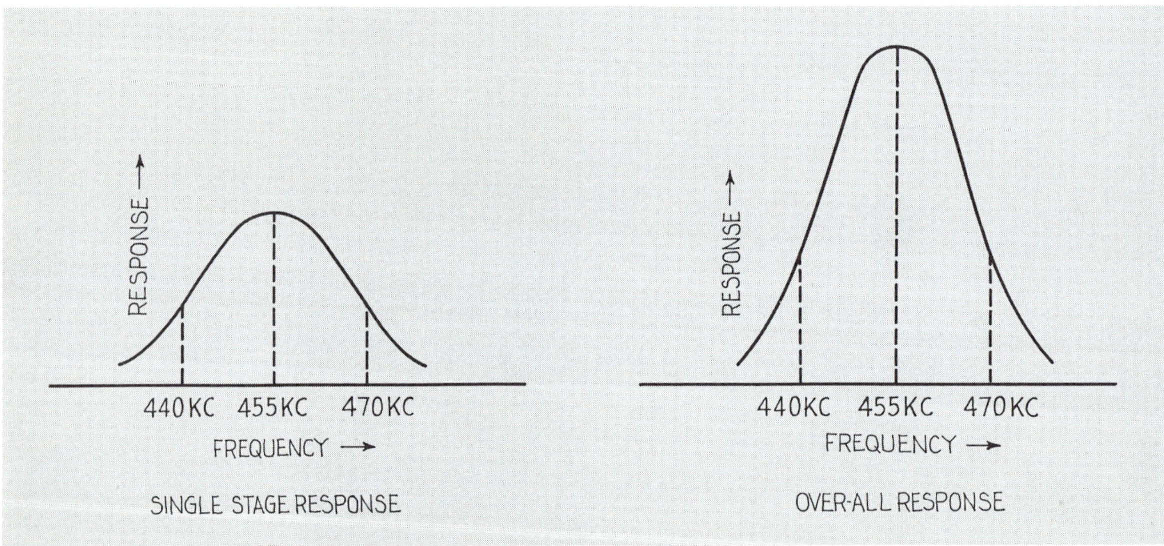

FIG. 15-14. Frequency-response curve of a high-fidelity i-f amplifier.

point, after which the gain is reduced and the selectivity becomes considerably broader, owing to the appearance of two peaks, one on each side of the resonant frequency. When the primary and secondary windings are closer than this critical point, the transformer is said to be "overcoupled." Figure 15-15 illustrates the effects of this overcoupling on the gain and selectivity of a transformer.

The design of the usual single-stage transformer makes some compromise as to coupling between the low and the critical points, so as to give high gain with good selectivity. Some receivers that feature broadband i-f amplifiers make use of overcoupled i-f transformers. Sometimes the coupling is made variable by a mechanical arrangement that raises and lowers one winding by turning a knob on the front panel of the receiver. The position of minimum coupling is labeled SELECTIVITY or SENSITIVITY, whereas the position of maximum coupling is labeled FIDELITY or TREBLE. This control is called a "fidelity" control.

Another method of broadening the response of the i-f amplifier is to load the tuned circuits with resistors, as shown in Fig. 15-16. The resistors may be placed across the primary winding, the secondary, or both; or they may be placed in series with the trimmer capacitor. In any case, the introduction of resistance loads the tuned circuit and results in a decreased gain and a broader response curve. The amount of broadening is determined by the amount of loading; that is, the ohmic value of the resistor. The single-stage curve of Fig. 15-14 is typical for a resistance-loaded transformer.

From the technician's point of view, a two-stage i-f amplifier presents few complications. The signal check is about the same as for a single-stage i-f amplifier, since the gain per stage is considerably lower. It merely adds another grid from which to check. The presence of two peaks close together is to be expected, especially where overcoupling is employed. The i-f transformers are subject to the same ills as with a single stage. They open and become noisy because of corrosion; the same checks are applicable. However,

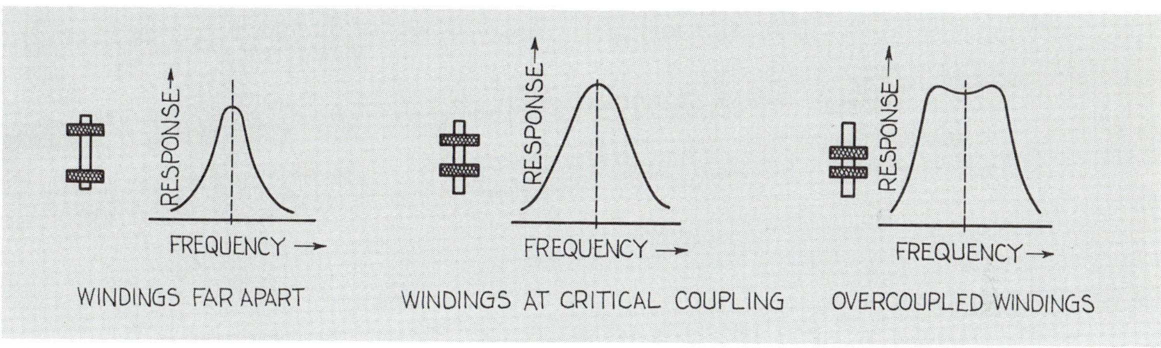

FIG. 15-15. Effect of coupling on the frequency-response curve of an i f transformer.

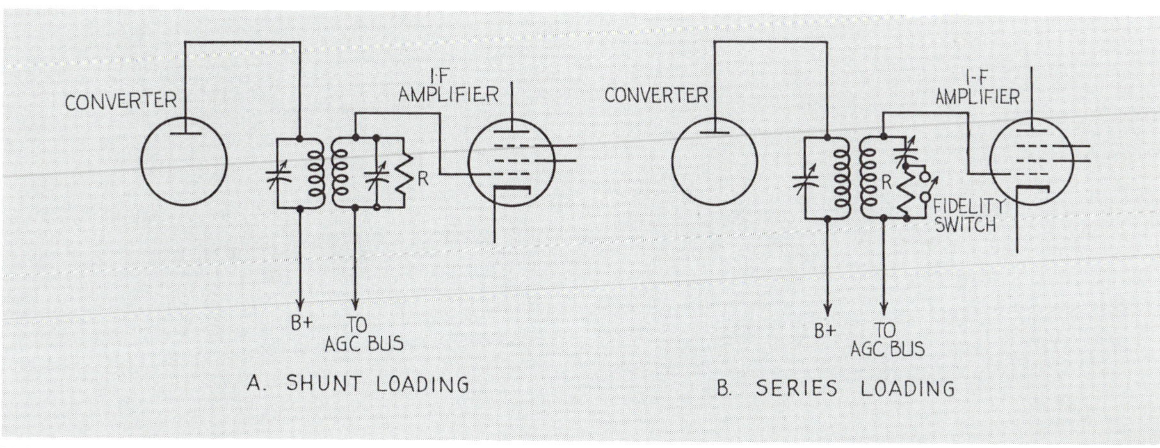

FIG. 15-16. Resistance loading to broaden response characteristic of an i-f amplifier.

when an i-f transformer is replaced in a two-stage i-f amplifier, it becomes more necessary to employ an exact replacement.

Because of an added stage, more decoupling filters will be used. However, the treatment of them will not vary from that given for our standard circuit.

The alignment of broad-band i-f amplifiers can best be performed with an oscilloscope, but satisfactory alignment can be obtained with a standard signal generator and output meter. It is extremely important to follow the manufacturer's service instruction. Where such instructions are not obtainable, a generalized procedure may be followed. Set the receiver for maximum gain position (not high-fidelity); that is, minimum coupling where a coupling control is used (shunt resistors switched out where this is the method), and align for maximum response, as usual. Then rotate the signal generator about 10 kc on each side of the intermediate frequency, noting the output-meter deflection. If it remains fairly constant for about 5 kc on each side of the intermediate frequency, the alignment may be considered good. If the output meter fails to remain constant, alignment adjustments should be repeated.

SUMMARY

Quick check.

Introduce a modulated signal to the signal grid of the converter tube and wobble through the intermediate frequency. When the i-f stage is functioning properly, the modulation note will be heard in the speaker at or near the intermediate frequency. The response will be much stronger than that heard when the detector stage is checked; that is, when the signal is applied to the i-f grid.

Diagram of typical i-f amplifier stage.

A diagram of the typical i-f amplifier stage is given in the accompanying figure.

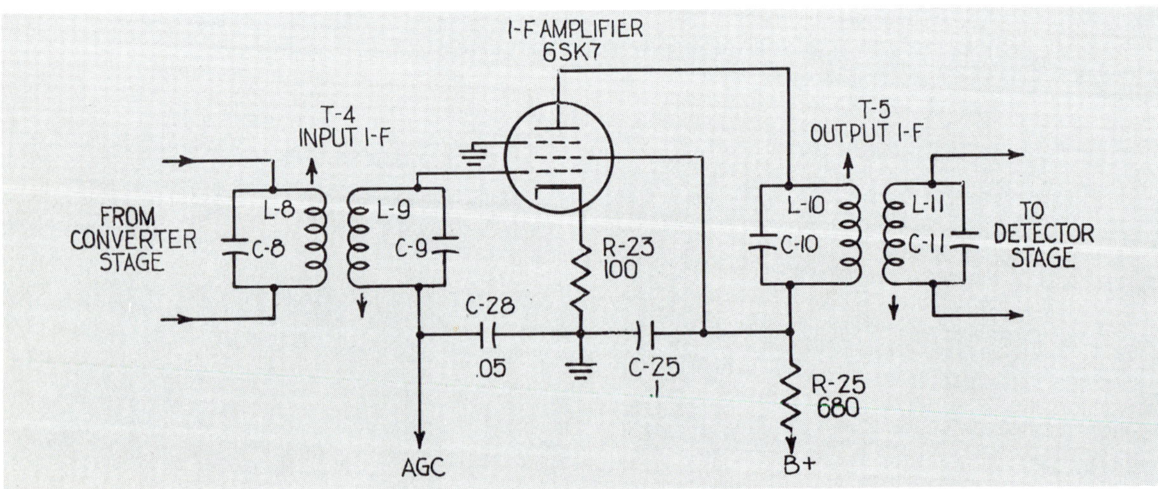

Voltage check.

Readings are taken from chassis or common negative terminal. Normal voltage data are given in the accompanying table.

Tube elements	A-c receivers, volts	6SK7 or 12SK7 pin No.	A-c/d-c receivers, volts	6BA6 or 12BA6 pin No.
Plate	90–135	8	90	5
Screen	90–135	6	90	6
Cathode	1	5	0–1	7

Normal resistance data.

Normal resistance data are given in the following table.

	Resistance ohms	
	Air core	Iron core
Across *L*-8, primary of *T*-4	30–50	5–15
Across *L*-9, secondary of *T*-4	30–50	5–15
Across *L*-10, primary of *T*-5	30–50	5–15
Across *L*-11, secondary of *T*-5	30–50	5–15
Cathode to chassis	0–100	
Control grid to chassis	2,700,000	

A wide divergence is given for the coils *L*-8, *L*-9, *L*-10, and *L*-11, to allow for differences between receivers. In any one receiver, owing to the common use of matched transformers, these coils should all check very close to the same value within the limits given.

Signal-substitution test procedure for an inoperative i-f amplifier.

The test oscillator, receiver, and output meter are connected as shown in Fig. 15-17. The signal generator is adjusted for modulated output on the i-f band. The receiver is adjusted for maximum volume, minimum bass response, and maximum selectivity (if there is such a control); the r-f portion is made inoperative by shorting the oscillator section of the gang tuning capacitor. Let us assume normal operation of the audio amplifier, as proved by a normal response when an audio test signal is applied to point ⑤, the input of the a-f amplifier, and no response or weak response when a modulated signal at the intermediate frequency is applied to point ①, the converter signal grid.

Step 1. The test lead from the signal generator is moved to point ②, the converter plate:

1. If a normal response results, the trouble may be
   a. A shorted converter signal grid (most likely a short in the gang tuning capacitor).
   b. A defective converter tube (substitute a good one).
   c. Open or shorted plate, screen, or cathode circuit in the converter tube (detected by voltmeter check).

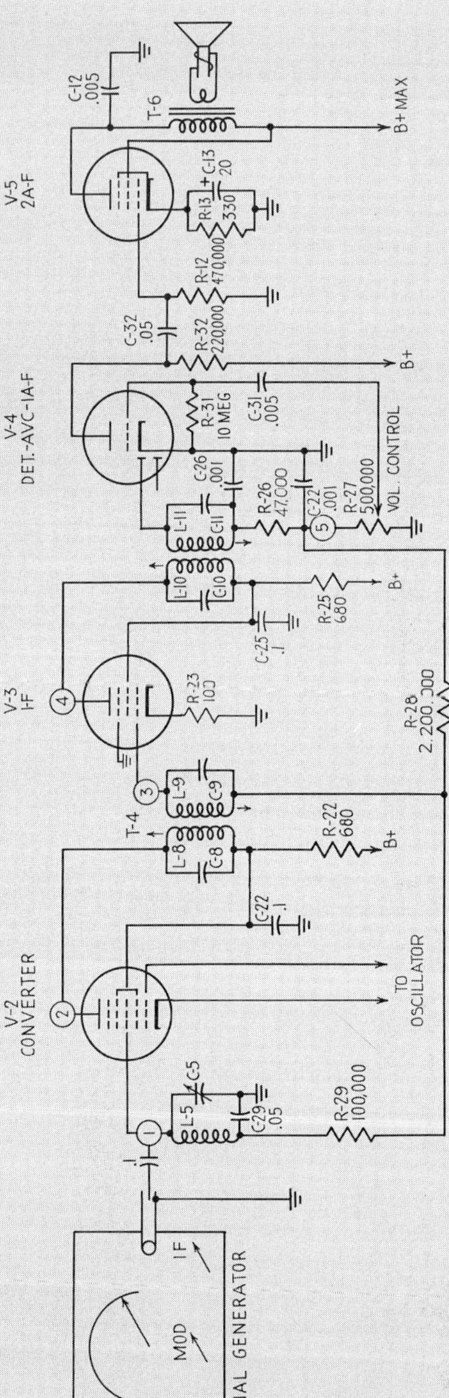

FIG. 15-17. Signal check for locating trouble in an i-f amplifier.

2. If the signal does not come through or remains very weak, move on to step 2.

Step 2. The test lead from the signal generator is moved to point ③, the i-f grid.

1. If a normal response (3,500 microvolts input for standard output) results, the trouble may be
   a. A defective input i-f transformer (detected by ohmmeter check).
   b. An open agc bypass capacitor *C*-28. (Bridge it with a good one and recheck from point ①.)
   c. Input i-f transformer *T*-4 badly misaligned (check alignment).
2. If the signal does not come through or remains very weak, move on to step 3.

Step 3. The test lead from the signal generator is moved to point ④, the i-f plate. The attenuator is advanced, and the frequency control is wobbled through the intermediate frequency.

1. If the signal comes through, the trouble may be
   a. A shorted i-f grid. (Detected by ohmmeter check. The short would most likely be between the grid wire or trimmer and the i-f shield can.)
   b. A defective i-f tube (substitute a good one).
   c. Open or short in the plate, screen, or cathode circuits of the i-f tube (detected by voltmeter check).
2. If the signal does not come through, check the detector stage (see Chap. 14).

SERVICE DATA CHART

Symptom	Abnormal reading	Look for
No reception	Plate voltage = 0	Open i-f output transformer. Shorted plate bypass capacitor *C*-25. Open plate circuit decoupling resistor *R*-25. Plate-to-ground short in i-f can
	High cathode voltage	Open minimum-bias resistor *R*-23
	All voltage checks are normal	Dead i-f tube *V*-3. Shorted trimmers in the i-f cans. Open i-f transformer secondaries. Open agc bypass capacitor *C*-28

Symptom	Abnormal reading	Look for
Weak signal	All voltage checks are normal	Weak i-f tube *V*-3. Open agc bypass capacitor *C*-28. Open cathode bypass capacitor *C*-23. Open plate circuit bypass capacitor *C*-25. Misalignment
Noise	All checks are normal	Noisy i-f tube *V*-3. Corrosion in the i-f transformer windings
Squeal or oscillation	All checks are normal	Open bypass capacitor *C*-25. Open ground connection to shielding. Open agc bypass capacitor *C*-28. Incorrect wire dress

QUESTIONS

1. A receiver does not play. Signal check shows normal operation when a test signal is applied to the i-f plate; no response when the test signal is shifted to the i-f grid. List the likely sources of trouble, and explain how you would check for each.
2. A receiver does not play. Signal check shows normal operation when the proper test signal is applied to the i-f grid; no response when the test signal is shifted to the converter plate. List the likely sources of trouble, and explain how you would check for each.
3. The a-m tuner of Fig. 14-13 is inoperative. A signal check shows that the trouble is in the i-f stage. A voltage check gives normal readings for the stage. List the likely causes of the trouble, and explain how you would check for each.
4. A receiver gives the following voltage readings for the i-f stage:

Plate	130 volts
Screen	130 volts
Cathode	50 volts

   What is the probable trouble, and how would you check for it?
5. A receiver gives the following voltage readings for the i-f stage:

Plate	0 volt
Screen	95 volts
Cathode	1 volt

   What is the probable trouble? What should the next check be?
6. An a-c superheterodyne receiver oscillates badly. The oscillation continues when the converter tube is removed but stops when the i-f tube is removed. This indicates that the cause of the trouble is probably in the i-f stage. What checks and adjustments should be made to track down the trouble?
7. What factors in the i-f stage can cause noisy reception? How would you check for each?

Converter: Mixer and Oscillator Stages

16

After the i-f check, the next area for investigation is the converter. This consists of two distinct stages: the mixer and the oscillator. Their functions are so closely interrelated that they are best handled as one unit—the converter. In most receivers, the two stages are combined in one pentagrid converter tube, although some receivers use separate mixer and oscillator tubes. Service analysis is similar for both types of receivers.

The modulated r-f signal from the stage before the converter is fed to the mixer grid of the converter tube, where it is mixed with the unmodulated r-f signal from the local oscillator stage. The signal on the mixer grid, regardless of frequency, is changed by the converter to a signal with the same frequency, the intermediate frequency of the receiver. The signal at the intermediate frequency retains the same audio modulation that is present in the r-f signal fed to the mixer grid. The i-f signal is then fed to the input of the i-f amplifier.

Many superheterodyne receivers do not incorporate an r-f stage. In receivers of this type, the antenna is coupled to the converter mixer grid. In the signal-substitution method of servicing, where the trouble shooter works from the speaker back to the antenna, the converter will be the last area of investigation for receivers of this type.

Quick check for the operation of the oscillator stage. Tune the receiver to 600 kc. Connect the signal-generator output to the converter signal grid through a 0.1 mfd capacitor, and rotate the signal-generator dial through 600 kc. If the signal-generator modulation note is heard in the speaker at or near 600 kc, the oscillator is functioning.

Quick check for the operation of the mixer stage. Tune the receiver to 1,400 kc. Connect the signal-generator output to the r-f grid (antenna if there is no r-f stage) through a 0.00025-mfd capacitor, and rotate the signal-generator dial through 1,400 kc. If the signal-generator modulation note is heard in the speaker at or near 1,400 kc, the mixer stage is functioning.

Function of the converter. The function of the converter is fourfold:

1. It tunes and amplifies the received signal.
2. It generates an unmodulated r-f signal of its own at a frequency different from the received signal.
3. It mixes the locally generated signal with the received signal to produce a new signal at the intermediate frequency.
4. It maintains a constant frequency difference (the intermediate frequency) between the locally generated signal and any signal to which the receiver is tuned.

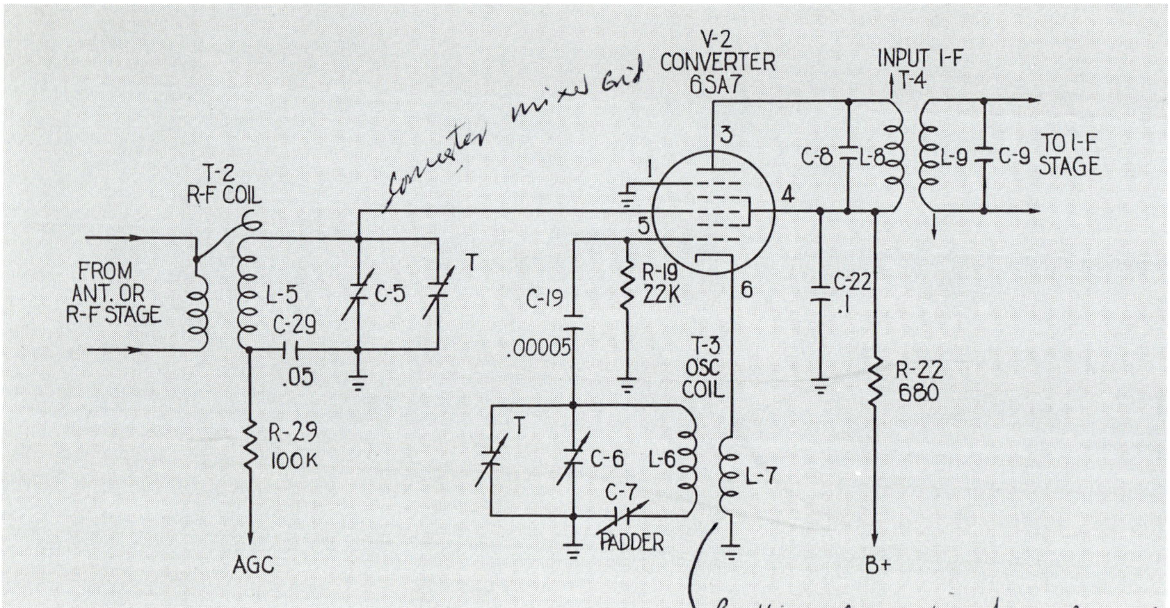

FIG. 16-1. Basic pentagrid converter circuit.

Standard circuit of a converter. This circuit is shown in Fig. 16-1.

Theory of operation of the converter. The theory of operation of the converter can be explained by elaborating the four functions listed above.

1. *It tunes and amplifies the received signal.* The input of the stage is r-f transformer *T*-2, which couples the preceding r-f stage or antenna to the converter tube. Tuning is accomplished by the circuit composed of *L*-5 and *C*-5. Capacitor *C*-5 is one section of the ganged tuning capacitor. The desired station signal is fed to grid *G*-3 which is the mixer grid or signal grid of the pentagrid converter. Grid *G*-4 is the mixer screen and grid *G*-5 is the mixer suppressor. Thus, grids *G*-3, *G*-4, and *G*-5 act as the three grids of an r-f pentode. The desired signal is therefore tuned and amplified and appears in the plate circuit of the tube. This idea is illustrated in Fig. 16-2A.

2. *It generates an unmodulated r-f signal of its own at a frequency different from the received signal.* The cathode and grids *G*-1 and *G*-2 act as a triode oscillator. This can be more easily seen by redrawing the oscillator stage of the converter, as shown in Fig. 16-2B. Grid *G*-1 acts as the oscillator grid while grid *G*-2 acts as the oscillator plate or anode. Coil *L*-6 and its associated capacitor *C*-6 are located in the oscillator grid circuit and make up the tuning section for the oscillator. Capacitor *C*-6 is the oscillator section of the gang tuning capacitor. Feedback is obtained by coupling between *L*-6 and *L*-7, the latter coil being in the oscillator cathode circuit. The feedback is in proper phase and of sufficient strength to maintain oscillation, the frequency of which is controlled by *L*-6, *C*-6, and *C*-7. The function of capacitor *C*-7 will be explained in more detail in the section on tracking.

3. *The converter mixes the locally generated signal with the received signal.* The electron stream coming from the cathode is caused to pulse by the oscillator action of grids *G*-1 and *G*-2, at a rate determined by the values

of *L*-6, *C*-6, and *C*-7. Since the oscillator anode is not a solid plate but a pair of rods, most of the pulsing electron stream will go right through the oscillator anode *G*-2 to the rest of the converter tube. The received signal is applied at *G*-3, where it contributes its own effect on the pulsing electron stream, thereby mixing the signal and oscillator output in the converter tube. Grid *G*-3 in the converter tube is sometimes called the "converter signal" grid, and sometimes called the "mixer" grid. The plate output circuit of the converter tube, therefore, will contain a signal with components at the received signal frequency, the oscillator frequency, the sum of these two frequencies and the difference of these frequencies. This type of mixing of two signals is known as "electron" mixing.

4. The locally generated signal must maintain a constant frequency difference (the intermediate frequency) with any signal to which the receiver may be tuned. Of all the signals present in the converter plate circuit, the i-f amplifier accepts only the one to which it is tuned. This is the signal that is at the difference frequency between the received signal and the locally generated signal. The oscillator frequency is usually higher than the frequency of the received signal. A few examples may clear this up. The most commonly used intermediate frequency is 455 kc. This will be used in the examples. Let us assume that the wanted station signal is 1,000 kc, approximately in the center of the broadcast band. Then the signal tuning circuit (*L*-5, *C*-5) will be at 1,000 kc. The frequency of the oscillator section, controlled by *L*-6, *C*-6, and *C*-7, will be 455 kc higher, or 1,455 kc.

The converter plate circuit will contain various frequency components:

1,000 kc Received signal
1,455 kc Oscillator signal
2,455 kc Sum of the above
455 kc Difference between the first two

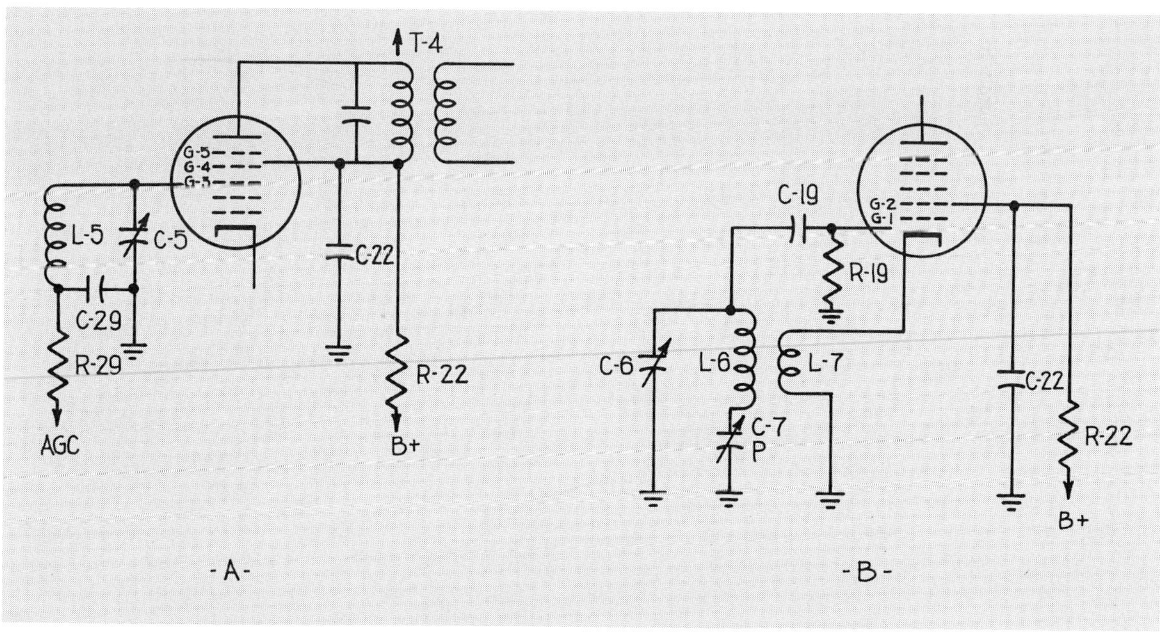

FIG. 16-2. (A) The r-f amplifier in the pentagrid converter. (B) The oscillator section in the pentagrid converter.

The sharply tuned i-f amplifier will accept the signal at 455 kc, amplify it, and pass it on to the detector.

If the desired signal is near the low-frequency end of the broadcast band at 600 kc, the signal input circuit (*L*-5, *C*-5) will be tuned to 600 kc, and the oscillator tuning circuit composed of *L*-6, *C*-6, and *C*-7 will be tuned to 1,055 kc, making the difference frequency 455 kc.

At the high-frequency end of the broadcast band, the oscillator must be adjusted to 1,955 kc to receive a signal at 1,500 kc. The two signals are mixed in the converter tube, giving, among others, the same difference frequency of 455 kc.

From the above examples, it can be seen that the prime function of the converter is to change any received signal to a signal at 455 kc, the intermediate frequency. It follows as a corollary that the oscillator frequency must be greater by 455 kc, the intermediate frequency, than the desired station signal frequency. An oscillator frequency 455 kc lower than the desired signal frequency could also be used. This is sometimes done in reception on the short-wave bands.

Tracking. In a receiver operating on the broadcast band, capacitor *C*-5 tunes coil *L*-5 from 550 to 1,600 kc in the received signal circuit (the mixer grid circuit). In the oscillator tuning circuit, capacitor *C*-6 tunes coil *L*-6 from 550 plus 455, or 1,005 kc to 1,600 plus 455, or 2,055 kc, where the i-f amplifier is tuned to 455 kc. Since capacitors *C*-5 and *C*-6 are parts of the same tuning gang, there is considerable design work needed to make these two tuning circuits always 455 kc (or the intermediate frequency) apart. The ability of a receiver to perform equally well on all parts of the tuning range is dependent on this factor, which is known as "tracking." Alignment instructions often include tracking adjustments on both ends of the tuning range and a tracking check in the center. The usual check points on the broadcast band are 600 kc for the low-frequency end, 1,000 kc for the middle, and 1,400 or 1,500 kc for the high-frequency end.

Oscillator tuning circuit. The r-f tuning circuits have a tuning range for the broadcast band of 550 to 1,600 kc. The oscillator tuning circuit for the same band must have a tuning range of 1,005 to 2,055 kc. The two tuning circuits must, therefore, be considerably different.

The oscillator tuning circuit can perhaps be better understood if it is redrawn, as in Fig. 16-3. It can now be recognized as an *L*-*C* circuit, the *L* being the oscillator coil.

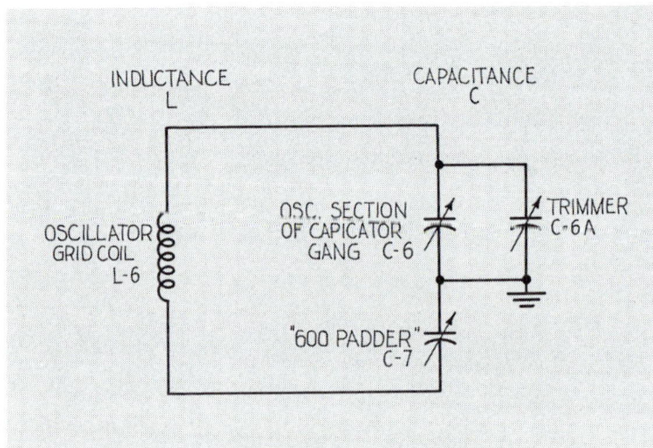

FIG. 16-3. Oscillator tuning circuit.

The *C* of the *L*-*C* circuit is composed of the main tuning capacitor *C*-6 with its shunt trimmer *C*-6*A*, both of which are in series with capacitor *C*-7. The latter is an adjustable capacitor of comparatively high capacitance. Trimmer *C*-6*A* is a low-capacitance unit. Now we need only remember that the capacitance of capacitors in series is lower than the individual capacitors, whereas the capacitance of capacitors in parallel is additive.

When tuning capacitor *C*-6 is in a low-capacitance position, the lumped *C* in the tuning circuit is small (series capacitors). Trimmer capacitor *C*-6*A* is an important cog at this position since its small capacitance is added to the small capacitance of the tuning capacitor. The setting of trimmer capacitor *C*-6*A*, therefore, controls the low-capacitance (high-frequency) end of the tuning range. This trimmer is often called the "high-frequency oscillator aligner."

When tuning capacitor *C*-6 is in a high-capacitance position, the lumped *C* in the tuning circuit is high since it is composed of two comparatively large capacitors in series. Trimmer capacitor *C*-6*A* has little effect in this position since its small capacitance is added to the large capacitance of the tuning capacitor. At this time, the setting of adjustable capacitor *C*-7 becomes of greater importance since its capacitance, now of about the same order as that of the tuning capacitor, will have a greater effect on the lumped *C* in the circuit. The setting of adjustable capacitor *C*-7, therefore, controls the high-capacitance (low-frequency) end of the tuning range. Since this adjustment is usually performed at 600 kc, capacitor *C*-7 is often called the "600 padder."

Cut-plate oscillator tuning capacitors. In some receivers, oscillator tuning capacitor *C*-6 has been designed to maintain the 455-kc difference without a low-frequency padder adjustment. In this case, the rotor plates of capacitor *C*-6 are smaller and differently shaped than the rotor plates of the other capacitors in the tuning gang, as shown in Fig. 16-4. When the oscillator rotor plates are shaped in this manner, the gang capacitor is known as one having a "cut-plate oscillator" section. The shape of the cut plates is so designed that tracking is automatic, in that the capacitance in the oscillator circuit maintains its frequency at a value 455 kc higher than the frequency of the received signal.

Functions and values of parts in the converter. From the above discussion, it can be seen that the values of the component parts in the tuning section of the receiver are an important part of the design of any receiver. The technician rarely, if ever, changes the values of any of these parts, since any such changes will seriously affect the operation of the receiver in selectivity, sensitivity, and dial calibration. Defective com-

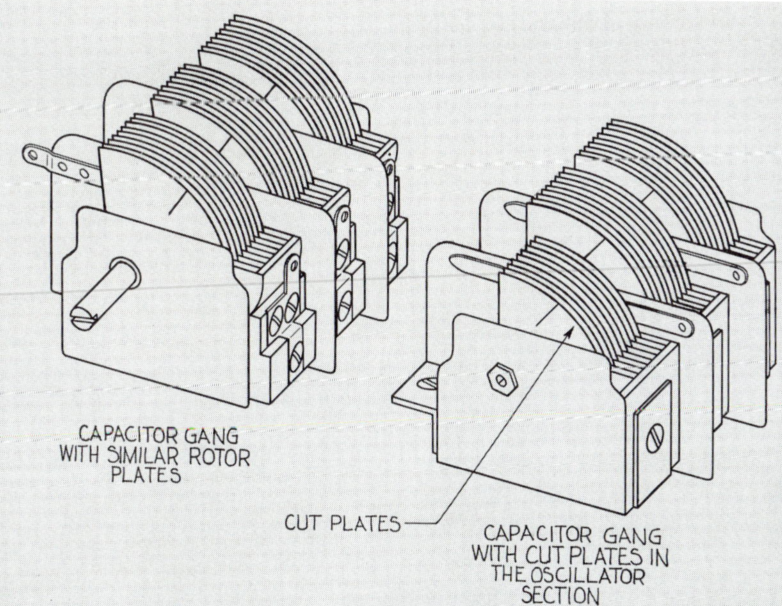

FIG. 16-4. Comparison between capacitor gangs with similar rotor plates and with cut-plate oscillator section.

ponents in the tuning circuit usually require the technician to obtain the original manufacturer's replacement parts. As a result, values of parts need not be given, and it merely remains to state the functions of parts not yet mentioned.

The oscillator grid leak and capacitor, *R*-19 and *C*-19, develop the oscillator grid-bias voltage. When a tube is in an oscillating condition, there is considerable grid current. This flows through *R*-19 and causes a voltage drop across it. The grid end of the resistor is negative, giving the bias voltage for the oscillator section of the tube.

The voltage developed across *R*-19 is also important from a service point of view, since it makes a good check as to whether the oscillator is operating.

Oscillator grid-leak resistor *R*-19 is usually a 22,000-ohm/½-watt resistor. Oscillator grid capacitor *C*-19 is usually a 0.0005-mfd mica or ceramic capacitor. Occasionally, a paper tubular capacitor is used for *C*-19.

The agc decoupling filter (*R*-29 and *C*-29) has been described in the detector and agc stage. Typical values are 100,000 ohms/½-watt for *R*-29, and 0.05 mfd/400 volts for *C*-29. When there is no r-f stage, *R*-29 and *C*-29 are usually omitted, and the signal grid return of coil *L*-5 is connected to the agc bus.

The tube used is the 6SA7 pentagrid converter. Another tube very commonly employed is the miniature 6BE6 pentagrid converter. In this case the circuit is slightly different. A circuit using the 6BE6 will be described later. A-c/d-c receivers use the 12SA7 or 12BE6 pentagrid converters.

The converter-plate-circuit decoupling filter consists of *R*-22 and *C*-22. Resistor *R*-22 is usually 470 to 1,000 ohms, while capacitor *C*-22 is 0.05 to 0.1 mfd. Capacitor *C*-22 also acts as the screen bypass for the pentode section of the converter tube. In receivers where decoupling filters are omitted, the screen bypass function is taken over by the filter capacitor across the *B*+ line.

The input to the mixer stage of the converter is the r-f transformer *T*-2. The primary *L*-4 is in the plate circuit of the r-f tube, or antenna circuit where no r-f stage is used. The secondary *L*-5, which is tuned by *C*-5 of the ganged variable capacitor, feeds the signal to the signal grid of the pentagrid converter tube. In some receiver circuits, r-f transformer *T*-2 is replaced by an untuned resistance-coupled stage.

In receivers that do not use an r-f stage, r-f transformer *T*-2 couples the antenna to the signal grid of the pentagrid converter tube. In this case, *L*-4 the primary of the transformer is connected to the antenna and ground. In loop-operated receivers that do not use an r-f stage, r-f transformer *T*-2 is replaced by the loop antenna. Coil *L*-5 is the main part of the loop, which is still tuned by capacitor *C*-5 in the usual way. Primary coil *L*-4 consists of two or three turns on the loop, which may be connected to an external antenna and ground when it is desired to obtain greater signal pickup for this type of receiver.

NORMAL TEST DATA FOR THE CONVERTER

Signal check for normal operation of the oscillator. When the operation of the oscillator is checked, the test signal is applied to the converter mixer grid (sometimes called the "signal" grid). This is the same point that was used in checking the i-f amplifier. Before the oscillator check is made, any short that had been placed on the oscillator tuning capacitor for previous tests is removed. The receiver is tuned to 600 kc. The signal-generator dial had been set at 455 kc for checking the i-f amplifier. At this position, the modulation note will still be heard in the speaker. The signal-generator dial is then rotated toward 600 kc. As the dial leaves 455 kc, the modulation note should die out, and it should be heard again, at about the same volume as before, when the signal-generator dial pointer passes 600 kc. This is the signal check for normal operation of the oscillator portion of the converter.

If the modulation note is not heard, the oscillator section is inoperative. If the note is considerably weaker than the note at 455 kc, the converter tube is probably weak. If the note is heard when the signal-generator frequency control is at a considerable distance from 600 kc, the oscillator circuit is probably out of alignment.

This check could be performed at any position in the tuning range. However, it is recommended that the check be performed at 600 kc, since oscillator action is normally weaker at the low-frequency end of the tuning range.

Signal check for normal operation of the mixer. When normal operation of the oscillator section has been found, the next step is to check for normal operation of the mixer portion of the converter. The test signal is applied through a 0.00025-mfd capacitor to the control grid of the r-f tube, which is most easily available at the stator connection of the r-f section of the gang tuning capacitor. Where there is no r-f tube, the test signal is applied to the antenna lead of the receiver. The antenna lead is, of course, readily available. Where the receiver has no r-f stage and where there is no antenna terminal on the receiver, mixer coil *L*-5 is in the form of a loop or a ferrite coil, either one of which is acting as the antenna for the receiver. In this case, the output of the signal generator is fed into a loop made up of a few turns of wire. This loop, known as an "injection loop," is then placed near the antenna of the receiver, as shown in Fig. 16-5.

The receiver dial is set to 1,400 kc. If an output meter is connected to the receiver, it should be switched to a high-voltage range. This is important since the amplification from the r-f grid is very high and even a moderate test-signal input may furnish sufficient output voltage to bend the output meter pointer.

The signal generator frequency control is rotated a few points each side of 1,400 kc. When the receiver is functioning normally, the signal generator modulation note should be heard in the speaker as its dial pointer passes 1,400 kc. It should be considerably

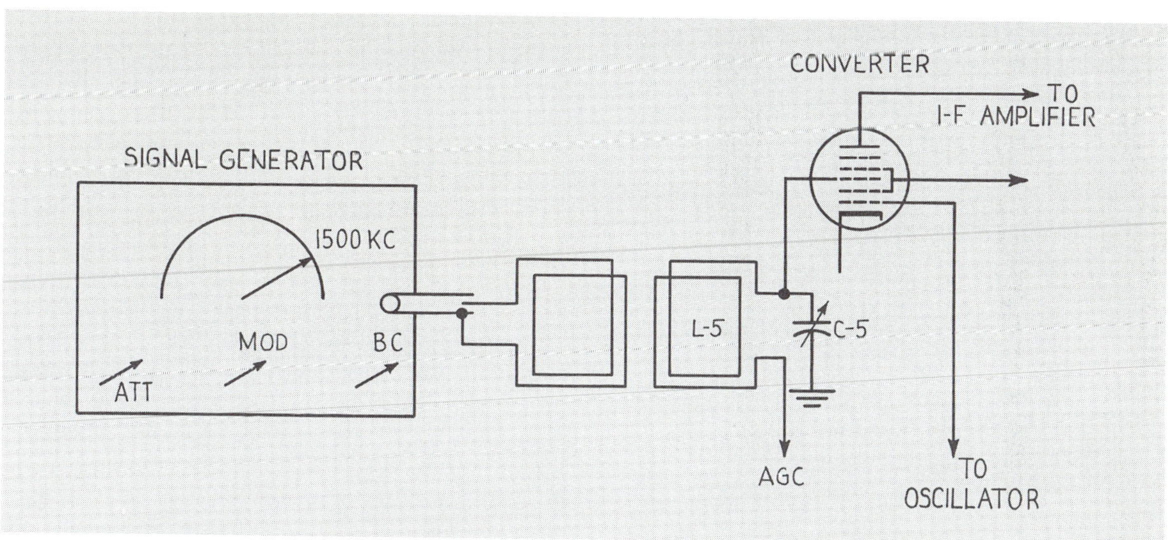

FIG. 16-5. Method of connecting a signal generator to a loop-operated receiver.

louder than the last check (the oscillator section), where the test signal was applied to the mixer grid.

If the signal-generator note is not heard, the mixer section must be checked. The same applies if the check shows no gain over the check from the mixer grid. If the note appears at a considerable distance from 1,400 kc on the signal-generator dial, alignment is indicated.

Normal voltage data for the converter. Readings taken from indicated terminals to the chassis or common negative terminal of the receiver are given in the accompanying table.

	6SA7 or 12SA7 pin No.	A-c receiver, volts	A-c/d-c receiver, volts	6BE6 or 12BE6 pin No.
Plate	3	90–135	90	5
Screen	4	90–135	90	6
Oscillator grid	5	−10	−10	1

Normal resistance data for the converter. Resistance data are given in the following table.

Across *L*-4, primary of the signal input transformer *T*-2	40 ohms
Across *L*-5, secondary of the signal input transformer	5 ohms
Across *L*-6, grid coil of the oscillator transformer *T*-3	5 ohms
Across *L*-7, feedback coil of the oscillator transformer	3 ohms
Signal grid (*G*-3) to chassis	2,700,000 ohms
Oscillator grid (*G*-1) to chassis	22K ohms

In receivers where the signal input transformer *T*-2 is a loop antenna, the grid coil of the loop will measure 1 to 3 ohms. The antenna winding will measure less than 1 ohm.

Sensitivity checks for the converter from the converter signal grid. The technician should run some checks on receivers known to be in perfect operating condition, so that he has a basis of comparative data on his bench test equipment, and normal gain data to be expected from the converter. In addition, he should tabulate his experience with each of the various types of converters.

There are two check points for the converter: the converter signal grid *G*-3, and the r-f grid or antenna, if the receiver does not use an r-f stage. The receiver, signal generator, and the output meter are connected, as shown in Fig. 16-6, to check from the converter signal grid. The receiver is adjusted as follows: The volume control is set to the maximum-volume position; the tone control to the minimum bass position; the selectivity control is set for the position of maximum selectivity; and the receiver dial is adjusted to 600 kc. Any short placed across the oscillator section of the tuning capacitor gang for previous tests should be removed. The output meter is switched to a high-voltage range. The signal-generator output leads are connected shield to chassis, and the "hot" lead through a 0.1-mfd capacitor to the converter signal grid. The signal generator is adjusted to give a modulated signal on the broadcast band. The attenuator setting is kept comparatively low, since approximately 50 microvolts will give standard output from the receiver.

The frequency-control dial on the signal generator is rotated through 600 kc for peak output from the receiver. When the peak position is found, the attenuator on the signal generator is adjusted to give the standard output of 50 mw from the receiver. When the output voltage is low enough, the range switch of the output meter is reduced so that the 16 volts which correspond to 50 mw can be read more accurately.

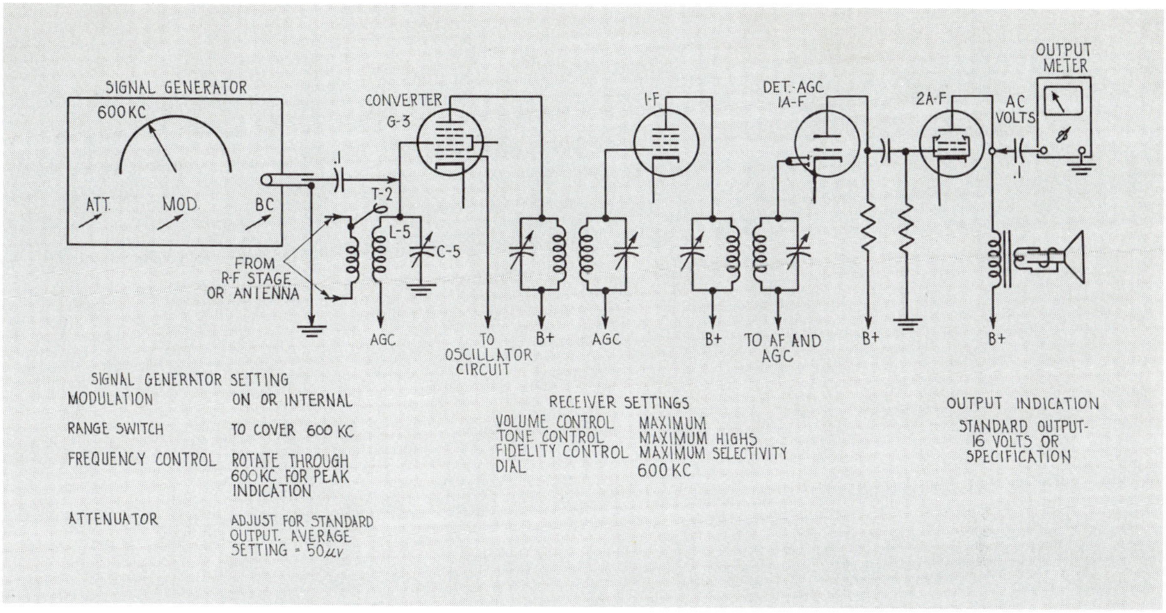

FIG. 16-6. Sensitivity measurements from the mixer grid.

The average 600-kc signal strength necessary to give standard output from the converter signal grid is 50 microvolts. In making sensitivity checks for the i-f amplifier (see Chap. 15), it was seen that the average i-f signal strength necessary to give standard output from the converter signal grid was also 50 microvolts. From the above it may be seen that the gain of the receiver from the converter signal grid should be approximately the same for a signal at the intermediate frequency as for the r-f signal to which the receiver is tuned. Any great difference in signal input (attenuator setting) for standard output would indicate a defective converter tube.

Sensitivity checks for the converter, including the tuned signal grid input. Since the capacitance of the signal generator will detune the converter signal grid circuit, measurements to include the tuned circuit must be made, as was done in all other checks, from a previous point in the receiver. Figure 16-7 shows the connections for a receiver with an r-f stage. Note that the capacitor in the "hot" lead of the signal generator is 0.00025 mfd. When the receiver has no r-f stage, measurements are made from the antenna terminal, as shown in Fig. 16-8. When a loop-operated receiver has no antenna terminal, coupling the signal to it through an injection loop, while satisfactory for signal checks, is unreliable for sensitivity measurements. In either case, the receiver is adjusted for maximum gain; that is, the volume control is set to the full ON position, tone control to the minimum bass, and the selectivity-fidelity control to the position of maximum selectivity. The receiver dial is tuned to 1,500 kc. (If a station is received at this frequency, it will interfere with the check. When this is the case, the receiver is tuned to a quiet part of the dial between 1,400 and 1,600 kc.) The output meter is set for a high-voltage a-c range. The signal generator is adjusted for a modulated output on the broadcast band.

The frequency-control dial on the signal generator is then carefully rotated through 1,400 to 1,600 kc for peak response from the receiver. When the peak position is found, the

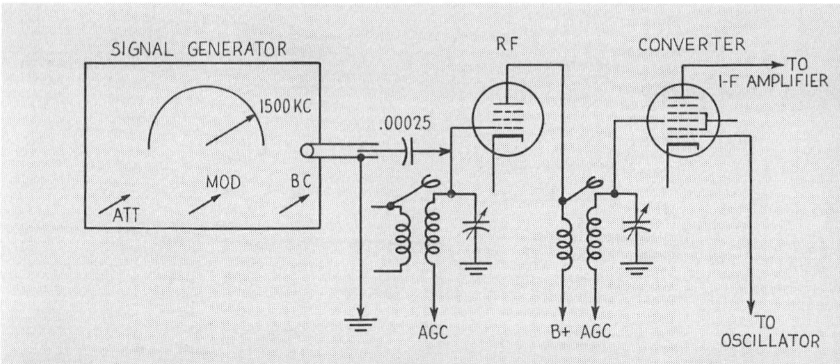

FIG. 16-7. Signal-generator connections for sensitivity measurements of the converter when the receiver incorporates an r-f stage.

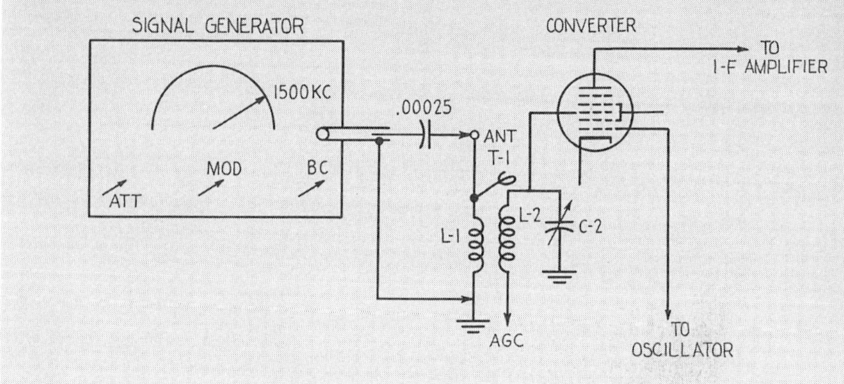

FIG. 16-8. Signal-generator connections for measurements of converter gain when the receiver does not include an r-f stage.

attenuator is adjusted to give the standard output of 50 mw in the speaker. When the standard output has been obtained, the signal-generator frequency control is again adjusted for peak response on the output meter, and the attenuator readjusted if a response greater than 50 mw was obtained. This repetition is necessary because a high-level signal input would bring agc action into play with a consequent broadening of the peak.

The average signal strength at 1,500 kc needed to give the standard output of 50 mw from the antenna of a receiver that does not use an r-f stage is 20 microvolts. When a receiver uses an r-f tube, the average input signal (at 1,500 kc) applied to the r-f grid is 5 microvolts for standard output from the receiver. The added gain is due to the amplification of the r-f tube.

Sensitivity checks at signal levels of 5 or 20 microvolts are unreliable because of leakage in the attenuator circuits, insufficient shielding, and noise pulses. Just look for a reduced attenuator setting from the previous 50-microvolts setting obtained at the converter grid. The reduced setting indicates the gain.

Having established comparative gain data with several good receivers, the technician is in a position to judge the gain characteristics of any converter stage. He should remember, however, that these checks are approximate and that there will be considerable variation shown when different receivers are checked.

COMMON TROUBLES IN THE CONVERTER

Troubles common to the r-f input transformer. The r-f input transformer, *T*-2, is likely to be an interstage r-f transformer coupling the r-f stage to the converter, an antenna coil coupling the antenna to the converter, or a coil loop acting as the antenna for the receiver, depending on the type of receiver. The three types of coupling units all have one common trouble—that is, the windings open—but they present different service problems and will be handled separately.

Service notes for an interstage r-f transformer. An open secondary winding of an interstage r-f transformer will be found on signal check. At such time, when a test signal, either at r-f or i-f, is fed into the converter signal grid, the signal will come through to the speaker, but the gain will probably be low. In addition, the modulation note of the signal generator will have a rough tone due to the open grid circuit. When the test signal is applied to the r-f grid, the response will be very low. The condition is then confirmed with an ohmmeter check.

When the primary of the interstage r-f transformer is open, the receiver will operate normally when the test signal is applied to the converter signal grid but will not operate at all when the test signal is shifted to the r-f grid. A voltage check will then show no voltage at the r-f plate, and a continuity check will confirm the trouble.

Before a defective interstage r-f transformer is replaced, it would be wise to examine the coil, since the break is often at or near a terminal lug and is easily repaired. Even removing a turn to effect a repair is permissible.

An exact replacement of the r-f interstage transformer is necessary, since tuning circuits will not bear wide tolerances. However, at times, the coil is beyond repair, an original replacement cannot be obtained, and a general replacement transformer is the only alternative. In this case, the technician should choose the replacement transformer carefully, so that it matches the original as closely as possible in physical characteristics. The important points to keep in mind are the size of the shield, length and diameter of the coil form, and size and location of the windings.

When the replacement transformer has coded leads, the color coding is the same as for an i-f transformer.

Blue wire	Plate
Red wire	B+
Green wire	Grid
Black wire	Grid return

The placing of the green grid lead can easily be altered to conform to the placing in the original transformer. For example, the replacement transformer has all leads coming out of the bottom and is to be used in an old receiver with a topcap 6A8 tube. Remove the

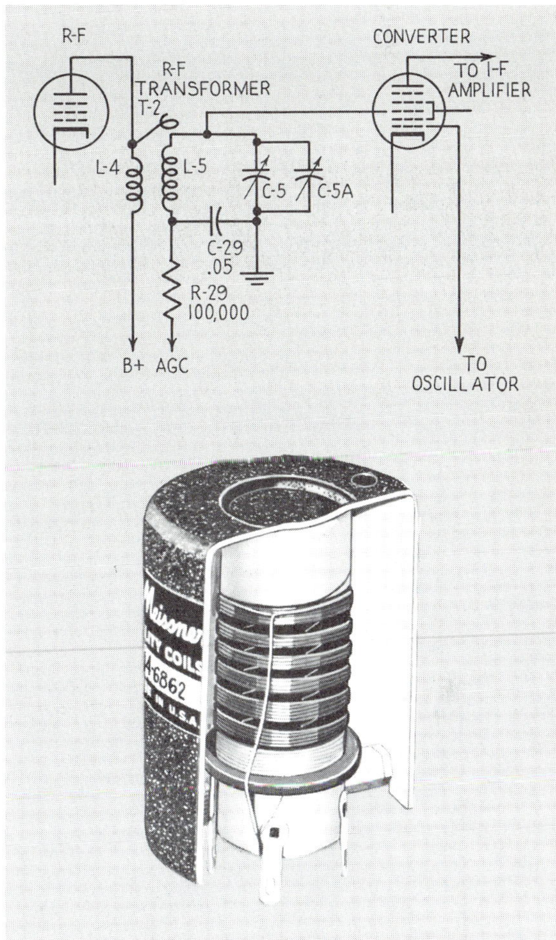

FIG. 16-9. A typical interstage r-f transformer and its position in the circuit.

coil from the can, drill a hole in the top of the can, and reroute the green lead when replacing the coil.

When the replacement-transformer coil leads are brought to unmarked soldering terminals, the terminals can be identified as described in the next section.

How to identify r-f transformer coil leads. For the identification of r-f transformer coil leads, see Fig. 16-10 and the following notes.

Plate lead. Look for the gimmick loop. Trace it to the coil terminal lug. This is the plate lead, which connects to the r-f plate.

B+ lead. Look for the leads on the primary coil. One goes to the plate lead. Trace the other lead to its terminal lug. This is the *B+* lead.

Grid lead. Look for the secondary coil leads. Trace the top end of the secondary winding (near the gimmick loop) to its terminal lug. This is the grid lead, which connects to the tuning capacitor stator and converter signal grid.

Grid return lead. Check to see that the remaining terminal lug goes to the bottom end of the secondary winding. This is the grid return lead, which connects to the agc circuit.

Service notes for an antenna r-f transformer. An open secondary of an antenna r-f transformer will be found by a signal check. The radio will operate at reduced gain and possible hum, when a test signal, either r-f or i-f, is applied to the converter signal grid, and at greatly reduced gain when the test signal is applied to the antenna terminal. An ohmmeter check then confirms the condition.

An open primary winding may or may not cause any appreciable difference in operation. The capacitance of the gimmick loop may transfer sufficient energy from the antenna to the secondary winding, so that operation is apparently normal for local reception, and the trouble would not be found unless sensitivity measurements or routine ohmmeter checks are made. In receivers where the open primary winding causes a large difference in reception, even a rough signal check will show a loss in gain between the antenna and the converter signal grid.

All the service notes pertaining to the r-f transformer can be applied to the antenna transformer, by making allowance for the fact that the primary connects to antenna and ground, instead of r-f plate and *B+*.

Antenna transformer color code. The E.I.A. color code for the antenna transformer follows:

Blue lead	Antenna
Red lead	Ground
Green lead	Grid
Black lead	Grid return

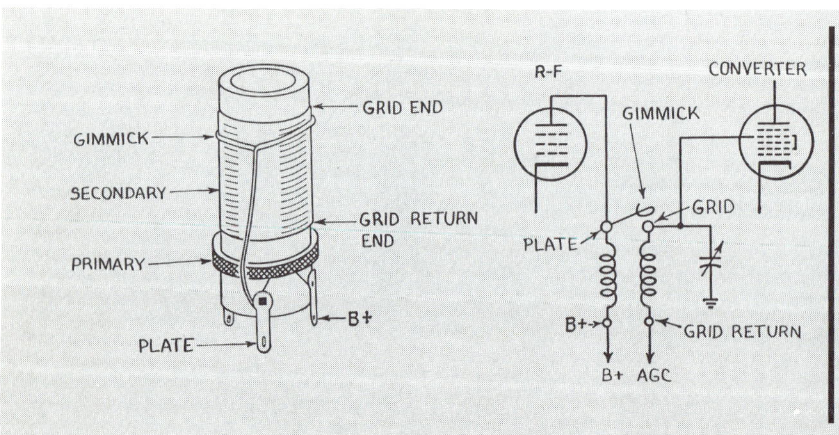

FIG. 16-10. Interstage r-f transformer leads.

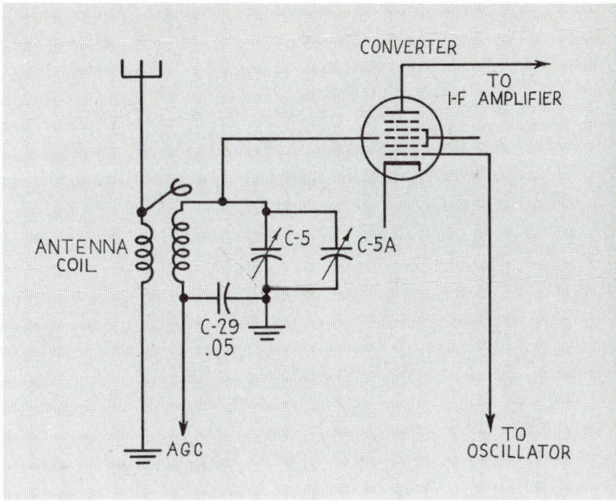

FIG. 16-11. A typical antenna coil and a circuit showing antenna input to the converter tube.

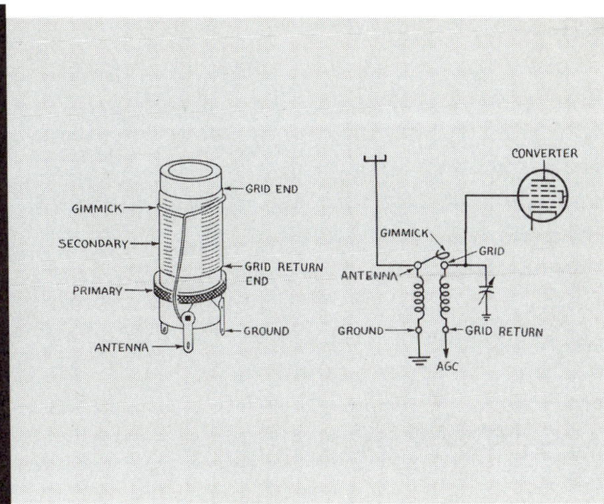

FIG. 16-12. Antenna coil leads.

How to identify antenna transformer leads. For the identification of antenna transformer leads, see Fig. 16-12 and the following notes.

Antenna lead. Look for the gimmick loop. Trace it to the coil terminal lug. This is the antenna lead, which connects to the antenna terminal of the receiver.

Ground lead. Look for the leads on the primary coil. One goes to the antenna terminal. Trace the other lead to its terminal lug. This is the ground lead.

Grid lead. Look for the secondary coil leads. Trace the top end of the secondary winding (near the gimmick loop) to its terminal lug. This is the grid lead, which connects to the tuning capacitor stator and converter signal grid.

Grid return lead. Check to see that the remaining terminal lug goes to the bottom end of the secondary winding. This is the grid return lead, which connects to the agc circuit.

Replacement notes for antenna and r-f transformers. When replacing an antenna or r-f transformer, the technician should be careful of the placement of the leads. Improper lead dress may cause oscillation. The leads should be routed as they were in the original transformer of the receiver. If the wiring has been disturbed, the following general rules should be observed. The blue (plate or antenna) and the green (grid) leads are the "hot" wires. The transformer should be so mounted that the green lead or grid terminal points to its connection point on the tuning capacitor stator or signal grid terminal of the converter tube. At the same time the blue lead (r-f plate or antenna terminal) points to its connection point, the plate terminal of the r-f tube socket or the antenna terminal. The leads are dressed close to the chassis and away from each other and all other wiring. The dress of the other two leads is not quite so important, but they should also be routed close to the chassis and directly to their connection points.

When an antenna transformer is being replaced in an a-c/d-c type of receiver, the transformer antenna terminal connects to the hank of wire that acts as the antenna or leadin of the receiver through a capacitor. The purpose of this capacitor is to insulate the receiver from accidental grounds through the antenna wire. The capacitor is usually a paper tubular type that almost never gives any service difficulties. However, the moving of leads, coincidental with the replacement of the antenna transformer, may have caused one of the capacitor terminal leads to break away from the tin foil of the plates, causing an intermittent or fading condition. It is a good idea, therefore, when replacing an antenna transformer in an a-c/d-c receiver to examine carefully the associated capacitor terminal leads. If they appear to move under the wax, or if a gentle pull causes the receiver to fade, the capacitor should be replaced. The capacitance of the capacitor is unimportant. Any capacitance over 0.002 mfd will be satisfactory.

When the antenna or r-f transformer is replaced, the circuit will have to be realigned as must be done when any component in any tuned circuit is changed. It is usual practice to realign the entire receiver.

When universal adjustable replacement antenna and r-f transformers are employed, it is necessary to alter the standard alignment procedure somewhat, so that the replacement transformer may be adjusted to work properly in the circuit in which it is being placed. The adjustable feature of these coils is permeability tuning with a screw adjustment similar to that used in i-f transformers, so that the inductance of the coil may be varied to suit the receiver. An adjustable replacement coil of this type is shown in Fig. 16-13.

FIG. 16-13. Universal adjustable replacement r-f or antenna coil.

The alignment procedure specified for the receiver being serviced, or the standard alignment procedure given in Chap. 18, is followed down to the adjustment of the oscillator trimmer and padder capacitors. At this point, the receiver dial is correctly calibrated. The hot lead of the signal generator is connected through a 0.00025-mfd capacitor to the antenna terminal of the receiver, the signal generator and receiver dials are both turned to 600 kc, and the permeability adjustment screw of the replacement transformer is tuned for maximum response on the output meter. The receiver and signal-generator dials are then turned to 1,400 kc, and the r-f or antenna trimmers on the gang capacitor are aligned for maximum output in the usual way. The permeability adjustment-screw setting is then checked at 600 kc and, if readjustment is required, the procedure at 1,400 kc is repeated.

Service notes for built-in antennas. Self-contained antennas are of many types. The two types most frequently found are the pancake loop and the ferrite loop or loopstick, both of which are illustrated in Fig. 16-14. The pancake loop is generally mounted on the back panel of the receiver cabinet. With printed wiring construction, the loopstick is generally mounted toward the rear of the printed board. Sometimes this placement leads to undesirable feedback effects, and the loopstick is mounted clear of the receiver on top of the gang tuning capacitor. The loop often includes a separate one- or two-turn winding for connecting an additional external antenna. The coils are usually wound with heavy wire and, as a result, are rarely troubled with corrosion, which is the main cause of trouble in all other coils in the receiver. However, the position of the loop in the back or top of the receiver makes it vulnerable to troubles of a mechanical nature. The leads connecting the coil to the receiver chassis become frayed and broken, various types of plug-in connectors lose contact, and sometimes the coil becomes partly unwound.

An open coil will be found on signal check. The radio will operate at reduced gain and possible hum when a test signal, either at modulated radio frequency or at modulated intermediate frequency, is applied to the converter signal grid. When the test signal is shifted to the antenna by means of an injection loop, the radio may operate at greatly reduced gain or not at all. An ohmmeter check then confirms the condition.

It is rarely necessary to replace the coil. The broken lead or loose contact is found by inspection and repaired. A partly unwound coil is rewound, and the wire is held in place with coil dope.

In the case of the loop, if several leads have broken away, there is likely to be some confusion as to where they should be replaced. The manufacturer's service notes are helpful in this regard, since they often include a wiring diagram of the antenna connections. When this information is not available, the servicing technician should examine the loop antenna to determine whether the primary antenna winding (one or two turns) is on the outside or the inside of the loop winding. After that, the conventional connections for both types are shown in Fig. 16-15. The outside and inside leads are always easily located. The two inner leads may not be so readily distinguished by visual inspection. A continuity check with the ohmmeter, however, will positively identify the inner leads. In the case of the a-c/d-c type of receiver, the technician should remember to check the insulating capacitor, which should be in the antenna or ground lead.

Troubles common to the tuning-capacitor gang assembly. Tuning capacitors usually develop troubles of a mechanical nature

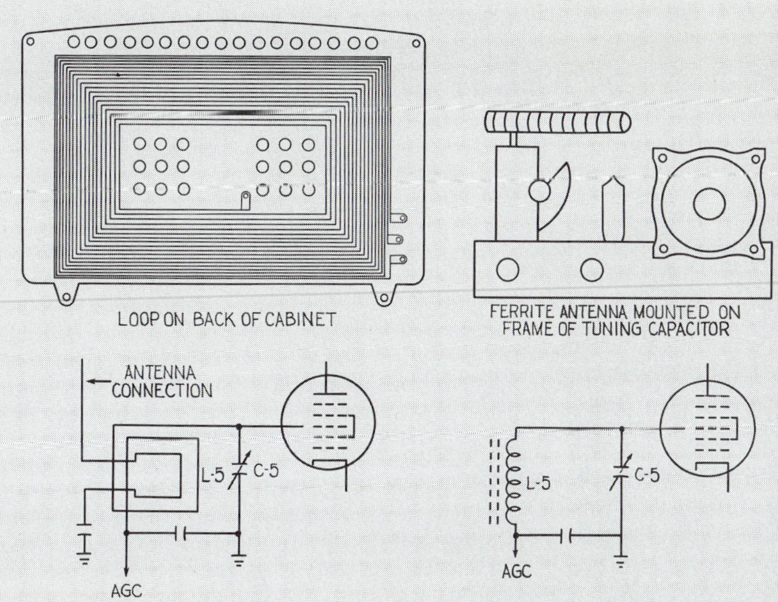

FIG. 16-14. Built-in antennas.

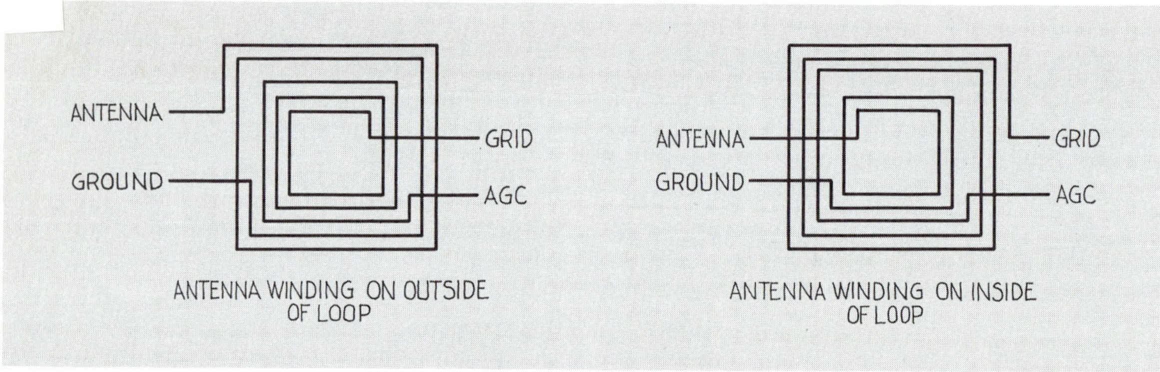

FIG. 16-15. Identifying loop antenna leads.

which may be repaired. Replacement of a tuning gang with anything but the original part would be extremely difficult, since the replacement would have to match the capacitor drive mechanism, and the dial and pointer. Also the plates would have to be so shaped that the dial calibrations would be reasonably accurate; in addition there are the usual considerations of size, capacitance, etc. For this reason, maintenance notes on tuning-capacitor gangs will be in considerable detail.

A very common trouble is slipping or failure of the capacitor and dial drive mechanism. Since there are such a large number of different types of drive assemblies in common use, the information under this heading will be generalized.

Sometimes the drive mechanism operates the dial pointer but the capacitor rotor plates do not turn, resulting in no stations or one station all over the dial scale, depending on the position of the rotor plates. This is usually due to loose setscrews between the capacitor drive and the rotor shaft. The cure is obvious—tightening the setscrew. Before doing so, however, the technician should refer to the receiver service notes to see if there are definite instructions about the positioning of the dial pointer. This information is usually given as part of the alignment instructions. If no reference can be found, the usual procedure is to rotate the gang tuning capacitor until the plates are fully engaged, set the dial pointer to the last calibration mark on the low-frequency end of the dial scale, and tighten the setscrew in this position (see Fig. 16-16).

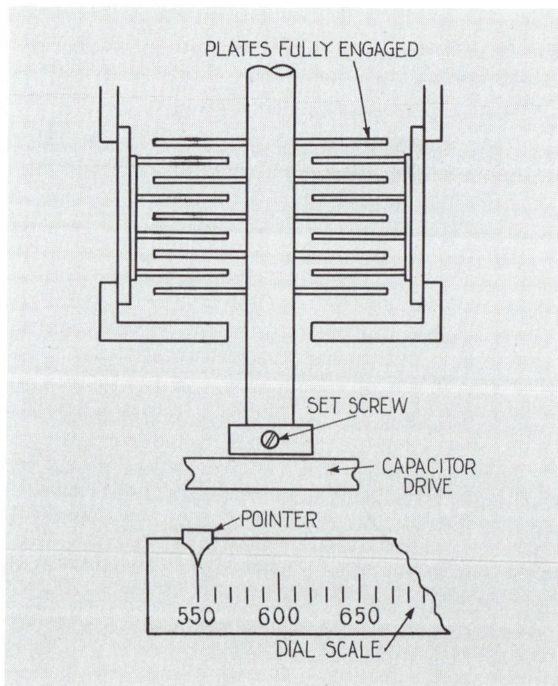

FIG. 16-16. Position of capacitor drive, setscrew, and dial pointer.

The trend in modern receivers is to use silk fish cord for the dial drive mechanism. Fortunately, receiver manufacturers are now issuing instructions for restringing the dial drive cords, as part of their service literature. Where this is not available, the technician must work out the mechanical details for himself. After some experience, a man with average mechanical ingenuity will have little difficulty with any of the multiplicity of dial drives in common use.

Servicing capacitor contact springs. Another common trouble with tuning capacitor gangs develops in the contact springs, often called "wipers." Figure 16-17 shows the location of this item. The wiper makes contact between the rotor plates and the capacitor shields that are grounded. Sometimes a ground wire is soldered to the contact spring, and sometimes no wire is connected. In either case, when dirt gets between the contact spring and the rotor, there will be resistance between the rotor plates and the ground. This may cause noisy reception, and even no reception over parts of the tuning range. In some receivers, poor contact at this point is a cause of oscillation. The cure is to remove the wipers, clean them, readjust the spring tension, and return them to their positions. When the wipers are riveted in place, the spring can be pried back at the point of contact with a screwdriver that has been dipped in carbon tetrachloride. The screwdriver is then removed, and the drop of cleaning solution is worked back and forth by rotating the capacitor gang quickly. When this procedure is repeated a couple of times for each wiper, the contact between the rotor plates and ground is reestablished.

At this time, it might be well to add a word about the general use of cleaning solutions, lubricants, and abrasives in radio service work. Wood alcohol makes a good general-purpose cleaning solution, since it dissolves grease, loosens dirt, dries quickly, and is not harmful to radio parts. Carbon tetrachloride is also used. A light machine oil should be used for lubricating bearings, pulleys, shafts, etc. Tuning-capacitor bearings should not be lubricated, since they are self-lubricating, and any oil at this point may work its way into the capacitor contact springs and insulate the rotor from the ground. Where an abrasive is needed, sandpaper should be used. Steel wool and emery cloth should not be used on or near a receiver. Although steel wool will do a good cleaning job on a capacitor contact spring, particles of it getting into the tuning gang or into the speaker will cause considerable trouble. The abrasive material in emery cloth is also a conductor and will cause similar trou-

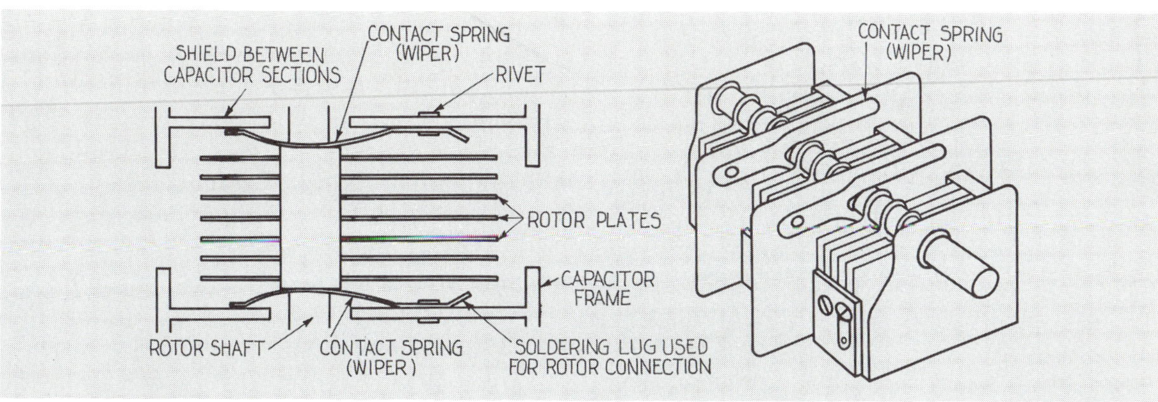

FIG. 16-17. Location of contact springs (wipers) for the rotor plate connection on a variable capacitor.

s. Special-purpose cleaning agents are prepared by some manufacturers and marketed under trade names. "Contactene" is used for switch contacts and capacitor wipers. "Lubriplate" is used where a grease-type lubricant is desired.

Shunt resistance and shorts in variable gang capacitors. Capacitors, especially when not covered by shielding, collect a considerable amount of dust and dirt. When a bakelite stator-plate support or a trimmer capacitor is dusty, the dust will act as a shunt resistor between the stator and the ground. The shunt resistance may cause very little effect on the operation of the antenna and r-f stages, but it can seriously impair the operation of the oscillator. Dusting with a soft brush usually takes care of the trimmer capacitor, and a wash with carbon tetrachloride cleans the stator insulator.

A short between the stator and the rotor plates of any capacitor in the tuning gang will cause noisy reception and dead spots in the tuning range. The short may be due to a number of causes: one or more bent plates (usually in the rotor), shifted stator plate, dirt and dust between the plates, and, in the case of plated capacitor plates, slivers of plating sometimes peel, causing shorts as the capacitor is rotated. A detailed procedure for locating and removing these shorts is given as part of the general overhaul procedure that follows.

General reconditioning procedure for variable gang capacitor.

1. Clean. Blow out the dust by applying a gentle air pressure as from a bicycle pump, to all parts of the variable gang capacitor. Go over the trimmer capacitors and stator supports with a soft brush. Wash the stator supports with carbon tetrachloride. Clean the capacitor contact spring as described previously. Clean and lubricate the dial drive mechanism.

2. Locate and remove any shorts. The cleaning of the trimmer capacitors and stator insulators removed some of the possibilities for shorts and shunt resistance in the variable gang capacitor. There are still dust between plates, bent rotor plates, and slivers of plating to be considered. To find these, remove the wiring from the capacitor stator soldering lugs, and then apply high voltage between the stator connection and the chassis while turning the rotor. The high voltage will show an arc at any shorting position. The arc will probably burn up any dust or sliver of plating that caused the short (thereby automatically removing it) and will show the position of a bent plate. The high voltage is most easily obtained from the rectifier plate terminal. Use a test lead with an alligator clip on one end and an insulated test prod on the other. Clip the alligator to one of the plate leads of the full-wave rectifier (socket terminal 4 or 6 for a 5Y3-G rectifier) and keep the test prod where it can reach the capacitor stator terminals. Switch the set on, touch the test prod to one of the stator lugs momentarily while watching for an arc either in the capacitor or at the stator terminal. Turn the rotor plates in and out of mesh while the prod is connected. If a bent plate is discovered, turn the current off, straighten the plate, and then resume the procedure until all signs of shorts have disappeared. The procedure is then repeated for the other capacitors in the tuning gang.

In this procedure, it must be emphasized that the technician is working with a live lead at 300 or more volts, which is quite dangerous. He should have the current on only when needed, the test lead should be well insulated, and he should exercise care and alertness in his movements. He should also remember that shorting the high-voltage winding may ruin the transformer. That is why the test prod is touched to the capacitor stator *momentarily* for checking the location of a short. It should not be left on a shorted capacitor for any length of time.

When an a-c/d-c receiver is serviced, the transformer high-voltage winding is not

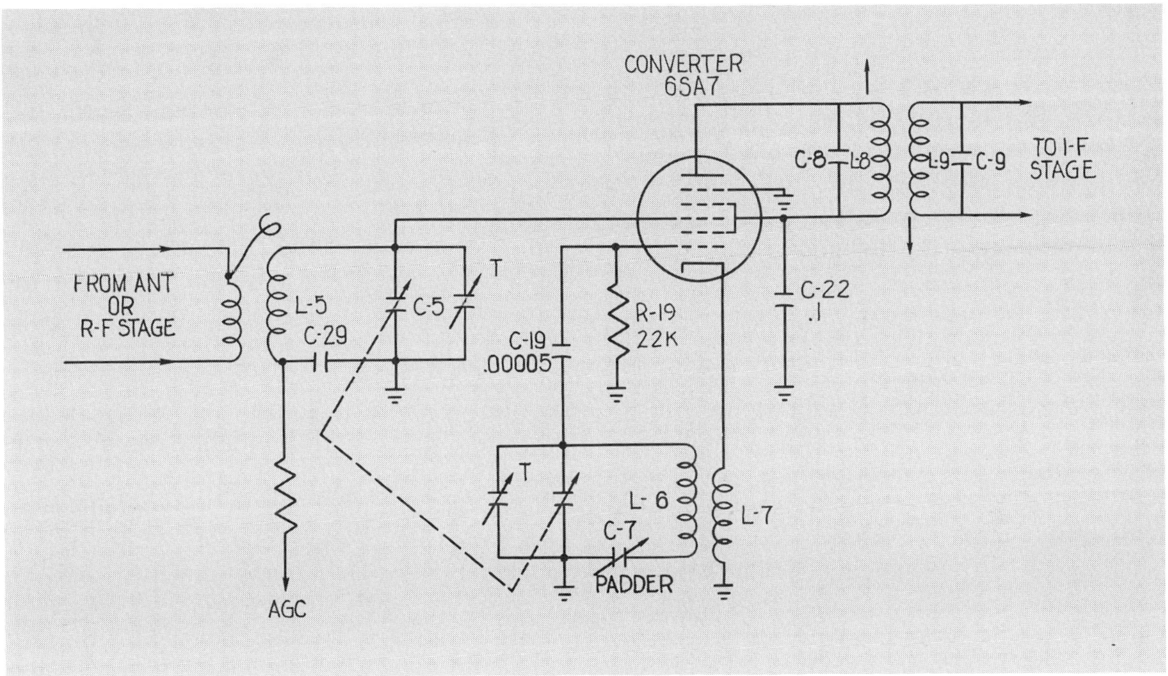

FIG. 16-18. A typical oscillator coil, *L*-6/*L*-7, and its position in the pentagrid converter circuit.

available right on the same chassis. In this case, a separate transformer may be used, connecting one high-voltage lead to the chassis, and the other to the test lead. The same procedure is then followed.

When all shorts have been removed, the stator leads are then resoldered.

3. *Check the dial drive.* The dial drive mechanism is then checked. If it is the cord or belt type and shows signs of wear, replace it. If it is the type that uses a friction rim drive, it will usually respond to a thorough cleaning. Finally check the position of the pointer, as described previously.

Troubles common to the oscillator coil. As with all other coils in the receiver, the main difficulty encountered with oscillator coils is open windings. If either winding opens, the oscillator will not function and the receiver will not pick up any stations. Signal check will isolate the oscillator stage as the field of trouble, since when a modulated test signal is applied to the converter signal grid at the intermediate frequency, there will be normal response from the receiver, whereas when the signal generator test frequency is shifted to radio frequency, there will be no response. An open feedback winding will be indicated in a voltage check, when high voltage appears at the oscillator cathode. An ohmmeter is used for the final check. The ohmmeter check will also show up an open grid winding on the oscillator coil.

If the oscillator coil proves defective, it is usually necessary to obtain an exact replacement of the original, since the oscillator circuit controls the calibration of the dial scale. When an exact replacement is used, it is necessary only to connect the wiring to the new coil without disturbing the lead dress and to realign. It is, of course, necessary to connect the leads correctly, since a reversal of the connections to either winding will reverse the phase of the feedback coil, and the circuit will not oscillate. A good way to make sure that the rewiring is correctly done is to follow the procedure suggested for replacing volume controls. The old coil is loosened with the wiring intact, the new coil is mounted, and the wiring is shifted one wire at a time to its corresponding soldering lug.

When an exact replacement oscillator coil is not obtainable, it is possible to use a universal adjustable replacement coil, where the inductance of the coil may be varied by means of a permeability adjustment screw. When this is done, it is necessary to follow the instruction sheet with reference to identifying the coil terminals, since oscillator coils are rarely color-coded. It is also difficult to determine the ends of the windings, since the windings are usually wax-impregnated and closely coupled.

The replacement coil is mounted in such a way that the oscillator grid and cathode leads are short. The replacement coil is then wired and the receiver is realigned. However, it is necessary to alter the alignment procedure, so that the universal replacement oscillator coil may be adjusted to work properly in the receiver in which it is placed.

Aligning a universal replacement oscillator coil. When the receiver is of the type that uses cut plates in the oscillator section of the variable gang capacitor and there is no 600 padder, the alignment procedure is as follows: First, the i-f transformers are aligned in the usual way. Then the "hot" lead of the signal generator is connected through a 0.00025-mfd capacitor to the antenna, and the generator is adjusted to give a modulated signal at 600 kc. The receiver dial is turned to 600 kc, and the permeability adjustment screw on the replacement oscillator coil is aligned to give maximum response. The signal generator and the receiver dials are shifted to 1,500 kc, and the high-frequency trimmer on the oscillator section of the gang capacitor is adjusted for maximum response. The procedure is repeated at 600 and 1,500 kc for optimum results. The r-f and antenna trimmers are then aligned in accordance with the standard alignment procedure.

When a universal adjustable replacement oscillator coil is placed in a receiver that uses a 600 padder, the alignment procedure is somewhat more involved.

Possibly, it would be best to review the function of each of the adjustments in the oscillator tuning circuit, in the hope that the procedure may become more understandable and usable. This is done in Fig. 16-19. Capacitor *C*-6 is the main tuning capacitor, which is the oscillator section of the gang. The capacitor *T* is the high-frequency trimmer. The capacitor labeled *P* is the series 600 padder. The adjustment screw on the universal replacement oscillator coil controls its inductance and is labeled *L*. It will be remembered that the high-frequency trimmer *T* controls the frequency of the oscillator tuning circuit at the high-frequency end of the dial, and the series padder controls the low-frequency setting of the oscillator tuning circuit. The actions of these controls are not entirely independent, since each will have some effect on the opposite end of the tuning range. This explains why alignment procedures always recommend repeating the setting of these adjustments until they are at their correct positions, as proved by no further need for readjustment.

When the inductance of the oscillator coil is also variable, as is the case when a universal replacement is used, its adjustment will control the frequency of the oscillator circuit all over the tuning range, since this depends on the inductance as well as the lumped capacitance of the circuit. As a corollary, any adjustment of the inductance by means of its permeability screw *L* will necessitate readjustment of the series padder and shunt trimmer. With all three controls variable and dependent upon each other, proper alignment will be extremely difficult, unless a planned procedure is followed closely.

If the 600 padder has been undisturbed,

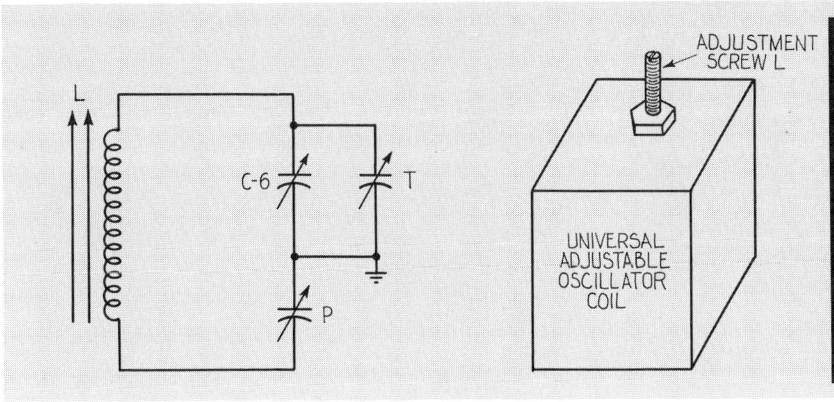

FIG. 16-19. Oscillator tuning-circuit adjustments when a universal adjustment replacement oscillator coil is used: *T*, high-frequency trimmer; *P*, series 600 padder; *C*-6, main tuning capacitor.

one of these variables will be eliminated, since the padder will be close to its correct setting. In this case, the 600 padder is neglected entirely, and the receiver realigned by the procedure just given for a circuit that uses cut plates in the oscillator section of the gang capacitor, and no 600 padding adjustment.

If the technician is not sure of the setting of the 600 padder, two alignment procedures may be used. In the first, the settings of the three adjustments are first made roughly and, then by repeated readjustments, are brought to their final positions. This procedure, although simple, is not always operative, owing to varying circuit constants and limited trimmer ranges in many receivers. The second procedure is more difficult, but it is always successful. In it, the alignment of the receiver is carried out at several prefixed positions of the 600 padder, each one is checked, and finally the position of best tracking is chosen.

Alignment procedure no. 1 for an oscillator circuit with variable trimmer, padder, and inductance

1. *Check i-f alignment.*

2. *Connect the "hot" lead of the signal generator to the antenna terminal of the receiver through a 0.00025-mfd capacitor.* Adjust the signal generator for a modulated output on the broadcast band.

3. *Set the trimmer and padder to center-capacitance range.* The average trimmer and padder capacitors require three full turns of the adjustment screw from full to low capacitance. For approximate center-capacitance setting, tighten the screws fully, then loosen one complete turn.

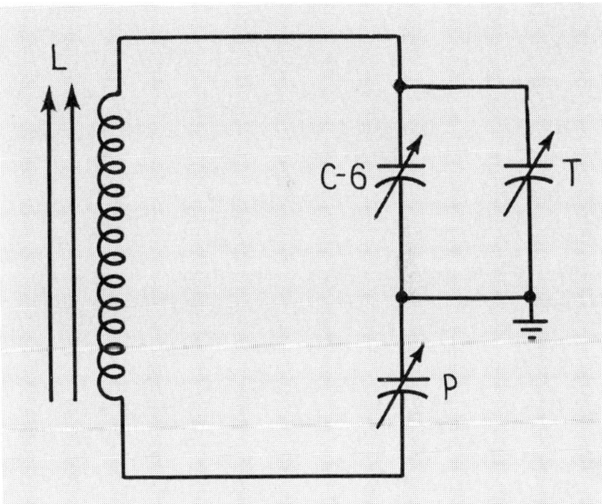

FIG. 16-20. Oscillator tuning-circuit adjustments.

4. *Adjust L at 1,000 kc.* Tune the signal generator and receiver to 1,000 kc, and adjust the permeability screw *L* on the oscillator coil for maximum output.*

* If the signal cannot be tuned in, the adjustment range is not large enough or the first rough setting for center capacitance is too far from the correct setting. Try the second alignment procedure.

5. *Adjust P at 600 kc.* Tune the signal generator and receiver to 600 kc, and adjust the 600 padder *P* for maximum output.*

6. *Adjust T at 1,500 kc.* Tune the signal generator and receiver to 1,500 kc, and adjust the high-frequency trimmer *T* for maximum output.*

7. *Repeat steps 4, 5, and 6 in sequence until each screw requires no further readjustment.*

8. *Align the r-f and antenna trimmers in the usual way.*

Alignment procedure no. 2 for an oscillator circuit with variable trimmer, padder, and inductance

1. *Check i-f alignment.*

2. *Connect the "hot" lead of the signal generator to the antenna terminal of the receiver through a 0.00025-mfd capacitor.* Adjust the signal generator for a modulated output on the broadcast band.

3. *Set P for minimum capacitance.* Examine the 600 padder and the action of its adjustment screw. Set the adjustment screw in the position where the plates begin to move together. This is the low-capacitance setting of the 600 padder.

4. *Adjust L at 600 kc.* Tune the receiver and signal generator to 600 kc, and adjust the permeability screw on the oscillator coil for maximum response. If the signal cannot be heard over the range of this adjustment, increase the capacitance setting of the 600 padder by a quarter turn and try again. Repeat this until the signal-generator note can be heard.

If the receiver does not have an r-f stage, and the signal-generator attenuator is well advanced, there is a possibility of tuning the oscillator stage to the second harmonic of the 600-kc signal. To make sure that this error does not spoil the alignment when the 600-kc note is first heard, tune the signal generator to 1,200 kc. If the signal is not heard at this point, *L* is correctly adjusted for 600 kc. If the signal is heard and with a stronger note, *L* has been adjusted for 1,200 kc. More inductance and probably more capacitance are needed in the circuit.

5. *Adjust T at 1,500 kc.* Tune the receiver and signal generator to 1,500 kc, and adjust the high-frequency trimmer on the oscillator section of the gang capacitor for maximum response. If the test signal cannot be heard, increase the capacitance of the padder *P* by an eighth turn of its adjustment screw. Then readjust *L* at 600 kc and try again to adjust *T* at 1,500 kc. Repeat this until the signal-generator note can be heard. The receiver is now tracking at 600 and 1,500 kc.

6. *Measure sensitivity at 1,000 kc.* Tune the receiver to 1,000 kc. Rotate the signal

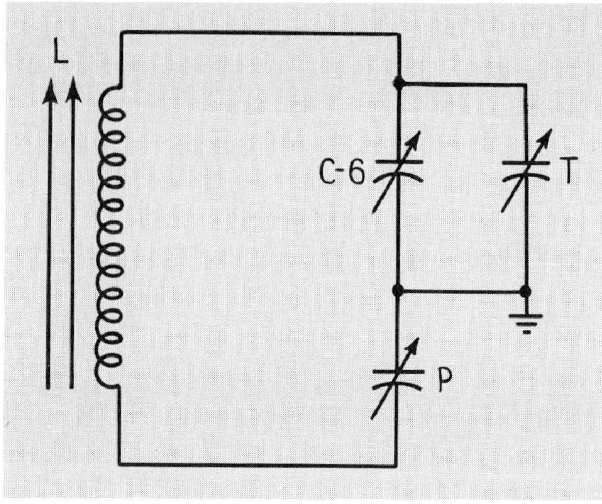

FIG. 16-21. Oscillator tuning-circuit adjustments.

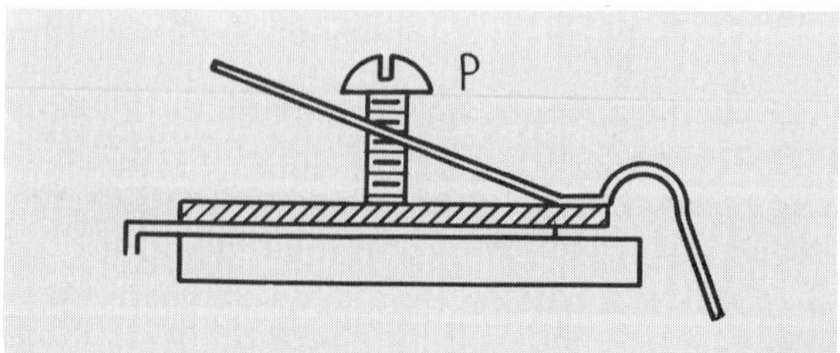

FIG. 16-22. Low-capacitance setting for padder *P*.

* If the signal cannot be tuned in, the adjustment range is not large enough or the first rough setting for center capacitance is too far from the correct setting. Try the second alignment procedure.

generator through 1,000 kc, while watching the output meter for maximum deflection. At peak response adjust the attenuator for standard output. Note the attenuator setting. This gives the sensitivity of the receiver at 1,000 kc.

7. *Adjust for maximum sensitivity at 1,000 kc.* Tighten the padder another eighth turn. Adjust *L* for maximum response at 600 kc. Adjust *T* for maximum response at 1,500 kc. Measure sensitivity at 1,000 kc. Note the reading. The receiver should show an improvement in the reading over the reading taken in step 6.

Repeat with another eighth turn on *P*. Readjust *L* at 600 kc and *T* at 1,500 kc, and measure sensitivity at 1,000 kc. Continue until the sensitivity decreases. The previous adjustment of the 600 padder was the correct one. Loosen *P* an eighth turn and complete the alignment.

Troubles common to the pentagrid converter tube. The converter tube is a common cause of trouble in the stage. Tube checkers are not very reliable in indicating an inoperative tube, since the tube may show adequate emission but still not oscillate. If the signal check shows normal response when a test signal at the intermediate frequency is applied to the converter signal grid, but no response or weak response when the test signal is shifted to 600 kc, there is a sure indication that the oscillator is not functioning. The most probable reason is the tube. The best check is to substitute another similar tube that is known to be good.

Another trouble often experienced with pentagrid converters is modulation hum, caused by cathode-to-heater leakage. Again the best check is substituting a similar tube known to be good.

Sometimes a receiver is encountered where conditions for maintaining oscillations are critical, and the oscillator circuit will not operate over the entire tuning range. Substituting another pentagrid converter tube usually clears this up. The original tube may not be defective and may operate perfectly in another receiver. This matter is treated in greater detail in the next section.

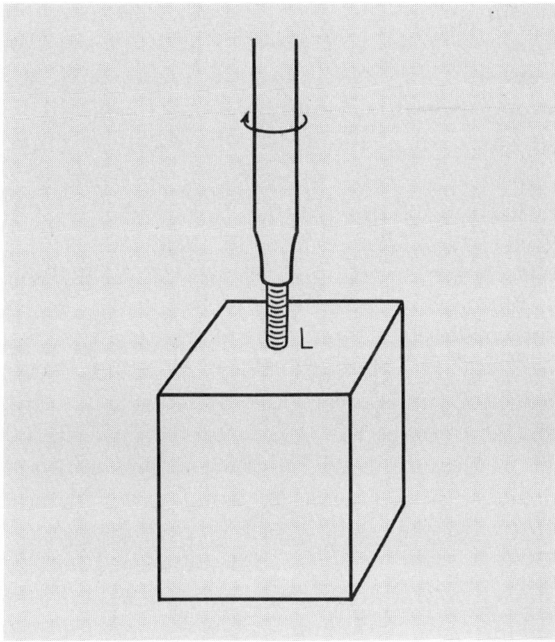

FIG. 16-23. Adjustment *L* at 600 kc.

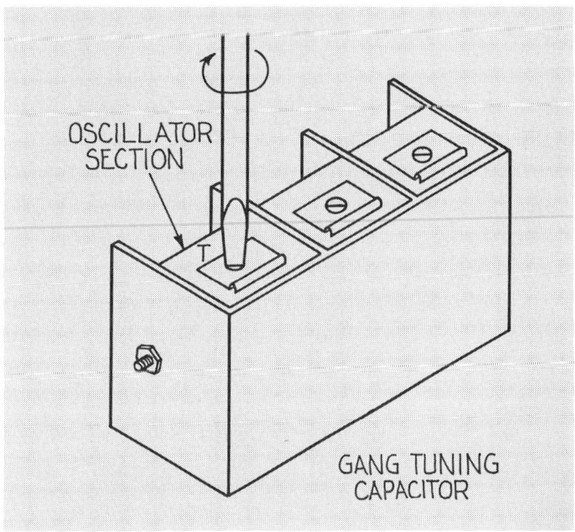

FIG. 16-24. Adjusting the high-frequency trimmer *T*.

Critical oscillator conditions. Superheterodyne receivers sometimes develop a peculiar trouble. Reception is normal on the high-frequency end of the tuning range, erratic at the middle frequencies, and dead on the low-frequency end. Such a condition could be caused by shorts in the gang tuning capacitor; more often it is due to failure of the oscillator at the low-frequency end of the tuning range. Which of the two possibilities is responsible can be quickly determined by the following procedure: Start at the low-frequency end of the tuning range, and tune toward the high-frequency end, noting the frequency of the first station received. Let us assume that it is at 1,100 kc. Then starting at the high 1,600-kc end, tune toward the 540-kc end, noting the stations as they are passed. If the 1,100-kc station comes in, followed by stations at 1,000 and 900 kc, and no stations after that, the trouble is sure to be in the oscillator circuit. The stations at 1,000 and 900 kc cannot be tuned in unless the radio is being tuned from high to low frequencies.

It is normal for oscillator operation to be more efficient at high frequencies than at low. Then if we assume, for example, an oscillator tube with weak electron emission, it may oscillate at the high-frequency end of the tuning range, but not at the low. Also, when the circuit is in an oscillating condition, the oscillation may continue as the operating frequency is reduced beyond the point of a normally nonoscillating position. Such operation might be called "critical oscillating condition."

A condition of critical oscillator operation may be caused by other factors than a weak tube. The tube, however, is the most easily checked, since we can substitute another that is known to be good. If the new tube does not entirely clear up the trouble but causes the oscillation to stop at a lower frequency than before, it may be advisable to try still another tube, with the hope of finding one that will continue to oscillate all over the tuning range.

At this point, it might be well to add that a condition of oscillation is easily determined by a check of the voltage between the oscillator grid and the chassis. When the circuit is oscillating, the oscillator grid voltage will be negative with respect to chassis. When oscillation stops, the oscillator grid will check zero or slightly positive.

When replacement of the tube fails to clear up the trouble, all components of the oscillator circuit should be carefully checked. This includes cleaning the oscillator section of the gang tuning capacitor, since a dusty shunt across the oscillator tank may be the cause of the condition.

If all components seem to be in good condition, refer to the receiver manufacturer's service notes, to see if later changes incorporated in the receiver include any change in the oscillator circuit. Often the condition is widespread for a particular receiver, and later changes include remedial measures. The change may be a different value for the oscillator grid resistor; or, the oscillator coil may have been changed, as indicated by a new part number. If any such alterations can be found, incorporating this same change in the receiver being serviced will clear up the difficulty.

Miscellaneous oscillator troubles. When the signal check indicates trouble in the oscillator section, any of the component parts might be at fault. The tube, oscillator coil, and tuning capacitors, which are the most common offenders, have been covered in previous sections. The resistors and capacitors in the oscillator grid and anode circuits remain as possible sources of faulty operation.

In the oscillator grid circuit, resistor *R*-19 and capacitor *C*-19 rarely give any servicing trouble. Occasionally, a paper tubular or ceramic grid capacitor may cause oscillator trouble because of leakage. When replacing capacitor *C*-19, use a mica capacitor of the proper size. In the oscillator plate circuit, the decoupling filter made up of resistor *R*-22 and capacitor *C*-22 is more often the cause of

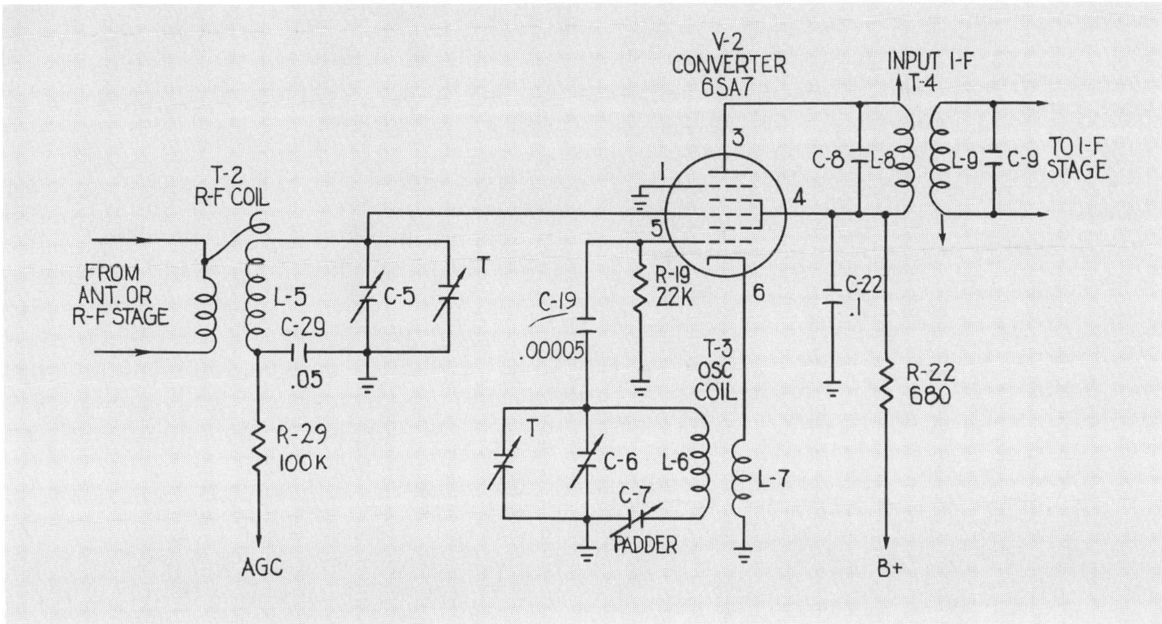

FIG. 16-25. Basic pentagrid converter circuit.

troubles. However, filter difficulties would have been found much earlier in the servicing procedure. A short in capacitor *C*-22 would cause a heavy overload on the *B*+ line and would be discovered after a routine voltage check of the power supply. If the capacitor were to open, the result would be a squeal from the receiver. The faulty capacitor would be found when the squeal stops when a test capacitor is connected across it, a standard servicing procedure for squeals. An open in resistor *R*-22 would make the converter section inoperative. The condition would be found when checking the i-f stage. An i-f test signal applied to the mixer grid would give no response. The two most likely causes are a dead converter tube and failure of the voltage supply to the tube. Then voltage analysis would show no voltage at the converter plate or screen, and a resistance check would confirm the open in resistor *R*-22.

CONVERTER VARIATIONS

There are many variations in the converter section. These are concerned with different tubes, different oscillator circuits, and different components. However, the functions of the stages remain the same so that the signal checks may be applied with little or no adaptation. The variations to be described are those commonly found or those requiring a special servicing technique.

Converter using a 12BE6 tube in an a-c/d-c printed-wiring receiver. The schematic diagram of a converter circuit using a 12BE6 tube is shown in Fig. 16-26. The circuit was taken from an RCA a-c/d-c receiver, as a typical example. The complete diagram of the receiver is given in Fig. 16-27.

The 12BE6 tube is the miniature equivalent of the 6SA7 tube of the basic circuit. Suppressor grid *G*-5 of the mixer portion of the 12BE6 is internally connected to the cathode. This makes it slightly different from the 6SA7 where suppressor grid *G*-5 is externally connected to the chassis. From the servicing point of view, this connection causes no complications and all signal and voltage checks given for the standard circuit

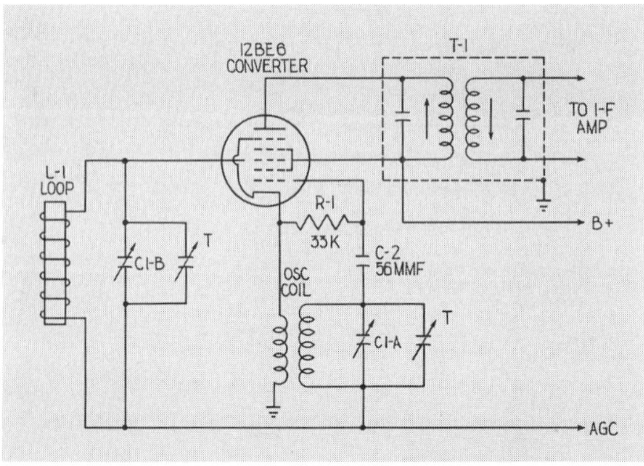

FIG. 16-26. Converter circuit using a 12BE6 tube.

may be applied. This is especially true when a 6BE6 tube is used in a transformer-type a-c receiver on a metal chassis.

This RCA receiver was selected also to point up the differences when working with a-c/d-c printed-board receivers. Like most a-c/d-c receivers, the circuit is a five-tube superheterodyne set with no r-f stage. Decoupling filters are at a minimum. Note that the *B*+ line goes directly to all plate and screen circuits, so that output filter capacitor *C*-10*A* is the only filter on the *B*+ line. The receiver shows one decoupler, *R*-13/*C*-13, in the agc feed line to the i-f stage. Most sets of this type omit even this decoupler.

The ground symbol in the schematic diagram does not indicate a chassis connection. This is because there is no metal chassis in this printed-board receiver. The ground symbol shown indicates the common negative wiring, which in the a-c/d-c circuit is one side of the power line. Because of the insulated-board mounting, the rotors of the tuning capacitors are not automatically grounded to the chassis. Therefore, the tuning capacitors may be connected directly across the coils in the tuning circuits without the need for intervening capacitors. The frame of the tuning capacitor is therefore connected to the agc bus. The agc bus is also wired to the mounting bracket for the volume control, so that both receiver control shafts are floated from the power line by agc bypass capacitor *C*-3.

When signal checks are made on receivers of this type, r-f and i-f signals are fed to the converter signal grid by means of an injection loop placed near the ferrite-coil antenna *L*-1.

FIG. 16-27. Schematic diagram of an RCA a-c/d-c receiver using a 12BE6 converter tube.

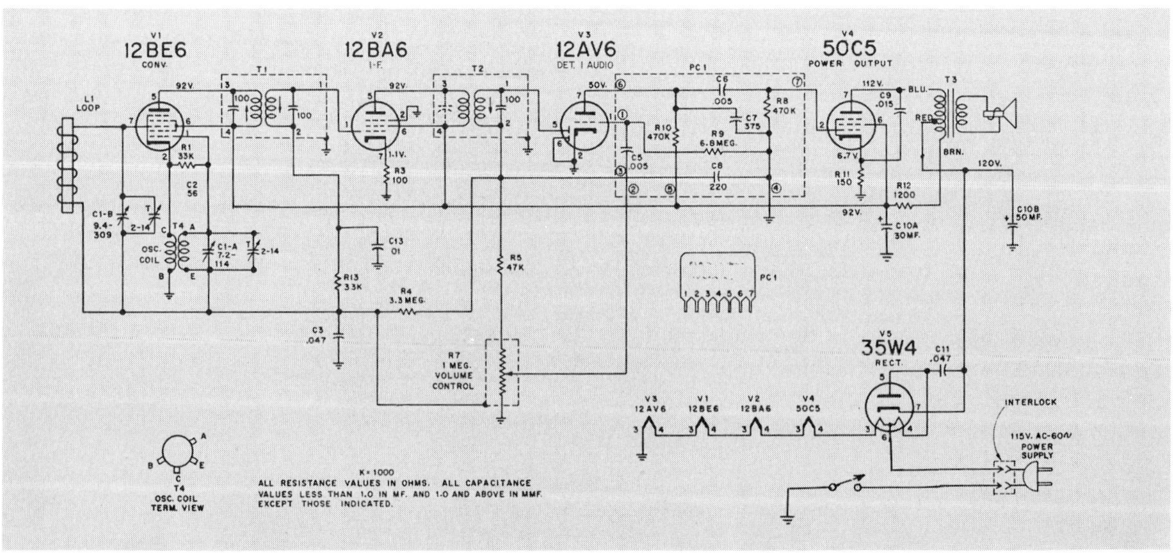

This method is also satisfactory for alignment purposes. When it is desired to make sensitivity checks or to feed signals to other parts of the receiver, the preferred method is to connect a 0.1-mfd capacitor in series with the grounded lead of the signal generator, and to connect it to common negative. Manufacturers' service notes recommend that the grounded lead of the signal generator should not be connected to the agc bus at either control shaft.

If the common negative connection creates a hum in the receiver, try reversing the receiver line plug. If the hum persists, an isolation transformer is necessary. The hot lead of the signal generator is then connected through its usual dummy antenna to the desired test point.

Converter circuits using a tapped oscillator coil. A common variation of the converter circuit uses a tapped coil in the oscillator circuit. Refer to Fig. 16-28. Oscillator coil *L*-6 is a tapped winding rather than the two-

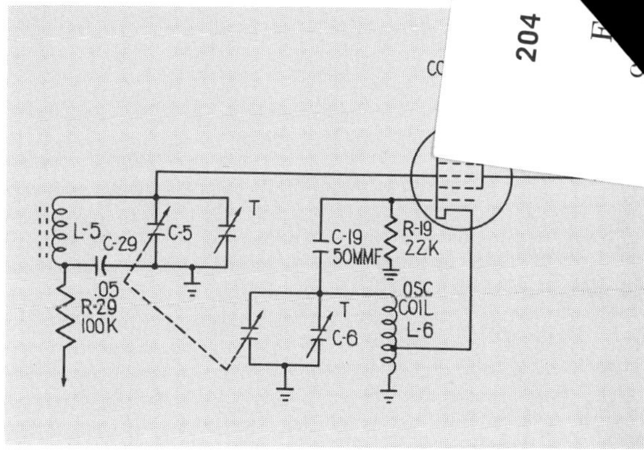

FIG. 16-28. Converter using a tapped oscillator coil in a Hartley oscillator circuit.

winding coil used in the standard circuit. The lower portion of the coil is the feedback winding between cathode and *B*– or ground. The entire coil is in the oscillator grid circuit. This type of circuit is called a Hartley oscillator. The schematic diagram of an Admiral clock radio is given in Fig. 16-29 as a typical example of this type of circuit.

FIG. 16-29. Schematic diagram of an Admiral receiver using a tapped oscillator coil.

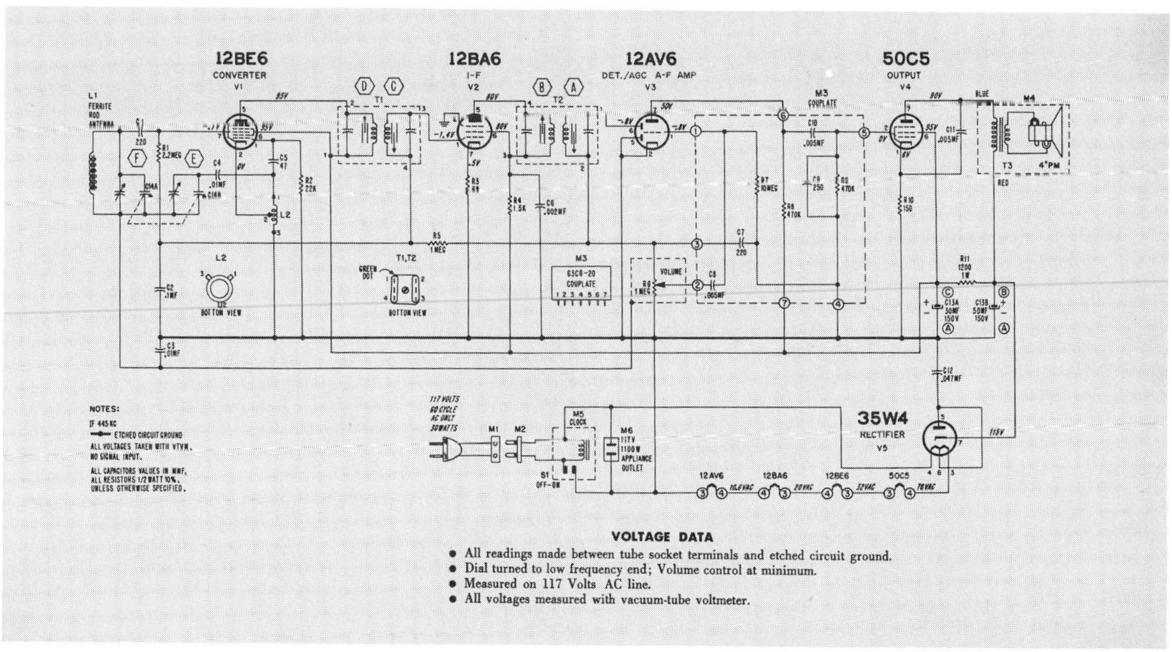

rom the servicing point of view, the receiver responds to signal checks in the same way as the standard circuit. Voltage checks are also the same. Resistance checks will differ because of the oscillator coil. The coil is not tapped at the center. The feedback winding generally measures 1 to 2 ohms, and the upper winding 4 or 5 ohms.

If the coil should become defective and require replacement, an exact replacement must be used. If this is impossible, general replacement tapped oscillator coils for use in Hartley circuits are listed in parts catalogues of supply companies, and may be used successfully.

Oscillator coils with a capacitive winding. Many receivers use a capacitive winding in the oscillator coil to replace the oscillator grid capacitor. A converter circuit with a winding of this type is shown in Fig. 16-30. The oscillator coil shown would normally have four leads for the two windings. The capacitive winding or gimmick places a fifth lead on the coil as indicated in the diagram. Capacitive windings are also used with the tapped coils of the Hartley oscillator circuit.

A capacitive winding, when used, adds no service problems. All signal, voltage, and resistance checks may be used without change. And, the capacitive winding itself is rarely the cause of trouble in the receiver. It becomes a problem when the oscillator coil becomes defective and requires replacement. If an exact replacement cannot be obtained, it is difficult to obtain a general replacement part that contains the capacitive winding. It now becomes necessary to use a tapped winding or two-winding general replacement oscillator coil, whichever the circuit calls for, and to replace the capacitive winding with an external mica or ceramic capacitor of 50 micromicrofarads (mmfd) as is done in the standard circuit.

Converters with separate mixer and oscillator tubes. Some receivers, particularly those meant to operate at high frequencies (f-m, TV, and short-wave receivers), use separate tubes for the mixer and oscillator functions of the converter. A basic circuit is shown in Fig. 16-31.

The mixer uses an r-f pentode like the 6AU6 in a conventional r-f amplifier circuit. Signal energy is fed to the mixer grid through the *L*-5/*C*-5 tuning circuit. The oscillator uses a triode like the 6C4. The Hartley oscillator shown in the diagram is often found in high-frequency applications. Energy from the oscillator is fed to the mixer grid through capacitor *C*-7. This is a small silver-mica capacitor of about 5 mmfd.

Note that, except for the use of two tubes and the addition of capacitor *C*-7, the circuit is almost the same as a standard pentagrid converter. The signal checks will be the same. With regard to the voltage check, the oscillator plate will be at the same *B*+ potential as the mixer screen and plate. The oscillator grid will be −10 volts, the same as oscillator grid *G*-1 in the pentagrid converter. All components have the same functions as in the standard circuit and are subject to the same troubles. The additional component, capacitor *C*-7, rarely gives any service difficulty.

Converter using a pentode tube. The pentagrid converter combines the twin functions of mixer and oscillator in one tube with five grids. In some receivers, the same result is obtained using an r-f pentode with three grids. Such a circuit is shown in Fig. 16-32.

The tube that is used is a 6AU6 or a 12AU6 r-f pentode. Oscillator coil assembly *T*-3 has three windings. Coil *L*-7 is in the cathode circuit. Coil *L*-17, in series with the plate circuit before the *B*+ bus, provides feedback. Coil *L*-6 and tuning capacitor *C*-6 form the oscillator tank which tunes the oscillator circuit to the proper frequency. Energy from the oscillator circuit is brought to the control grid by capacitor *C*-19, and resistor *R*-19 functions as the oscillator grid leak. Signal energy is also fed to the control grid through capacitor *C*-5A. This circuit is used in small a-c/d-c receivers and as the a-m converter in some a-m/f-m receivers.

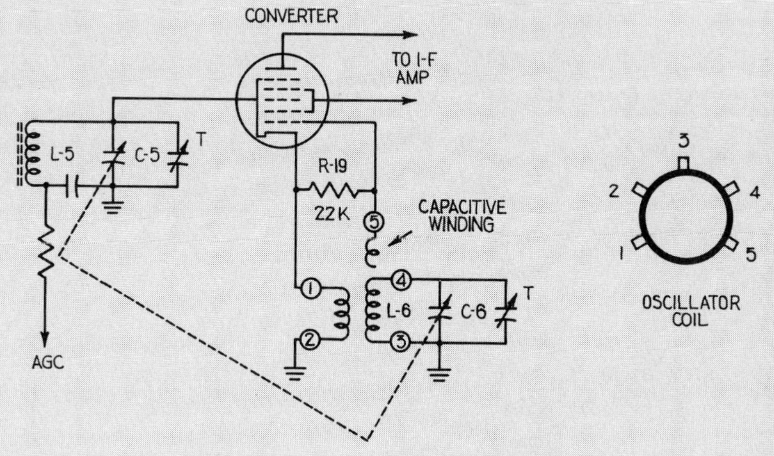

FIG. 16-30. Oscillator coil with a capacitive winding and its connections in a converter circuit.

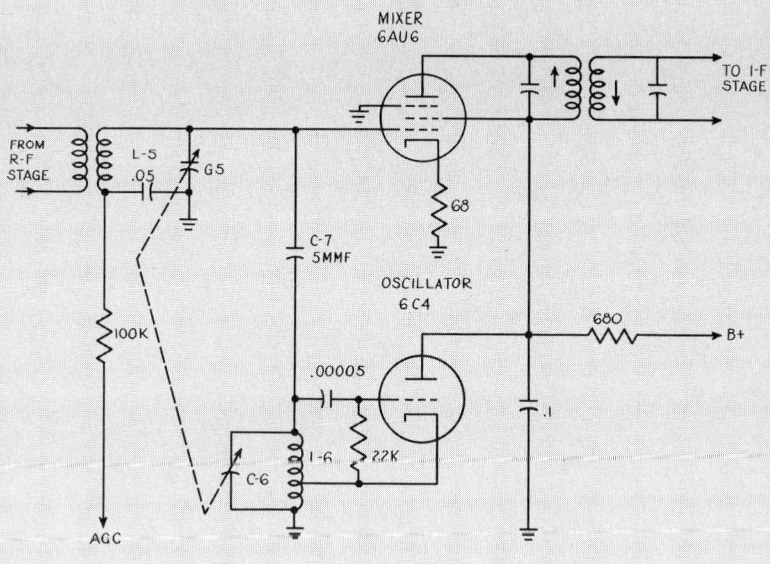

FIG. 16-31. Converter with separate mixer and oscillator tubes.

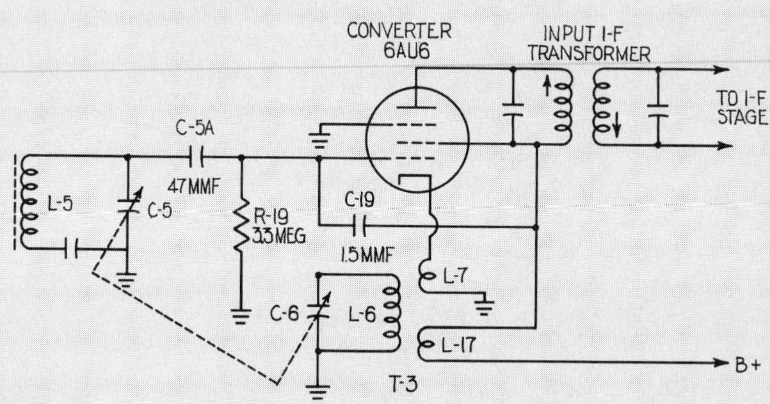

FIG. 16-32. R-f pentode as a converter.

In servicing a converter of this type, signal checks may be applied in the usual way. Normal voltage readings will be somewhat different. Plate and screen voltages should read the *B*+ value for the receiver. The control grid should have a negative reading of 4 to 6 volts. If the signal check indicates oscillator malfunction, the absence of this negative reading confirms that the trouble is is in the oscillator. Resistance data for the oscillator coil assembly are as follows: 1 to 2 ohms for windings *L*-7 and *L*-17, and approximately 6 ohms for winding *L*-6. If the oscillator coil assembly should become defective, an exact replacement will be required.

Multiband receivers. In multiband receivers, the inductances in the tunable circuits are switched, so that the receiver may operate over more than one band of frequencies. A simplified circuit indicating how this may be accomplished is shown in Fig. 16-33.

The gang tuning capacitors *C*-5 and *C*-6 are permanently connected across the signal grid and oscillator grid circuits. Switch *S*-2 is the wave-band switch. In the broadcast (*BC*) position shown, coil *L*-5 with its trimmer *C*-5*A* is connected in the signal grid circuit. In the oscillator grid circuit, for broadcast, the *B* section of the switch throws coil *L*-6 with its low-frequency padder *C*-7 and its high-frequency trimmer *C*-6*A* across the main oscillator tuning capacitor *C*-6.

When the switch is thrown to the short-wave (*SW*) position, coil *L*-105 with its associated trimmer *C*-105*A* is connected in the signal grid circuit, while coil *L*-106 with its associated low-frequency padder *C*-107*A* and high-frequency trimmer *C*-106*A* is connected across the oscillator tuning capacitor *C*-6. An arrangement such as the above makes it possible to align each wave-band position individually.

By using more positions on the wave-band switch, each one throwing in a different set of coils with their associated trimmers, it is possible to have an all-wave or communications receiver. The common bands used for radio receivers are as given in the accompanying table.

The *B* or police band has either of the two

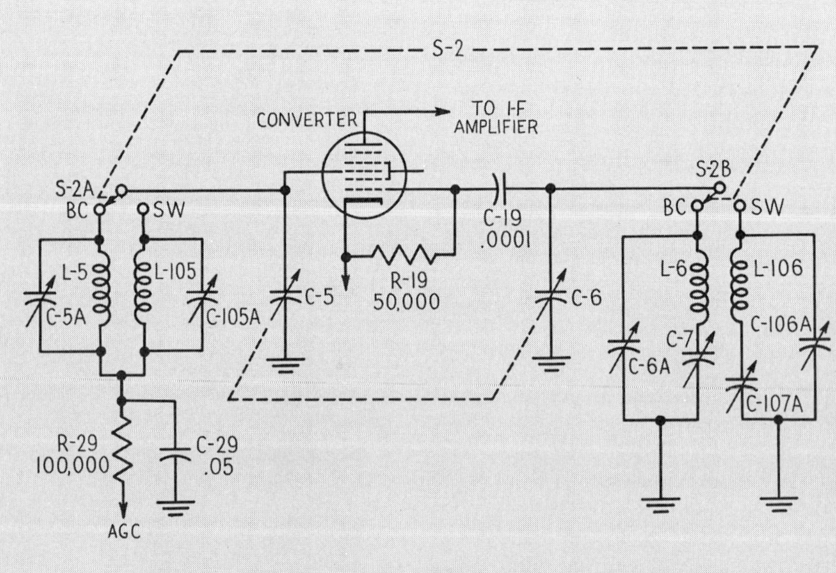

FIG. 16-33. Simplified switching circuits for a multiband receiver.

Band	Approximate frequency range	Type of program
X or *LF*	150–400 kc	Maritime and aircraft
A or *BC*	540–1,600 kc	Standard broadcast
B or police	{1.5–4.6 mc, 2–6.2 mc}	Police, amateur
C or *SW*	5.8–18 mc	U.S. and foreign short wave

frequency ranges shown, depending on whether it is desired to include state police at 1,600 to 1,800 kc or the United States and foreign broadcast stations at 6 to 6.2 megacycles. When the latter range is chosen, the broadcast band is usually extended to 1,750 kc to include the state police broadcasts.

Multiband receivers usually include two or three of the frequency ranges listed above. Communications receivers include all bands, and sometimes add a fifth which extends the high-frequency range to approximately 40 megacycles.

The circuit of Fig. 16-33 is simplified in that it makes no provision for shorting the unused coils (a usual procedure), and does not show the primary of the coil in the signal grid circuit or the feedback winding of the oscillator coils, either or both of which may also require switching.

Practical multiband receivers use a variety of switching and coil arrangements. In addition to changing coils, the switch may include extra sections which accommodate auxiliary functions. For example, pilot lights may be switched on and off so that the proper frequency range on the dial scale is illuminated. Another common practice is to increase sensitivity and to alter the tone response, when the receiver is switched to short-wave reception.

Figure 16-34 is a two-band superheterodyne receiver. The range switch has four decks or wafers, labeled *A*-1 to *A*-4. The switch has three positions: broadcast band (antenna), broadcast band (loop), and short-wave band. In the position shown, broadcast band (antenna), range switch sections *A*-1 and *A*-2 connect the antenna to the center tap of antenna coil *L*-3; antenna tuning capacitor *C*-4 is connected across *L*-3 with its trimmer *C*-1; and the tuning circuit, composed of *L*-3 and *C*-4, is connected in the grid circuit of the r-f amplifier tube, which is a 6SK7 type. Oscillator tuning capacitor *C*-5 is connected through switch sections *A*-3 and *A*-4 to broadcast oscillator coil *L*-6 and the oscillator grid circuit of the 6SA7 tube. The cathode of the 6SA7 tube connects through the tap on short-wave oscillator coil *L*-7 to the tap on broadcast oscillator coil *L*-6. Terminals 6 and 7 on switch deck *A*-3 throw a short across capacitor *C*-28*A* in the grid circuit of the lower 6V6-GT tube through the wires labeled *Y*-*Y*. This is a tone-compensation circuit which will be open on short-wave reception.

When the range switch is moved one position to broadcast (loop), switch section *A*-2 opens the antenna circuit, and *A*-4 shorts out the minimum-bias resistor in the cathode circuit of the r-f 6SK7 tube. The other connections remain the same in the broadcast (antenna) section.

When the range switch is moved to the short-wave position, the following changes take place: The antenna is connected to the center tap of short-wave antenna coil *L*-4 through terminals 9 and 10 on range-switch section *A*-2. Terminals 1 and 2 on the same section connect short-wave antenna coil *L*-4 to the 6SK7 grid circuit. Terminals 7, 6, and 5 on the *A*-1 section of the range switch short the lower halves of the loop and broadcast antenna coil. Terminals 2 and 3 on the same section connect antenna tuning capacitor *C*-4 through *C*-6 to the grid circuit of the 6SK7 tube. Capacitor *C*-6 is for band-spread purposes. In the oscillator section, terminals 8, 9, and 10 of section *A*-4 keep the r-f tube in the sensitive position by shorting out the cathode resistor, and by grounding the bottom

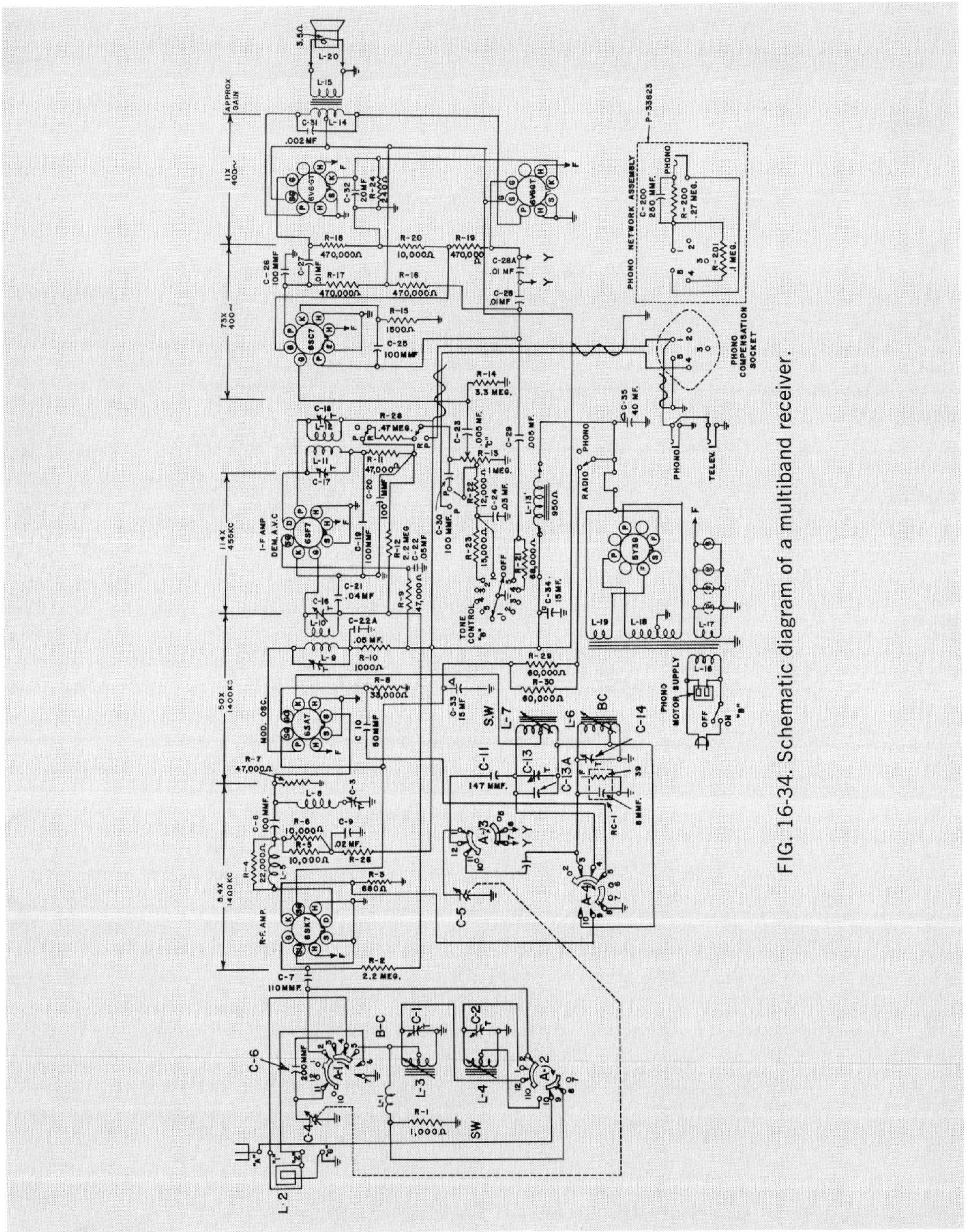

FIG. 16-34. Schematic diagram of a multiband receiver.

lead of the short-wave oscillator coil *L*-7 as well as the bottom half of the broadcast oscillator coil *L*-6. Oscillator tuning capacitor *C*-5 is connected through band-spread capacitor *C*-11 to the top lead of short-wave oscillator coil *L*-7 and the oscillator grid circuit of the 6SA7 tube. Switch terminals 6 and 7 of section *A*-3 connect the capacitor in the grid circuit of the 6V6-GT tube for tone compensation.

Note also the following points of general interest in this diagram. Coupling between the r-f and converter tubes is untuned. The *L*-8/*C*-3 tuning device in the converter signal grid circuit is a wave trap adjusted to 455 kc and is meant to keep incoming signals at or near the intermediate frequency from causing interference. Trimmers *C*-1, *C*-2, *C*-13, and *C*-14 adjust the high-frequency end of each band, and the permeability adjustments on the associated coils adjust the low-frequency end of each band to ensure accurate dial readings and tracking. Gain data are included above the check points for each stage.

Servicing multiband receivers. Although multiband receivers look more complicated than a single-band unit and may take a little longer to service, they are no more difficult. As a matter of fact, the range switch opens a possibility of faster diagnosis in some ways. When a receiver is dead on the broadcast band but operates normally on other bands, the defective condition is more quickly narrowed down to defective coils in the r-f or oscillator portions of the receiver.

Servicing procedures for the multiband receiver are the same as for any other, until the r-f portion of the receiver is reached. At this point the technician need only make sure that the range switch is in the broadcast position in order to continue in the usual way.

There are, of course, some service problems connected with multiband receivers that will not be present in single-band radios. These include the short-wave coils, the range switch, and alignment.

Short-wave coils are usually wound as a single-layer inductance using heavy wire. This type of winding rarely gives any service trouble. The technician, however, should be able to check the windings with an ohmmeter. This may not be so easy as it sounds, since it is sometimes difficult to determine which lead is which on a multiunit coil assembly. Also, the technician may not be sure as to whether the winding is being shorted by the range switch. The suggested method is to work with the schematic diagram and check the coils and switch at the same time. Because of the diversity of range switches, no standardized procedure can be arranged for checking coils and switches.

When an open coil is found, the technician should first make sure that the defect is not due to a broken lead wire. If the coil must be replaced, it is necessary to obtain an exact duplicate from the manufacturer of the receiver. Even if it is necessary to replace an entire coil assembly for one open winding, this must be done since the chances of finding a usable section are very slim.

The lead dress between the range switch and the short-wave coils is very important. In high-frequency circuits, stray capacitance of the wiring represents a considerable portion of the total capacitance of the circuit. Very often, the wiring in these circuits makes use of heavy bus bar, so that the positioning will be maintained. Any replacement of switch or coils should be accomplished with a minimum of bending or rearranging of the wiring, so as to avoid undesirable coupling or changes in the stray capacitance.

Range switches become covered with a layer of dust and dirt, resulting in poor contact and sometimes leakage between terminals. Dusting with a soft brush and then cleaning with carbon tetrachloride comprise the usual service procedure. A good way to clean the contacts is to wet them and the contact arms with carbon tetrachloride and then rotate the switch rapidly.

If a switch contact or wafer becomes broken, it is necessary to replace the entire switch. Again, an exact duplicate must be obtained from the manufacturer.

Aligning multiband receivers. In realigning a multiband receiver, the manufacturer's instructions should be followed to the last detail. This advice is given whenever the word "alignment" is mentioned. However, in the case of multiband receivers, it is of more than usual importance for two reasons. One is that the alignment on one wave band may affect the alignment on the other ranges, and the proper alignment sequence should be followed. The other is the fact that some receivers are so designed that the oscillator frequency should be 455 kc lower than the signal frequency on the short-wave band, while the conventional 455 kc higher signal on the other bands is maintained. Often both frequencies are within the scope of the trimmer adjustment, and it is important to use the peak at the lower or higher capacitance setting, as instructed, in order to maintain proper tracking. If alignment instructions are not obtainable, the following suggestions may be of value.

I-f alignment. First turn the range switch to the broadcast position, short the oscillator section of the gang tuning capacitor, and align the i-f trimmers in accordance with the general alignment instructions given in Chap. 18. Then remove the short from the oscillator section of the tuning capacitor and check the position of the dial pointer, by turning the gang tuning capacitor to full capacitance (full mesh). The dial pointer should be in line with the last calibration mark on the low-frequency end of the dial scale.

As for the sequence of range alignment, when trimmer settings of one band affect another, it is suggested that the broadcast band be aligned last, because the cumulative effect will be most noticeable on the lowest frequency band. In addition, the owner of a receiver is usually most concerned about the operation of the broadcast band. For these reasons, when more exact instructions are not available, it is advisable to align the broadcast band last.

Short-wave alignment. Connect the signal generator through a dummy antenna of 400 ohms to the antenna and ground of the receiver. Turn the range switch to the highest frequency band on the receiver, and set the dial at a convenient mark near the high-frequency end of the scale. Adjust the signal generator to a modulated output at the same frequency as shown on the receiver dial. Adjust the short-wave oscillator coil trimmer for maximum response. If only one peak is obtained, the oscillator is adjusted to the proper frequency. If two peaks are obtained, choose the peak at minimum capacitance. This sets the oscillator to a higher frequency than the signal.

If the receiver has an r-f stage, the interstage coil trimmer is next to be aligned. If there is no r-f stage, the antenna-coil trimmer follows the oscillator adjustment. In either case, the trimmer is adjusted for maximum response. If two peaks are obtained, the one that maintains the oscillator at the higher frequency is the peak that is nearer the maximum capacitance setting of the trimmer. If the peak is unobtainable or if the receiver does not track at the low-frequency end of the dial, the alignment should be tried with the oscillator set for a lower frequency than the signal. This is done by choosing the maximum capacitance peak for the oscillator trimmer and the minimum capacitance peak for the antenna trimmer.

If the receiver has more than one short-wave band, the next lower frequency range should be aligned. The same procedure is followed as for the highest frequency band except that the oscillator circuit should be adjusted for a higher frequency than the signal. However, it is doubtful if two peaks are obtainable on any but the highest frequency band. This band may or may not include a low-frequency adjustment.

Broadcast alignment. Set the range switch at the broadcast position and adjust the tuning capacitor until the dial reads 600 kc. Connect the signal generator to the antenna and ground, using a 0.00025-mfd capacitor as the dummy antenna. Adjust the signal generator for a modulated signal at 600 kc. Adjust the oscillator 600 padder for maximum response. Then set the receiver dial to 1,500 kc and adjust first the oscillator-coil high-frequency trimmer, then the r-f coil trimmer (if present), and finally the antenna-coil trimmer for maximum response. Return both dial and signal generator to 600 kc and reset the oscillator 600 kc adjuster, if necessary. Then align the r-f low-frequency adjuster, if present. Finally return to 1,500 kc and check alignment. If readjustment is necessary, repeat the alignment at 600 kc and check at 1,500 kc until further readjustment is unnecessary.

Permeability tuning systems. In the conventional receiver, tuning is accomplished by changing the capacitance of a variable capacitor connected across a coil, thereby changing the resonant frequency of the combination. The same effect can be brought about by allowing the capacitor capacitance to remain fixed and changing the inductance of the coil. This is the basis of permeability tuning systems, where the inductance of the coil is changed by varying the position of an iron-core plug in the coil.

Coils with adjustable cores are used as i-f transformers, and for the fixed-tuned antenna and oscillator coils of the circuit-switching type of push-button tuners. In these cases, the inductances of the coils are adjusted during the alignment procedure or when setting up the push buttons, and remain undisturbed thereafter. In some receivers, instead of using a variable capacitor, the antenna and oscillator core plugs are ganged, and their adjustment is brought out to the front control panel as a continuously variable adjustment of the tuning range of the receiver. Such a tuning system is known as a "permeability" tuner. Figure 16-35 shows a tuner of this type, where the positions of the core plugs are varied by means of a drive cord.

The coils, drive pulley, and idler pulley are fastened to a subassembly. The coil mounts are so arranged that either coil may be shifted slightly to the right or left for tracking purposes. Ordinary dial-drive cord is used to vary the position of the core plugs. Note that when the drive shaft is rotated in the direction shown, both core plugs are pulled into their respective coils.

From the servicing point of view, any trouble in the permeability tuner would be found in the same way that a similar trouble would be found in a conventional tuner, since the circuits are alike. The alignment procedure is somewhat different, and the manufacturer's service notes should be followed closely in this regard. Another difference lies in the

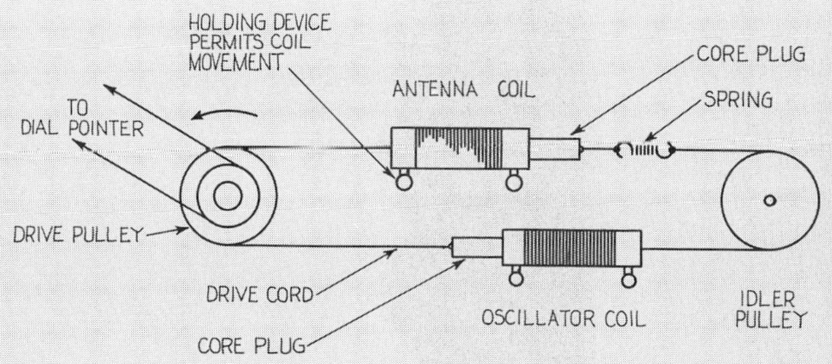

FIG. 16-35. Drive-cord type of permeability tuner.

fact that restringing the drive cord calls for realignment, since the restringing process will slightly alter the relative positions of the tuning slugs.

If specific restringing and alignment notes are not available from the receiver manufacturer, the following generalized procedure should be of help. The skeleton schematic diagram of Fig. 16-36 is included as an aid in locating and identifying the trimmers.

Restringing and alignment procedure for drive-cord permeability tuners

1. Restring the tuning slugs, using the frayed or torn pieces of the old drive cord as a guide, so that the relative positions of the tuning slugs are as close as possible to their original settings.
2. Set the antenna coil in the center of its positioning range, so that it may be shifted either to the right or left.
3. Rotate the drive so that the antenna core plug is completely out of the winding.
4. Set the oscillator coil so that its core plug is in the same relative position as the antenna coil and its core plug. Set the dial pointer to the highest frequency division on the dial scale.
5. Rotate the drive to make sure that the dial pointer and tuning slugs move together and cover the entire tuning range.
6. Check the alignment of the i-f amplifier and, if necessary, align in the usual manner.
7. Connect the signal generator to the receiver antenna, using a 0.00025-mfd dummy antenna. Rotate the drive to the high-frequency end of the dial scale, and adjust the signal generator for a modulated output at the same frequency. Adjust the oscillator trimmer *C*-2 to a maximum response. Then adjust the antenna trimmer *C*-1 to a maximum response.
8. Rotate the drive to 1,400 kc, and adjust the signal generator frequency control to a peak. This should occur at 1,400 kc. If it is too far off, the starting position of the oscillator core plug was incorrectly adjusted. This should be corrected and steps 7 and 8 repeated.
9. When the peak at 1,400 kc in step 8 has been obtained, the antenna coil is shifted to the right or left for a maximum tracking peak.
10. Return the dial and signal generator to the highest frequency reading on the dial scale, and check the adjustment of the antenna trimmer *C*-1. If no appreciable change is needed, the antenna coil is in track. If a considerable change has been made, repeat steps 9 and 10.

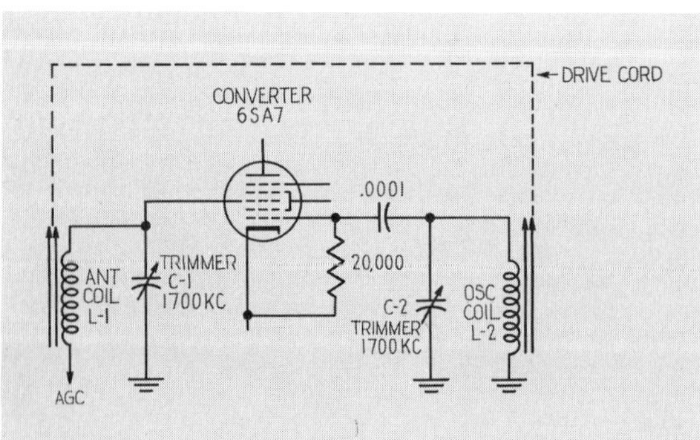

FIG. 16-36. Skeleton diagram of drive-cord type of permeability tuner.

Screw-drive permeability tuners. Another type of permeability tuner uses a screw for driving the ganged tuning slugs. Figure 16-37 shows a tuner of this type. The proportions have been altered to permit viewing the operation of the unit.

The coils *L*-1, *L*-2, and *L*-3 are mounted on the back plate of the carriage. They are generally in individual shield cans to prevent feedback effects. The bakelite strip is threaded to take screws attached to the core plugs. Adjusting these screws permits adjustment of the relative positions of the individual core plugs with respect to their coils. Rotating the drive shaft causes the bakelite strip to move in and out, carrying the core plugs with it, and thereby changing the inductance of the coils. The drive shaft is the tuning control for the radio. The gear ratio is usually chosen to allow several turns of the drive shaft for complete coverage of the tuning range, thereby giving vernier tuning. A similar ratio on a drive pulley (not shown in the diagram) operates a conventional dial pointer from the same tuning shaft.

Tuners of this type, employing three coils, may be used for a receiver with a tuned r-f stage. It is widely used in automobile radios. The diagram of Fig. 16-38 shows the r-f and converter sections of a Motorola auto radio as a typical example of a permeability tuner of this type.

The ganged cores tune antenna coil *L*-1, mixer coil *L*-3, and oscillator coil *L*-5. Trimmers across each coil provide the high-frequency adjustment for each circuit. The individual core adjustments on the bakelite strip are used for sensitivity and tracking at the low-frequency end of the tuning dial.

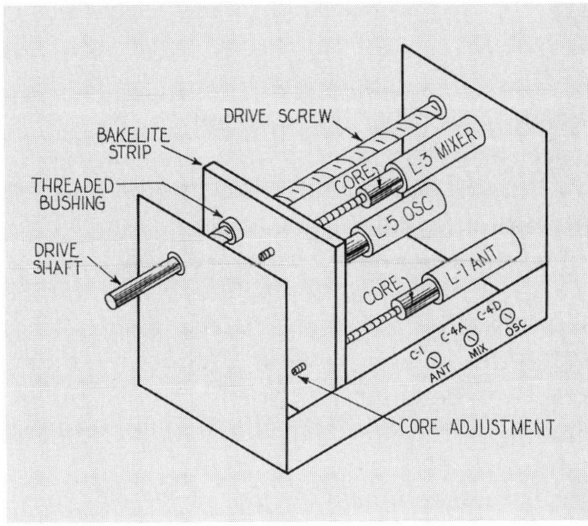

FIG. 16-37. A permeability tuner of the screw-drive type.

Note some peculiarities common to this type of receiver. The antenna and mixer coils are not transformers. Impedance coupling is most often employed. The suppressor of the r-f amplifier is connected to the agc (labeled AVC) bus, rather than its usual connection to cathode or ground. In the oscillator circuit, the feedback coil is in the oscillator plate and screen circuit, rather than in the

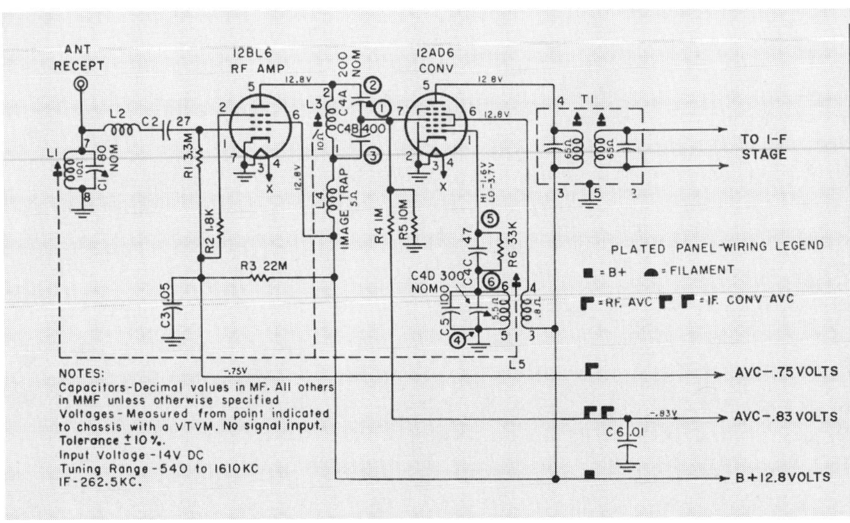

FIG. 16-38. R-f and converter section of a Motorola auto radio using a three-coil permeability tuning system.

cathode circuit. The plate and screen voltages measure only 12.8 volts, since the receiver is powered directly from the 12-volt storage battery in the automobile. Note the negative voltage at the oscillator grid and how it varies, depending on the frequency of the tuning dial.

From the servicing point of view, signal checks may be applied to receivers of this type just like any other. If the voltage and circuit differences are kept in mind, the tuner offers no new service problems, except for alignment. A generalized alignment procedure follows.

Alignment procedure for three-coil permeability tuners

---

1. Check the intermediate frequency of the receiver. Auto receivers are often designed to work with an intermediate frequency near 260 kc, rather than 455 kc. Then align the i-f amplifier in the usual way.
2. Check the dial pointer setting and positioning of the core plugs by the following steps.
   a. Rotate the tuning shaft to the low-frequency stop.
   b. See that the core plugs are fully engaged in their respective coils.
   c. Set the dial pointer at the lowest frequency calibration mark on the dial scale.
3. Connect the signal-generator output lead to the antenna connection through a 0.00025-mfd dummy antenna.
4. Set the signal generator to feed a modulated signal at 600 kc, rotate the tuning shaft to read 600 kc on the receiver dial, and align the oscillator-coil core plug for maximum output.
5. Tune the receiver to a quiet point near the high-frequency end of the tuning range, adjust the signal generator to feed the same frequency as that shown on the receiver dial scale, and align the oscillator high-frequency trimmer for maximum output. This trimmer is labeled *C-4D* in Figs. 16-37 and 16-38.
6. Check the 600-kc adjustment by repeating step 4. If considerable readjustment is necessary, realign the high-frequency adjustment by repeating step 5 and then recheck at 600 kc.
7. Tune the receiver and signal generator to a peak near 600 kc, and adjust the core plugs on the mixer and antenna coils for a maximum response.
8. Tune the signal generator and receiver to a peak near 1,500 kc, and align first mixer and then the antenna trimmers for maximum output.
9. Check the 600-kc adjustment by repeating step 7. If considerable readjustment is needed, realign the high-frequency adjustment by repeating step 8.
10. If the receiver is an auto radio, the high-frequency antenna trimmer is finally readjusted in the car. The receiver is connected to its own antenna and tuned to any weak station at the high-frequency end of the dial. The antenna trimmer is then adjusted for best reception from this station.

Receivers with push-button tuning. Many receivers, particularly auto sets, use push buttons for tuning favorite stations. There are two general types of push-button arrangements, mechanical and switching. In the mechanical type, when the button is pushed, the tuning dial is moved to predetermined positions by means of an electric motor, a solenoid system, or lever action. In the switching method, when the button is pushed, the dial-operated tuning circuit is switched out, and a separate tuning circuit, preset to the desired station, is switched in.

There is a large number of each type of arrangement. For best servicing procedures, refer to the manufacturer's service notes. Some typical examples will be described, together with applicable notes.

Switched push-button tuning circuits. Figure 16-39 shows the schematic diagram of a Zenith receiver featuring broadcast, short-wave, and push-button tuning. The latter is inaugurated by turning the range switch to the automatic tuning position. This disconnects the gang tuning capacitor, all three sections of which are labeled *C*-1, throws a row of preset trimmers across the antenna tuning circuit, eliminates the converter signal-grid tuning circuit by converting it into an untuned resistance-coupled circuit, and throws any one of a row of permeability-tuned coils in the oscillator grid circuit.

The predetermined station is then tuned in by depressing the proper push button. The buttons are sprung so that they are normally in the OFF position. Then as any button is depressed, a catch holds this button in place while automatically releasing any button previously depressed. Each button controls a double-pole switch, one pole of which connects one of the permeability-tuned coils in the oscillator grid circuit, while the other pole connects the proper associated trimmer in the r-f grid circuit.

The trimmers and coils in the automatic tuner have a limited range (approximately 400 kc), so that each button cannot tune many desired stations in the broadcast band. However, the values of coils and capacitors are staggered, so that any station can be tuned in on some one button. The tuning range of each button is usually marked near the adjustment screws.

Figure 16-40 shows a simplified drawing of the tuning circuit of the Zenith receiver of Fig. 16-39, when the range switch is in the automatic position. One push button is depressed, showing one preset trimmer connected across the r-f tuning circuit and one preadjusted permeability coil in the oscillator tuning circuit. The coupling between the r-f and converter tubes is of the resistance-capacitance type. This coupling also remains the same for the short-wave position of the

FIG. 16-39. Schematic diagram of a Zenith receiver featuring broadcast, short wave, and push-button reception.

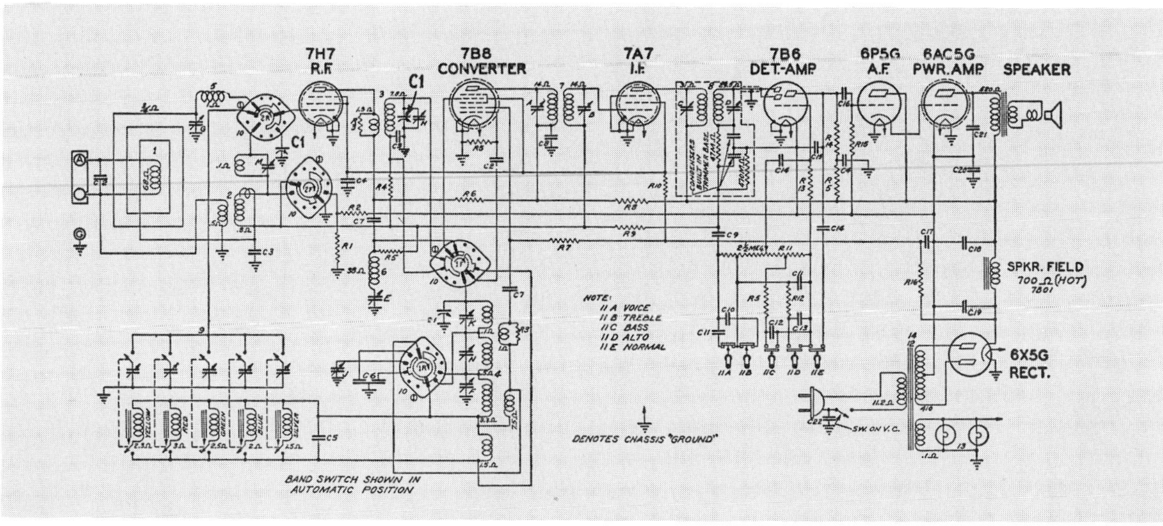

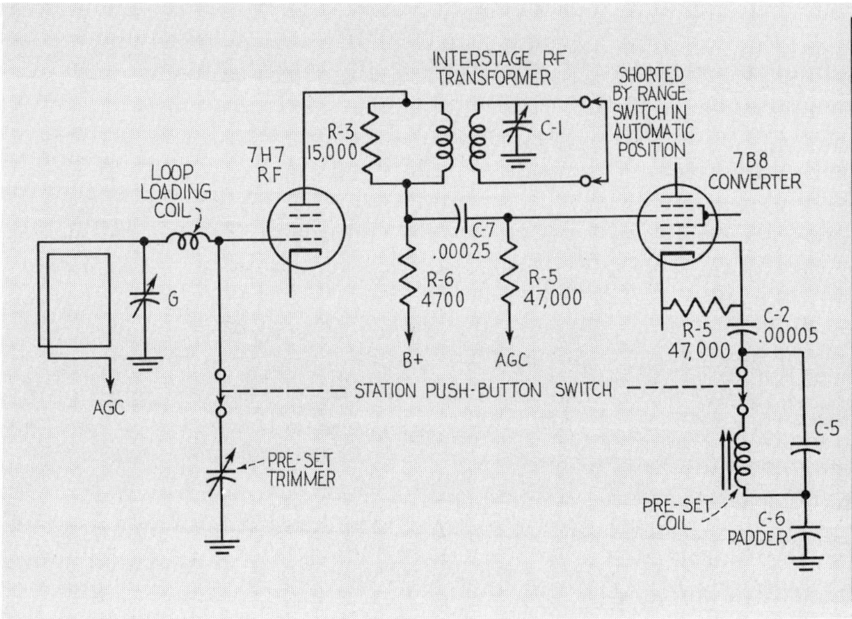

FIG. 16-40. Simplified diagram of the tuning circuit of the Zenith receiver shown in Fig. 16-39.

range switch. The circuit is tuned only in the manually operated broadcast position of the range switch.

The system just described is typical for push-button tuners of the switching type. These systems differ mainly in the number of preset stations available. In some cases, switching from manual to automatic tuning is taken care of by an extra similar push button, rather than a position on the range switch. Often the radio-phonograph switch is also an extra similar push button. In addition, some types provide two sets of trimmer capacitors, instead of one set of trimmers and one set of permeability-tuned coils. In these types, the regular broadcast oscillator coil is used, the oscillator tuning capacitor is switched out of the circuit, and one of the preset trimmer capacitors is substituted for it.

Servicing push-button tuners of the switching type. Push-button systems of the switched tuning-circuit type give very little service difficulty. Occasionally, the switches do not make good contact. When this happens, the following cleaning procedure is effective: Dust the entire switch assembly with a soft brush. Depress the first switch, and apply carbon tetrachloride to its contacts and also the arm and contacts of the next switch. Then depress the two switches alternately: first the second, then the first. Repeat the procedure for the first and second switches, this time depressing the second button before applying the carbon tetrachloride to it. Repeat on the next pair, making sure that each switch has been washed in both the open and the closed position.

Another service problem is resetting the adjustment screws, which may change their position with time. When doing this, the receiver should be allowed a warm-up period of about 15 min, to allow all components to reach normal operating temperature. The oscillator control is adjusted first, followed by the antenna adjustment. If the adjusting screws are not marked, the technician can identify them by checking the wiring diagram or by the operation of the adjustments. The oscillator adjustment is critical—a fraction of a turn will bring the station in or out. The antenna adjustment is broad in comparison. If the receiver is equipped with a magic eye, it should be used to indicate exact resonance. An output meter cannot be used for this purpose, since the reading will vary with the modulation of the program. A vacuum-tube voltmeter, if available, connected to the agc bus, can also be used as the resonance indicator. If neither the magic eye nor a vacuum-tube voltmeter is available, the adjustments are set for best volume and tone by ear. A good check for correct settings is to tune to the same station with the switch set for manual operation, and then switch from manual to push button and note any difference. Operation should be the same, except in the case of a receiver like that of Fig. 16-39, where the manual switch throws in an extra tuning circuit.

When push buttons are set up or when the adjustment screws are far from their correct alignment positions, it would be timesaving to use the signal generator for finding the desired stations.

Figure 16-41 shows the method of connecting the signal generator to the receiver. Adjust the signal generator for a modulated output at the frequency of the first desired station. Depress the first push button, and adjust the associated oscillator control until the signal-generator note is heard. It will be accompanied by a squeal, caused by the beating action between the generator signal and the signal from the desired station. Disconnect the signal-generator "hot" lead. If the squeal does not stop because of leakage, detune the signal generator. Then readjust the oscillator control for maximum response from the station. Finally, adjust the antenna trimmer. Repeat the procedure for the other buttons.

Mechanically operated push-button tuners. Figure 16-42 shows two views of a typical mechanically operated push-button tuning system. This type is known as a "rocker-bar" mechanism. Each button depresses a preset pawl, which turns the rocker bar as far as the pawl setting will allow. A gear connected to the rocker bar rotates the gang tuning capacitor. The tuning knob and dial pointer rotate with the capacitor gang. The return spring maintains the push button in its normal out position and, at the same time, keeps the pawl away from the rocker bar.

When a button is set up, the locking screw is loosened. A screwdriver is kept pressed against the loosened locking screw, thereby depressing the push bar and pushing the pawl against the rocker bar. The desired station for each button is tuned in manually, thereby pushing the pawl to its proper setting. The

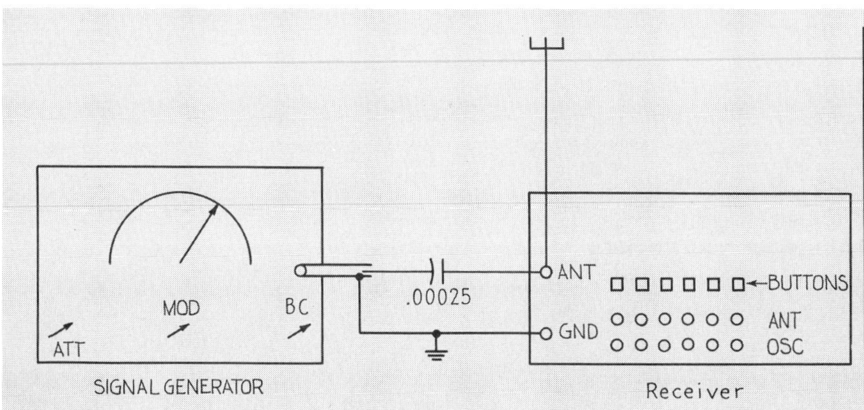

FIG. 16-41. Using a signal generator as an aid in quickly presetting pushbuttons.

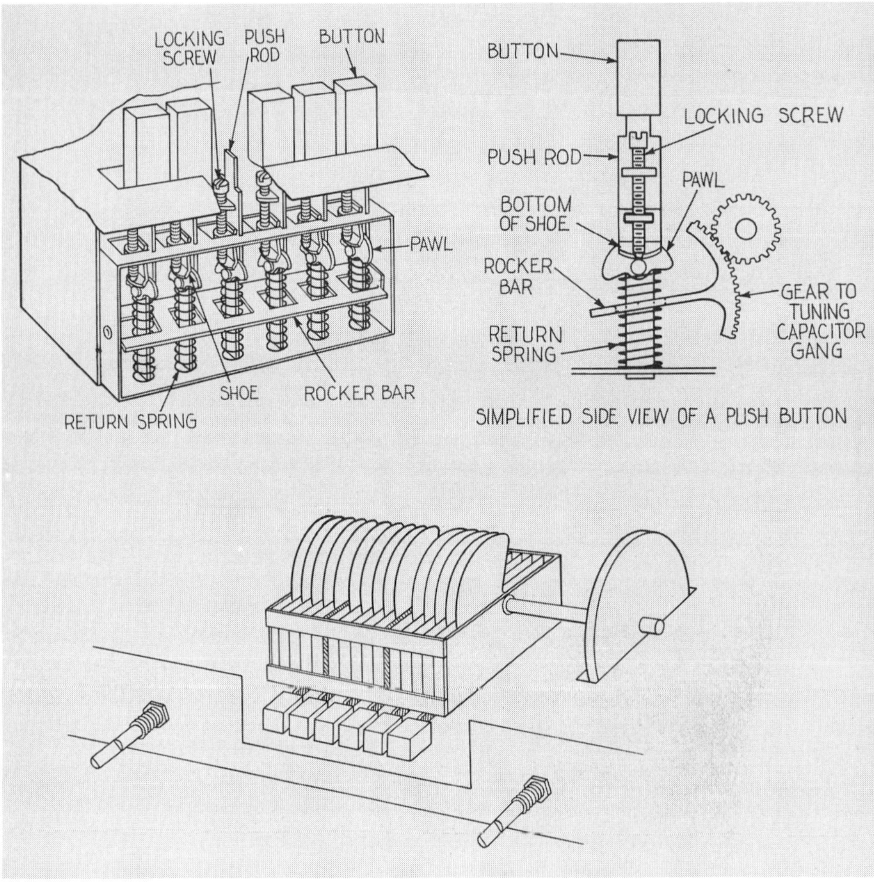

FIG. 16-42. Mechanically operated push-button tuner of the rocker-bar type.

locking screw is then tightened, fixing the pawl firmly between the shoe and push rod. Subsequently, when the button is depressed, the pawl pushes the rocker bar to its set position, thereby bringing in the desired station.

From the servicing point of view, loosened adjustments are about the only difficulty experienced with mechanical buttons of this type. A complete adjustment procedure follows.

Adjustment of push buttons for rocker-bar tuners. Rotate the range switch to the broadcast position. Select the stations desired for automatic tuning. Choose one of these stations and any button to be adjusted for it. Follow the procedure outlined:

1. Grasp the button firmly and remove it from its shaft by pulling straight out (see Fig. 16-43*A*).
2. Insert a screw driver into the slot of the locking screw. Press in and loosen the screw 1 to 1½ turns (see Fig. 16-43*B*).
3. With the screw driver seated in the screw slot, press the screw in as far as possible. Hold it in firmly with one hand, and tune in the desired station with the other hand by pressing in and rotating the selector knob (see Fig. 16-43*C*).
4. Release the selector knob and tighten the screw firmly.
5. Check the adjustment by tuning

well past the station, using the selector knob and then pushing in the button shaft. The station should come back in again clearly and with maximum volume. After the adjustment is tested, check to see that the locking screw is tightened firmly. Replace the button on its shaft.

6. Adjust the remainder of the buttons in the same manner as outlined above.

Figure 16-43*D* shows a common method of inserting station tabs.

Mechanical automatic permeability tuners. A Motorola Tuner, illustrated in Fig. 16-44, is a push-button control of associated permeability tuners. It is similar to the rocker-bar arrangement just described, except that powdered-iron cores are inserted to different extents in the antenna, r-f, and oscillator coils by the depression of each of the buttons.

Like the rocker-bar system, about all the trouble that develops is a falling out of adjustment. The tuner is then reset as described below:

1. Turn on the receiver and allow it to warm up for a few minutes.
2. Pull out any station-selector push button to unlock the tuner.

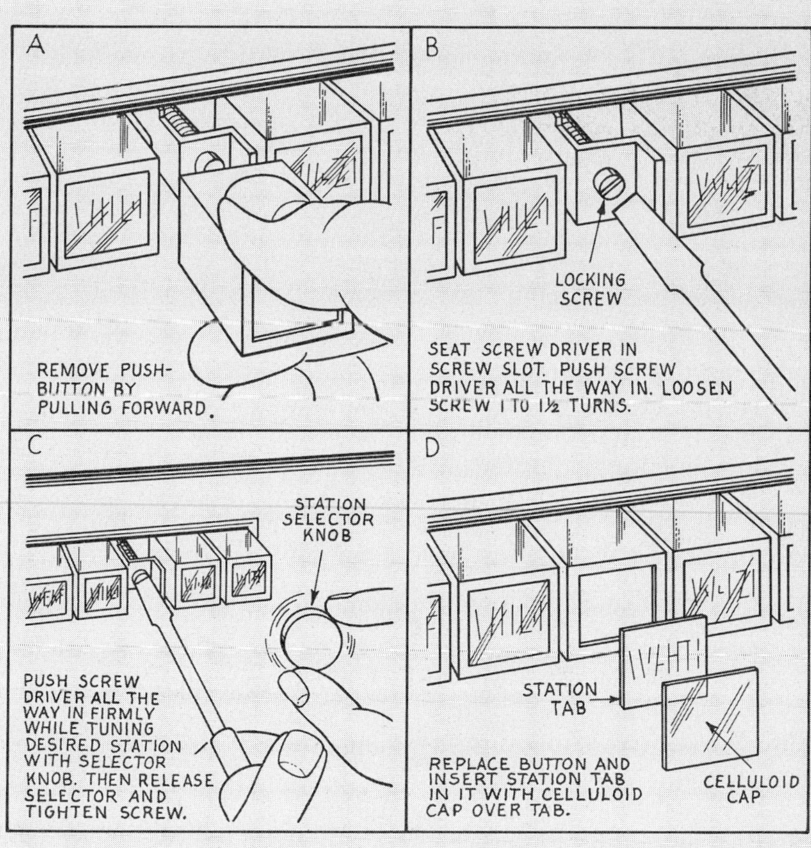

FIG. 16-43. Adjustment of pushbutton for mechanical automatic tuners.

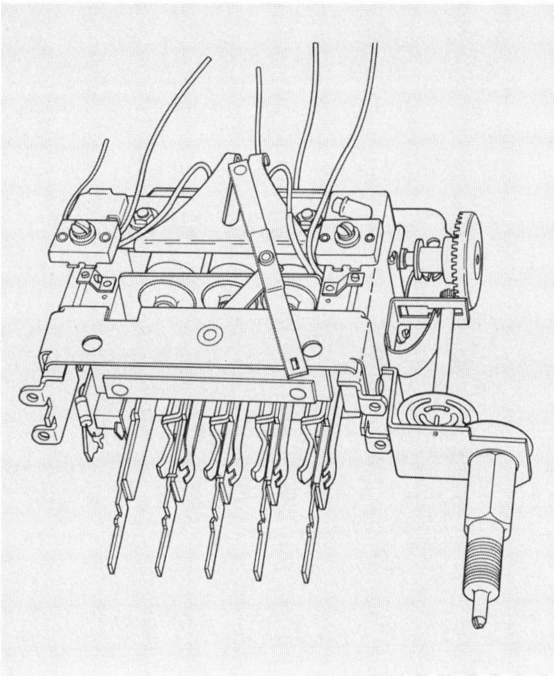

FIG. 16-44. Mechanical automatic permeability tuner.

3. Tune manually to the station you desire to set up. Be sure you tune for clearest reception.
4. Push the button back on and depress to lock the tuner. This push button is now set for the station you just selected.
5. Follow the same procedure for the remaining push buttons and other desired stations.
6. It is usual to set the push buttons for higher-frequency stations as you move from the left to the right.

Mechanically operated push-button tuners of the motor-driven type. Motor-driven push-button tuners are too varied in their operation, adjustment, and service problems for any generalized treatment in a book of this nature. The technician is referred to the manufacturer's service notes when he experiences difficulty with any of these devices.

SUMMARY

Quick check for normal operation of the oscillator.

Tune the receiver to 600 kc. Connect the signal-generator output to the converter signal grid (mixer grid) through a 0.1-mfd capacitor, and rotate the signal-generator dial through 600 kc. If the signal-generator modulation note is heard in the speaker at or near 600 kc, the oscillator is functioning.

Quick check for normal operation of the mixer.

Tune the receiver to 1,400 kc. Connect the signal-generator output to the r-f grid (antenna if there is no r-f stage) through a 0.00025-mfd capacitor. Rotate the signal generator dial through 1,400 kc. If the signal-generator modulation note is heard in the speaker, at or near 1,400 kc, the mixer stage of the converter is functioning.

Standard diagram.

This circuit is shown in the accompanying figure.

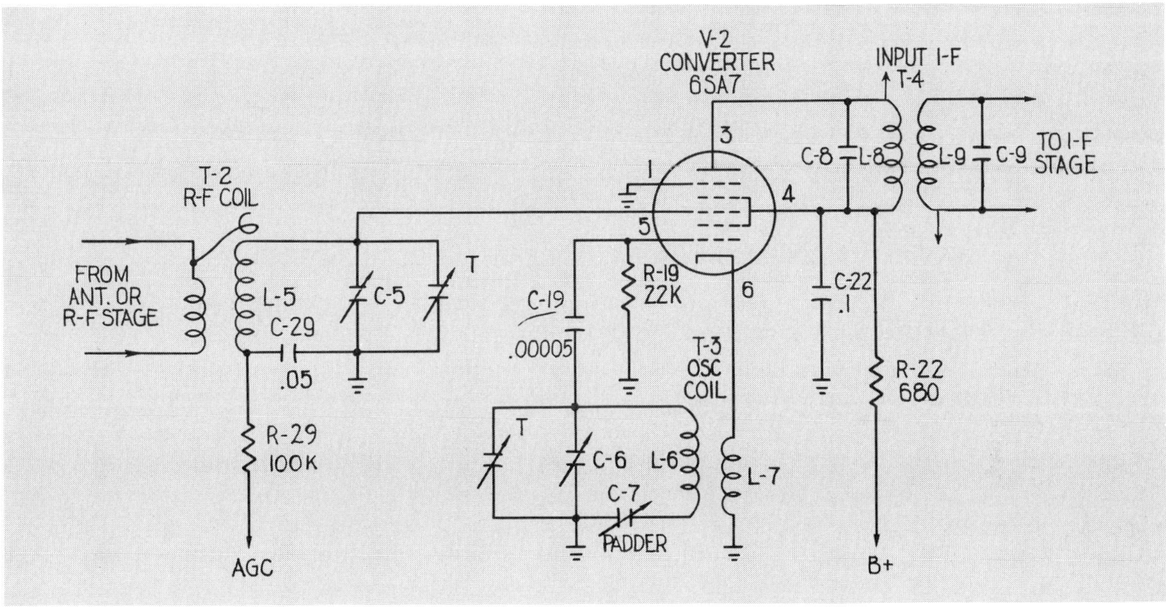

Normal resistance data for the converter.

Across *L*-4, primary of the signal input transformer *T*-2	20–40 ohms
Across *L*-5, secondary of the signal input transformer	3–5 ohms
Across *L*-6, grid coil of the oscillator transformer *T*-3	4–5 ohms
Across *L*-7, feedback coil of the oscillator transformer	1–2 ohms
Signal grid (*G*-3) to chassis	2,700,000 ohms
Oscillator grid (*G*-1) to chassis	22,000 ohms

In receivers where the signal input transformer *T*-2 is a loop antenna, the grid coil of the loop will measure 1 to 3 ohms. The antenna winding, if present, will measure less than 1 ohm.

Normal voltage data for the converter.

Readings taken from the indicated terminals to the chassis or common negative terminal of the receiver are given in the accompanying table.

	6SA7 or 12SA7 pin No.	A-c receiver, volts	A-c/d-c receiver volts	6BE6 or 12BE6 pin No.
Plate	3	90–135	90	5
Screen	4	90–135	90	6
Oscillator grid	5	−10	−10	1

SERVICE DATA CHART FOR AN INOPERATIVE OSCILLATOR STAGE

Assume an inoperative oscillator section in a dead receiver, as shown by normal response when an i-f test signal is applied to the mixer grid, and no response when the test signal frequency is changed to r-f. The following test procedure is recommended.

Procedure	Reading	Trouble and subsequent check
Make a voltage check.	Oscillator grid reads zero or positive	Nonoscillating converter tube. Substitute one that is known to be good. If the trouble still persists, it is in the oscillator grid circuit. Make ohmmeter check

In the ohmmeter check look for the following conditions:

Open oscillator coil grid winding *L*-6.
Shorted oscillator section of the gang tuning capacitor *C*-6.
Leakage across the oscillator tuning capacitor *C*-6 or its shunt trimmer.
Open padder capacitor *C*-7.
Open or leaking oscillator grid capacitor *C*-19.
Open or wrong resistance value for oscillator grid leak *R*-19.

SERVICE DATA CHART FOR AN INOPERATIVE MIXER STAGE

Assume an inoperative mixer in a dead or weak receiver, as shown by normal response when an r-f test signal is fed to the mixer grid, and no or weak response when the r-f test signal is fed to the r-f grid.

Procedure	Reading	Trouble and subsequent check
Apply the test signal to the r-f plate.	Normal response	Trouble is in the r-f tube or its associated circuit. Substitute an r-f tube known to be good. Make a voltage check and a resistance check of the r-f stage to find the defective component

Procedure	Reading	Trouble and subsequent check
	No or weak response	Trouble is in the mixer grid circuit. Check the mixer input transformer *T*-2 for opens. Check the agc decoupling filter *R*-29 and *C*-29 for opens

SERVICE DATA CHART FOR THE CONVERTER

Symptom	Look for
Hum	Open mixer grid coil *L*-5
Modulation hum	Defective converter tube. (Cathode-to-heater short or leakage)
No reception on LF end of the tuning range	Weak converter tube. Check oscillator circuit for critical oscillator conditions
Distortion	Short-circuited agc capacitor *C*-29
Weak reception	Open plate bypass *C*-22. Misalignment
Weak reception—high noise level	Open agc capacitor *C*-29
Squeal	Open plate bypass *C*-22. Shielding improperly grounded. Leads improperly dressed
Squeals or birdies when tuning certain stations	Image frequency interference. See Chap. 17
Noisy, intermittent operation	Defective converter tube. Corrosion in input transformer *T*-2 and oscillator transformer *T*-3. Check gang tuning capacitor for shorts and poor wiper contacts. Check wiring for bad connections and rosin joints
One station all over the tuning range	Capacitor gang not tuning. I-f amplifier tuned to the wrong frequency

QUESTIONS

1. Outline a procedure for determining the cause of a defective oscillator circuit in a dead receiver.
2. Outline a procedure for determining the cause of a defective mixer circuit in a dead receiver.
3. A receiver operates on the high-frequency end of the broadcast band but not on the low-frequency end. List the probable causes and state how you would check for each.
4. The receiver of Fig. 16-27 does not operate. Signal check shows normal response when checking with an i-f signal at the 12BE6 mixer grid, and no response when checking with an r-f test signal from the same point. Voltage check shows no reading at the oscillator grid and normal readings for the 12BE6 plate and screen. Where is the trouble likely to be? How would you check for it? How would you remedy the condition?
5. The receiver of Fig. 16-34 operates normally on the short-wave range but not on the broadcast band. Outline a procedure to locate the cause of the trouble.
6. The receiver of Fig. 16-34 gives no reception. Signal check shows normal operation when a test signal, either r-f or i-f, is applied to the 6SA7 signal grid, and no reception when an r-f test signal is applied to the r-f grid. Voltage check on the r-f tube shows a reading of zero at the plate terminal. What are the probable causes of the trouble? How would you check each?
7. The receiver of Fig. 16-39 operates normally on the manual and short-wave positions of the range switch, but gives no reception on any push button. What is the most likely cause of the trouble? How would you check for it? How would you remedy the condition?
8. What are the important points to remember when replacing an oscillator coil?
9. A receiver has a tunable hum. The line filter is checked and found to be O.K. What else is likely to cause this condition? How would you check for it?
10. A superheterodyne receiver squeals all over the tuning range. How would you check the converter stage for this?
11. The receiver of Fig. 16-27 does not operate. When a test signal, either r-f or i-f, is applied to the converter signal grid, the response is heard weakly and with a rough note. What is likely to be wrong? How would you check for it?
12. The receiver of Fig. 16-38 operates, but reception is weak, and the noise level is high. Signal check shows normal operation from the converter signal grid, but reduced gain when the r-f test signal is shifted to the r-f grid. What factors are likely to be at fault, and how would you check for each one?

R-f Amplifier Stage 17

Many receivers incorporate a stage of r-f amplification ahead of the converter. It is called the "r-f" stage, or sometimes the "antenna" stage. It requires no quick check for operation. Since it is last in a line of stage checks, if all others check perfect for a defective receiver, the r-f stage must be defective by a process of elimination.

Of course, the entire receiver may be normal and the trouble may lie in the antenna system, which is the first link in the signal chain. This possibility, however, should not occur on a test bench setup. Only at the customer's home will such trouble arise, and the alert technician will recognize the condition. Checking the antenna is usually a routine part of the home service call. Service notes relating to antennas will be included later in this chapter.

Function of the r-f stage. The r-f stage receives energy from the antenna, tunes the desired signal (station), amplifies the signal, and passes it on to the converter. Because of these functions—tuning and amplification—it increases the selectivity and sensitivity of the receiver. The r-f stage provides other advantages. One is the reduction of the noise level when a stronger signal is fed to the converter. Another is the improvement of the agc action, since another controlled tube is added in the r-f and i-f chain. A third is the elimination of image-frequency interference—peculiar to superheterodyne receivers.

Image-frequency interference. Examine the operation of a superheterodyne receiver. The antenna picks up station signals broadly at all frequencies. Where a receiver has no r-f stage, the antenna energy is fed through a tuned circuit to the signal grid of the converter tube. For example, suppose the tuning circuit is set to receive a desired signal at 1,000 kc. The local oscillator of the receiver will then have an output at 1,455 kc if the i-f amplifier of the receiver is fix-tuned to 455 kc. The station and oscillator signals are mixed in the converter tube and the output from the latter is many frequencies. The i-f amplifier, usually sharply tuned by four resonant circuits, accepts only the signal that is the difference in frequency between the station and the oscillator signal—in this case 455 kc, the intermediate frequency. This i-f signal is then amplified and sent on to the detector and a-f amplifier.

Any signal at 455 kc in the converter plate circuit will be accepted by the i-f amplifier and passed on. We have just seen how one 455-kc signal is developed at the input of the i-f amplifier. Is it possible for a second one to be present at the same time?

Reexamine the converter signal grid. It is tuned to 1,000 kc, by means of one tuned circuit that is somewhat broad. As a result, the grid may receive signals of widely different frequencies picked up by the antenna. Normally, these signals mix with the local oscillator signal (1,455 kc) and produce difference frequencies which are rejected by the sharply tuned i-f amplifier. However, there might be one station signal at 1,910 kc on the converter grid which, after mixing with the local oscillator signal of 1,455 kc, will produce a difference frequency at 455 kc. This 455-kc signal will be accepted by the i-f amplifier and result in two stations at the speaker output – 1,000 and 1,910 kc.

Similarly, if the receiver is tuned to a station at 600 kc, the oscillator will be tuned to 600 + 455, or 1,055 kc, and an intermediate frequency of 455 kc will be produced at the i-f amplifier grid. And a station signal at 1,510 kc at the converter signal grid will mix with the oscillator signal and again produce a difference, or i-f frequency, of 455 c which will appear as an interfering station. Thus, any desired station is likely to experience interference from another station that happens to have a frequency which is higher than that of the desired station by twice the intermediate frequency (as much above the oscillator frequency as the desired station is below the oscillator frequency). This defect of superheterodyne receivers is known as "image-frequency interference."

Since the two stations are rarely exactly twice the intermediate frequency apart, they will beat with the oscillator signal to produce intermediate frequencies very close to 455 kc and to each other. These two intermediate frequencies will be accepted by the i-f amplifier, where they will beat with each other to form a difference frequency which will be a-f and result in a high-pitched squeal. Thus, image-frequency interference appears in the receiver in a form called "birdies" or "whistles," which mar reception on certain stations.

Early superheterodyne receivers employed an intermediate frequency of 175 kc. Here, image-frequency interference would occur from stations 350 kc from the desired ones (twice the intermediate frequency, or 2 × 175 kc). If there were only one tuned circuit before the converter, that tuned circuit would not be sufficiently sharply tuned to eliminate image frequencies 350 kc from the desired signal, and interference would be troublesome. Modern practice employs an intermediate frequency of 455 kc, thereby placing the image frequency much farther away, 910 kc from the desired station. As a rule, with such a wide spread between desired and image frequency, even one r-f tuning circuit is adequate to keep a station 910 kc away from affecting the converter signal grid.

A preselector circuit is an added tuning circuit between the antenna and the converter signal grid, the purpose of which is to sharpen the tuning to reduce image-frequency response. Figure 17-1 shows a typical preselector circuit together with its over-all selectivity curve.

Preselector circuits are rarely used in modern receivers, since they reduce the sensitivity, and the same reduction in image-frequency response is possible by increasing the intermediate frequency.

A tuned r-f stage, ahead of the converter, combines the added selectivity of the preselector in reducing image-frequency response and, because of the amplification of the tube, adds to the sensitivity of the receiver.

Standard circuit. The standard circuit is shown in Fig. 17-2.

Functions and values of component parts. The antenna transformer *T*-1 couples the energy picked up by the antenna to the grid of the r-f tube. The secondary winding *L*-2 is tuned by *C*-2, the antenna section of the gang tuning capacitor. As explained before in connection with the components of the converter stage tuning system, the values of parts in any tuning system are important parts of the design of the receiver and cannot be changed without altering the calibration and tracking.

When the receiver is of the loop-operated type, antenna transformer *T*-1 is replaced by the loop antenna. In this case, the main

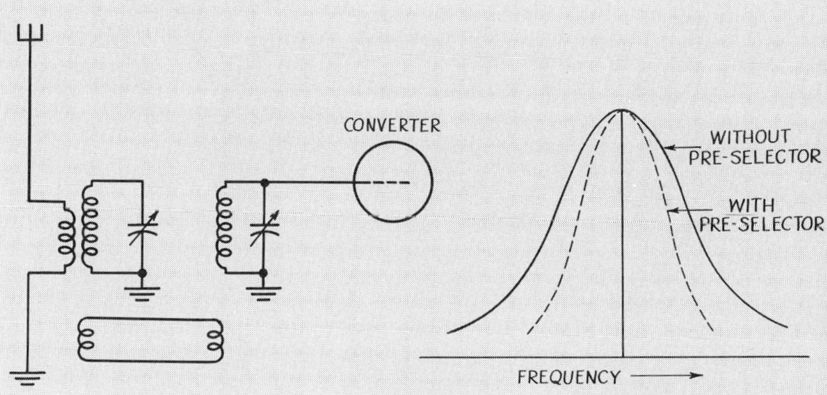

FIG. 17-1. A preselector circuit and its effect on selectivity before the converter signal grid.

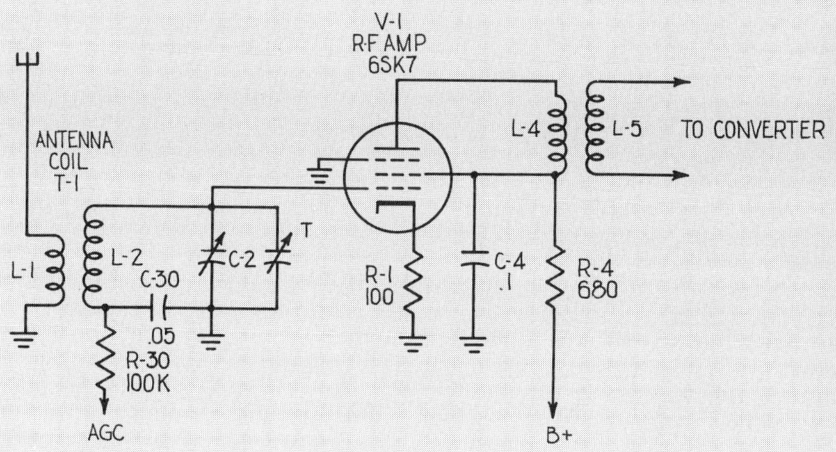

FIG. 17-2. Schematic diagram of a typical r-f stage.

portion of the loop winding acts as the antenna for the receiver and is tuned by capacitor *C*-2. Should it be desired to connect an external antenna for greater sensitivity, the loop is equipped with a primary winding of one or two turns which is connected to the antenna and ground.

Decoupling filters in the r-f stage. When a receiver incorporates an r-f stage, there is more likelihood of undesirable coupling. As a result, there must be more decoupling filters not only in the r-f stage, but also throughout the receiver.

The r-f cathode may be tied to the i-f cathode. Decoupling is achieved by using separate cathode resistors for each stage. Omitting a bypass capacitor provides degeneration. In some receivers, a 0.1-mfd bypass capacitor is used for added gain. Sometimes r-f cathode resistor *R*-1 is made variable. This provides a variable minimum bias for the tube, giving a control of its gain and acting as a sensitivity control for the entire receiver. When used, a sensitivity control of this type provides a range of 100 to 25,000 ohms, and it is generally bypassed by a 0.1-mfd capacitor.

The plate supply also usually includes decoupling filters for one or more of the r-f, i-f, and converter plate circuits. The decoupling resistor *R*-4 for the r-f plate is usually 600 to 1,000 ohms. The plate bypass capacitor *C*-4 is usually 0.05 to 0.1 mfd/600 volts.

Tubes commonly used in the r-f stage. Tubes of the 6SK7 remote-cutoff pentode type are commonly used as r-f amplifier tubes. The miniature equivalent is the 6BA6. A-c/d-c receivers use the 12SK7 and 12BA6 tubes. The r-f tube is often enclosed in a closely fitted metal shield.

NORMAL TEST DATA FOR THE R-F STAGE

Normal signal check for the r-f stage. The signal generator is connected to the antenna and ground through a 0.00025-mfd capacitor, and adjusted for a modulated output at 1,500 kc, with the attenuator set for a very low output. The receiver is adjusted for maximum gain; that is, volume control full on, tone control at maximum high a-f response, and fidelity control in the selective position. The receiver dial is set for a quiet point between 1,400 and 1,500 kc. If an output meter is connected to the receiver, it should be on a high-voltage range. The signal generator dial is then rotated through 1,400 to 1,500 kc.

When the receiver is operating normally, the signal generator modulation note will be heard in the speaker very strongly as the signal-generator dial passes the point at which the receiver is tuned. The r-f stage is then known to be functioning. Usually, the output in the speaker will be greater than the standard output of 50 mw, even with both attenuation controls set at zero. In addition, with the receiver gain controls set at maximum, random noise pulses, picked up by the receiver, cause considerable output meter deflections, so that sensitivity measurements for the r-f stage cannot be made to get the usual standard output.

However, if the signal check does not show some gain over a signal measurement from the r-f grid, it may be assumed that there is trouble between the antenna and the r-f grid.

Normal voltage data for the r-f stage. Readings are taken from the indicated terminal to the chassis or common negative. Normal voltage data for the r-f stage are given in the accompanying table.

Tube terminal	6SK7 or 12SK7 pin No.	A-c receiver, volts	A-c/d-c receiver, volts	6BA6 or 12BA6 pin No.
Plate	8	90–135	90	5
Screen	6	90–135	90	6
Cathode	5	1	1	7

Normal resistance data for the r-f stage. Normal resistance data are given in the following table:

Antenna to ground, or across *L*-1 (primary)	30–50 ohms
Across *L*-2, secondary of the antenna transformer *T*-1	5 ohms
Cathode to chassis	100 ohms
Plate to *B*+	40 ohms plus the resistance of a decoupling filter, if used
Control grid to chassis	2,700,000 ohms

In receivers where the antenna transformer is replaced by a loop antenna, antenna to ground will measure less than 1 ohm and the grid coil of the loop will measure 1 to 3 ohms. A ferrite loopstick, when used, measures 1 to 2 ohms.

COMMON TROUBLES IN THE R-F STAGE

Most of the parts used in the r-f stage are similar to those used in other stages in the receiver. In this section, the common troubles and how they are found will be covered, but the reader will be referred to other parts of the text to avoid repetition of replacement notes.

Troubles common to the antenna transformer. Either winding of the antenna transformer *T*-1 is likely to open. An open secondary would cause weak, noisy reception, possibly accompanied by hum. The trouble would be localized on a signal check since there would be no gain between the r-f grid and the antenna. An ohmmeter check would then confirm the difficulty.

An open primary winding might not be so easily found, owing to the fact that the capacitive gimmick winding may transfer sufficient energy to the secondary for fair operation of the receiver. The nature of the open in the winding will affect the type of trouble experienced with the radio. Usually it will be noisy, and a noisy receiver always calls for a routine check of coils that will locate the trouble.

Replacement notes on antenna transformers will be found in Chap. 16.

Troubles common to the antenna tuning capacitor. The antenna tuning capacitor develops troubles in common with any section of the gang tuning capacitor. The plates may touch and cause a short and no operation or very noisy operation over parts of the tuning range. Also the plates and the associated trimmer capacitor collect dust and cause noisy operation and a partial short. Sometimes the wipers make poor contact, again causing noisy operation or weak reception over parts of the tuning range.

A shorted tuning capacitor or trimmer would be found on signal check. When the converter stage is checked from the r-f grid, the defective capacitor would short the signal-generator output and cause no signal. An ohmmeter check would then disclose the short. Since this might be in the trimmer or the plates of the tuning capacitor, the technician must determine which unit is defective. An easy way of doing this is to rotate the tuning capacitor to the full open position and check again. If the short remains, it is probably due to cracked mica in the trimmer capacitor.

When the tuning capacitor is causing noisy reception over part of the range, a thorough overhaul of the tuning-capacitor gang will be necessary. A procedure for doing this is given in Chap 16.

Troubles common to the r-f stage decoupling filters. The resistors in decoupling filters usually give no service troubles unless the associated capacitor shorts. If *C*-4 should short, there would be an overload of current through *R*-4 which might damage it. When this happens, the resistor is usually replaced. Capacitor *C*-30 in the agc circuit rarely shorts and, if it should, there is insufficient current in this circuit to harm resistor *R*-30.

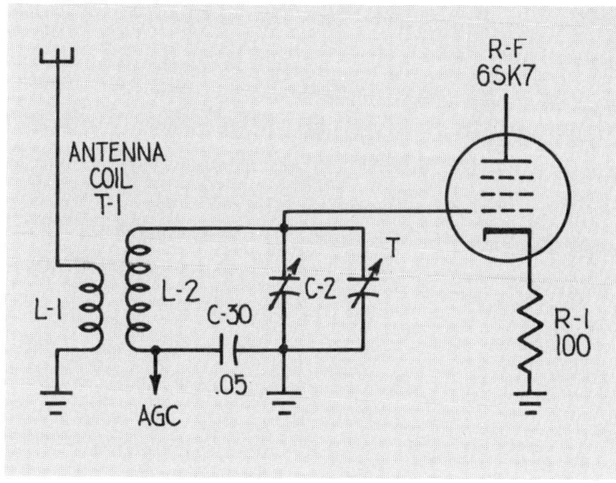

FIG. 17-3. An antenna coil and its position in the r-f circuit.

If capacitor *C*-4 shorts, the stage becomes inoperative owing to lack of plate voltage. The condition would be found much earlier in the test procedure, however, since the short would reduce the total *B* voltage and affect the operation of stages previously tested. The shorted capacitor would be found by voltage and resistance checking.

If capacitor *C*-30 shorts, the agc voltage applied to the stage would be shorted out. This would cause the stage to be operating at maximum volume with consequent overloading of itself or succeeding stages. The overload would cause poor tone on all but weak signals, a symptom of defective agc operation, which would focus attention on this circuit.

If any of these capacitors open, the trouble would be found by checking for the symptom that ensues. If plate capacitor *C*-4 opens, the gain of the stage would be reduced with possible oscillation also resulting. Signal check and sensitivity measurements would show normal response from the converter grid and insufficient gain from the r-f grid. This could be caused by a weak r-f tube, improper operating potentials, a defective interstage r-f transformer, or an open plate bypass capacitor *C*-4, all of which would have to be investigated. The condition would be found when a test capacitor, bridged across *C*-4, restores normal operation.

If agc bypass capacitor *C*-30 opens, the tuning circuit in the r-f stage becomes ineffective. This causes a weakening of the received signal, which in turn causes the agc to step up the gain of the controlled tubes. The net result is that strong locals come in like weak stations, that is, with a high noise level, and weak ones do not come in at all. The trouble would be found on signal check, since operation would be normal from the r-f grid and show no gain or a loss when checking from the antenna. An open antenna coil primary may give the same results. The trouble would be confirmed when a test capacitor bridged across *C*-30 restores normal operation. A second check that may be used for confirmation is that the trimmer capacitor across the antenna section of the gang tuning capacitor is ineffective.

Code interference. Any superheterodyne receiver is especially sensitive to any signal at its intermediate frequency. With an intermediate frequency adjusted to 455 kc, a signal at or near this frequency may get to the converter grid, be present in the converter plate circuit, and be accepted by the i-f amplifier. This condition will cause interference with the desired station. In areas near the seacoast, many powerful shore-to-ship transmitters operate at frequencies close to 455 kc. These cause interference that may blanket the low-frequency half of the broadcast band. Since the shore-to-ship transmission is generally in code, its interference is called code interference.

Code interference is a problem in some seacoast areas, and is most pronounced in receivers which do not use an r-f stage, since the converter grid is that much closer to whatever is being used for antenna pickup. Some receivers have built-in protection from such interference in the form of a wave trap. This trap is placed in the converter grid circuit and is tuned to give minimum response at the intermediate frequency. When intermittent code interference is the complaint, the installation of such an i-f wave trap is about the only remedy.

If the receiver does not have an r-f stage and uses a loop or ferrite loopstick for the antenna, the trap *L*-3/*C*-3 is placed in the converter grid circuit, as shown in Fig. 17-4A.

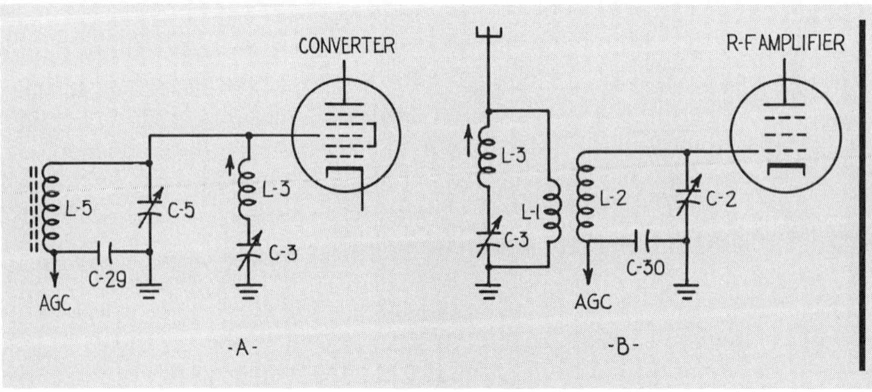

FIG. 17-4. Installation of i-f wave trap for reducing code interference.

Small permeability-tuned i-f wave traps are available. The lead to the converter-mixer grid should be as short as practicable.

Where a receiver uses a shielded antenna coil, the trap could still be located in the converter grid circuit, but it might be more feasible to wire it into the antenna circuit, as shown in Fig. 17-4*B*.

To adjust the trap, feed the signal generator output to the antenna of the receiver and wobble the frequency of the generator around the intermediate frequency to get a peak response from the receiver. Then adjust the trap while feeding in the same test frequency until there is minimum response from the receiver.

Reduction of image-frequency interference. Interference caused by the normal response of superheterodyne receivers to stations operating at the image frequency causes service difficulties in relatively few instances. Receivers that incorporate a tuned r-f stage are not troubled. Loop-operated receivers are likewise little affected, since the tunable loop antenna is less responsive to a station 910 kc off resonance than are the ordinary antenna and antenna coil which respond to a very large band of frequencies. The usual offenders, as regards image-frequency interference, are the types of receivers that employ an antenna and no r-f stage, or an r-f stage followed by an untuned converter stage. Also, the trouble is a local one, since another requirement is the presence of a strong image-frequency signal at the high-frequency end of the broadcast band (twice the intermediate frequency above the frequency of the desired station).

Image-frequency interference can be recognized as a whistle, or "birdie," which mars reception on one station at the low-frequency end of the broadcast band while reception is normal for all other stations. In the metropolitan New York area, the stations affected are either WMCA – 570 kc, or WNBC – 660 kc. In the case of station WMCA, the interference will be prevalent in the vicinity of station WHOM operating at 1,480 kc (570 + 910 = 1,480). Station WNBC may be troubled with image-frequency whistles caused by the presence of a signal from WQXR operating at 1,560 kc (660 + 910 = 1,570). In addition, reception at many points in the tuning range may on occasion experience image-frequency interference, if the receiver should be in the vicinity of police or amateur stations operating on frequencies ranging from 1,700 to 2,400 kc.

When the service job is to reduce image-frequency interference, various methods can be employed by the technician. A simple yet sometimes effective method is to reduce the signal input to the receiver. Modern superheterodyne receivers are usually more sensitive than needed for normal requirements of local reception, and may perform satisfactorily with very little antenna pickup. When this is the case, a reduced antenna may receive so much less signal from the interfering station that the whistle disappears. It is always worth while, therefore, to try the effect of a short indoor antenna on the interference. If it is effective, the technician should then check carefully to see that reception from all desired stations is satisfactory with regard to both signal strength and freedom from noise.

Another expedient is the installation of a wave trap tuned to the frequency of the interfering station. This frequency may be determined by adding twice the receiver intermediate frequency to the frequency of the station experiencing the interference. The wave trap chosen should have a range which includes the frequency of the interfering station.

Still another method of reducing image-frequency interference is to change the intermediate frequency of the receiver. The operation of a receiver is not greatly altered in respect to sensitivity, selectivity, tracking, etc., if the intermediate frequency is shifted about 10 kc, provided the receiver is completely realigned. The change, however, may reduce image-frequency interference. For example, assume a receiver with an i-f amplifier tuned to 455 kc and experiencing an image whistle when tuned to a station at 570 kc. The whistle is caused by a station operating at 1,480 kc (1,480 − 910 = 570). Suppose now that the i-f amplifier is retuned to 465 kc, and the entire receiver is realigned to operate at this intermediate frequency. The station at 1,480 kc will still be present and will cause image-frequency interference when a station at 550 kc is tuned in (1,480 − 930 = 550). But there is no local station at 550 kc and the image-frequency interference that marred reception from the station at 570 kc will be greatly reduced or entirely eliminated.

Variations in the r-f stage. Multiband receivers will, of course, cause changes in the r-f stage. These will be in the tuning circuit, switching arrangements, etc. The servicing of these components has been dealt with in connection with multiband circuits in the converter stage, and it is felt that the technician will be able to apply the same techniques to similar situations in the r-f stage.

Function of the antenna. The antenna is the first link in the receiver signal channel. Radio waves from all stations sweep across the antenna. Radio waves are one of the variety known as "electro-magnetic" waves. When such waves sweep across a wire conductor like an antenna, they generate a voltage in the wire, which in turn drives a feeble current through the antenna-ground system in the receiver. The current produced has the same frequency as that in the broadcasting-station antenna and possesses the same modulation that represents the desired intelligence.

Types of antennas used. The oldest type of antenna was simply a length of copper wire, about 75 feet in length and held horizontally by insulators at each end. A lead-in wire was then brought from one end of this antenna to an input terminal on the receiver. The antenna was mounted as high as possible to give the greatest signal pickup.

A later variety was the pancake loop antenna consisting of a flat loop of wire that was attached at the rear of the radio cabinet. It was placed in the receiver as the inductor for the input tuning circuit. Many pancake loop antennas are still being used.

The latest development is the ferrite loopstick antenna. Here, a small coil of wire is wound on a powdered-iron core rod. It, too, is the inductor for the input tuning circuit.

The old elevated antenna wire type is still found in fringe areas where station signals are weak. But it is usually connected to a separate winding on the pancake loop or ferrite loopstick. This winding has a small length of wire extending from the rear of the receiver. The leadin is connected to this short length of wire. When using an outdoor elevated wire antenna, it is advisable to take precautions against damage resulting from lightning. A commercial lightning arrester should be connected from the leadin wire to a metal pipe driven into the ground where the soil is somewhat damp. Many communities require such protection.

There are special antenna types used for f-m receivers. These will be described in the chapter on f-m receivers. We are concerned at this time only with a-m receivers and their antennas that feed signal to the r-f amplifier stage or to the converter.

Directional characteristics of antennas. It is common knowledge that antennas have directional characteristics. An elevated wire antenna will have best signal pickup in a direction along the length of the wire, and particularly from the direction opposite to the end at which the leadin is connected.

Pancake loop antennas have best pickup when the receiver is turned so that the desired station is coming from a direction along the horizontal axis of the loop. The ferrite loopstick will also have best signal pickup from one particular direction. Such reception behavior should not be looked upon as a defect in the receiver.

Trouble common to the built-in antenna. A common difficulty in most receiver coils is that windings open. This occurs in the ferrite-rod or pancake loop antenna.

A receiver with an open antenna winding may still play. But if it does, reception will be weak, erratic, and noisy, and the receiver may hum. On signal check, there will be a normal response when the i-f amplifier is tested. When the test signal is then shifted to the converter grid, the modulation note in the speaker will be at a reduced volume, rather than the normal increased volume, and it probably will be accompanied by hum. The open antenna winding is then found by a resistance check of the converter, when the reading across *L*-2/*C*-2 in Fig. 17-3 is open rather than its normal reading of 1 ohm.

The break in antenna winding *L*-2, regardless of whether it is of the ferrite-rod or pancake-type loop, may be in the winding itself, in the connections at the soldering lugs, or in the connecting leads to tuning capacitor *C*-2. Often enough, the break is located at the connection points or in the leads, either of which is easily repaired. Therefore it is wise to check for these possibilities before trying to find a very-hard-to-get exact replacement for antenna coil *L*-2.

Carefully examine the coil or loop to see if you can pick up a broken end near one of the connection lugs. If you can, the repair is accomplished by cleaning, tinning, and reconnecting the broken end. The wire may be enamel-covered; clean with fine sandpaper.

When checking a lead for the break, clip an ohmmeter to each end, and yank on the wire. Often enough, the break is inside the insulation and cannot be seen. It may even be making temporary contact at the time of the ohmmeter check and hence the need for the yank. If a defective lead is found, replace it. Use stranded hookup wire for the purpose, especially on pancake loops, since these are generally attached to the back of the cabinet and are subject to strain every time the receiver is opened for checking tubes, etc. If the break is in the winding itself and cannot be repaired, it must be replaced with an exact manufacturer's replacement part.

SUMMARY

Quick check for normal operation of the r-f stage.

If all previous stages checked showed a normal response, the trouble must be in the r-f stage.

Standard r-f stage diagram.

The accompanying figure shows the standard r-f stage.

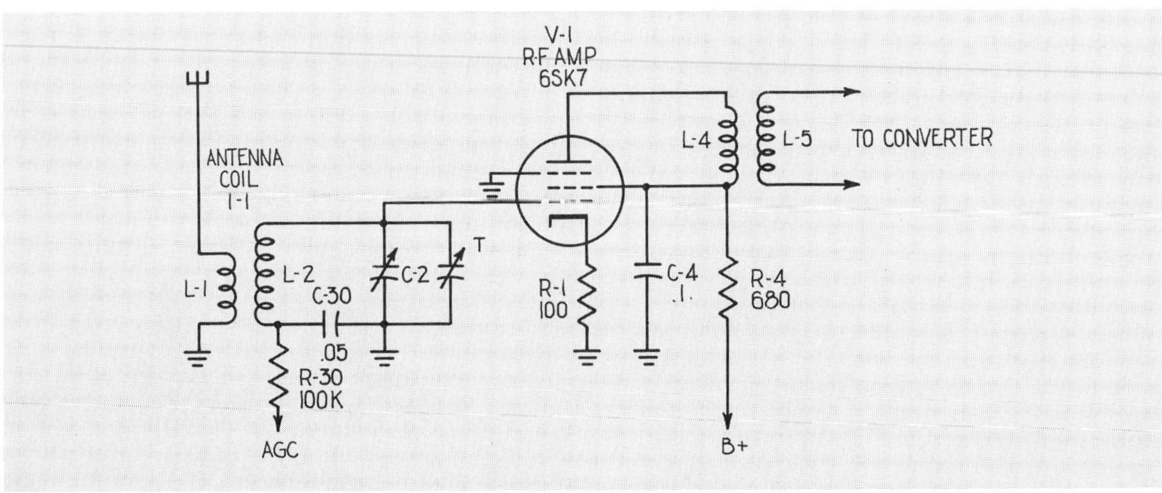

Normal voltage data for the r-f stage.

Readings are taken from the indicated terminal to the chassis or common negative. Normal voltage data are given in the accompanying table.

Tube terminal	6SK7 or 12SK7 pin No.	A-c receivers, volts	A-c/d-c receivers, volts	6BE6 or 12BE6 pin No.
Plate	8	90–135	90	5
Screen	6	90–135	90	6
Cathode	5	1	1	7

Normal resistance data for the r-f stage.

Across *L*-1	30–50 ohms
Across secondary of the antenna transformer	5 ohms
Cathode to chassis	100 ohms
Across primary of interstage r-f transformer	30–50 ohms
Control grid to chassis	2,700,000 ohms

SERVICE DATA CHART

Symptom	Abnormal reading	Look for
No signal from the speaker	Plate voltage = 0. Other voltages normal	Open primary of interstage r-f transformer *T*-2
	Plate and screen voltages = 0	Open plate decoupling resistor *R*-4
	Plate and screen voltages = 0. Other voltages low	Plate-chassis short circuit in the r-f circuit. Short-circuited decoupling capacitor *C*-4

Symptom	Abnormal reading	Look for
	Cathode voltage high. Other voltages normal	Open cathode resistor *R*-1
	Cathode voltage = 0. Other voltages normal	Dead r-f tube *V*-1
	All voltages normal	Short circuit in gang tuning capacitor *C*-2
Weak signal	All voltages normal	Weak r-f tube *V*-1. Open antenna transformer *T*-1. Open plate bypass capacitor *C*-4. Open agc bypass capacitor *C*-30. Misalignment
Oscillation	All voltages normal	Open screen bypass capacitor *C*-4. Shielding improperly grounded. Incorrect lead dress
Noisy operation	All voltages normal	Open or corroded antenna transformer *T*-1. Open agc bypass capacitor *C*-30 Corrosion in the interstage r-f transformer *T*-2. Defective r-f tube *V*-1. Defective gang tuning capacitor
Poor tone quality	All voltages normal	Short-circuited agc bypass capacitor *C*-30
Whistles on one or two stations	All voltages normal	Image-frequency interference

QUESTIONS

1. Outline a procedure to find the source trouble in a receiver that has a defective r-f stage.
2. A weak a-c superheterodyne receiver gives a normal response when the proper test signal is applied to the r-f grid and a weaker response when the same test signal is applied to the antenna. What are the likely sources of the trouble, and how would you check for each?
3. A dead a-c superheterodyne receiver gives a normal response when the proper r-f test signal is applied to the converter signal grid and no response when the same test signal is shifted to the r-f grid. Use the standard circuit of Fig. 1-1 and list the possible causes of the trouble. How would you check for each?
4. A superheterodyne receiver gets a station operating at 570 kc all over the dial. What is likely to be wrong? How can you check for this condition?
5. A receiver operates normally on all local stations except one at 660 kc. On this station, there is a persistent whistle that cannot be tuned out. What is the most likely cause of the difficulty? Outline a procedure to be followed in an attempt to minimize the condition.
6. A receiver with an r-f stage like the standard circuit of Fig. 1-1 picks up local stations but the reception is below normal in strength and is coupled with considerable noise. A signal check shows normal response when the proper test signal is applied to the r-f grid, and a loss when the same test signal is applied to the antenna. Name the parts that may cause this condition. How would each one be tested?

Alignment of a Superheterodyne Receiver 18

What is receiver alignment? The average superheterodyne receiver has seven or more tuned circuits, each of which has to be in resonance at its proper frequency for best operation of the receiver. When they are not in resonance, the receiver responds with a loss of sensitivity and selectivity to an extent depending on the degree to which the tuned circuits are out of resonance. The procedure for bringing these circuits to resonance at their operating frequencies is called "alignment."

Best alignment procedure for any particular receiver is an individual process. In each case, it is best to follow the exact method of alignment recommended by the receiver manufacturer in his service notes. Nevertheless, all methods are sufficiently similar to permit the presentation of a generalized procedure for use where specific service notes are not available. The purpose of this chapter is primarily to present such a generalized procedure.

The alignment procedure presented will be that to be used for a standard superheterodyne receiver operating in the broadcast band. Where special receiver variations occur, such as multiband receivers or broadband high-fidelity receivers, alignment notes have been given in the sections where such variations were described. They will not be repeated in this chapter.

In most receivers, alignment adjustment for tuned circuits is performed by varying small semivariable capacitors in parallel or in series with the main tuning capacitors of the tuned circuits. These capacitors are known as "trimmers" when in parallel, and as "padders" when in series. They remove or add capacitance to the main tuning capacitor. In some cases, as in wave traps or i-f transformers, the trimmers are across the coils, and no other capacitor is present.

In many modern receivers, the alignment adjustment is performed by varying the position of powdered-iron core plugs in the coils of the tuned circuits. This procedure varies the inductance and therefore the resonant frequency of the tuned circuit. These are known as "permeability-tuned coils." Regardless of whether the adjustment screw varies the capacitance or the inductance of the tuned circuit, the alignment procedure is the same.

When does a receiver require realignment? The technician is often confronted with the question of whether or not to realign the receiver. He must be guided by the symptoms of trouble, by his own purposes, and by a few general rules.

A receiver with a complaint of lack of sensitivity or selectivity, or oscillation, may be in need of realignment. However, other factors may cause the same complaint. These include weak tubes, open bypass capacitors, etc., which must first be investigated. Be sure that the receiver is perfect in all other respects before resorting to realignment, which should not be performed unless it is necessary.

Alignment can be used as a service tool to find the cause of trouble in a receiver. For example, when the signal check shows a broad, weak response at an off-frequency setting, the circuit involved may be out of alignment. Then, when the associated trimmer is adjusted and fails to give a peak response, the indication is trouble in that circuit, rather than in alignment. Further investigation will show an open bypass capacitor, an open coil, or something similar. If adjusting the trimmer causes an improvement in response, the indication is that the alignment is at fault, and the receiver should be completely realigned.

As a general rule, when a coil or capacitor, which is part of one of the many tuning circuits within the receiver, is replaced, realignment should become a routine step. This rule is important because the replacement will rarely have the same value as the original unit.

Alignment of a superheterodyne receiver involves the adjustment of the tuning circuits of three main units within the receiver. These are the i-f stages, the local oscillator, and the r-f stages. The order presented is the order in which the units should be realigned.

Equipment used for receiver alignment. When a receiver is aligned properly, it gives maximum output for any signal introduced at any point within the receiver. The equipment required, therefore, for alignment is that necessary to introduce a signal into the receiver and that necessary to measure the output for maximum.

To introduce a signal, the signal generator is a prime requirement. It furnishes any frequency needed at high or low magnitude. As indicated in Chap. 7, where r-f signals are delivered to the receiver, as at the antenna, r-f amplifier, mixer, and oscillator stages, a 0.00025-mfd/600-volt capacitor should be connected in series with the "hot" lead of the signal generator. Where i-f signals are delivered to the receiver, a 0.1-mfd/600-volt capacitor should be used. Since automobile radio receivers have a special antenna-input circuit for matching purposes with the short antenna used, a special coupling is needed between the signal generator and receiver for alignment of the r-f stage. The circuit is shown in Fig. 18-1. When aligning the i-f section of the auto receiver, the same 0.1-mfd/600-volt capacitor may be used.

When the receiver being aligned uses a loop or ferrite loopstick antenna, the signal generator output is fed to the receiver as illustrated in Fig. 18-2. A two-turn loop, with a diameter of about 6 inches, is constructed with hookup wire, and the ends are connected to the output terminals of the signal generator. The two-turn coil, known as an injection loop, is placed in close proximity to the antenna of the receiver. The strength of the signal fed to the receiver can be adjusted by varying the distance between the two loops, or by adjusting the signal-generator attenuator.

The second requirement is a resonance or output indicator. The purpose of this device is to give not a numerical output value, but an indication of maximum output, regardless of its numerical value at that point. When a tuned circuit is at resonance, the output indicator will read maximum.

The resonance adjustment for most tuned circuits is screw-controlled. The head is usually slotted for screwdriver adjustment. Sometimes the head is hexagonal in shape, with or without a slot. Therefore, all that is required for adjustment is an aligning wrench or a screwdriver. However, an ordinary screwdriver should not be used. All resonant circuits in a receiver are quite sensitive. The capacitance introduced by the metal shank of an ordinary screwdriver throws off the true resonant point when the screwdriver is removed. To overcome this difficulty, alignment screwdrivers of polystyrene or bone fiber have been made which eliminate the capacitance effect. Socket wrenches for adjusting hexagonal screw heads or nuts are also available in nonmetallic materials.

Connecting an output indicator. If a receiver to be aligned is equipped with an electron-ray tuning-indicator tube, the tube gives a satisfactory indication of when a circuit is tuned to resonance, and no other output meter is needed. Maximum output is indicated when the "eye" closes to the greatest extent. In this case, the signal-generator output need not be modulated.

Various other methods of obtaining an indication as to when resonance is reached may be used. Probably the most satisfactory method is to feed a modulated r-f or modulated i-f signal to the receiver, and to measure the output at the speaker with an a-c voltmeter. If the service multimeter has a low a-c voltage range of 2.5 or 3 volts which will give a good indication at 0.5 volt, it may be connected as shown in Fig. 18-3A. If the meter does not give a good reading for such a low voltage, it is connected as shown in Fig. 18-3B, where a large signal voltage is obtainable. The purpose of the capacitor marked "0.1-mfd/600-volts" is to insulate the meter from the d-c plate potential of the second a-f tube. The capacitor offers low impedance to the audio signal voltage. The latter is the modulation note of the signal generator—in most cases about 400 cycles per second. The relative output-signal strength is then obtained from the reading of the meter.

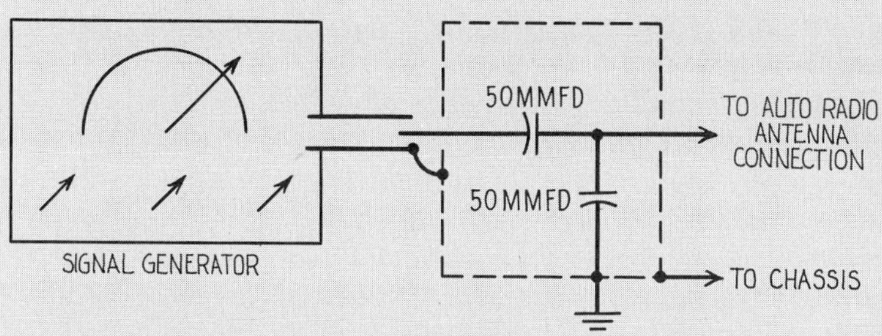

FIG. 18-1. Automobile radio coupling device.

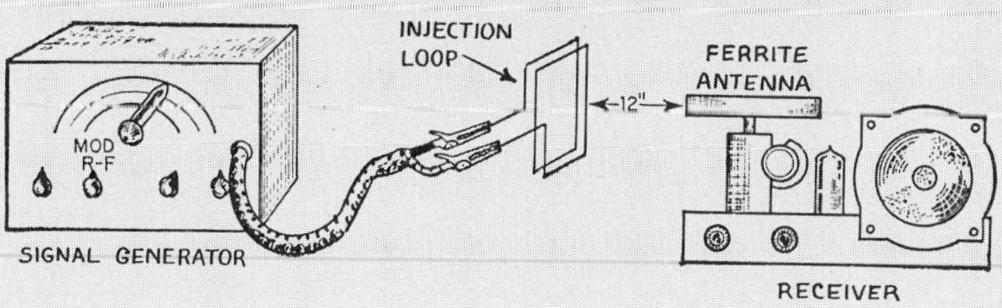

FIG. 18-2. Coupling the signal generator to a receiver with a ferrite loopstick antenna.

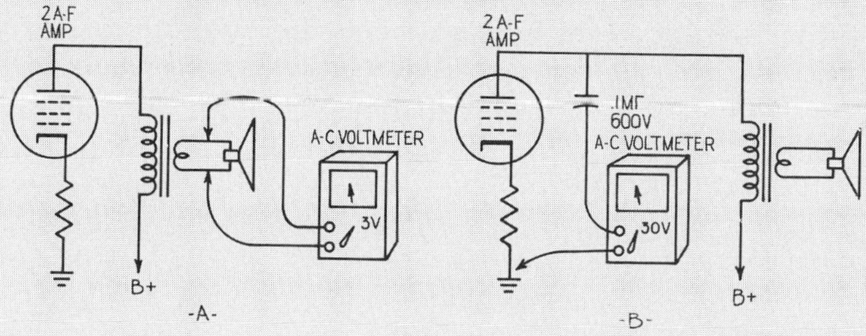

FIG. 18-3. Connecting an output indicator for alignment.

Strength of signal input. When the output indicator described above is used, it is extremely important to utilize as little output from the signal generator as is necessary to give a reading. The reason for this is the operation of the agc circuit. When a large signal is introduced to the receiver and a tuned circuit is adjusted for resonance, as the circuit is detuned, the agc voltage that is developed will drop. As a result, the controlled stages will have increased gain in an effort to keep the signal output of the receiver constant, and no drop in output at the output indicator will be seen. When a very small signal input is used, the agc circuit will not become operative, and the controlled tubes will operate at maximum gain. As a result, with detuning, the output indicated on the indicator will drop, and the agc circuit will not be functioning to raise the output to the constant level. The detuning will therefore be evident.

The standard output of 50 mw for receivers will require a signal input that is small enough to be below the level at which agc action will interfere with alignment adjustments, and should therefore be used for this operation. In an average receiver, 50 milliwatts corresponds to an output meter reading of 0.4 volt in the circuit of Fig. 18-3A, and a reading of 16 volts when the connections of Fig. 18-3B are used.

Receiver adjustments when aligning. When a receiver is aligned, its controls should be set in such positions as to give maximum gain. Below is a list of the positions for maximum gain. Of course, when a receiver does not have one or more of the listed controls, the technician will not make those settings.

Volume Control. Turned on full.

Tone Control. Set to the minimum bass position.

Fidelity Control. Set to low fidelity, or maximum selectivity position.

Sensitivity Control. Set to maximum sensitivity.

The tuning dial should be set up to the frequency required and indicated in the alignment procedure.

Location of i-f trimmers in a receiver. Before attempting alignment, the technician would be wise to check service notes for diagrams showing the location of trimmers and padders. If such diagrams are not available, the following suggestions will be helpful. The i-f trimmers are usually located in the i-f cans. Figure 18-4 shows a common arrangement. Look for the detector, i-f, and converter tubes, and visualize the block diagram of the superheterodyne receiver. The second i-f can is between the detector and the i-f tubes. The first i-f can is between the i-f and the converter tubes.

Manufacturer's instructions always advise the adjustment of the trimmers of the secondaries of the i-f transformers before those of the primaries of the i-f transformers. The location of which is primary and which is secondary is not easily determined. However, it does not matter too much which we adjust first, since the alignment of them is repeated.

Sometimes the i-f cans look like those in Fig. 18-5. Here, only one adjustment screw is visible from the top of each can. This arrangement is common for permeability-tuned i-f transformers. The top adjustment screw will be for either the primary or secondary of the i-f transformer. The adjustment screw for the other half of the transformer extends from the bottom end of the i-f can and is adjusted from the underside of the chassis.

Sometimes there will be a hexagonal hole instead of a screw slot under the hole in the i-f can. In these transformers, both permeability adjustments are made from the top of the can. Figure 18-6 shows such an i-f transformer assembly with special aligning tool.

The primary and secondary coils are wound on an internally threaded form. The tuning slugs are threaded on the outside and fit in the coil forms. The adjusting slots are hexagonal and extend through the cores. The alignment tool has a hexagonal head mounted

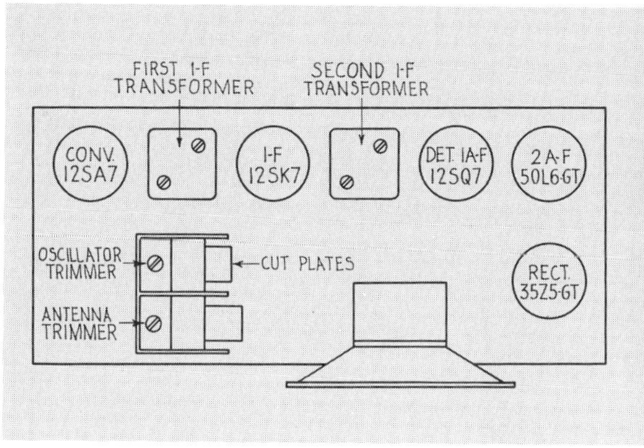

FIG. 18-4. Layout diagram of a typical small receiver showing trimmer locations.

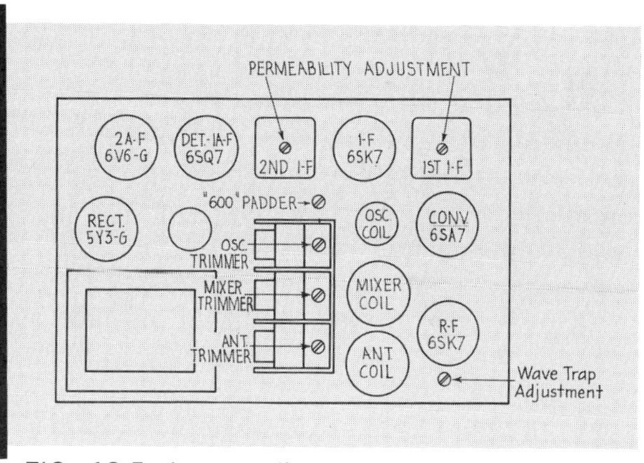

FIG. 18-5. Layout diagram of a six-tube superheterodyne receiver.

on a thin nylon shank. When adjusting the lower core, the aligning tool is slipped through the upper core slug and engages the lower slug. Turning the tool then causes the lower slug to move up or down in its threaded form without affecting the position of the upper slug, because the thin nylon shank does not engage the upper hexagonal slot. To adjust the upper slug, the tool is drawn up so that the head engages the upper slot.

Location of oscillator trimmers and padders. The two controls for the oscillator-tuned circuit are the 600 or low-frequency padder, and the 1,500 or high-frequency trimmer. If the tuning gang capacitor is of the variety with cut plates for the oscillator section, as in Fig. 18-4, the oscillator trimmer may be easily located. Receivers using this type of capacitor usually are not equipped with a 600 padder.

When the capacitors of the tuning gang all look alike, as in Fig. 18-5, the oscillator section can easily be found as follows: Tune in a station and touch the stator plates of each capacitor section. When the antenna and r-f sections are touched, reception will be little affected. But touching those of the oscillator section will detune the receiver. The trimmer mounted on this oscillator section is the high-frequency oscillator adjustment. The 600 padder will be located close to this oscillator section.

Some receivers do not use a padding capacitor, but a permeability-tuned oscillator coil instead. The alignment procedure is the same in either case. The oscillator coil is usually close to the oscillator section of the gang tuning capacitor. The permeability adjustment is usually located in the top of the coil, which is so mounted as to make the screw

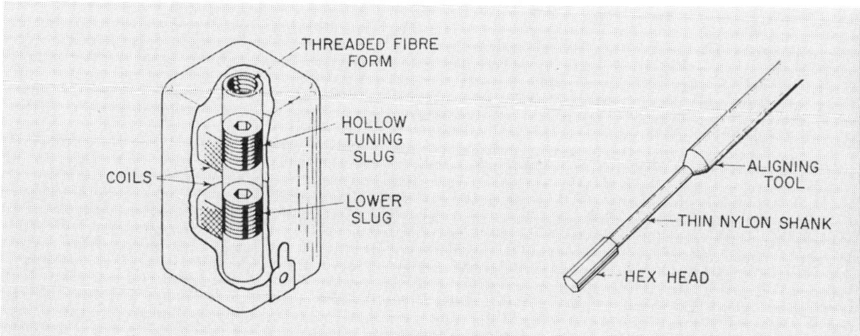

FIG. 18-6. Permeability-tuned i-f transformer in which both core slugs are adjusted from the top.

accessible from the top, side, or rear of the chassis.

Location of r-f, antenna, and wave-trap adjustments. Trace the antenna wire to the antenna coil. The wave trap will be close to the antenna coil. The antenna coil will also lead to the antenna section of the capacitor tuning gang. This locates the antenna trimmer. The r-f trimmer is the only one to be located. It may be found by tracing the stator lead to the mixer coil.

When the receiver is of the multiband type, the i-f trimmers are identified as before. The r-f trimmers, however, are usually mounted on the coil assemblies rather than on the gang capacitor. This makes the location of the trimmers somewhat more difficult, but it can be done.

Figure 18-7 shows the layout of a two-band superheterodyne receiver. All the coil cans, r-f and i-f, look alike. However, the second i-f can is between the detector and the i-f tubes. The first i-f can is between the i-f and the converter tubes. The oscillator coil can is in one line with the rear gang-capacitor section and the converter tube. The 600 padder is between the gang capacitor and the oscillator coil. The r-f coil can lines up with the center section of the 3-gang capacitor, and the antenna coil lines up with the front section of the gang capacitor and the r-f tube. Somewhat between the antenna lead and the r-f coil can, the wave trap will be found. In Fig. 18-7, the wave-trap adjustment is the trimmer near the antenna post. It can be confirmed by tracing the antenna wiring. The only thing left is to determine which trimmer on each coil is for the broadcast band and which is for short wave. This is not too difficult. Start the alignment procedure with the adjustment of the i-f amplifier in the usual way. When the first coil trimmer adjustment is reached, set the receiver for broadcast reception and give either trimmer a half turn. If it has no effect, return it to its original position. The other trimmer is the broadcast coil trimmer. To verify, give it a half turn and note the effect.

Adjusting a trimmer to a peak response. When making a trimmer adjustment, turn the trimmer screw back and forth slowly on each side of the peak output position. Note the peak position reading. The final adjustment is always one of tightening the screw, stopping at the peak response.

Sometimes, when the alignment tool is removed from the adjusting screw, the response falls below the peak obtained while the tool was in position. This effect occurs because the weight or capacitance of the tool affects the trimmer capacitance. When this happens, make the final tightening adjustment to the peak position, and tighten the screw an extra fraction of a turn. When this is done, the output meter reading will fall below the peak reading, but it should return to peak when the tool is removed.

In some receivers, peak response position occurs at the low-capacitance setting of the trimmer capacitor (when it is wide open), as shown in Fig. 18-9. A trimmer should not be left with this type of adjustment since, in this position, the spring tension of the top plate is at its weakest. Any jar or vibration, like that from the speaker, will cause the upper plate of the trimmer to vibrate, with accompanying noise and microphonics. When peak response position is at minimum capacitance, remove the adjustment screw and bend the top plate back, as shown in Fig. 18-10. Microphonics will thereby be eliminated.

Determination of the intermediate frequency of a receiver. Before attempting to align the i-f trimmers, the technician must know the intermediate frequency for the particular receiver to be aligned. This information is always found in the manufacturer's service alignment notes, or on the schematic diagram.

Where the usual information is not available, the technician can assume that the intermediate frequency is probably 455 kc, if the receiver is of modern vintage. The i-f amplifiers of the past few years have been designed to peak at some value between 440 and 480 kc. If, because of lack of information, a receiver designed to peak its i-f amplifier at 465 kc is aligned at 455 kc instead, no great harm will have been done. The receiver will operate normally and satisfactorily, although the tracking of the tuning dial may be slightly off.

If the technician is entirely uncertain of the intermediate frequency, he can determine it approximately by connecting the "hot" lead of the signal generator to the mixer grid. Then, by rotating the signal-generator frequency control from 500 to 150 kc, he can see at what frequency a response is heard. Usually, a misaligned receiver is not too far from its correct setting and a broad response, with possibly two peaks, will be obtained. For example, if a receiver shows a broad response centering at approximately 270 kc, it may safely be assumed that the receiver was originally designed to operate at 260 kc, the nearest commonly used intermediate frequency. The most commonly used intermediate frequencies are 175, 260, 455, and 465 kc.

When the procedure just described is used, it is important to make sure that the oscillator of the receiver is made inoperative. If the receiver oscillator is operating, there will be many squeals and responses heard, depending on the position of the receiver tuning control. The oscillator is made inoperative by placing a short across the oscillator tuning capacitor.

It is also important in the procedure to make

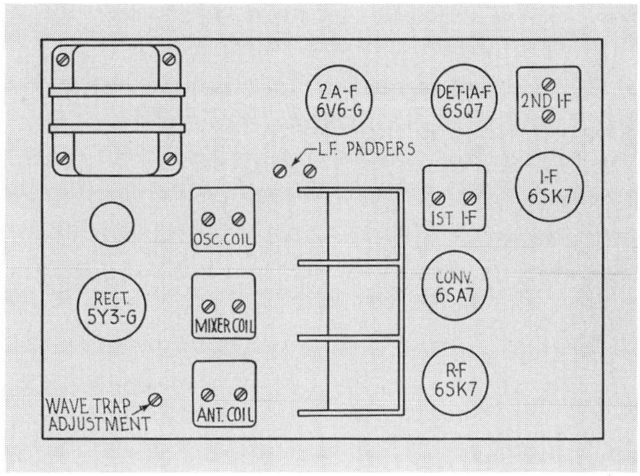

FIG. 18-7. Layout diagram of a six-tube two-band receiver.

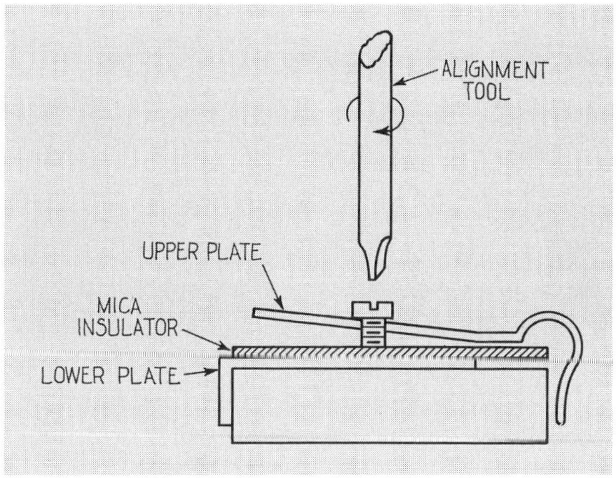

FIG. 18-8. Adjusting a trimmer capacitor.

FIG. 18-9

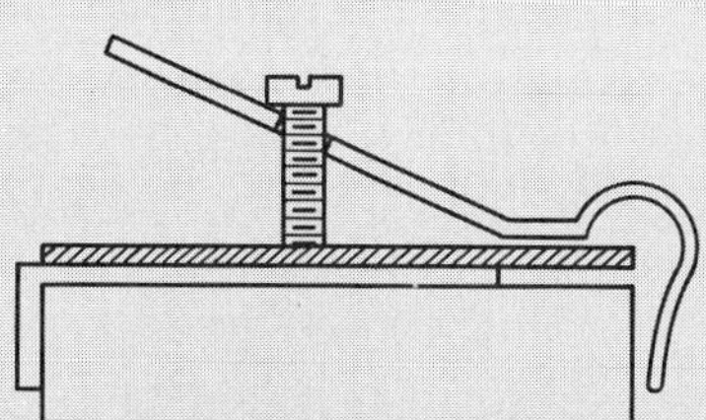

FIG. 18-10

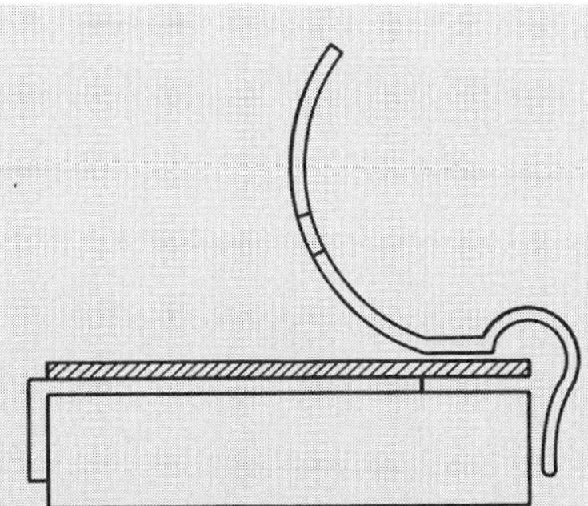

FIG. 18-9. A trimmer capacitor at its low-capacity setting.

FIG. 18-10. Provision to avoid microphonics when the trimmer capacitor peaks at the low-capacity setting.

sure that the receiver is not responding to a harmonic of the signal-generator output. This unwanted response can be avoided by starting the search for the intermediate frequency at the high-frequency end of the i-f spectrum, that is, at 500 kc. Then, when a receiver shows a response at 460 kc, the technician should expect another weaker response at 230 kc, the second harmonic of which is 460 kc. Still another very weak response should occur at 153 kc, whose third harmonic is about 460 kc. Were the technician to make the check at random frequencies, he might stop at the 230-kc response, assume an intermediate frequency of 260 kc, and try to align the receiver at that frequency, with very poor results.

Aligning the i-f amplifier. The receiver, signal generator, and output meter are hooked up as shown in Fig. 18-11. An isolation transformer should be used if the receiver is of the a-c/d-c type. The output meter is adjusted for a high-voltage a-c range. The signal generator is adjusted for a modulated output at the intermediate frequency of the receiver. The signal-generator output is connected through a 0.1-mfd/600-volt capacitor to the mixer grid of the receiver. The stator terminal of the oscillator capacitor is shorted to the capacitor frame. The receiver controls are set for maximum gain. Both the receiver and the signal generator are allowed a warm-up period of 15 min.

The signal-generator attenuator is adjusted to give an output as low as possible, with the modulation note heard faintly in the speaker of the receiver. Either trimmer in the output i-f can is then adjusted for the loudest note from the speaker. The attenuator is then reduced until the note can just be heard again. Then the other trimmer in the second or output i-f can is adjusted for the loudest note from the speaker. The procedure is repeated for the two trimmers in the input or first i-f shield can—first reducing the output to a faint note and then adjusting the trimmer for a maximum note.

The alignment is then repeated for the four trimmers in turn, starting with either trimmer in the second i-f can. This time the output meter setting is reduced to the 3-volt range if the meter is across the speaker or to the 30-volt range if it is connected to the plate of the output tube. The attenuator of the signal generator is adjusted to give a low reading of about 0.3 volt on the 3-volt range or 10 volts if the meter is in the plate circuit. Then, each trimmer is adjusted for peak voltage reading on the output meter. If the meter climbs to a high reading while any one trimmer is being aligned, reduce the attenuator of the signal generator once again and continue the adjustment. Remember that the output signal is to be kept below 50 milliwatts to prevent operation of the agc circuit. Approximately 0.4 volt at the speaker voice coil or 16 volts at the plate of the output tube corresponds to 50 milliwatts of output power.

When all four trimmers have been readjusted to give peak response on the output meter, the i-f alignment is complete.

Setting the receiver tuning-dial scale adjustment. The next step in the alignment procedure is to adjust the receiver tuning-dial pointer. The most common adjustment is to rotate the gang capacitor to maximum capacitance (plates fully engaged) position. Then set the dial pointer to the last calibration mark on the low-frequency end of the dial scale, as shown in Fig. 18-12.

Aligning the oscillator circuits. After the i-f trimmers have been adjusted, there remain the oscillator and r-f or antenna adjustments. Normally, the oscillator will have two adjustments: the high-frequency trimmer and the 600 padder. However, when a receiver uses cut plates in the oscillator section of the tuning gang capacitor, and no 600 padder, the adjustment of the oscillator circuit is relatively simple. The remainder of this section will give the alignment procedure for such an oscillator circuit, and it will be followed by the r-f and antenna circuit adjustments. The section after that will present the alignment

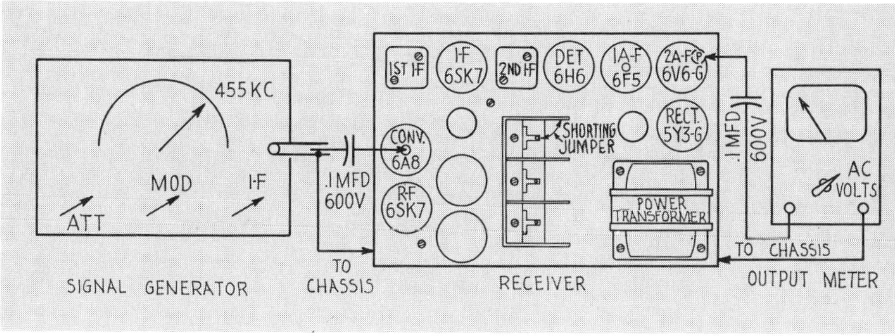

FIG. 18-11. Connections to a receiver for aligning the i-f amplifier.

procedure for an oscillator stage with a 600 padder.

The short (used in the i-f alignment) is removed from the oscillator section of the gang tuning capacitor. The signal-generator output lead is shifted to the antenna post of the receiver, and a 0.00025-mfd/600-volt mica capacitor is added as the dummy antenna in series with the generator output lead. The output meter is switched to a high-voltage a-c range. The receiver is tuned to a quiet point on the tuning range, between 1,500 and 1,700 kc. The receiver controls are kept at the maximum gain positions.

Now, the signal-generator frequency control is adjusted to deliver the same frequency as that shown on the receiver tuning dial, let us say 1,500 kc. The oscillator trimmer is then loosened all the way and tightened carefully until the signal is heard. This adjustment is very critical. When it is reached, the signal-generator output is reduced, and the output meter is switched to the lower a-c range. Then the trimmer adjustment is repeated for peak voltage output. If the reading goes down after removing the aligning tool, follow the procedure suggested in the section on adjusting a trimmer.

Adjusting the r-f and antenna trimmers. There remain now only the r-f and antenna trimmer adjustments. The signal-generator dial is turned to 1,400 kc. Then the receiver is tuned for maximum or peak response at or near 1,400 kc. The generator attenuator is reduced to give a low reading on the output meter, and the mixer trimmer is adjusted for peak response on the meter. If an antenna trimmer is present, it is then adjusted for peak response. Alignment of the receiver is now complete, except for the adjustment of a wave trap, if present.

Adjusting the oscillator and r-f circuits when a 600 padder is present. The alignment procedure is somewhat different when a 600 padder or permeability adjustment is present for alignment at 600 kc. The signal-generator output is fed to the antenna of the receiver through a 0.00025-mfd mica capacitor, acting as a dummy antenna. The generator is adjusted for a modulated output at 600 kc. Then the receiver is tuned to 600 kc, and the 600 padder is adjusted for peak response.

Next, the receiver is tuned to a quiet point near the high-frequency end of the dial. The signal-generator frequency control is adjusted for the same frequency as that shown on the

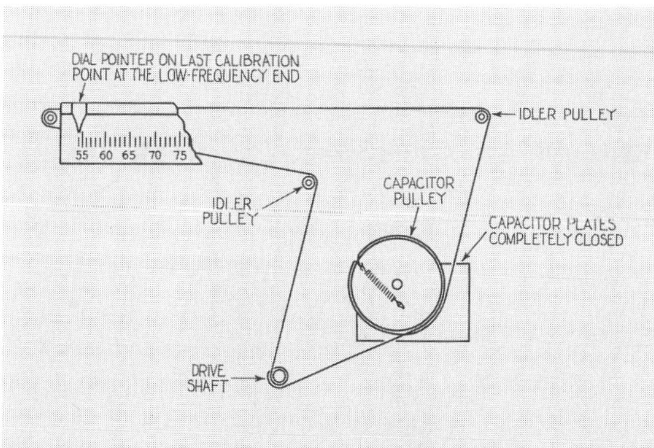

FIG. 18-12. Setting the dial pointer.

receiver tuning dial. The high-frequency trimmer on the oscillator section of the gang capacitor is then carefully adjusted for peak response. This last adjustment is critical. In making it, the trimmer screw should first be loosened and then slowly tightened until the note is heard. The attenuator is then reduced for a low reading on the output meter, and the adjustment repeated for peak voltage on the output meter.

The next step comprises the r-f and antenna trimmer adjustments. The oscillator is set to 1,400 kc. The receiver tuning knob is tuned back and forth near 1,400 kc and left at the position of greatest response. The r-f and antenna trimmers are now adjusted for peak response, as described previously.

Alignment of the oscillator circuit is not yet complete. The gang tuning capacitor must be rocked at 600 kc on the tuning dial, and the 600 padder adjusted for peak response. This rocking procedure is described below.

Rocking the gang tuning capacitor at 600 kc. In the rocking procedure, performed step by step, the receiver and signal generator are both tuned to 600 kc, and the 600 padder is readjusted for peak response. The attenuator is then set to give an output meter reading of 50 milliwatts. The signal, of course, is being fed to the antenna.

The receiver is then tuned slightly higher than 600 kc, such as 605 kc. The signal generator, however, is left at 600 kc, and the 600 padder is readjusted for peak response. If the output meter reading increases above the previous reading, the maneuver is repeated until a maximum voltage is obtained. If the output meter reading does not increase, the receiver tuning capacitor is rocked in the other direction; that is, the receiver is tuned slightly lower than 600 kc, such as 595 kc, and the 600 padder is adjusted. The output is noted. If there is an increase, the capacitor is rocked still lower, the padder is adjusted again, and the peak output is noted. The rocking and padding adjustments are made and repeated until maximum output is reached.

The adjustment of the high-frequency oscillator trimmer is then checked at 1,500 kc for peak response at the high-frequency end of the broadcast band.

Adjusting the wave trap. The last step in receiver alignment is adjustment of the wave trap. The signal generator is connected to the antenna and ground of the receiver, the "hot" lead fed through a 0.1-mfd/600-volt capacitor. The generator is then adjusted to give a strong response at the intermediate frequency of the receiver, say 455 kc. The receiver is tuned to 1,000 kc, approximately the center of the tuning range. Then the wave trap is adjusted to give minimum response in the output meter. The receiver is now completely aligned.

SUMMARY OF ALIGNMENT PROCEDURE

1. Set receiver controls for maximum gain.
2. Connect ground lead from signal generator to receiver chassis (or *B*− in the case of an a-c/d-c receiver).
3. Connect output meter to plate pin of second a-f tube and chassis.
4. Allow signal generator and receiver to operate for 15 min as a warm-up period before aligning.

Signal generator		Receiver			Notes
Frequency setting, kc	Dummy antenna mfd	Generator "hot" lead to receiver terminal	Tuning capacitor setting, kc	Adjustment	
455 or I-f	0.1	Mixer grid	Any	I-f trimmers	1, 2
			Maximum capacitance	Tuning dial pointer	
600	0.00025	Antenna	600	600 padder	
1,500	0.00025	Antenna	1,500	High-frequency oscillator trimmer	
1,400	0.00025	Antenna	Tune for maximum output	R-f and antenna trimmers	
600	0.00025	Antenna	600	600 padder	3
1,500	0.00025	Antenna	1,500	High-frequency oscillator trimmer	4
455	0.1	Antenna	1,000	Wave trap	5

Notes:

1. Short oscillator section of gang tuning capacitor.
2. Align i-f trimmers in the following order:
   a. Detector input trimmer (secondary of second i-f transformer).
   b. I-f plate trimmer (primary of second i-f transformer).
   c. I-f grid trimmer (secondary of first i-f transformer).
   d. Converter plate trimmer (primary of first i-f transformer).
   e. Repeat the adjustment of the trimmers in the same order.
3. Rock the tuning capacitor during this adjustment as follows: Turn the rotor of the gang capacitor back and forth and adjust the 600 padder until a peak is obtained.
4. This step is a check to see whether the previous adjustment of the 600 padder has affected the setting at the high-frequency end. If the high-frequency trimmer requires considerable readjustment, the 600 padder must also be readjusted by repeating the previous step.
5. Adjust for minimum output.

Phonographs and Home Audio Systems

19

Phonographs and home audio systems are simply expansions of the audio section of a standard receiver. However, the wide interest in good reproduction of music from records, particularly in hi-fi systems, has led to a complexity and refinement of audio systems that need special consideration. This chapter is designed to promote the understanding of such systems and to teach the procedures for servicing them.

We may classify phonograph players and home audio systems in various ways. One classification divides them into monophonic, or monaural, and stereophonic systems. Different types of records are prepared for each of the two systems. The monaural is the older of the two systems. However, phonographs of this type are still being manufactured because of their relatively low cost, and they will probably continue to be made.

Another classification may be made with respect to degree of fidelity of the reproduced sound. Factors to be considered here are tonal response and wattage. In the smaller phonographs, the audio amplifiers are very similar to those used in radio and TV receivers. Larger phonographs, designed for a wider tonal response, generally use more audio stages and include bass and treble controls and push-pull circuits.

The monaural phonograph system. A monaural phonograph consists basically of an electric motor that drives a turntable, a pickup, a single-channel audio amplifier, and a loudspeaker. The arrangement is shown in block form in Fig. 19-1.

The motor is generally designed for constant speed within wide limits of applied voltage and load. In some turntables, the motor drives a spindle, which in turn drives a rubber-rimmed wheel. This wheel acts as a friction drive applied to the inside rim of the turntable. The spindle diameter is graduated so that a manual adjustment may be used to set the wheel at different diameters, thereby causing the wheel to rotate at different speeds. In this way, the turntable may be made to rotate at speeds of 33, 45, and 78 revolutions per minute (rpm). Manufacturers often provide a fourth speed of 16 rpm. In other turntable systems, speed change is accomplished by engaging different idler driving pulleys.

The turntable systems in most common use, even in inexpensive players, are fully automatic. Several records are stacked for playing. The final positioning of the pickup arm near the end of the record activates a changing mechanism. In the changing cycle, the pickup is automatically lifted and moved to the edge of the turntable, the next record is dropped into place on the turntable, and the pickup is gently placed on the starting grooves of this record. After the last record is played, the changer mechanism operates a switch

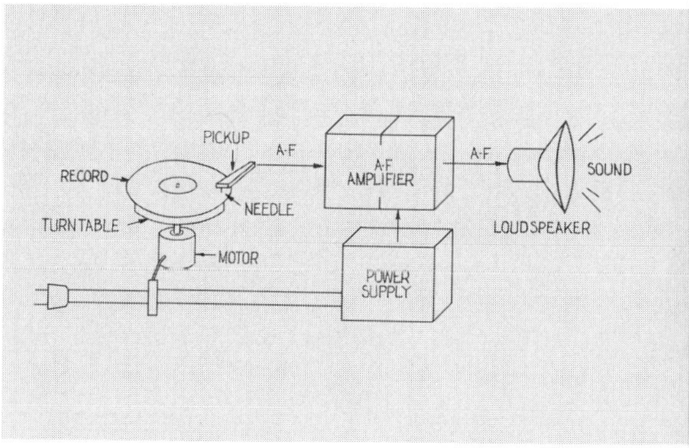

FIG. 19-1. Block diagram of a monaural phonograph.

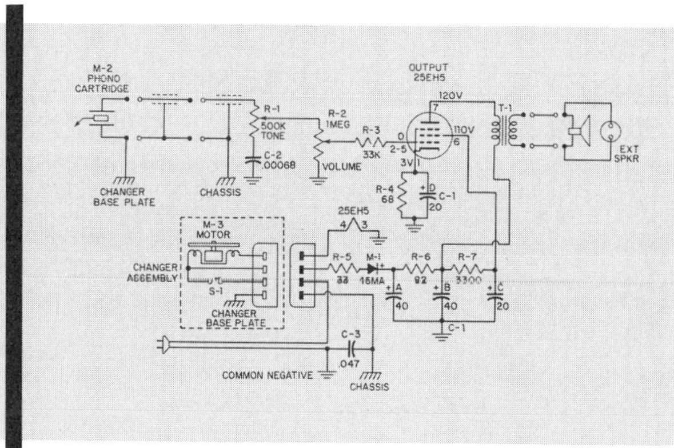

FIG. 19-2. Schematic diagram of a one-tube phonograph.

which turns off the motor. The same switch is often used to turn off the amplifier as well.

At the recording studio, records are made by a cutting stylus that cuts grooves into the record, beginning at the outer edge and slowly spiraling in to the center. Sounds fed to the studio microphone produce electric currents that pass through an a-f amplifier to the cutting stylus, which is forced to vibrate horizontally at the varying audio frequencies. This action produces a record groove with slight variations in wall shape, the variations representing the original sounds.

In phonograph playback, the needle in the pickup head rides the groove and is forced into horizontal vibration by the irregularities in the groove walls. The vibrating needle acts on a magnetic, crystal, or ceramic cartridge which converts the vibrations into a-f electric currents at the same frequencies as those produced in the studio microphone. When these a-f currents are amplified and fed to the loudspeaker, the original sounds at the recording studio are recreated. Of course, when there is a pause or no sound at the studio, the cutting stylus simply cuts smooth grooves; these do not activate the phonograph cartridge, so no sound is produced from the speaker.

The crystal and the ceramic pickup cartridge of the phonograph produce an average audio signal output of about 1 volt. This is comparable to the audio output level of a detector stage and is used in most radio-phonograph combination receivers. There are special high-output crystal cartridges which produce up to 3 volts of audio output. These require only one stage of audio amplification and are used in inexpensive record players. The magnetic, or variable-reluctance, pickups have a wide frequency response and exert a very light pressure on records. This gives a full tone and reduces needle scratch and record wear. The audio output is very low, so three or more stages of audio amplification are required. Magnetic pickups are generally used in audio systems designed for high-fidelity reproduction.

One-stage phonograph. The schematic diagram of a one-tube monaural phonograph with an automatic record changer is given in Fig. 19-2. The pickup cartridge is of the high-level type, followed by a single stage of amplification powered by an a-c/d-c power supply. The small number of parts and the simplicity of the circuit make this type very popular in inexpensive phonographs.

Note the number of connector plugs and jacks indicated between the amplifier chassis, the changer, the phonograph (phono) cartridge, and the speaker. These are for easy disassembly of the units for service work.

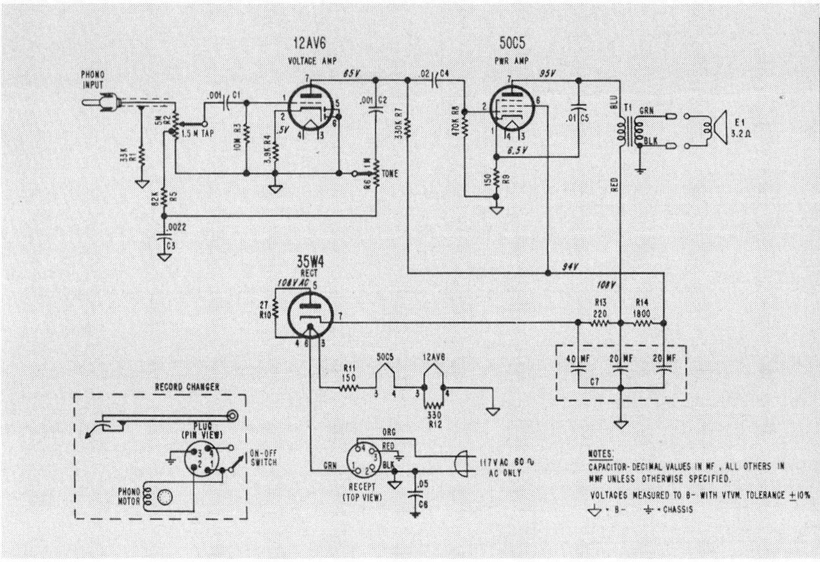

FIG. 19-3. Schematic diagram of a two-stage phonograph.

The wiring between the cartridge and the amplifier is shielded, as indicated by the dotted lines around the ungrounded ("hot") phono lead. The shielding is grounded to the changer base plate and the chassis. Most a-c/d-c radio receivers are enclosed in an insulated plastic cabinet, and precautions are taken so that the user does not have accidental contact with the "hot" chassis. In a phonograph, the user unavoidably comes into contact with metal parts on the changer base plate or the pickup arm. To avoid shock hazard, these components must be insulated from common negative, which is one side of the line in the a-c/d-c power-supply circuit. The changer base plate and the chassis are floated from the line by capacitor *C*-3. This capacitor insulates these metal parts at the power frequency, thereby preventing the shock hazard. The capacitor also provides a conductive path to ground for the higher-frequency a-f signals. In some phonographs, the chassis-floating circuit includes a resistor of high ohmage value that is bridged across the isolation capacitor.

The tube used in the amplifier is a 25EH5 type, which is rated to deliver 1.4 watts to the speaker when a signal of 3 volts is applied to the grid. Since a high-level cartridge delivers an average of 3 volts of audio signal, there is ample volume. Trace the "hot" line lead through the switch, phono motor, and the heater of the 25EH5 tube to common negative. The motor is acting as the series resistor to drop line voltage for the tube heater. The on-off switch *S*-1 is part of the changer assembly. It is turned on by the reject mechanism of the changer and automatically turned off after the last record is played. Note that the switch wiring also controls the plate power supply.

The latter circuit starts at surge resistor *R*-5 and selenium rectifier *M*-1. This is followed by a two-section *R-C* filter. Practically all phonographs use at least a two-section filter for better hum elimination. The filter capacitor unit, *C*-1, has a fourth section labeled *C*-1*D* in the diagram. This capacitor bypasses the cathode resistor of the output tube. This bypassing increases the gain of the tube, an important consideration in a one-stage amplifier.

Phonographs generally are equipped with tone controls, even in inexpensive models. In the circuit of Fig. 19-2, the tone control consists of potentiometer *R*-1 and capacitor *C*-2 connected across the input signal. Its

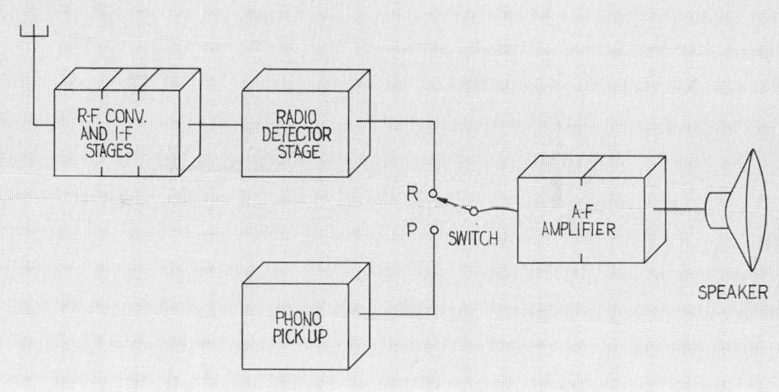

FIG. 19-4. Block diagram of a radio-phonograph combination.

action cuts the response in the treble register. When the arm of the tone control *R*-1 is moved toward the capacitor end, capacitor *C*-2 is across the input circuit and bypasses high a-f signals out of the amplifier, making the bass notes seem louder by comparison. This gives the user a control of needle scratch, as well as of tone. Needle scratch is in the high audio frequencies; turning the control toward bass reduces the high audio response and consequently reduces scratch noise.

Two-stage phonograph. The schematic diagram of a three-tube, two-stage phonograph is given in Fig. 19-3. The pickup is the standard crystal or ceramic type that produces an average audio signal of approximately 1 volt. Two audio stages are therefore required.

The amplifier circuit is fairly standard and is very similar to that used in radio receivers. Note the special phonograph details. There are plug-in arrangements for the changer, pickup, and speaker. The changer base and the chassis are insulated from common negative by capacitor *C*-6 in the power supply. The ON-OFF switch is located on the changer. Hum elimination is taken care of by a two-section *R*-*C* filter. There is a treble-cut tone control consisting of capacitor *C*-2 and potentiometer *R*-6 in the plate circuit of the voltage-amplifier stage.

Further, note that volume control *R*-2 is tapped for a bass compensation circuit, consisting of resistor *R*-5 and capacitor *C*-3. Bass compensation capacitor *C*-3 is shorted out when treble-cut tone control *R*-6 is in the maximum treble position.

Radio-phonograph combinations. Many receivers are equipped with a phonograph in a radio-phonograph combination. The receiver may also come equipped with a phonograph switch and an input jack so that the phonograph turntable and pickup unit may be added when desired. The phonograph will utilize only an audio amplifier and therefore will use only the audio stages and speaker of the receiver. At such time, it would be undesirable to have the r-f portion of the receiver in an operating condition. Therefore, a switch is used to block the radio signals from entering the audio stages. Likewise, when radio signals are being received, it is desirable that the phonograph pickup be disconnected from the audio stages. The setup is shown in the block diagram of Fig. 19-4, together with a simplified switching arrangement. The switch is shown in the RADIO (*R*) position used for the reception of radio signals.

The switching is usually arranged in the coupling between the detector and a-f stages before the volume control so that the latter is operative for either the radio or the phonograph. A typical radio-phonograph switch hookup is shown in Fig. 19-5.

The switch is shown in the RADIO (*R*) position, which is normal operation for the receiver. When the switch is changed to the PHONOGRAPH (*P*) position, the pickup feeds the audio amplifier through the volume control. Since some radio signals may leak through the switch and spoil the phonograph reception, provision is made to kill the radio when the switch is in the phonograph position. This is accomplished by opening the cathode, screen, or plate circuits of one or more of the tubes in the r-f section of the receiver. The lower half of the double-pole switch in Fig. 19-5 opens the plate circuit of the i-f tube when the switch is adjusted for phonograph operation.

The complete schematic diagram of a Motorola radio-phonograph combination is shown in Fig. 19-6, as a typical example of this type of unit. The radio is a standard 5-tube a-c/d-c receiver. It is mounted on a printed-wiring board. The usual precautions are taken to keep the user away from contact with common negative. The record-changer base is isolated from common negative by capacitor *C*-8, bridged by resistor *R*-11. These are shown in the diagram connected to the shielded lead from the phono cartridge as it enters the radio.

Note some interesting features of radio-phonograph operation used in this typical combination. The ON-OFF switch for the entire unit is ganged to the volume control, so this must be turned on first, regardless of the mode of operation. The switch on the record player is activated in the usual way, and it will turn off the phonograph motor but not the receiver. Now examine the radio-phonograph switch circuit. It is ganged to the tone control. The extreme counterclockwise position places the switch in the radio mode of operation, the position shown in the diagram. Here the detector output is fed to the volume control. The extreme clockwise position of the tone control throws the switch to the phono mode of operation. The upper portion of the switch now opens the detector circuit and feeds the output of the phono cartridge to the volume control. The lower portion of the switch breaks the cathode lead of the i-f amplifier tube, thereby making the radio inoperative and simultaneously throwing common negative to the phonograph motor so that it can operate. Note that the phonograph motor will not turn if the switch is in radio mode of operation.

Troubles common to phonographs and radio-phonograph combinations. As we have seen, the audio amplifier in a phonograph is very similar to the audio amplifier in a radio, with possibly a little more attention to tone quality and tone controls. Servicing these amplifiers therefore presents few new problems. The main problems to be considered are troubles with the motor, the pickup, the wiring, and the switches in radio-phonograph combinations.

The servicing of phonograph motors and record changers is a field in itself. The radio technician, however, should be able to check a motor for proper operation and to make a proper installation of a replacement unit, as well as minor repairs.

Phonograph-motor maintenance notes.

FIG. 19-5. Typical radio-phonograph switching circuit.

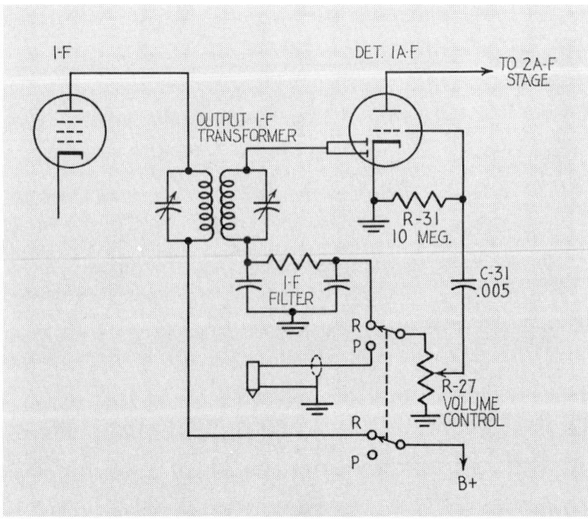

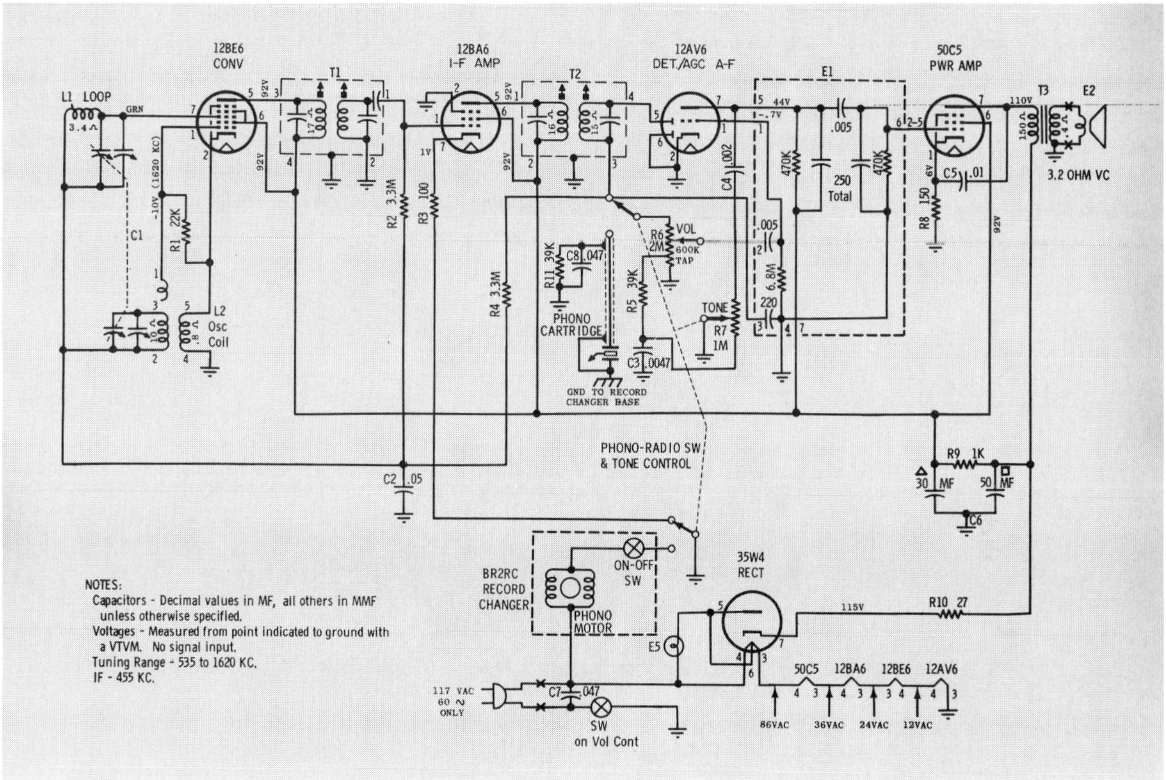

FIG. 19-6. Motorola radio-phonograph combination.

When a phonograph motor fails to operate, the line switch and wiring should be checked before condemning the motor. If the turntable speed is incorrect, the tone quality of the recording will suffer. The speed can be checked by means of a stroboscopic disk, such as the one shown in Fig. 19-7. The turntable is operated under a single fluorescent or neon lamp. One of the circles of dots will appear to stand still. Reference to the number above the stationary circle of dots will give the number of revolutions per minute (rpm) of the turntable. If the turntable has a speed adjustment, it can be properly set with the aid of the disk.

If the proper circle of dots remains stationary for the most part but shows a periodic jump for some of the dots, erratic action of the motor or drive mechanism is indicated. A worn spot on a rubber-rim friction-drive wheel could cause such an effect.

When the phonograph motor is supplied with oil cups, they should be filled with light machine oil. In lubricating a motor, the technician should be careful that oil is not smeared on the motor spindle or the rubber-tired drive wheel of a rim-drive motor (see Fig. 19-8). These should be washed with alcohol to remove any oil or grease. The same applies to the inner rim of the turntable.

The phonograph motor and turntable should float freely on rubber or springs. In some cases the motor mounting is floating; in others, it is solid and the entire motor board is floated. Figure 19-9 shows the mounting details for a typical phonograph motor.

If the spring or rubber suspension is inadequate, rumble may ensue. This is particularly important in combinations where the phonograph motor and speaker are housed in the same cabinet. The rumble is caused by a

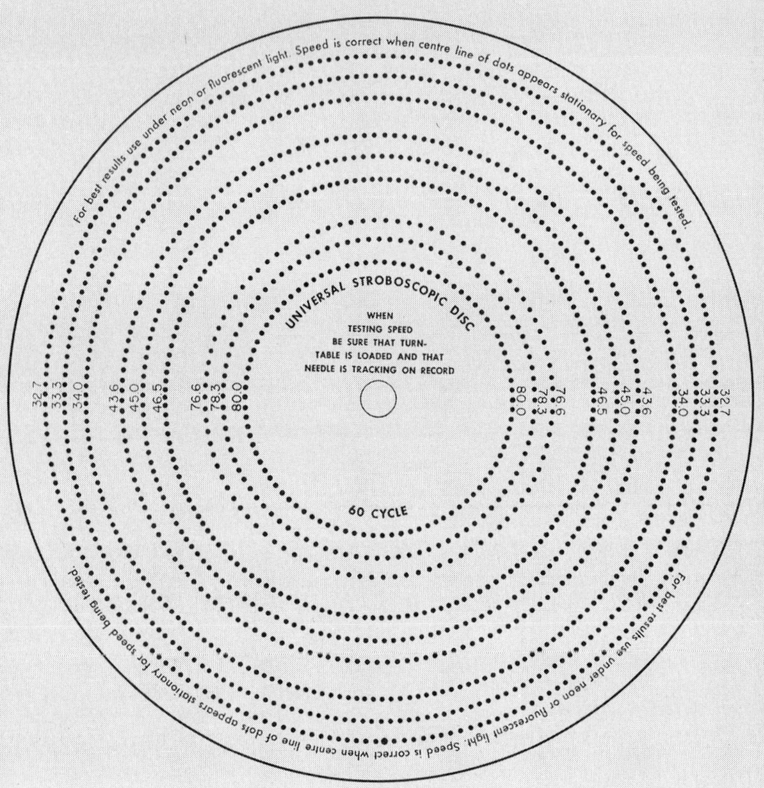

FIG. 19-7. Stroboscopic disk used for regulating speed of phonograph motors.

sort of mechanical feedback between the speaker and the pickup. Speaker vibrations cause the turntable and the pickup to vibrate. The vibration is in turn amplified and builds up the rumble.

Troubles common to the pickup unit. Phonograph pickup units develop various troubles: no output, weak output, and distorted output. In a radio-phonograph combination, the pickup is suspect when operation is normal in the radio mode.

In the case of a phonograph, it becomes necessary to look for trouble in either the pickup or the audio amplifier. If the defect is distortion, check the loudspeaker, tubes, and tube voltages in the usual way. If these are normal, the pickup should be checked by substitution, as shown in Fig. 19-10*B*. Try operation with pickup that is known to be good. The test unit can be a replacement part rated at the proper output level. A pair of clip test leads will serve for connections. The leads are clipped to the cartridge and the volume control. Then the cartridge is held lightly on a turning record by hand, as shown in the figure. Held in this manner, the cartridge will give almost normal operation as regards volume and tone quality.

If the above checks prove that the original cartridge is defective, it must be replaced. The replacement unit should be the same type as the original one, or one listed as a proper replacement for it. If the original type or a listed replacement is not obtainable, the replacement chosen must match the original one as nearly as possible in output level, weight, type of needle used, and mounting details. The output level is very important,

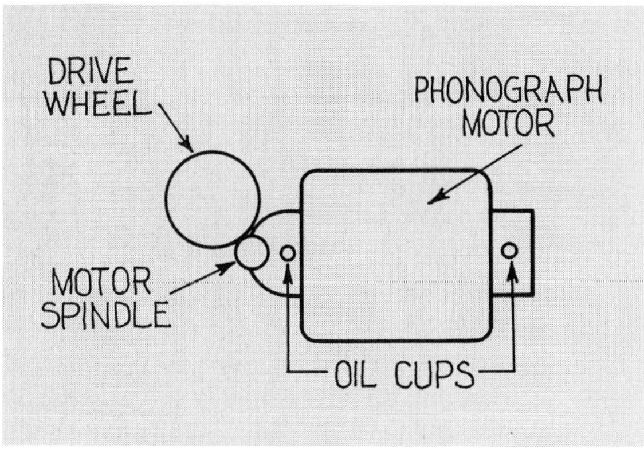

FIG. 19-8. Rim-drive phonograph motor showing location of oil cups.

since a 1-volt crystal will hardly be heard through a single-stage amplifier. On the other hand, a 3-volt crystal will overload the voltage amplifier tube in a two-stage amplifier, causing distortion at nearly all volume levels. Manufacturers of phonograph cartridges list most of this data in their literature as an aid in choosing a proper replacement.

For no output or weak output from a phonograph, apply an audio test signal to the hot end of the volume control. If the normal response is heard in the speaker, the pickup unit is at fault. Remember that the normal response with a one-stage player will be low. If the response is not heard, do not immediately conclude that the amplifier is defective. Shorts in the pickup or in its leads are quite common. Such a short will short out the test signal. Unplug the pickup lead or open its wiring and check again. Also check across the pickup leads for a short with an ohmmeter. If there is no short in the leads and the test signal is still not heard, the trouble is in the amplifier, its power supply, or the speaker. These are checked by the same procedures previously given for a radio receiver.

If the hot end of the volume control gives a

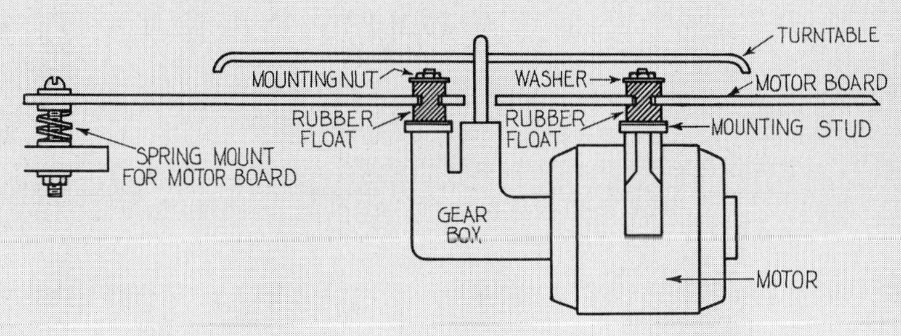

FIG. 19-9. Phonograph-motor board assembly showing rubber suspension mounting for the motor and spring mounting for the motor board.

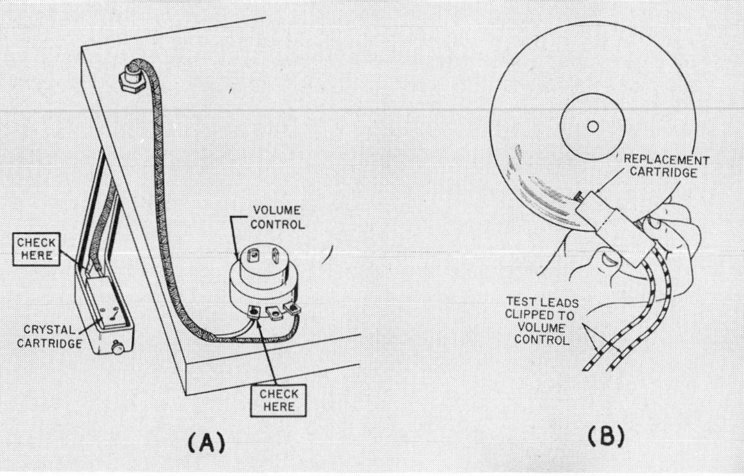

FIG. 19-10. Test procedure for determining location of trouble in pickup, leads, or amplifier.

normal test-signal response, the indication is that the amplifier is operative and that the trouble is in the pickup cartridge or the connection leads. To check the leads, apply the audio test signal to the hot end of the pickup cartridge (ungrounded end). The test points are illustrated in Fig. 19-10*A*. If the test signal is not heard from this cartridge end, the lead wire is open and must be replaced. If the test signal is heard from the cartridge end, then the cartridge is at fault. The final substitution test of the cartridge previously described should be made. If it is defective, it must be replaced, keeping in mind the factors mentioned.

How to work with shielded wire. In the previous section, a detailed procedure was given to find shorts or opens in the leads between the pickup cartridge and the amplifier. These leads are usually in the form of a flexible shielded cable. When one of the leads proves to be defective, it is best to replace the entire cable. Provide yourself with a roll of good-quality shielded phono cable and make up a new cable in the following manner:

1. Cut off the proper length. Include allowance for connections.
2. Push back the shielding to loosen the weave, as shown in Fig. 19-11*A*.
3. Bend loop in the lead (Fig. 19-11*B*).
4. Work the weave back and forth with a pointed instrument, such as a scriber, at the top of the loop until the hole made is large enough to pull the wire through. See Fig. 19-11*C*.
5. Slide the scriber under the wire and pull the end through the hole, as shown in Fig. 19-11*D*.
6. Pull out the empty piece of sleeving, as shown in Fig. 19-11*E*.
7. Repeat steps 1 to 6 at the other end of the shielded wire.
8. Strip and tin the ends of the lead ready for connecting (Fig. 19-11*F*).

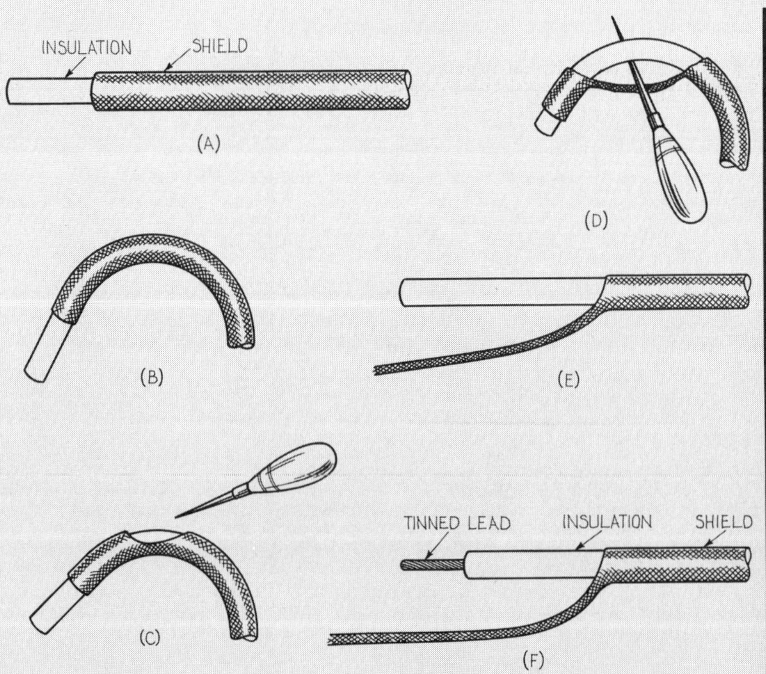

FIG. 19-11. How to prepare shielded wire for use.

When replacing shielded cable, use the prepared ends for connections. Do not attempt to solder the shielding in the middle of the lead to the chassis. The heat of the soldering tool will probably ruin the insulation inside the shield.

Troubles common to radio-phonograph switches. Radio-phonograph switches are subject to erratic action due to dirt between the contacts. This is almost always the case when it is necessary to flip the switch two or three times before positive contact takes place. A cleaning with carbon tetrachloride usually takes care of this difficulty. The usual procedure is to wet the switch arms and contacts with the carbon tetrachloride and then to rotate the switch quickly. The procedure may be repeated if necessary.

Sometimes a switch contact or the entire assembly is broken with use. When this happens, the switch must be replaced. Owing to the large variety in radio-phonograph switches, it is essential that one similar to the original switch be obtained. In replacing the switch, considerable care must be exercised to make sure that the wiring is correct and that the heat of the soldering iron does not draw the temper from the spring contacts. For correct wiring on an identical switch, it would be best to remove the old switch with its wiring intact, install the new one, and then change the wires, one at a time.

In soldering the switch terminals, it would be best to solder "uphill" where possible so that the solder or resin does not roll down to the switch contact. Production speed soldering will not draw the temper from the spring contacts. In this method of soldering, the resin-core solder is applied to the joint first, as shown in Fig. 19-12. Then the iron tip is pressed to both the joint and the solder. This makes a fast, clean joint that will not heat the contact unduly.

Precautionary check for a-c/d-c amplifiers. After completing the servicing of a phonograph or radio-phonograph combination with an a-c/d-c transformerless power supply, be sure to make the following precautionary check. With an ohmmeter, measure the resistance from chassis or changer base plate to common negative on either side of the line plug. This checks the floating-chassis circuit. The reading should be open circuit or a high-ohmage reading of 50,000 to 100,000 ohms, since the isolating capacitor is sometimes bridged by a resistor of approximately that value. If the ohmmeter reading is low, look for and clear an undesirable short so that your repair job does not create a serious shock hazard.

PUSH-PULL OUTPUT STAGE

Function of the push-pull amplifier. The push-pull amplifier is generally used as the last audio or power stage in high-fidelity home audio systems. It receives its signal voltage from the first a-f stage, amplifies it, and drives the loudspeaker. This function is similar to that of the second a-f stage in the standard receiver.

In addition, the push-pull stage serves in other functions. It makes it possible to deliver large amounts of power to the speaker system

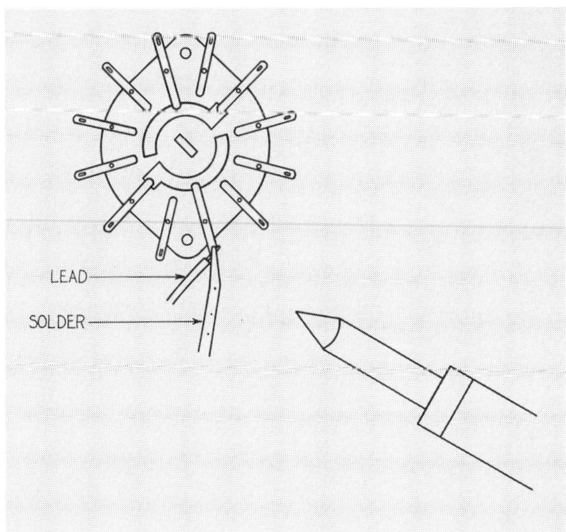

FIG. 19-12. Method of supplying solder when speed-soldering wafer-type switch connections.

with low distortion, while using small tubes and low *B* voltages. Another characteristic of the circuit is the reduction of hum. To summarize, the push-pull stage amplifies at audio frequencies with a large power output and with little distortion and hum.

Basic requirements of a push-pull amplifier. The theory underlying the operation of the push-pull amplifier can best be explained by means of a step-by-step development. Figure 19-13 is a block diagram showing the requirements of a typical push-pull second a-f amplifier. The audio signal from the first a-f amplifier is broken up into two voltages of equal magnitude but opposite phase, and they are then fed to the two grids in the push-pull stage. The output from the two tubes is coupled to the speaker.

A simple method of obtaining the 180-deg phase difference for the two grids, as well as equal voltages, is by means of a center-tapped transformer. Such a circuit is shown in Fig. 19-14*A*. At any one instant, the signal voltage at one end of the secondary will be positive, and at the other end negative in polarity. Since the center tap is grounded, one end of the secondary will be as much above ground in the positive direction as the other will be below ground in the negative direction.

Each of the terminals of the transformer is connected to the grids of the tubes as shown in Fig. 19-14*B*. As a result, the grids are 180 deg out of phase. When one grid is being driven positive with respect to ground by the signal, the other grid is driven equally negative with respect to ground. In the first tube, as a result, plate current will increase; in the second tube, plate current will decrease to the same extent.

A negative bias voltage is obtained in the usual manner, as shown in Fig. 19-15, by connecting resistor *R*-7 in the directly heated cathode circuit of both push-pull tubes. The plate currents of both tubes flow through this resistor and establish the bias voltage across it. The grids of both tubes are connected to

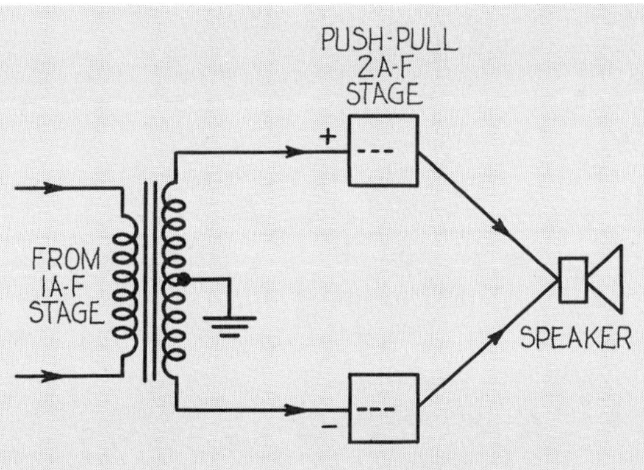

FIG. 19-13. Block diagram of a push-pull amplifier.

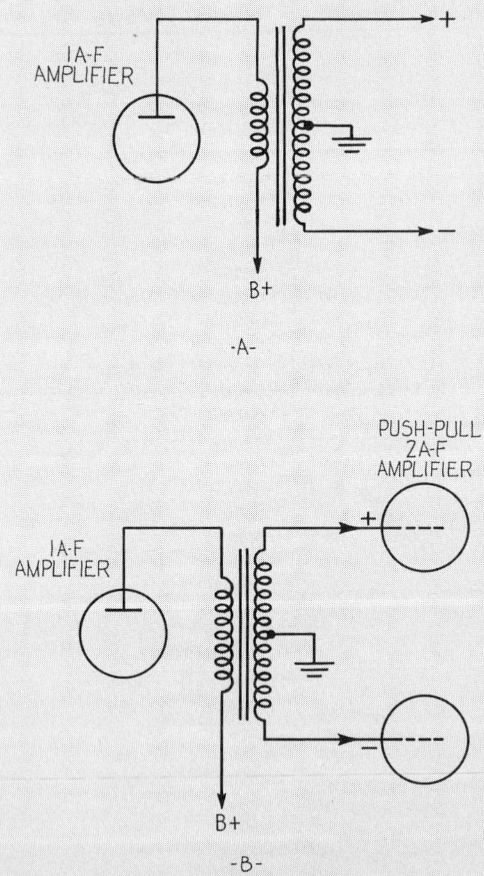

FIG. 19-14. The center-tapped transformer for obtaining input voltages for a push-pull stage.

the negative end of this bias resistor, thereby establishing the same no-signal bias voltage on each tube. Capacitor *C*-7 bypasses the bias voltage. In some circuits, it may be omitted, since its function is not so important in push-pull operation as it is in a single-tube second a-f stage.

The output from both tubes is fed into a push-pull output transformer with center tap connected to *B*+, as shown in Fig. 19-16. It should be noted that the flow of plate current through the two halves of the primary is in opposite directions. But coupled with this condition is another factor. Because one grid is driven more negative while the other is being driven less negative, we have a decreasing current through the first tube and an increasing current through the second one. The combined effect is to add the outputs from the two tubes in the secondary of the output transformer where a considerable voltage is induced. The outputs from the two tubes then drive the speaker.

A splendid characteristic of the push-pull stage is that it reduces or eliminates even-harmonic (primarily second-harmonic) distortion, a characteristic of single-tube operation. For sine-wave input, by way of illustration, even harmonics have a tendency to flatten one-half the cycle, as shown in Fig. 19-17*A* and *B*. In the second tube of the push-pull amplifier, since the signal is 180 deg out of phase, the distorted curve would look like Fig. 19-17*C*. These two distorted signals are combined in the output transformer, with a canceling out of the even harmonics, as shown in Fig. 19-18.

Because even-harmonic distortion is reduced or eliminated in a push-pull stage, the tubes may be overloaded somewhat without distortion. This accounts for the great power output from such a stage. For example, maxi-

FIG. 19-15. A method of obtaining bias for a push-pull stage.

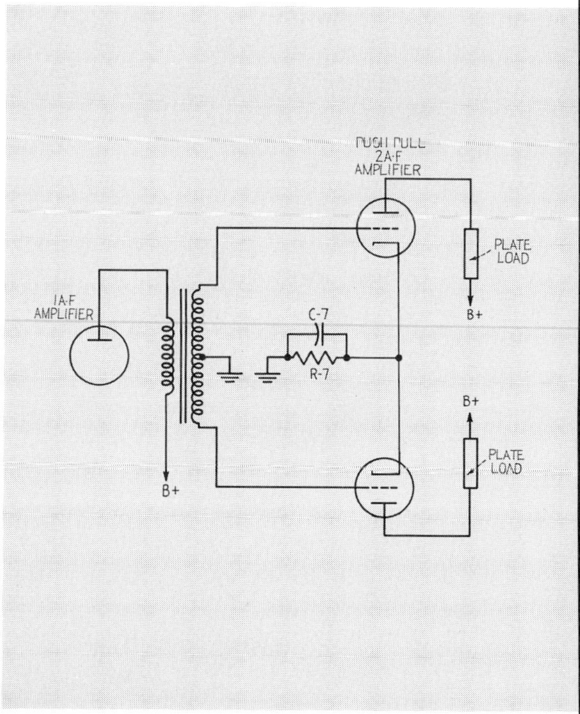

FIG. 19-16. The output circuit of a push-pull amplifier.

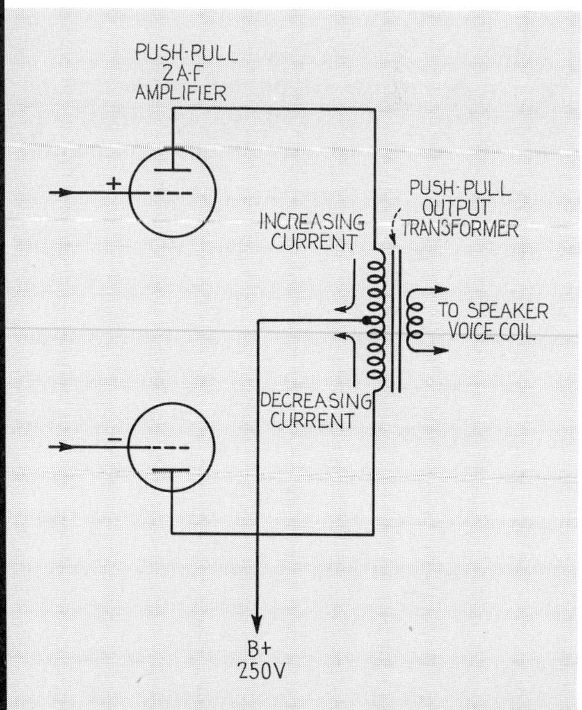

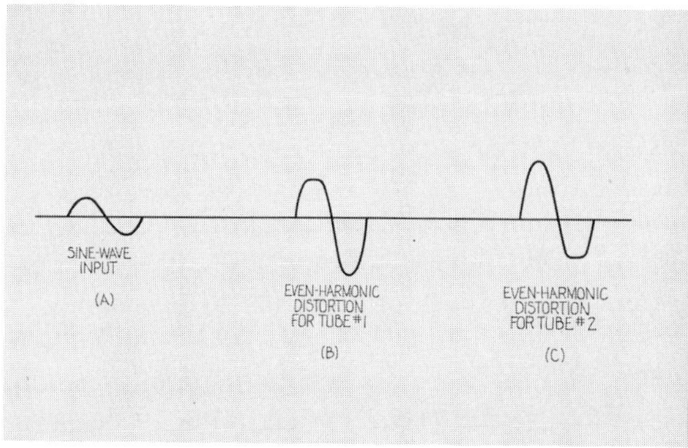

FIG. 19-17. The waveform of even-harmonic distortion developed in single-tube amplifiers.

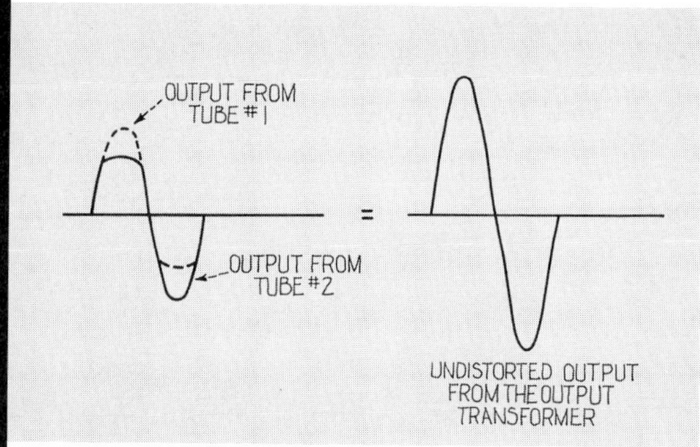

FIG. 19-18. Waveform showing cancellation of even-harmonic distortion in the output circuit of a push-pull amplifier.

mum undistorted output for a single 6V6 will be 5.5 watts, while a pair of push-pull 6V6 tubes will deliver a maximum undistorted output of 14 watts.

A push-pull amplifier will also reduce or eliminate any hum due to hum ripple fed to its plates. The reason is obvious. Hum ripple from the power supply will be fed to the center tap of the primary of the output transformer. Here it will move in opposite directions through the primary, but (in contrast with signal plate currents through it) will be rising or decreasing at the same time in the primary halves. As a result, the hum ripple cancels out.

An important point should also be considered. Since each tube requires the same signal-driving voltage, practically, as if it were operating alone, the total signal voltage delivered from the first a-f stage must be twice as great as for a single tube.

Figure 19-19 now shows the complete transformer-coupled push-pull stage. Transformer *T*-1 is the push-pull input transformer, which couples the output from the first a-f amplifier to the grids of the two tubes in the push-pull second a-f stage. For high-fidelity reproduction, the push-pull stage is usually operated in class *A* or AB_1, where the grids are always negative and the tubes always or nearly always pass plate current. Resistor *R*-7 and capacitor *C*-7 make up the common self-bias system for the push-pull tubes. Transformer *T*-2 is the push-pull output transformer. It combines the output from the two tubes and couples them to the speaker.

Transformer-coupled push-pull amplifiers are very common in older receivers and are still used in transistorized receivers. However, most modern tube amplifiers replace the input push-pull transformer with a resistance-capacitive type of coupling in conjunction with a phase-inverter tube.

Push-pull amplifier with phase inverter. Transformers are costly, and it is desirable to substitute for them where possible. Thus, resistance-capacitance coupling between the first a-f stage and the push-pull stage would be cheaper and more desirable. But it introduces a new problem. How can we get the 180-deg phase difference of signal voltage fed to the push-pull grids? The practical solution requires the use of another tube, known as a "phase-inverter" tube.

Figure 19-20 is a block diagram of a push-pull stage with a phase inverter. The principle of the phase inverter is that, for any tube, the grid voltage variations are 180 deg out of phase with the plate voltage variations. Hence, the positive signal voltage going to one of the second a-f grids is also directed to the inverter grid. The out-of-phase negative plate voltage pulse of the inverter is then fed to the grid of the other second a-f amplifier.

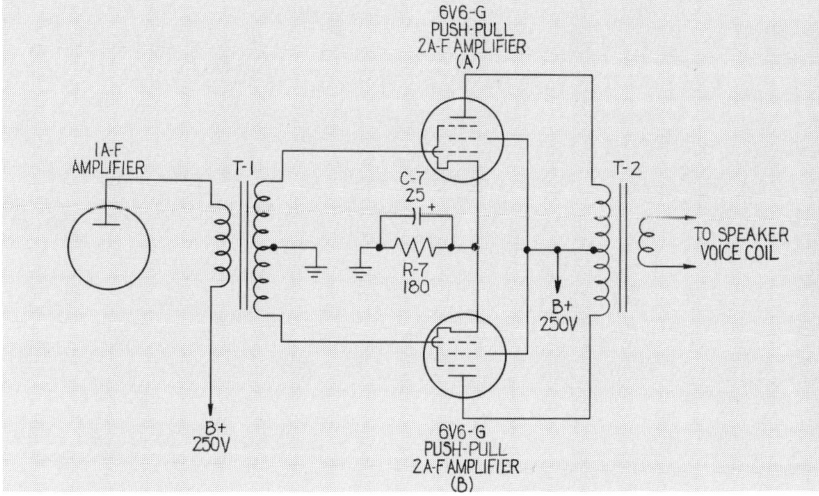

FIG. 19-19. Schematic diagram of a transformer-coupled push-pull stage.

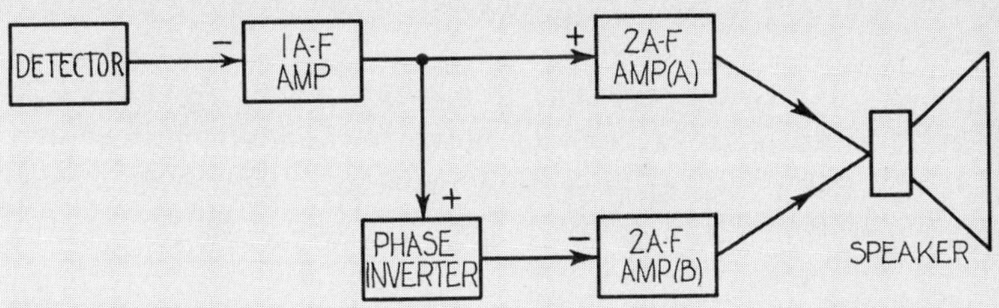

FIG. 19-20. Block diagram of a phase-inverter type of push-pull amplifier.

Another job that must be performed by the inverter is that of delivering the same voltage to the second second a-f grid as that which the first second a-f grid receives directly from the first a-f stage. With transformer coupling, a center tap to ground neatly takes care of this requirement. When using a phase-inverter tube, some provision must be made to compensate for the normal amplification of the added tube.

Let us examine the development of such a circuit. Figure 19-21 shows one of the push-pull tubes coupled by normal resistance-capacitance coupling to the first a-f amplifier. Note that a negative signal fed to the first a-f grid produces a decrease in current through *R*-2. This results in a lowered voltage drop across the resistor and, as a result, a rise in positive voltage on the plate. This rise in positive voltage feeds a positive pulse through capacitor *C*-2 to the control grid of the second a-f (*A*) tube. The first a-f tube has thus shifted the signal phase by 180 deg. The grid-leak resistor for the second a-f (*A*) tube is shown as two separate resistors, *R*-4 and *R*-6.

To obtain an equal but opposite voltage on the control grid of the other second a-f (*B*) tube, a portion of the positive output voltage pulse of the first a-f tube is tapped off from the grid leak of the second a-f (*A*) tube. Then this positive voltage is placed on the grid of a phase-inverter tube. Figure 19-22 shows the circuit described above. The self-bias circuit for the first a-f tube, made up of *R*-1 and *C*-1, is common to the inverter tube.

The positive signal pulse on the inverter

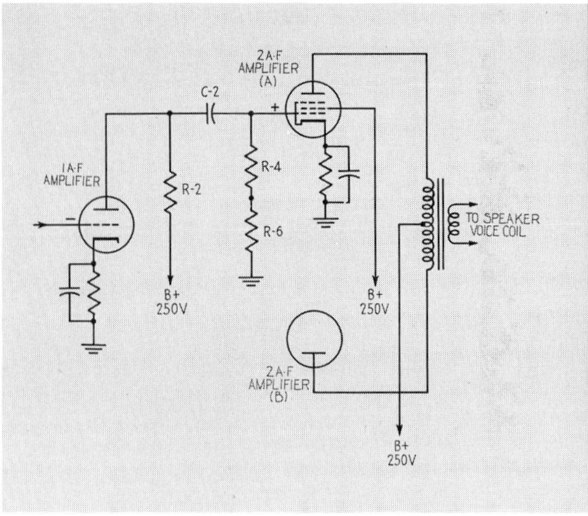

FIG. 19-21. Obtaining a positive signal on one push-pull tube grid from a negative signal on the first a-f amplifier grid.

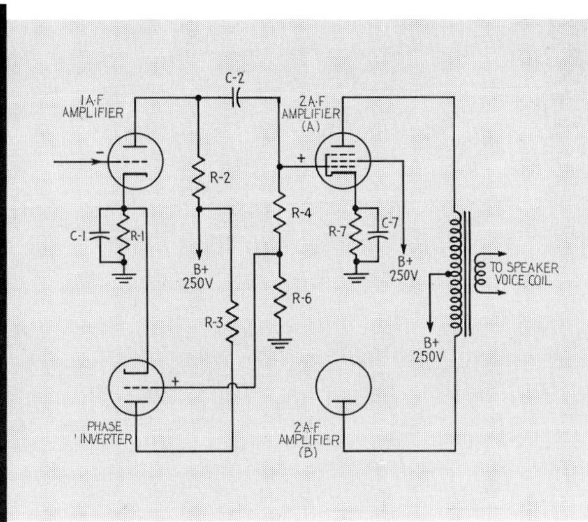

FIG. 19-22. Applying a portion of the positive output signal of the first a-f amplifier to the phase-inverter grid.

grid causes a rise of plate current through the plate load resistor *R*-3. The voltage drop across *R*-3 becomes greater and, as a result, the plate voltage of the inverter drops and produces a negative pulse. This latter pulse is fed by resistance-capacitive coupling to the second a-f (*B*) control grid, as shown in Fig. 19-23. Capacitor *C*-7 and resistor *R*-7 make up the common self-bias system for both push-pull tubes.

Note that the overall effect is to place opposite voltages on the control grids of the two output tubes, a condition that is desired. There still remains the problem of making these two voltages equal in magnitude. The output from the first a-f tube is fed to the second a-f (*A*) control grid. If this same output were fed to the inverter grid, the inverter gain would furnish a much larger voltage to the second a-f (*B*) control grid. This is not desirable. Hence, if the gain of the inverter is 20, a tap on the grid leak (*R*-4 and *R*-6) is taken so that only 1/20 of the voltage across it is fed to the inverter grid. Then, the voltages fed to the grids of the two push-pull tubes will be equal. For example, if *R*-4 is 10,000 ohms and *R*-6 is 190,000 ohms, the total resistance in the grid circuit of the second a-f (*A*) tube is 200,000 ohms. Then, since the voltage drop divides in direct proportion to the resistance,

$$\frac{10{,}000}{200{,}000} = \frac{1}{20}$$

and 1/20 of the voltage fed to the second a-f (*A*) tube is fed to the inverter grid. Here the voltage gain of 20 gives the same input voltage for both push-pull grids.

There are many methods for obtaining the necessary phase inversion, but the use of a phase inverter tube is the most common. Considerable variation exists in the tube used, in the gain of the stage, and in the circuit values. The schematic diagram of a typical a-c monaural radio-phonograph combination using a push-pull amplifier is given in Fig. 19-24. Since the power transformer isolates common negative from the line, there is no need for an elaborate isolating circuit. The chassis and changer base plate do not present a shock hazard and may be connected to an earth ground for most stable operation and minimum hum. The ON-OFF switch is ganged with the tone control and, as in most combinations, breaks the line supply to the

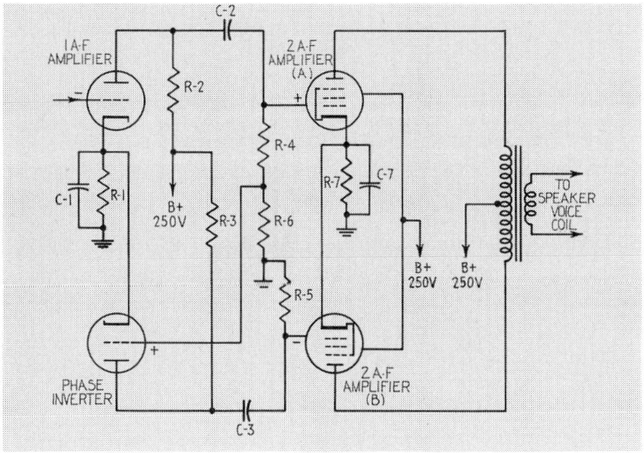

FIG. 19-23. Obtaining a negative signal on the other push-pull tube grid from the output of the phase-inverter tube.

changer. Note that the i-f amplifier, diode detector, and agc circuit are combined in a single 6SF7 tube. The twin triode 6SC7 combines the functions of first a-f amplifier and phase inverter. The signal from the detector is tapped from the volume control (*R*-13) and fed to the grid of the first a-f section of the 6SC7 tube. The output from this driver is fed by resistive-capacitive coupling to the control grid of the lower 6V6-GT output tube. The grid leak for this latter tube is made up of resistors *R*-19 and *R*-20. The combined resistance of these two is 480,000 ohms. The tap between the two furnishes 10,000/480,000 or 1/48 of the voltage to the grid of the inverter section of the 6SC7 tube, the gain of which is 48 under these operating conditions. The output of the inverter is then resistive-capacitive coupled to the control grid of the upper 6V6-GT output tube. In this way, equal but opposite signal voltages are obtained for the control grids of the two output tubes. Note capacitor *C*-31 across the primary of the output transformer (*L*-14). Its function is to reduce the tendency of the push-pull stage to fall into oscillation.

Signal check for normal push-pull stage operation. The signal check for the stage is shown in Fig. 19-25. The signal generator is adjusted to give an a-f signal, and the attenuator is set for maximum output. The signal-generator ground lead is connected to the receiver chassis and the "hot" lead is connected through a 0.1-mfd/600-volt capacitor in turn to the plate terminals of each second

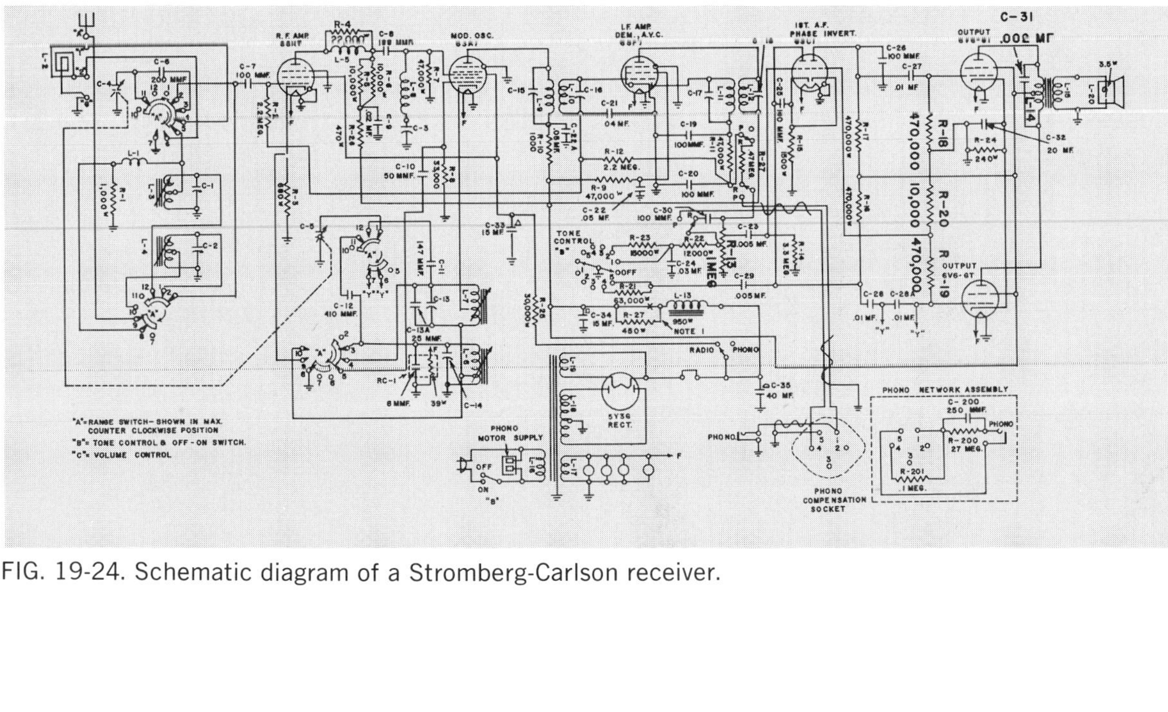

FIG. 19-24. Schematic diagram of a Stromberg-Carlson receiver.

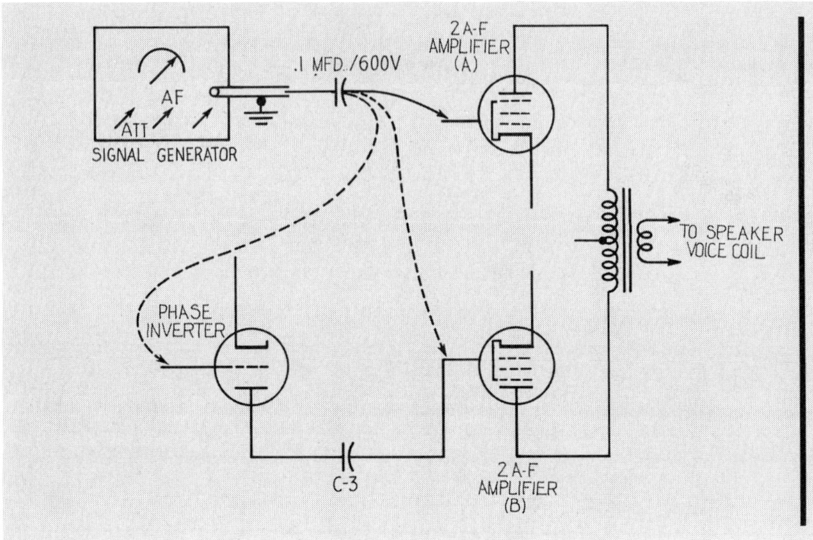

FIG. 19-25. Signal check of a push-pull amplifier.

a-f tube. If the signal generator has sufficient output, the audio note will be heard faintly each time in the speaker. The "hot" lead of the generator is then shifted to the control grid terminal of each tube. The signal-generator note should be heard in the speaker at a much greater volume, owing to the gain of each of the tubes.

Any audio input signal, like that obtained from the tip of a plugged-in soldering tool, may be used for quick checks.

To determine if the inverter is functioning, the "hot" lead of the signal generator is shifted to the inverter grid, and the attenuator is adjusted for less output. The signal-generator note should be heard clear and loud in the speaker.

Standard circuit. We shall consider the circuit shown in Fig. 19-26 as the standard circuit for a push-pull amplifier. The interstage coupling values given are for a high-gain voltage amplifier and phase inverter tube. Except for signal voltage divider resistor *R*-6, these values are fairly representative for any phase inverter circuit. Normal voltage and resistor data will refer to this circuit.

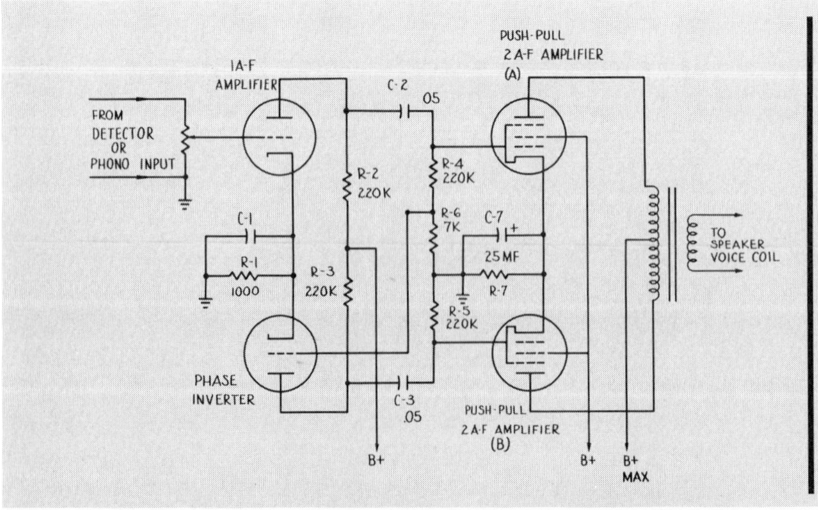

FIG. 19-26. Schematic diagram of a typical push-pull amplifier.

Normal voltage data.

	a-c	a-c/d-c
Either push-pull control grid to ground	0 volts	0 volts
Either push-pull plate to ground	200–300 volts	120 volts
Either push-pull screen to ground	135–200 volts	80–100 volts
Either push-pull cathode to ground	15–19 volts	5–7 volts
Inverter plate to ground	100–170 volts	50–70 volts
Inverter grid to ground	0 volts	0 volts
Inverter cathode to ground	2 volts	1 volt

Normal resistance data. The normal resistance data in the following table refer to Fig. 19-26.

Second a-f (*A*) plate to second a-f (*B*) plate	300–500 ohms
Push-pull cathodes to ground	60–200 ohms
Either push-pull control grid to chassis	220,000 ohms
Inverter plate to *B*+	220,000 ohms
Inverter grid to chassis	7,000 ohms
Inverter cathode to chassis	1,000 ohms

Troubles common to the push-pull amplifier. The troubles common to the push-pull amplifier are similar in many ways to those of the single-ended second a-f stage described in Chap. 12. Only those troubles that apply to the push-pull amplifier will be discussed here.

Troubles common to the output transformer. The output-transformer primary may open. As a result, one plate is open and no plate current flows from that second a-f tube. The voltage drop across the common self-bias resistor becomes less, and the bias for the other second a-f tube becomes too small. As a result, the operative second a-f tube distorts the signal badly. The distorted signal will also be weak because only one tube is functioning. A voltage check will show no plate voltage on one second a-f tube. Finally, an ohmmeter check will confirm the condition.

An open output-transformer secondary is not usual. When it does occur, the output signal will produce no sound from the speaker. For a defective transformer, an exact replacement is recommended. Where such a replacement is not possible, a universal push-pull output transformer may be used. Care must be taken to obtain proper impedance match between the second a-f tubes and the voice coil of the speaker. The transformer should be about the same size as the original in order to assure proper wattage dissipation. And the feedback connection, if present, must be properly connected. The reader is referred to the replacement notes on output transformers in Chap. 12 for a more detailed explanation.

Troubles common to the tubes. The second a-f tubes may become weak or inoperative. The result would be very similar to that of an open output-transformer primary. The operative second a-f tube will be improperly biased. The sound from the loudspeaker would be weak and distorted. Replacement with a good second a-f tube clears up the condition.

Troubles common to the coupling capacitors. The coupling capacitors *C*-2 and *C*-3 may become leaky. As a result, a positive voltage will be placed on the control grid of the second a-f tube to which it is coupled, causing bad distortion. Replacement with a capacitor of similar capacitance will remedy the defect.

Make sure that the voltage rating is as great as or greater than the original.

The inverter. The operation of the inverter is almost always foolproof. Only when its coupling capacitor *C*-3 becomes leaky does trouble arise. This has been described in the preceding section.

Troubles in the inverter tube or its associated circuit will also affect the first a-f tube and be found in a check of the first a-f stage. When the inverter is a separate tube, it operates in a manner very similar to the first a-f tube. A complete analysis of the first a-f stage is given in Chap. 13.

High-fidelity loudspeakers. To obtain full advantage of the elimination of distortion in a receiver with push-pull output, many manufacturers use high-fidelity loudspeakers. This setup usually consists of two loudspeakers with different response characteristics. One responds especially to the high frequencies and is known as a "tweeter." The other responds especially well to the low frequencies and is known as a "woofer." To obtain the desired low-frequency characteristic, the woofer uses a large-diameter paper cone with soft suspensions. It is also ruggedly built because low notes are associated with high power. The tweeter uses a small-diameter stiff cone. In some cases, horn speakers, which have a very good high-frequency response, are used for tweeters.

A typical circuit using this speaker system is shown in Fig. 19-27. The lower frequencies of audio signal are fed to the woofer voice coil and are blocked by the reactance of capacitor *C*-8 from getting into the tweeter voice coil. On the other hand, higher audio frequencies meet little opposition from the capacitor and feed through the tweeter voice coil. Capacitor *C*-8 has an approximate value of 4 mfd. It is either a paper capacitor or a nonpolarized electrolytic unit. Since this is a low-voltage circuit, capacitor *C*-8 rarely gives any service difficulties.

A three-unit high-fidelity speaker system is also quite common. In some cases, the third speaker is a second tweeter connected in parallel with the first tweeter. This is done because high-frequency sounds are very directional and may not be heard unless the listener is standing directly in front of the tweeter. The two tweeters are angled in the speaker compartment so as to give a wider distribution of the high-pitched sounds. In other cases, the third speaker is a mid-range unit. It is comparable to an ordinary radio loudspeaker, which gives good mid-range reproduction but does not respond very well to either the very low or the very high audio frequencies. All three speakers are connected in parallel; but a capacitor like *C*-8 in Fig. 19-27 is placed in series with the mid-range speaker, as well as with the tweeter to keep the high-power, low-frequency audio currents from affecting these other units.

Four-unit speaker systems are also found. These generally consist of a woofer, a mid-range speaker, and two tweeters.

Servicing high-fidelity speaker systems adds only the precaution to see that sound is coming out of each speaker. Repairs can be made on any individual speaker in accordance with the speaker service notes given in Chap. 11. However, if one of the speakers is to be replaced, it is important to obtain an exact replacement in order to ensure proper tonal balance.

The replacement speaker must also be connected properly. In a single-speaker set, the polarity of the speaker is unimportant. However, when two or more speakers are causing adjoining air spaces to vibrate, it is important that all paper cones push the air in unison and pull the air in unison. If the speakers are improperly polarized, one will push forward, compressing the air in front of it, while the others will pull back, rarefying the air in front of them. The undesirable effect will be a partial cancellation of sound in the area between the speakers.

Connecting a multispeaker system correctly is known as "phasing." Speakers meant to work in unison often have an identifying

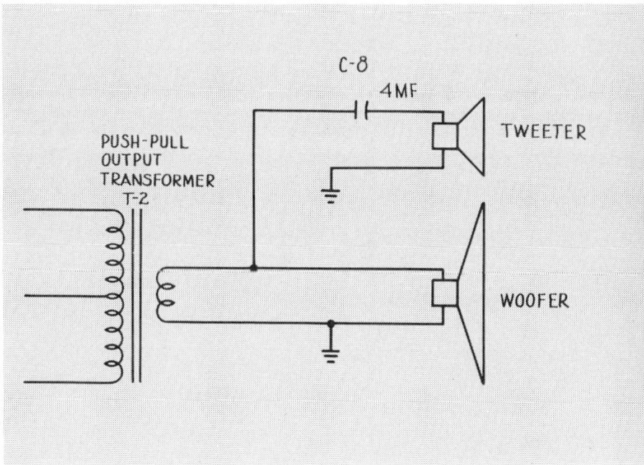

FIG. 19-27. A high-fidelity loudspeaker system.

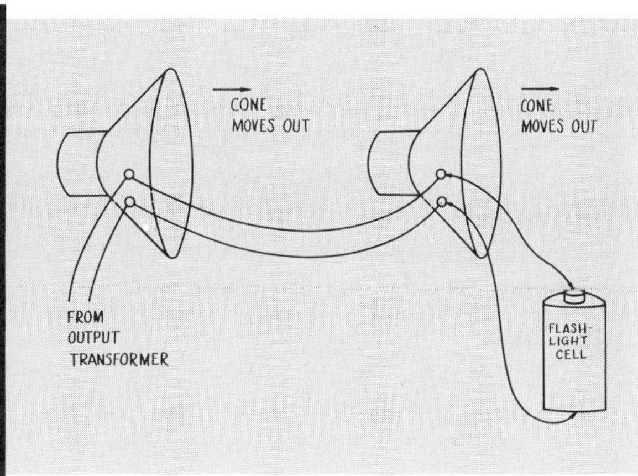

FIG. 19-28. How to check a speaker system for proper phasing.

phasing dot at one terminal. One output-transformer lead is connected to each of the marked terminals of each speaker. If the speakers have no identifying mark, or if there is any doubt about the proper connection, proper speaker phasing may be checked with a flashlight cell, as shown in Fig. 19-28. It is not necessary to disconnect the output-transformer lead. Connect the cell across any pair of speaker terminals, as shown, and break one cell lead while watching the paper cones. If the speakers are properly phased, all cones will move in the same direction, either all in or all out, as the connection is broken. If one moves in while the others move out, it is out of phase. Reverse the connecting leads to this speaker and check again to make sure that all speakers are now properly phased.

Tone controls. Almost all high-fidelity amplifiers provide manual controls for the high-frequency or treble response and for the low-frequency or bass response. There are many reasons for these controls. For example, if the pickup or speaker operates more efficiently in the bass register, cutting bass response with the control will restore the tonal balance. But it is not always desirable to have such flat overall response. If the listener's hearing is deficient in the treble area, or if his preference is to hear brilliant tones from the high-frequency instruments, he will probably prefer a tonal setting which gives maximum gain for the highs and reduced response for the lows.

Tone controls are generally placed between a pair of stages in the amplifier. Some common tone control circuits are shown in Fig. 19-29. In all three circuits, point *X* is the output of the first audio stage. It carries signals at all audio frequencies as they are produced by the pickup and amplified by the first audio tube. In most a-m radio receivers, point *X* is directly connected to the grid of the second audio stage. Here, all signals are further amplified and reproduced as sound from the loudspeaker. There is no control of tone. In the three circuits of Fig. 19-29, discriminating resistor-capacitor combinations are placed between point *X* and the following amplifier grid to give a control of the tonal response of the entire audio system.

Treble control. Those radios that employ tone controls and most inexpensive phonograph players use the treble tone control circuit of Fig. 19-29*A*. When the arm of tone control *R*-2 is toward the top, in the position marked CUT, capacitor *C*-2 bypasses signals out of grid-load resistor *R*-4. Since the reactance of a capacitor decreases with frequency, high-frequency audio signals are bypassed and reduced more than low-fre-

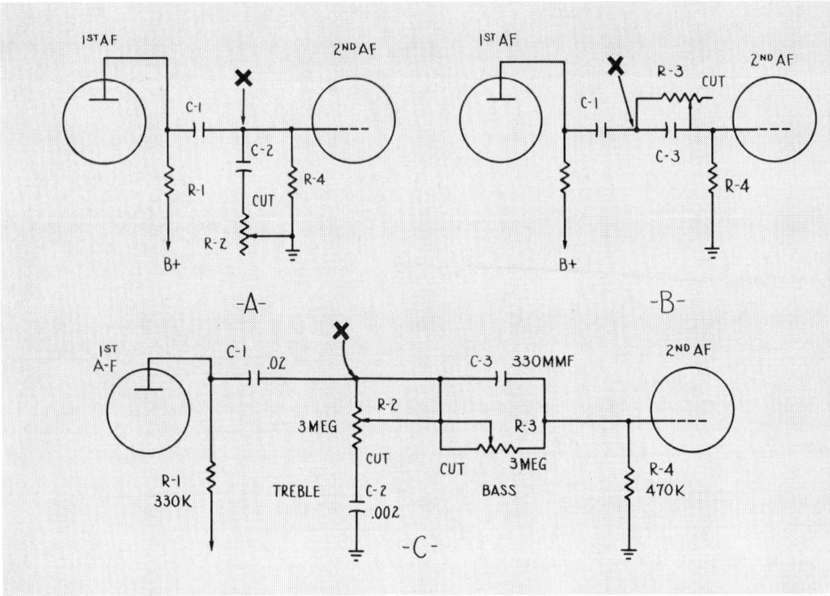

FIG. 19-29. Tone controls: (a) treble; (b) bass; (c) combined treble and bass.

quency signals, resulting in a loss of treble notes. This is the position called "mellow," where the bass notes boom through unaffected. Needle scratch, which is in the high-frequency register, is also reduced. As the control is rotated toward the bottom position, the bypass action of capacitor *C*-2 is gradually reduced, and more and more of the treble response is restored.

Capacitor *C*-2 generally has a capacitance of 0.0005 mfd or 0.001 mfd. The larger capacitance cuts the treble response more drastically and extends the effect further into the middle-frequency range. Tone control *R*-2 has a value of 0.5 or 1.0 megohms.

Bass control. The circuit of Fig. 19-29*B* shows a method of controlling the bass response. Coupling capacitor *C*-1 has a capacitance of 0.02 to 0.1 mfd, calculated to pass signals at all audio frequencies. Capacitor *C*-3, with a much smaller capacitance, is placed in series with the coupling capacitor *C*-2 and the next amplifier grid. Again, the reactance of the capacitor decreases with increasing frequencies, so that high a-f signals are passed unimpaired. The reactance to lower a-f signals is greater with the result that bass note signals are not readily passed on to the next amplifier grid. When the arm of the tone control *R*-3 is toward the position marked CUT, it has little effect on the action of capacitor *C*-3, and the bass response of the amplifier is reduced. As the control arm is rotated toward the other end, capacitor *C*-3 is gradually shorted out of the circuit and the bass response is restored.

Capacitor *C*-3 generally has a capacitance of 0.00025 to 0.0005 mfd. The smaller capacitance cuts the bass response to a greater extent and extends the effect further into the middle-frequency range. Bass tone control *R*-3 generally has a value of 2 or 3 megohms.

Combined treble and bass tone control. The circuit of Fig. 19-29*C* shows a combined treble and bass tone control circuit, taken from a high-fidelity Admiral receiver. High-frequency a-f signals go through capacitor *C*-3 to the next audio grid, regardless of the position of *R*-3. Low-frequency a-f signals are blocked from the next audio grid until the effect of capacitor *C*-3 is shorted out by the bass control *R*-3. Treble control *R*-2 permits more or less bypass action by capacitor *C*-2.

Servicing tone controls. There are few servicing problems in the tone control circuits. Since they operate in a low-voltage signal circuit, there is little danger of burned out capacitors or resistors. If one of the controls should become noisy or erratic in operation, the cause of the trouble is self-evident. If the faulty control does not respond to a cleaning, the replacement should match the original. Similarly, if a capacitor should open, its associated control would have no effect, thereby focusing attention on the tone control circuit. Again, a replacement capacitor should have the same values as the one replaced.

In some bass-treble control circuits, although the basic action of Fig. 19-29*C* is followed, more resistors and capacitors are added, generally to give a more gradual control. Often, the more complex resistor-capacitor network is combined in a single couplate component. When trouble is localized to the couplate, an exact replacement is required.

Tone controls in a feedback circuit. In some amplifiers, the tone controls are placed in a degenerative feedback line. The complete schematic diagram of an Airline phonograph player is given in Fig. 19-30 as a typical example of this type of control. It will also serve to review some of the features of high-quality audio amplification.

First, observe the switch on the changer and the plugin connections which are common to most phonograph players. Next note the floating chassis arrangement needed for an a-c/d-c type of power supply.

The amplifier uses a dual-triode 12AX7 tube and a pair of 50C5 output tubes in a push-pull circuit. The first triode section of the 12AX7 tube is a voltage amplifier, feeding the upper 50C5 tube and the phase inverter. The phase inverter is the second triode section of the 12AX7 tube. In turn, the phase inverter plate feeds its out-of-phase signal to the lower 50C5 output tube, completing the input to the push-pull circuit. Grid resistors *R*-8 and *R*-9 make up the voltage divider, feeding reduced signal to the phase inverter grid.

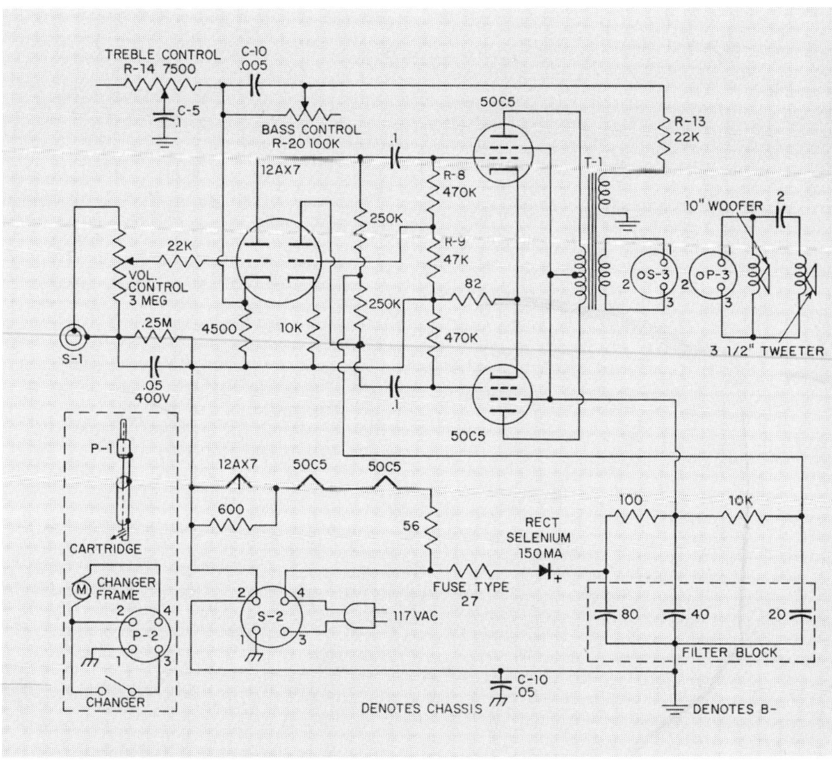

FIG. 19-30. Schematic diagram of a phonograph featuring push-pull amplification, a dual speaker system, and bass and treble tone controls in a feedback line.

The dual speaker system includes a 10-inch woofer for low- and mid-frequency audio signals and a 3½-inch tweeter for the higher frequencies. Note the 2-mfd capacitor in series with the tweeter to block low audio-frequency signals.

A basic audio degenerative or negative feedback circuit is shown in Fig. 19-31A. Some of the output voltage is fed back to the input of the amplifier through resistor *R*-13 in an out-of-phase relationship to the incoming signal. The feedback is to the unbypassed cathode of the input tube. Sometimes, the feedback is to the grid of the input tube. The effect of the degeneration is to cancel distortion created by the amplifier. At the same time, it causes loss of gain; therefore the amplifier is designed for higher gain to compensate for this loss. It follows, therefore, that if the degeneration were removed, the volume of the amplifier would increase. Frequency discrimination in the feedback line would therefore cause increased gain for the frequencies that were discriminated against. This is the reason why there is a resistance-capacitance network for control of tonal balance in the feedback line.

The feedback circuit of the amplifier of Fig. 19-30 is redrawn in Fig. 19-31*B* to show how tone control is obtained. Series capacitor *C*-10 in the feedback chain, like all series capacitors, offers high impedance to low-frequency signals and low impedance to the highs. By itself, it would keep low a-f signals out of the feedback chain, thereby giving full amplification and high gain for the bass notes. Variable resistor *R*-20 is across capacitor *C*-10. When the arm is turned toward the open end of the resistor, it has little effect on the action of the capacitor and we have maximum bass response. When the control arm is rotated toward the other end, capacitor *C*-10 is shorted out, low notes are fed back in the degenerative chain, and we have reduced amplification for these low notes. Variable resistor *R*-20 therefore gives a manual control of the bass response. High a-f signals come through the

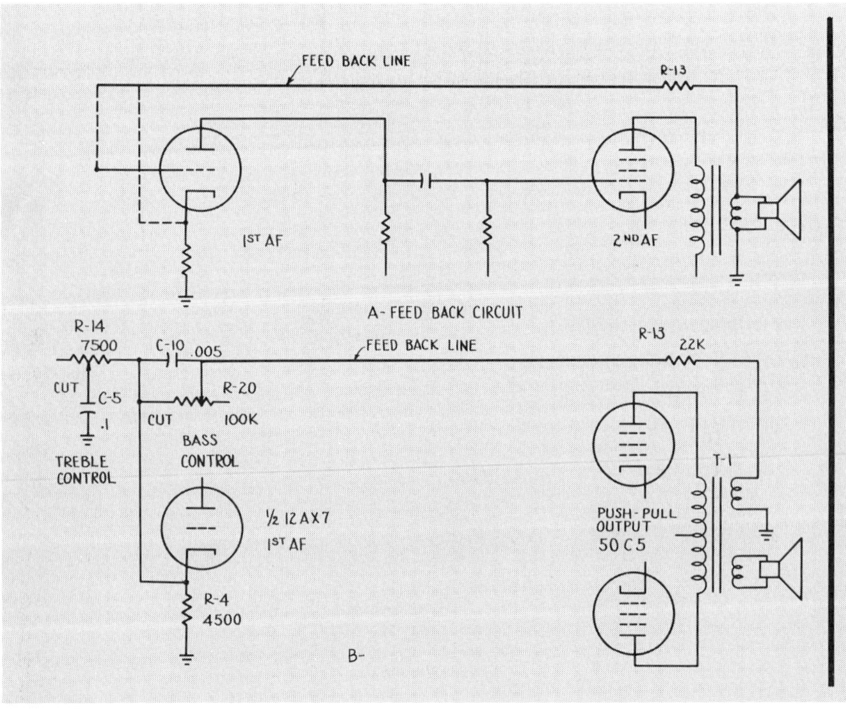

FIG. 19-31. Bass and treble tone controls in a feedback chain.

feedback chain at all settings of the bass control *R*-20; and so this control has little effect on the treble response.

Treble control *R*-14 and its associated capacitor *C*-5 control the treble response as follows. At the left or open end of the control, capacitor *C*-5 is 7,500 ohms away from the cathode, and so has little effect on the signals fed to it through the feedback line. As the control is advanced, capacitor *C*-5 bypasses high-frequency signals out of the cathode, thereby reducing the degenerative effect, permitting more gain, and increasing the treble response. Capacitor *C*-5 will have little effect on the bass response, since its capacitance is too small to give any bypass action at low audio frequencies.

The only new servicing problem connected with the amplifier of Fig. 19-30 would be encountered if the fault were to be a burned out output transformer. The replacement should be an exact duplicate, so as to get the proper feedback signal voltage. Also, the transformer must be wired in with due regard to the correct polarity. Reversal of the polarity would make the feedback signal in phase with the input signal. This is regeneration or positive feedback and results in oscillation. If the amplifier oscillates after changing the output transformer, reverse the two plate leads or the feedback winding, whichever is more convenient. This will restore correct polarity to the feedback line.

Stereophonic phonograph systems. Thus far, we have been concerned with monaural phonograph systems. Let us now turn to the more recent stereophonic system. A stereophonic (usually abbreviated to stereo) recording and playback system is illustrated in Fig. 19-32. When the recording is made, two microphones are spaced several feet apart. Each is connected to a separate amplifier. The amplifier outputs are fed to the stereo recording head, shown in Fig. 19-32*A*, where they operate at right angles to each other on the single cutting stylus. The latter is set at an angle of 45 deg to the record blank. Since both amplifiers are operating simultaneously, the cutting stylus will simultaneously cut both walls of the groove in a complex pattern. Generally the left microphone is used for the vocalist, while his accompaniment is placed near the other microphone.

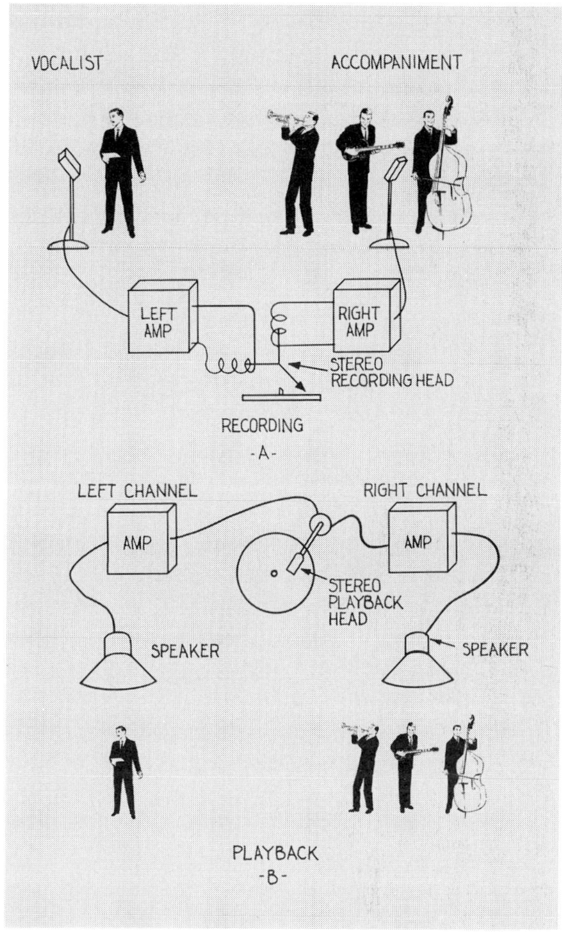

FIG. 19-32. The stereophonic phonograph system.

In the playback, illustrated in Fig. 19-32*B*, we start with a special stereo pickup, where a single needle set at 45 deg rides the stereo record groove and activates two cartridge units set at right angles to each other in the pickup head. Each crystal picks up one of the original sound components. This is the reverse action of the recording process. Each output signal is fed to a separate phono amplifier. The amplifiers are generally called left channel and right channel or channel 1 and channel 2, each channel feeding a separate speaker. When the two speakers are spaced far enough apart to approximate the conditions at the recording studio, a listener placed between the speakers seems to hear an added space factor in that the music from different instruments in a recorded selection seems to be coming from different parts of the room. In the case of a vocalist and his accompaniment, the vocalist seems to be coming from the left speaker and the accompaniment from the right. This is the stereophonic effect, adding another dimension to sound reproduction. The stereophonic effect seems to add a depth and richness to the reproduced sound even in inexpensive players.

Simple stereo phonograph. The schematic diagram of a small stereo phonograph is shown in Fig. 19-33. Note the symbol *M*-2 for the stereo cartridge, where the two crystal units are shown at 90 degrees to each other, and the output of each is fed to a separate phono plug. Sometimes a single three-connection plug is used (two hot leads and one common), but the separate plug arrangement is more common for convenience in reversing channels. In diagrams of stereo systems, the left channel or channel 1 is generally placed above the similar right channel, with the power supply below as shown.

Note that the tone controls are ganged, but separate volume controls are provided for each channel. This is a common arrangement in that the tone control is usually left at the position giving the desired mellowness and freedom from needle scratch. The same setting will do for both channels. The separate volume or loudness controls allow the user to equalize the volume from each speaker or adjust them differently if preferred. The controls are sometimes one unit with concentric shafts and a dual tapered pair of knobs. By this arrangement, volume is easily adjusted for both channels at once while allowing freedom to reset either if desired.

The cathodes of the two power output tubes are tied together. Resistor *R*-9 together with bypass capacitor *C*-1*D* provide bias for both

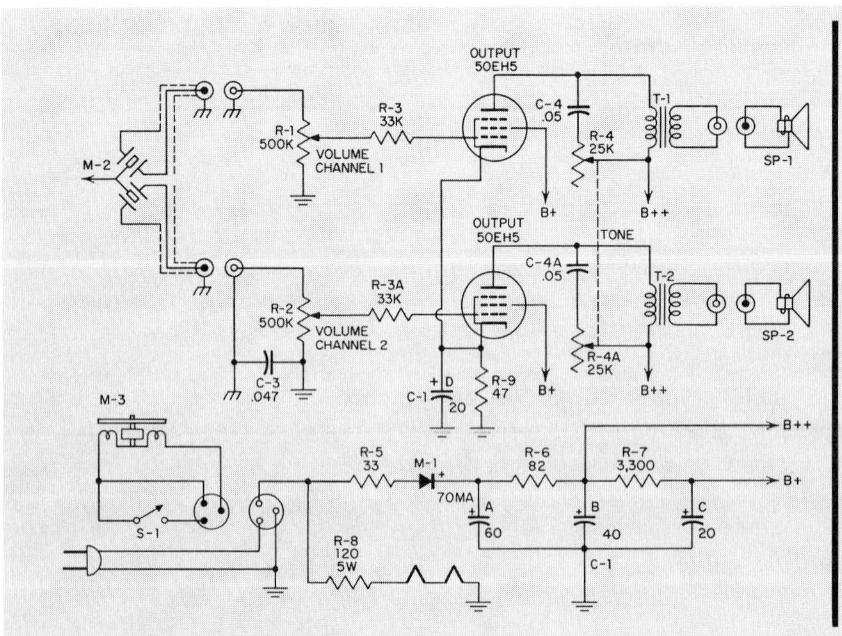

FIG. 19-33. Schematic diagram of a typical stereo phonograph using a single stage in each channel.

of the tubes. Since the plate and screen voltages are provided by one power unit, our two amplifiers share a common power supply, but their signal circuits are separate. We have a separate input source and speaker output for each channel. Note the continued use of plug-in arrangements, two-section filter, and floating-chassis construction, common for most phonographs. The channel 1 speaker is often contained in the cabinet with the phono turntable and amplifier. A separate speaker enclosure and extension cord is provided for the channel 2 speaker, so that it can be placed a distance away. A popular arrangement is to hinge the two speakers to the main phonograph cabinet, thereby making the entire unit portable. The hinges are separable so that either speaker can be lifted off and placed further away for a better stereophonic effect. A distance of 6 to 10 feet between speakers is recommended as being correct for an average room.

No special provisions are made for playing monaural records on a stereo phonograph. A monaural groove will vibrate a stereo needle and activate both crystals equally so that both channels will be in operation. Of course, there will not be any stereo effect. Approximately the same signal will be coming from each speaker, and the net result will be that the music seems to be coming from some point between them. Incidentally, the reverse is not true. Stereo records should not be played on a monaural system, even though there will be almost normal monaural reproduction initially, because a monaural head will quickly ruin stereo records by gouging out the sides of the grooves.

Servicing this stereo unit will bring no new problems. We simply have one power supply, two amplifiers, and two speakers to check. If anything, because of the dual amplifiers, service work on the stereo unit is even easier than a monaural phonograph. If the set is dead, the trouble is almost sure to be in the common power supply, since it is very unlikely that both tubes or both speakers will go dead at once.

Suppose the complaint is weak output or distortion. The procedure is simple. Put on a monaural record. A well-balanced stereo pickup head should pick up about equal signal in each crystal. Turn the channel 1 volume control to the off position, making that channel dead. Listen to reproduction through the other channel on speaker 2. Now reverse the controls; that is, turn the channel 1 volume control up and the channel 2 control down, so that you can listen to the reproduction through channel 1 and speaker 1. If both channels show the weakness or distortion, the trouble is in a common unit, such as a weak rectifier or leakage in filter capacitor *C*-1. Check for these in the usual way. If the above check shows that one channel is giving normal operation and the trouble is confined to the other, check only that channel for the weak tube, for an off-center speaker, or whatever is causing the complaint. Do not fail to check that the chassis is still floating from the line before you complete your job.

Stereo balance control. The schematic diagram of a Philco stereo phonograph is given in Fig. 19-34 as a typical example of a larger unit. The transformer in the power supply isolates the chassis and common negative from the line. This allows the changer base to be grounded to the chassis without incurring short and shock hazards. Each channel contains a two-stage amplifier. The voltage amplifier is one triode of the twin 12AX7 tube and the power amplifiers for the two output stages are separate 6AQ5 tubes. At the voltage delivered by the power supply, each channel delivers about 5 watts of audio power to its own speaker system.

The gain of each amplifier can be varied by means of potentiometer VR4 connected from grid to grid of the output tubes. This potentiometer is called the balance control. When its arm is in the center or normal position, both amplifiers are working at about the same

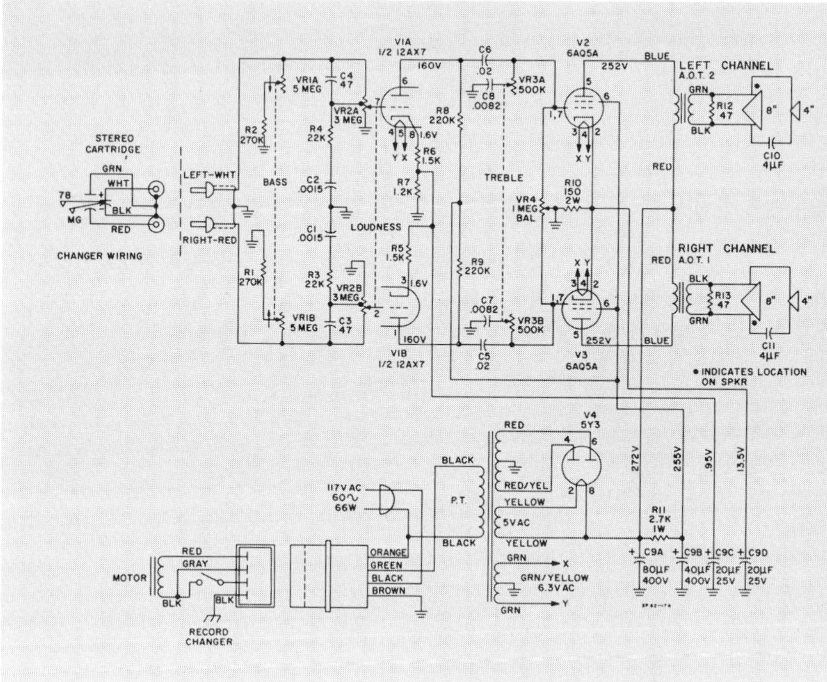

FIG. 19-34. Schematic diagram of a Philco stereo phonograph.

gain at a volume level determined by the ganged loudness control in the grid circuits of the two voltage amplifiers. As the grounded arm of the balance potentiometer is shifted toward either grid, let us say the top one, the gain of the left channel is cut down and the gain of the right channel is increased. As a result, there is less volume from the left speaker system and more from the right. Yet, there will not be too great a change in the overall volume level. This method of balancing outputs is considered superior to separate volume controls where balance procedures affect the overall loudness.

The circuit of Fig. 19-34 also illustrates a type of bass tone control not previously mentioned. The frequency response of a crystal pickup is greatly affected by the load across it. The bass response increases with high resistance loads. This effect is used to give a control of the bass. Bass tone controls VR1A and VR1B are connected across the crystals to vary the load on each. Resistors *R*-2 and *R*-1 provide minimum loads. Without them, the crystals could be shorted out at the minimum positions of the variable bass tone controls. The treble control is conventional.

If an amplifier of this type should go dead or develop a hum, the fault would most likely be in the common power supply. This is checked like any transformer-type power supply, as described in Chap. 8.

When working in the amplifier channels, the same procedure mentioned before of deactivating first one and then the other channel will prove helpful in locating the cause of the trouble. This can be done by means of the balance control. When the arm is in either extreme position, the associated channel is shorted out.

If balance control VR4 should become defective and require replacement, it is necessary to choose a replacement control carefully. If a control with a nonlinear taper is used, the balance area will not be at the center of rotation, but rather toward one end. If an exact replacement cannot be obtained, use a linear control of approximately the correct resistance value. In the circuit of Fig. 19-34, a 0.5-megohm linear control could be used if a 1-megohm replacement is not available.

When servicing stereo systems with a balance control, there is one type of complaint which may be baffling. This is that the balance control must be reset for every position of the loudness control, especially at low volume levels. This complaint is almost always due to unequal tapers in the loudness controls. To confirm the defect, unplug the line cord and check with an ohmmeter from ground to arm contact of each of the ganged loudness controls at several low-volume positions. If there is an appreciable difference, the loudness control must be replaced. Check the new control in the same way to make sure that the tapers run uniformly.

Stereo center channel. The schematic diagram of an Arvin stereo phonograph is given in Fig. 19-35 to illustrate the use of a center speaker in stereo systems. The center speaker is housed in the phonograph cabinet. The left and right channel speakers are hinged to the cabinet. Separable hinges and patch cords are provided so that these latter speakers may be extended. Note the representation in the diagram for the hinge connections and patch cords.

The best position for listening to stereophonic recordings and programs is between the two separated speakers and across the room from them. When the listener is closer to the equipment, the sound from the speakers may become dissociated, leaving an impression which is called a "hole in the middle." The purpose of the speaker in the center is to fill this "hole." Note how the center speaker

FIG. 19-35. Schematic diagram of an Arvin stereo phonograph, with a system having a balance control and a center speaker.

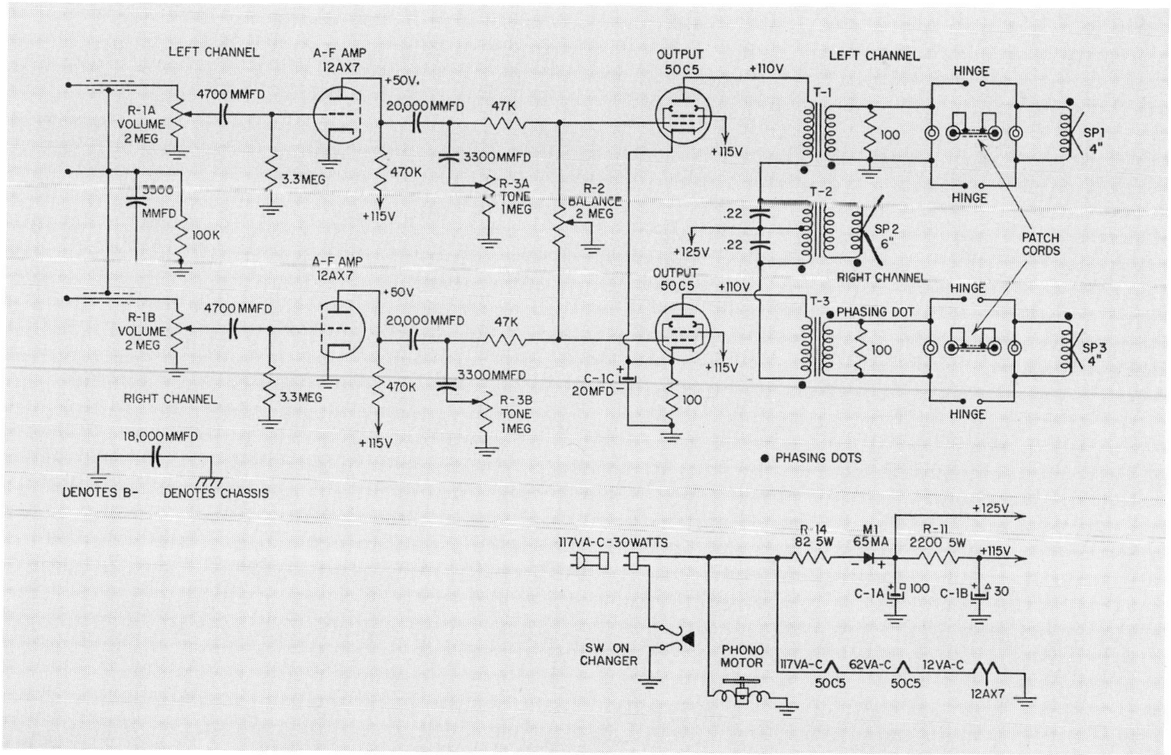

is connected through a tapped transformer to receive signals from both channels.

The center speaker is made to serve another function. Lower-frequency sounds are not directional and do not require separation to obtain the stereophonic effect. The center speaker, mounted in the cabinet, is therefore used for the low-frequency sounds, while the high-frequency sounds come from the hinged speakers which may be extended by their patch cords to enhance the stereo effect.

Separation of the audio signals into highs and lows is accomplished by means of the 0.22-mfd capacitors across each half of the primary winding of center transformer *T*-2. In the left channel, signals at high audio frequencies build up across the primary winding of transformer *T*-1 and are bypassed out of the primary of *T*-2. This forces the high audio frequencies to activate the left hinged speaker. The primary winding of transformer *T*-2 has a high impedance to low-frequency a-f signals, and so these activate the center speaker. In the right channel, the same separation is taking place to force high-frequency a-f signals into the right hinged 4-inch speaker and lows into the center 6-inch speaker. Phasing of the transformers and speakers is necessary to get the proper effects. Note the phasing dots indicated for each transformer and speaker.

Another method for obtaining center channel stereo operation is illustrated in Fig. 19-36. The diagram shows the output stages and speaker arrangement of a Westinghouse a-m/f-m receiver with stereo amplifier and automatic record changer. The large console cabinet houses five speakers: two 4-inch tweeters, two 6-inch mid-range units, and one 12-inch woofer. The tweeters and mid-range speakers are placed to the left and right in the console with the woofer in the center.

The upper part of the diagram showing the left amplifier channel feeds mid-range speaker *SP*-2 and tweeter *SP*-1 through capacitor *C*-41. The capacitor has a value of 70 mfd, which

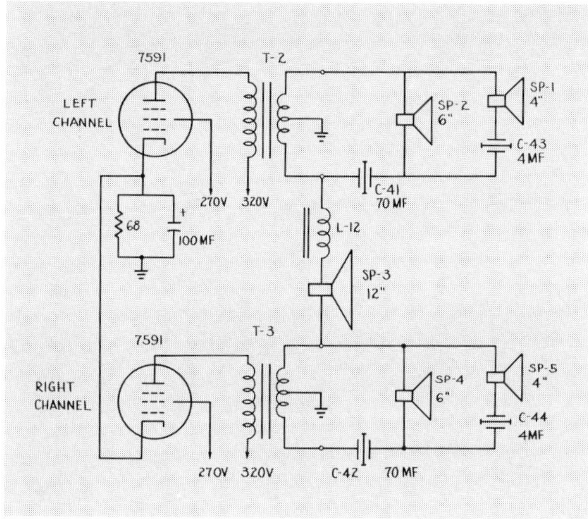

FIG. 19-36. Speaker system in a Westinghouse stereo receiver.

will pass all but the very lowest a-f signals. Note the symbol for the nonpolarized capacitor. Capacitor *C*-43 with a capacitance of 4 mfd passes only the very high a-f signals to the tweeter. A similar arrangement feeds speakers *SP*-4 and *SP*-5 from the right channel amplifier. Both channels feed the center woofer. The reactance of coil *L*-12 serves to keep high-frequency audio signals out of the woofer.

When doing service work on a stereo amplifier with a center speaker, it is not so easy to determine which channel is at fault, since the center speaker receives signals from both channels. If there is any doubt, it becomes necessary to disable the center speaker. This can best be done by opening one of its leads. The channels are then working separately through their small speakers. After the defective channel is found, there is then only one audio amplifier to check.

Small radio-stereo-phono combinations. The diagram of a typical small a-m radio-stereo-phono combination is shown in Fig. 19-37. First note that we have a standard a-m superheterodyne receiver, an a-c/d-c power supply, and a second audio channel. Observe the floating chassis and the plug-in connection arrangements for the changer, the pickup and the speakers. Note the loudness, treble, and bass controls. The loudness control uses tapered concentric knobs for single or separate control of the volume in each channel.

Changeover to the different modes of operation is made by switch *M*-4, called the selector switch in the diagram. It is a 12-point wafer switch, using only contacts 3, 4, 5, 6, 7, 8, 10, and 11. Points 3 through 8 are shown at the middle left side of the diagram. Points 10 and 11 are shown separately at the lower right, to simplify the drawing.

The selector switch has three positions: RADIO, HI-FI, and STEREO. The direction of turn of the switch is shown by the indication at its center. In the diagram, the switch is shown in the RADIO position for radio operation. In this position, the detector output from the radio is connected to pin 6 on the switch. Pin 6 makes contact in the switch with pins 4 and 5. These latter pins feed the hot ends of the two loudness controls *R*-1*A* and *R*-1*B*. Thus, the detector is feeding both audio channels simultaneously, and the radio will play through both speakers.

The ON-OFF switch is shown near the line cord. It is ganged to the selector switch. In the RADIO position, common negative is connected to one side of the power line so that it is in the ON position. Note the other part of switch *M*-4 at the right. The *B*+ line goes to pin 11, through the switch to pin 10, and from

FIG. 19-37. Schematic diagram of a typical radio-stereo-phonograph combination.

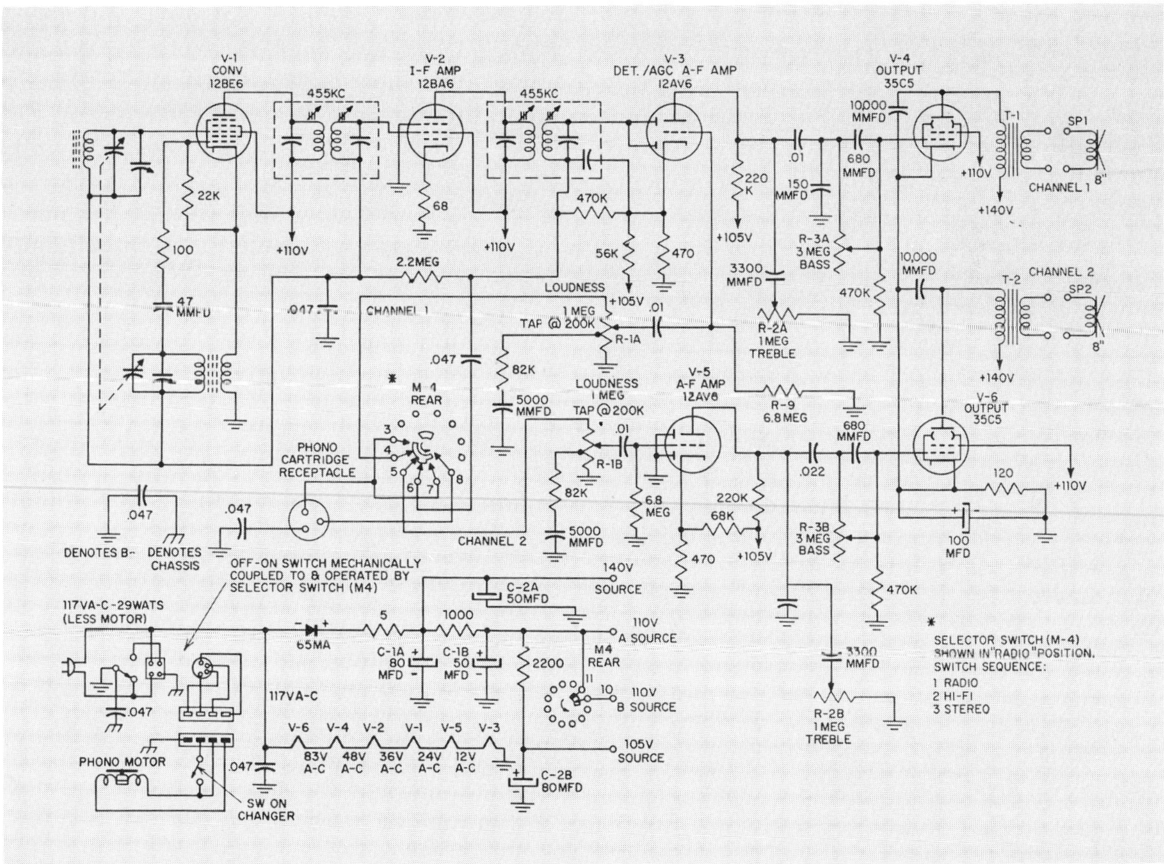

there (110V B SOURCE) to the converter plate and screen, energizing that tube.

When the selector switch is turned one notch, it goes to the HI-FI position for phonograph operation. Now, pins 4, 5, 7, and 8 are connected together in the switch, so as to feed the outputs of both cartridges in the phono head to both audio channel inputs. This will give monaural reproduction through the two speakers. In this position, pin 10 loses connection to pin 11 in switch *M*-4, removing *B*+ voltage from the converter and thereby silencing radio reception during phono operation. Common negative at the ON-OFF switch, being ganged to the selector switch, is now thrown to the automatic switch on the changer, so that the entire set goes off after the last record has been played. Note the insulation of the turntable and pickup arm from the line to avoid shock hazard.

One more turn of the selector switch brings us to the STEREO position. The switch on the changer is still in control and *B*+ is still broken at pins 10 and 11 to the converter plate, silencing radio reception during the phono operation. One output from the stereo cartridge is fed to the input of channel 1 through the interconnection of pins 3 and 4 in the selector switch. The output from the other stereo cartridge is fed to the input of channel 2 through the interconnection of pins 5 and 8 of the switch. We now have true stereo sound.

Large a-m/f-m radio-stereo-phono combinations. Large combinations are generally built up of several individual chassis interconnected by patch cords and power cables. There is generally an a-m/f-m stereo tuner chassis delivering outputs for both the right and left amplifier channels. In older combinations, there will be a separate multiplex chassis. Next, there will be a preamplifier chassis containing the first audio stages for each channel and most of the controls: the selector switch, and the loudness, treble, bass, and balance controls. Then there will be the heavy chassis containing the power supply and dual main amplifier. The output stages are likely to contain a pair of push-pull tubes for each channel, delivering 25 watts or more per channel to each of the two speaker systems.

Sometimes the components are grouped differently. For example, the preamplifier may be combined with either the tuner or the main amplifier. Sometimes the units are self-powered, receiving line voltage from the preamplifier section. Sometimes the main amplifier furnishes heater current and *B* voltage to the different sections through plug and socket arrangements. Sometimes all units are housed in one large console cabinet. In other cases, they may be spread over one or two walls, built into different compartments in book cases.

Obviously, no single example can be representative of all the possible combinations. When servicing such large units, it becomes necessary to first make a careful analysis of the individual arrangement and interconnections. Then, basic principles and servicing techniques for each section may be applied, as given elsewhere in this book.

SERVICE DATA CHART

(An a-m/f-m radio-phono combination is assumed.)

Symptom	Monaural	Stereo
Dead on all functions	Check power supply, amplifier, and speaker	Check power supply
Distortion on all functions	Check amplifier and speaker	Operate on one channel to isolate trouble. (Open center channel, if present.) Check defective amplifier and speaker
Hum on all functions	Check power supply, amplifier, and patch cords for open	Same as for monaural operation
Malfunction on phono, radio is good	Check pickup and input circuit	Same as for monaural operation
Malfunction on radio, phono is good	Check tuner	Check tuner

Servicing Transistor Radios 20

The servicing of transistor radio receivers presents no more difficulty than the servicing of tube radios. We have the same superheterodyne circuit in both types of receivers, with transistors replacing tubes. However, because circuit operation is somewhat different, we shall first study the transistor as a single unit, and then follow through with a study of stages where the transistor functions as a converter, an intermediate-frequency amplifier, and as an audio-frequency amplifier. This stage analysis will be made from a typical transistor receiver. Then we shall discover how to service a typical transistor receiver. The final study will be that of variations from our typical set. Be sure to note how the information you learned when studying the tube receiver is used with minor variations with transistor receivers.

The transistor. A transistor, as we find it in a receiver, is essentially a composite solid, made up of specially prepared semiconductor materials. Semiconductors are a group of elements whose electrical conductivity lies between those of good conductors and good insulators. Examples of such solid elements are germanium and silicon.

The type of transistor commonly used today, known as the junction transistor, came into existence in 1949. Essentially, it consists of three distinct specially prepared semiconductor solids placed side to side, thereby producing two contact areas known as junctions. Each of the three solid sections is connected to an external lead.

In many ways, these three leads correspond to the tube pins connected to the cathode, grid, and plate of a triode tube, as shown in Fig. 20-1. In contrast to the action in a tube, however, all the electronic action in a transistor occurs within a solid substance. The transistor, like the tube, is capable of achieving all the functions of oscillation, detection, mixing, and amplification. As with tubes, assigned voltages are applied to each of the three leads of the transistor. However, where a tube requires a heater for emitting electrons, a transistor uses no heater; it functions as soon as voltages are applied.

How the transistor works. To understand the operation of a transistor, let us begin with a small piece of pure germanium. If a voltage is applied across this material, an extremely minute current will flow because pure germanium is a relatively poor conductor. However, when this pure germanium is "doped" by adding small but definite amounts of certain other elements, great changes take place in the electric conductivity of the germanium. Its conductivity increases, with resulting useful properties. Let us see how this occurs.

When controlled amounts of such elements as arsenic or antimony are added to the germanium, a great number of electrons are in-

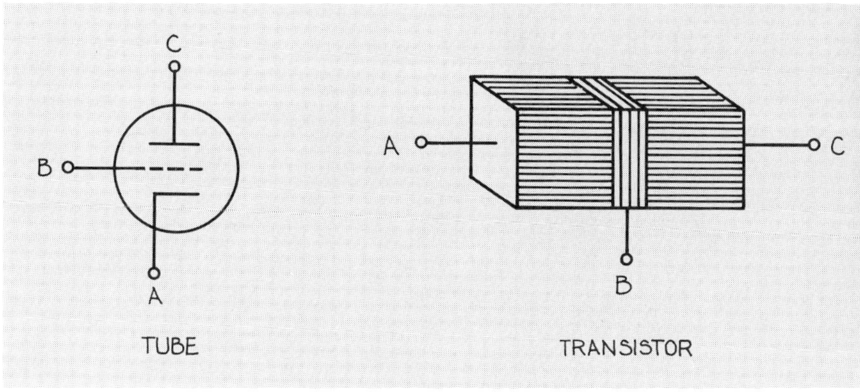

FIG. 20-1. The transistor resembles a triode tube; equivalent electrodes are labeled similarly.

troduced. These electrons are capable of easy movement through the treated germanium. The doped germanium is now known as the N-type, because of its ready ability to conduct negative charges. If we place a voltage across such N-type germanium, a relatively large current flows.

When controlled amounts of certain other elements, such as indium, gallium, or aluminum, are added to the pure germanium, a different effect occurs. Now we create a condition among the atoms where, instead of electrons free to move about, there are electron deficiencies. Where electrons normally would be found in the germanium atomic structure, there now occur a scattering of spaces known as holes. Such holes may be considered to be positive in nature because electrons from other germanium atoms tend to jump into these holes. But when an electron does so, it leaves a new hole at the position which it just left. The jumping of electrons in one direction in such germanium creates holes which appear to be drifting in the opposite direction. Germanium doped in this manner is known as the P-type because of the great abundance of these "positive" holes. The flow or drift of holes may be considered as much a current flow as the drift or flow of electrons. The conductivities of pure, N-type, and P-type germanium are shown in Fig. 20-2.

Now let us take a slab of N-type germanium

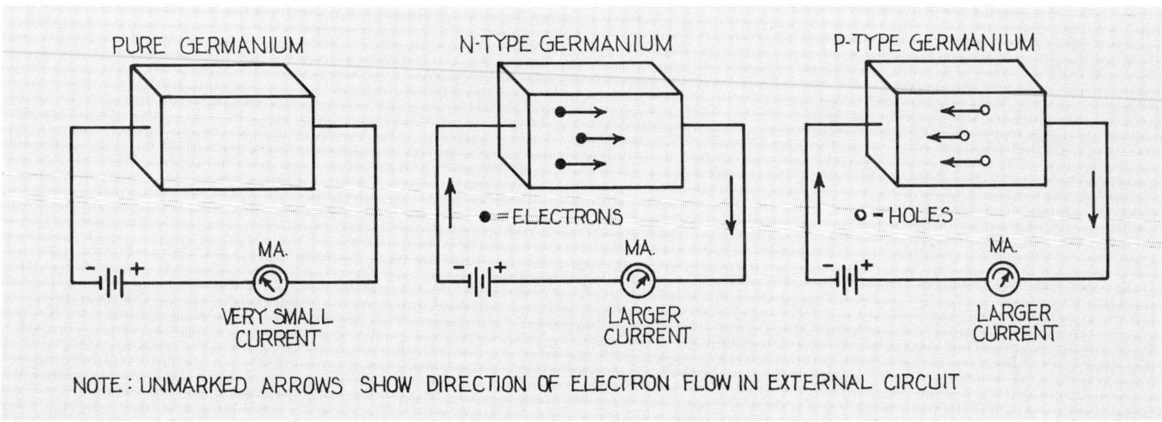

FIG. 20-2. Current flow through pure and doped germanium.

and place it adjacent to a slab of P-type germanium, producing a solid with a single junction. Connect a battery across this combination as shown in Fig. 20-3A. Practically no current flows in the circuit because the holes and electrons in the P-type and N-type, respectively, tend to move away from the junction,—the holes toward the negative lead and the electrons toward the positive lead. The junction presents a barrier to the passage of current carriers, electrons or holes, across it. This P-N junction diode is said to be reverse biased.

Now reverse the battery as shown in Fig. 20-3*B*. The battery draws electrons across the junction toward the positive lead and draws holes across the junction toward its negative lead. The junction barrier is thus reduced, and the meter shows a relatively large current flowing in the circuit. The P-N junction diode is said to be forward biased. It should be noted that the movement of holes as current carriers takes place only within the doped germanium. Once they arrive at the negative lead from the battery going to the junction diode, they pick up electrons delivered by the battery. Only electrons flow in the external conducting wires.

Now let us substitute an a-c generator for the battery and place a load resistor in the circuit, as shown in Fig. 20-4. The generator alternately gives forward and reverse biasing to the diode. The output is pulsating direct current. This diode could be used as a detector or rectifier.

These understandings may be expanded to the understanding of the transistor. A transistor is essentially two such germanium units placed back to back, with a common central slab. This relationship is shown in Fig. 20-5A. The first P-type slab is known as the emitter; the other P-type slab is known as the collector. The N-type slab, common to both diode units, is extremely thin and is known as the base. In all practical circuits, the emitter-and-base diode is biased in the forward direction and the base-and-collector diode is biased in the reverse direction. This may not at first be obvious with respect to the latter diode. But note that the collector is highly negative with respect to the emitter. However, the base is only slightly negative with respect to the emitter. Hence, the collector is quite negative with respect to the base and is reverse biased, as it should be.

Thus, there is low junction resistance between base and emitter and high junction resistance between base and collector. At

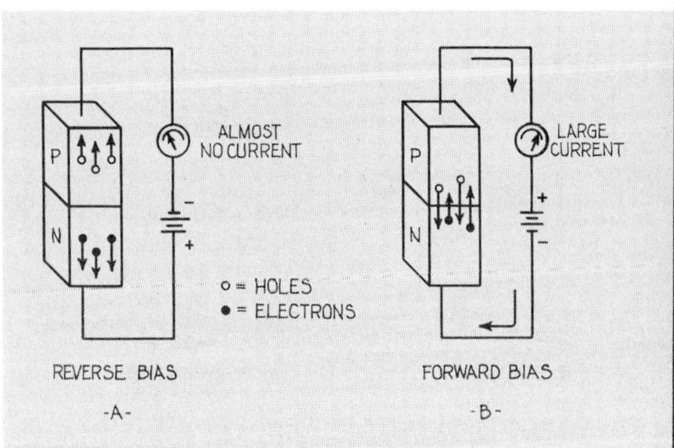

FIG. 20-3. Reverse and forward biasing in a junction diode.

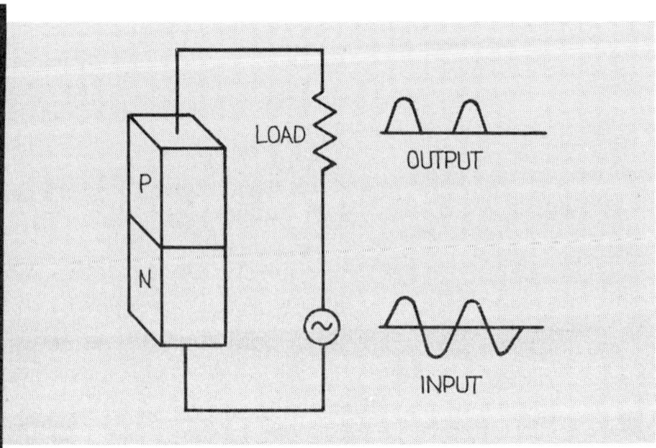

FIG. 20-4. How the junction diode rectifies an alternating current.

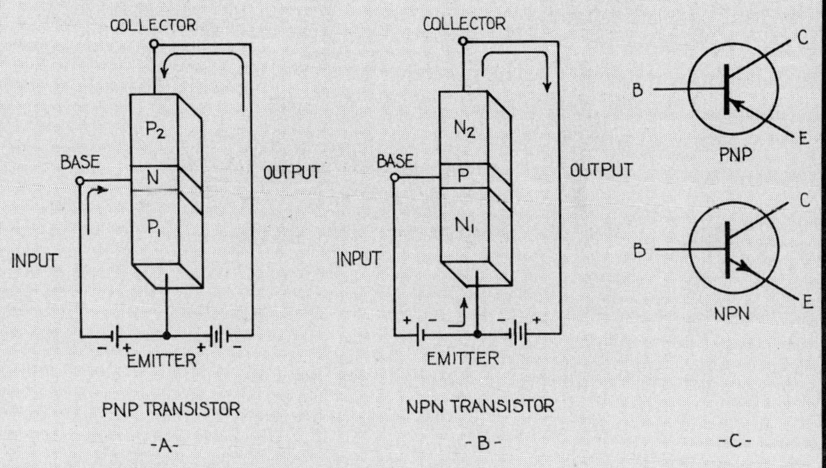

FIG. 20-5. A transistor is two junction diodes, back to back.

first, holes readily drift through P_1 to the P_1N junction (emitter junction). Under the influence of the batteries, the holes diffuse through the thin base region (generally less than 0.001 inch wide) and then across the NP_2 junction (collector junction). At the collector terminal, the holes draw electrons from the battery. Thus, a flow of electrons takes place from the battery to the P_1-germanium.

Sometimes, the three doped germanium slabs are united in a different manner: two N-types are made to sandwich a thin P-type, as in Fig. 20-5*B*. The first transistor described is known as a PNP junction transistor; this latter type is known as an NPN junction transistor. The circuit difference between the two types is primarily in the polarity of the batteries. Also note the different directions of electron flow in the external circuit.

Examine the diagram of the NPN transistor. The N_1P diode is forward biased, with the base slightly more positive than the emitter. The collector is much more positive than the emitter. Therefore, the collector is more positive than the base and the collector junction is reverse biased, as it should be. Under the influence of the small battery, electrons drift through the N_1 section across the emitter junction and into the P slab. Then under the influence of the large battery, the electrons almost entirely diffuse through the thin P slab, across the N_2 slab to the collector lead.

In both types of junction transistors, current carriers (holes for the PNP type and electrons for the NPN type) pass through the three germanium sections from emitter to collector. Remember, however, that for both types only electrons flow in the external circuits.

Figure 20-5*C* shows the symbols for a PNP and an NPN junction transistor. The direction of the arrow on the emitter lead identifies the type of transistor. A simple rule enables us to remember proper battery polarity. If we take the first letter of PNP or NPN, it tells us the polarity of the emitter voltage with respect to the base. If we take the second letter, it tells us the polarity of the collector voltage with respect to the base. The N stands for negative and the P for positive.

The common emitter circuit. The circuits just presented are common emitter circuits. In these circuits when the input signal is to the base circuit, the output is in the collector circuit, and the emitter circuit is common to the input and output circuits. Most transistor receivers use transistors in a common emitter circuit because it has high voltage gain and very high power gain.

The common emitter circuit most nearly resembles the triode tube circuit. The emitter may be considered equivalent to the cathode, the base may be considered equivalent to the signal grid, and the collector may be considered equivalent to the plate. This relationship is shown in Fig. 20-6.

Note that the circuits are similar, except for their different voltage and polarity requirements. Even the 180-degree phase shift between input and output signals is similar. Signal is fed to the base, varying the voltage between base and emitter, just as signal varies voltage between signal grid and cathode in a triode tube.

Transistor amplifiers. The transistor amplifier has three main circuits corresponding to the three doped germanium slabs: the emitter circuit, the base circuit, and the collector circuit. Examine the basic PNP transistor amplifier in Fig. 20-7A. The currents flowing in the three external circuits are shown: I_e is the emitter circuit current, I_c is the collector circuit current, and I_b is the base circuit current. Note that almost all, 95 percent, of the emitter current flows through the collector circuit. The remaining 5 percent of the emitter current flows in the base circuit.

The input signal is applied to the base. A small negative voltage, applied to the base, increases the forward biasing of the emitter junction. As a result, there is an increase of

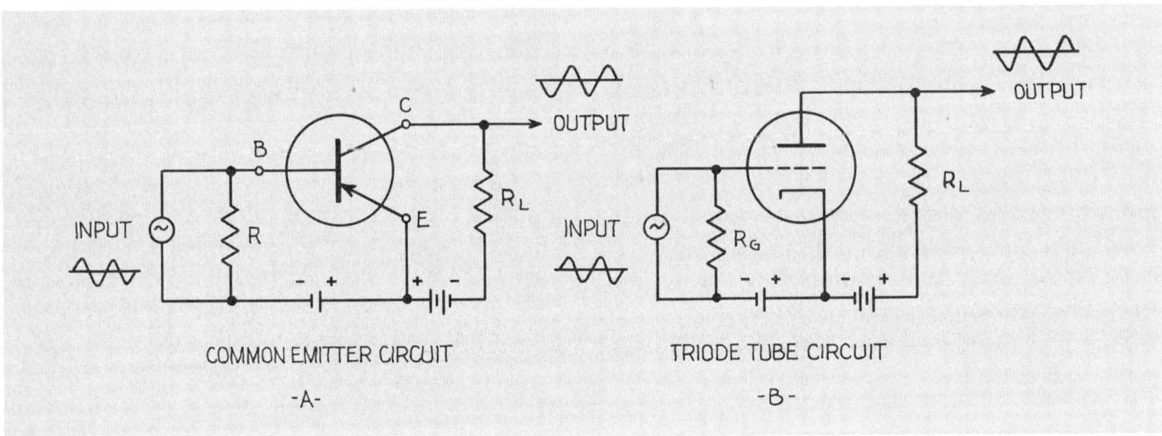

FIG. 20-6. The common emitter circuit resembles a triode tube circuit.

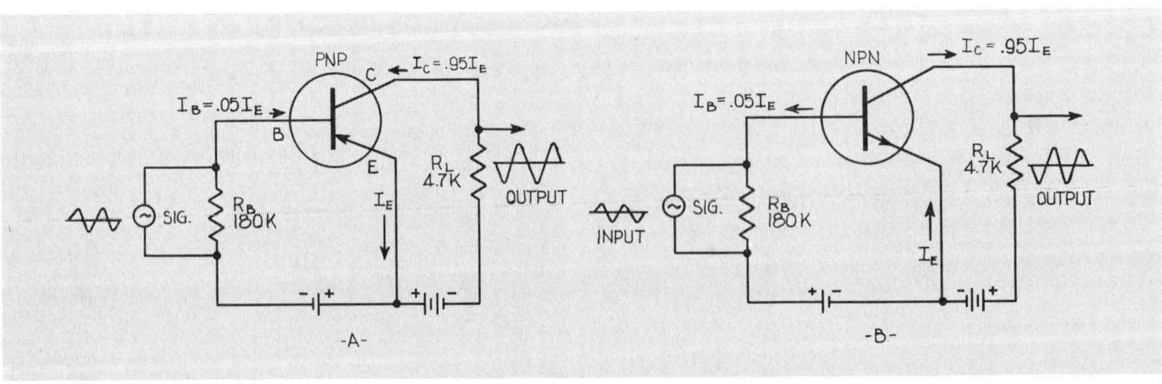

FIG. 20-7. The transistor as an amplifier.

base, emitter, and collector currents. Conversely, a small positive signal voltage to the base decreases the forward biasing of the emitter junction and reduces all circuit currents.

The transistor is a current-operated device because it involves internal current carriers. Its gain involves the gain of current from input to output.

$$\text{Current gain} = \frac{I_c}{I_b}$$

Thus, if the emitter current is 5 ma, base current will be 0.25 ma (5 × 5%), and collector current will be 4.75 ma (5 × 95%). The current gain will then be 4.75/0.25 or 19.

A more important consideration for the transistor amplifier is the voltage gain obtained. This may be expressed by the formula:

$$\text{Voltage gain} = \frac{E_{\text{out}}}{E_{\text{in}}} = \frac{I_c \times R_L}{I_b \times R_{\text{in}}}$$

The value of I_c as 4.75 ma and of I_b as 0.25 ma is already known. A little thought must be given to the other factors. The emitter junction, being forward biased, has a relatively low input impedance of about 2,000 ohms. It is so low that it is unaffected by the base resistor R_b (180,000 ohms). This resistor serves primarily to keep the base current at a desirable low level. So we may consider the input impedance, R_{in} as 2,000 ohms. The collector junction is reverse biased and therefore has a very high output impedance, even up to 500,000 ohms. However, the load resistor R_L is only 4,700 ohms, so much smaller that it shunts the junction output impedance; and we may consider its resistance as the value of the output impedance. If the proper values are now substituted,

$$\text{Voltage gain} = \frac{4.75 \times 4{,}700}{0.25 \times 2{,}000} = 44.65$$

Another important consideration is power gain which may be expressed as below:

$$\text{Power gain} = \frac{\text{Power}_{\text{out}}}{\text{Power}_{\text{in}}} = \frac{I_c^2 \times R_L}{I_b^2 \times R_{\text{in}}}$$

Substituting values, power gain is found.

$$\text{Power gain} = \frac{(4.75)^2 \times 4{,}700}{(0.25)^2 \times 2{,}000} = 858$$

So it is seen how the common emitter transistor circuit can give a good voltage gain and an even better power gain, the requirements of an amplifier. The same results would be obtained with the NPN common emitter circuit shown in Fig. 20-7*B*.

Transistor bias. Biasing a transistor means setting the d-c operating point for proper operation of the transistor. Bias is stated as the d-c no-signal values of collector voltage and emitter current. The proper bias is obtained by means of the emitter-base and collector-base batteries and the accompanying circuit elements. The operating point is primarily determined by the value of emitter current, which in turn depends on the value of emitter voltage. It varies with different transistors.

In commercial receivers, it is desirable to avoid the use of two biasing batteries; only one is preferred. Such a one-battery transistor amplifier is shown in Fig. 20-8. Examination will show that the emitter is positive with respect to both base and collector and that the collector is considerably more negative than the base for proper bias. For an NPN transistor, the battery would simply be reversed.

It is very important that the operating point be kept fairly constant, rather than drift. Three conditions may cause such a drift: the change of bias voltage between base and emitter, a rising temperature within the transistor, and a change of collector voltage. The last factor has the least effect. Thus, some circuit design must be made to control the first two factors in order to stabilize the d-c operating point. Let us examine these controls.

Stabilizing circuits. The net effect of temperature rise in a transistor is to cause a run-

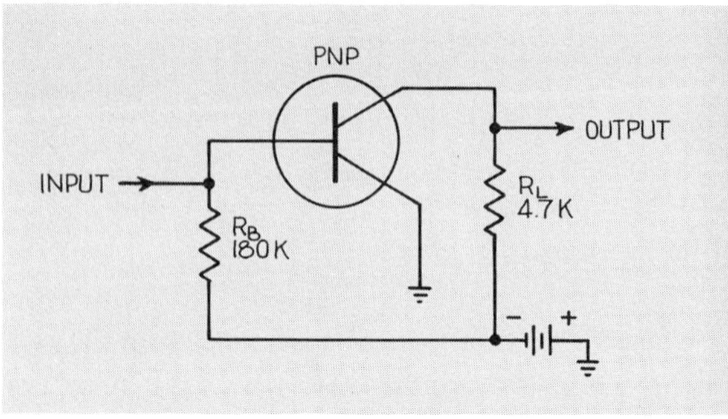

FIG. 20-8. Using a single battery for biasing a transistor.

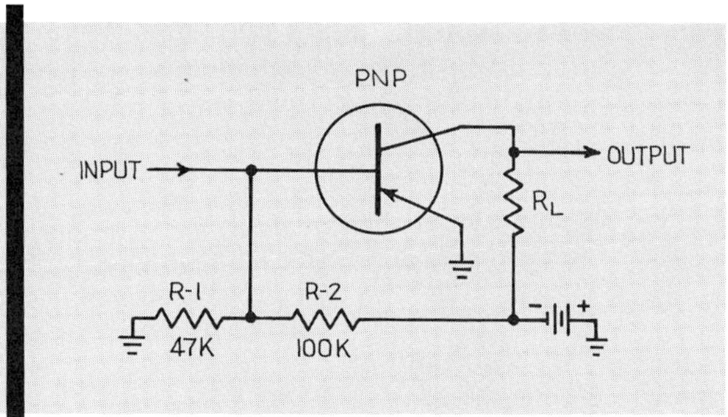

FIG. 20-9. Base bias by means of a voltage divider.

away rise of collector current. Severe distortion would occur, or the transistor could become inoperative, or even destroy itself. The primary cause is that a temperature rise reduces the emitter junction resistance. The forward-biased emitter-base diode is said to have a negative coefficient of resistance. To get stable bias, external compensating circuits must be employed.

One simple way to give a little more stability is to supply base bias via a voltage divider, as shown in Fig. 20-9. Note that resistors *R*-1 and *R*-2, the voltage divider, are in series across the battery. Base bias is taken from the tap between the two resistors. This device alone will not give sufficient stability.

A more effective stabilizing device is that of d-c current feedback. This is achieved by placing a resistor in the emitter lead as shown in Fig. 20-10. This resistor *R*-3 is in many ways similar to the cathode resistor in a tube. When the circuit becomes unstable and collector current starts to increase, the additional current across *R*-3 produces an additional voltage drop across the resistor in the direction shown. The polarity of this voltage is such that it reduces the forward bias of the emitter junction and tends to restore the stable operating point. A capacitor is often placed across the emitter resistor, as with a tube, to prevent signal degeneration.

Let us explore the circuits of Fig. 20-10 a

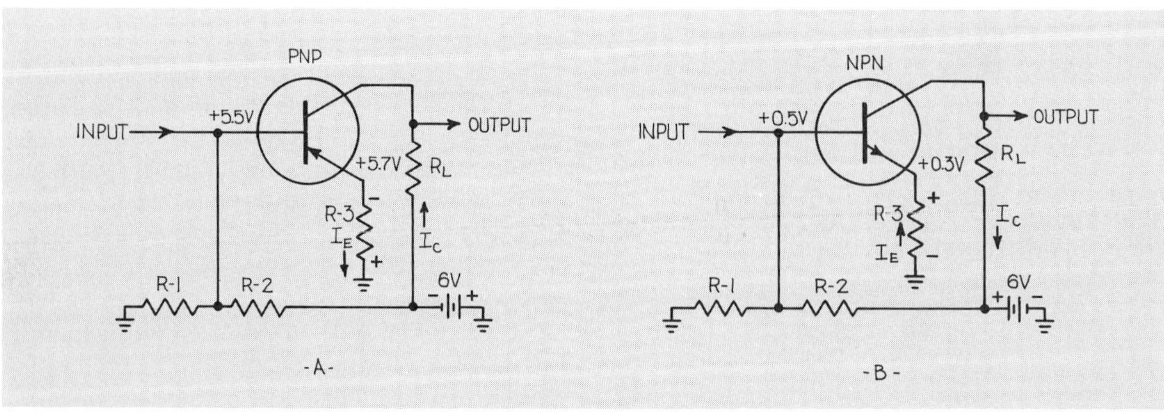

FIG. 20-10. Stabilizing operation by means of an emitter resistor for current feedback.

little further. Note that in the PNP amplifier, the negative end of the 6-volt battery is applied to the collector through the load resistor. The positive end of the battery is applied to the emitter through resistor *R*-3. As a result, the bias voltage on the emitter is 5.7 volts positive. Battery current through the voltage divider furnishes the base with a positive bias of 5.5 volts. The effect is to make the emitter 0.2 volt positive with respect to the base. Thus the emitter junction is forward biased, while the collector junction is reverse biased. The amount of forward biasing of the emitter junction usually ranges from 0.1 to 0.3 volt.

With the NPN amplifier, the positive end of the battery is connected to the collector through the load resistor. The emitter is connected to the negative end of the battery through resistor *R*-3. As a result, the bias voltage on the emitter is slightly above ground level (−6v) at 0.3 volt positive. Current through the voltage divider furnishes the base with a positive bias of 0.5 volt positive. Thus the base is 0.2 volt positive with respect to the emitter and is again forward biased. The collector, at a high positive bias, is positive with respect to the base. The collector junction is thus reverse biased, since it is more positive than the low positive base.

Return now to the stabilizing circuits. Another device to give operating stability is to use d-c voltage feedback. As seen in Fig. 20-11, the feedback is obtained from a voltage in the output. Resistors *R*-2 and *R*-3 give the base and emitter their proper positive biasing for forward bias. Collector load resistor allows the negative end of the battery to make the collector negative for reverse biasing of the collector junction. Stabilization occurs in the following manner. If the collector current starts to rise because of thermal effects, the collector becomes less negative (more positive) because of the larger voltage drop across the load resistor. As a result, a positive pulse is fed through feedback resistor R_F to the base, and reduces the forward biasing of the emitter junction. This effect reduces the collector current and stabilizes the amplifier.

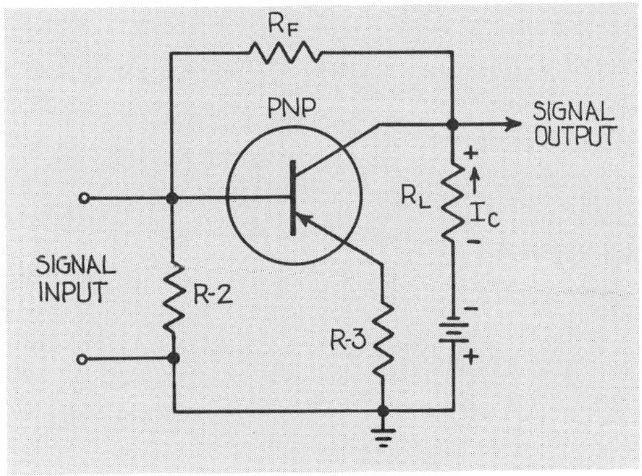

FIG. 20-11. Stabilization by voltage feedback.

When the collector load resistor is large, as in R-C coupled stages, current and voltage feedback are equally effective. When the load resistor is low, as in transformer-coupled stages, current feedback is more effective.

The previous stabilization circuits involved the use of standard resistors as circuit elements. Operation may also be stabilized by employing temperature-dependent resistors or PN junction diodes. Some of these circuits will now be described.

One such circuit, shown in Fig. 20-12, employs a thermistor. The resistance of a thermistor decreases with a rise in temperature; that is, it has a negative temperature coefficient of resistance. There are two voltage dividers across the battery. One is made up of resistors *R*-1 and *R*-4. The voltage across *R*-1 furnishes the base bias. The other voltage divider is made up of resistor *R*-2 and thermistor *R*-3. The voltage developed across *R*-2 furnishes the emitter bias. With the rise of temperature, the collector current tends to run away. However, the temperature rise also causes the resistance of the thermistor to decrease with a consequent increase of current through resistor *R*-2. This increase of current feeds a negative pulse to the emitter, thereby decreasing the forward bias of the emitter junction. As a result the collector current is decreased to normal operation.

The previous circuit employed a thermistor to control emitter voltage. The circuit shown in Fig. 20-13 uses a thermistor to control base voltage. Resistor *R*-1 and thermistor *R*-2 make up a voltage divider across the battery. Base bias is taken from the junction of the two. With the rise of temperature, collector current tends to rise. The rise of temperature makes the resistance of the thermistor decrease, increasing the current through the voltage divider. As a result, less battery voltage is dropped across the thermistor and the

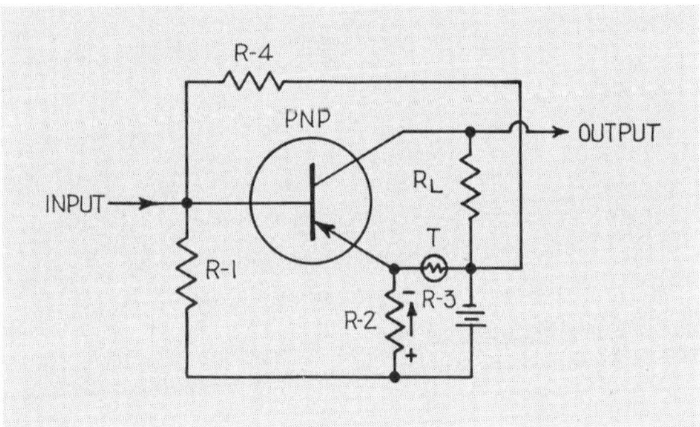

FIG. 20-12. A thermistor-stabilized transistor circuit.

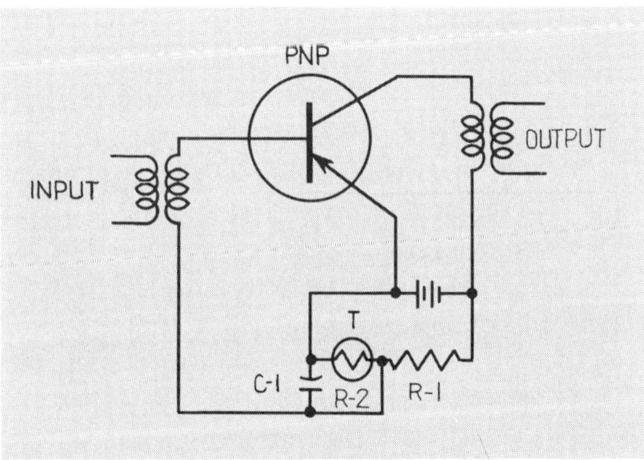

FIG. 20-13. A thermistor control of base voltage.

emitter junction forward biasing is decreased. This brings the collector current back to normal operation. It is always advisable to provide for good thermal contact between the transistor and the stabilizing thermal-sensitive component, to keep the temperature of the two similar. Although thermistors do tend to stabilize a transistor amplifier, they do not do so perfectly because thermistor resistance change does not track with the thermal effect change in the transistor.

A more effective thermal-sensitive device for stabilizing a resistor amplifier is a PN junction diode. Like a thermistor, a junction diode has a negative temperature coefficient of resistance regardless of whether it is forward or reverse biased. The main advantage of the junction diode is that it can be made of the same material as the transistor. As a result, thermal resistor changes of the diode tend to track better with the thermal effect changes in the transistor.

A circuit employing a forward biased PN diode as a stabilizing component is shown in Fig. 20-14. The circuit is very similar to that with the thermistor described in Fig. 20-13. It works in the same manner. It is well stabilized, except at higher temperatures. To get better stabilization at higher temperatures, use is made of two PN junction diodes, as shown in Fig. 20-15. Resistor *R*-1 and forward biased junction diode *CR*-1 serve the same purpose as in the previous circuit; namely, to compensate for the change in emitter junction resistance, particularly at temperatures below 50 deg C. Resistor *R*-2 and junction diode (reversed bias) serve to stabilize the circuit at higher temperatures. As with the previous circuit, capacitor *C*-1 bypasses the a-c signal around the junction diodes.

Some amplifiers employ a single reverse bias junction diode to stabilize operation. Such a circuit is shown in Fig. 20-16. It is used if the stage is resistance-capacitance coupled to the previous stage. Here is how it works. Emitter resistor *R*-1 prevents the

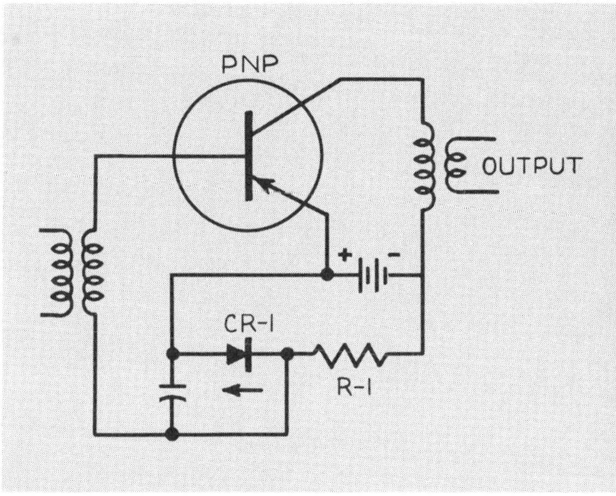

FIG. 20-14. Stabilization by a forward-biased junction diode.

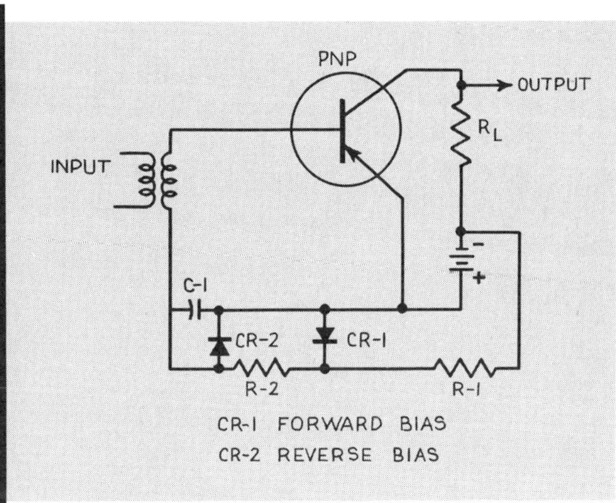

FIG. 20-15. Stabilized amplifier with two junction diodes.

build-up of a large emitter current at the lower temperatures. At the higher temperatures, the base region tends to overaccumulate current carriers, holes in the P-type and electrons in the N-type. Such an accumulation tends to increase the emitter current and in turn the collector current. The small reverse bias current that flows through the junction diode *CR*-1 drains the current carriers from the base, preventing the accumulation and stabilizing the circuit.

Another system for stabilizing amplifiers

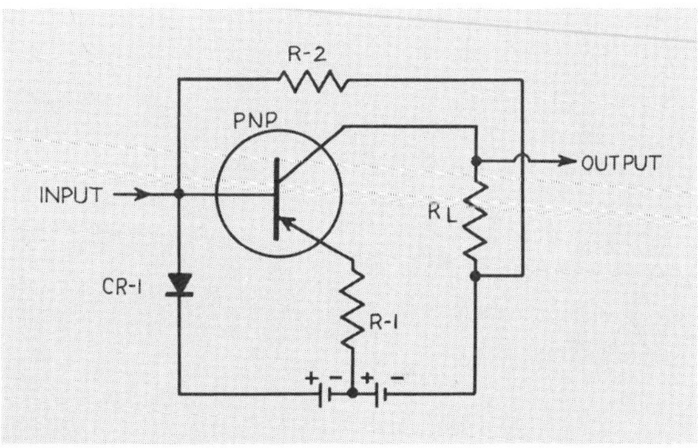

FIG. 20-16. Stabilized amplifier using one reverse-biased junction diode.

is that of using currents and voltages of one temperature-stabilized transistor to temperature-stabilize another transistor. A circuit showing this system is shown in Fig. 20-17. Note that the base of the NPN transistor is connected to the emitter of the PNP transistor and that the emitter of the first is connected to the base of the second. This gives proper forward biasing to the emitter junctions of both transistors. The NPN transistor is stabilized by having an emitter resistor *R*-1 and by having virtually no resistance in the base lead. As the temperature increases, the emitter junction resistance decreases because the base and emitter sections form a junction diode with a negative temperature coefficient of resistance. However, since the NPN transistor is stabilized, the current across the emitter junction remains constant. With constant current and reduced resistance, the voltage across the emitter junction drops. Since this emitter junction voltage furnishes the forward bias voltage for the PNP transistor, the latter is given reduced forward bias and its runaway collector current is brought back to normal. Capacitor *C*-1 bypasses the a-c signal around resistor *R*-1 and the emitter battery. This

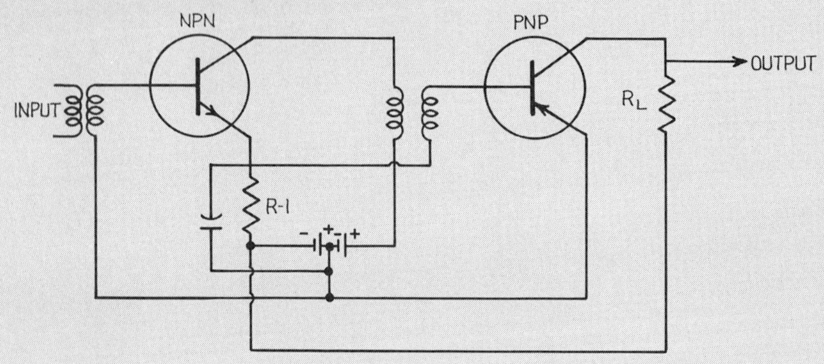

FIG. 20-17. Temperature stabilization of a transistor by another transistor.

circuit relies on the common emitter-base voltage for the two transistors.

Figure 20-18 shows a stabilization circuit in the same system, employing a common emitter-collector current for the two transistors. The first transistor is stabilized by placing a resistor in the emitter lead and by keeping resistor *R*-1 at a low ohmage. The stabilized collector current of the first transistor is made to flow through the emitter-collector of the second transistor by connecting the collector of the first directly to the emitter of the second. This type of stabilization eliminates the need for a stabilizing resistor in the emitter lead of the second transistor, an important consideration in power amplifiers that draw heavy emitter currents. Note that the first transistor amplifier is not that of a common emitter circuit, but rather that of a common collector circuit where the output signal is from the emitter. Capacitor *C*-2 places the first collector and the second emitter at ground potential for the a-c signal.

The circuits just described that stabilize the currents also tend to stabilize transistor voltages. Figure 20-19 shows a circuit for further stabilizing the collector supply voltage. Junction diode *CR*-1 is a reverse biased diode. At a certain reverse bias voltage, the current

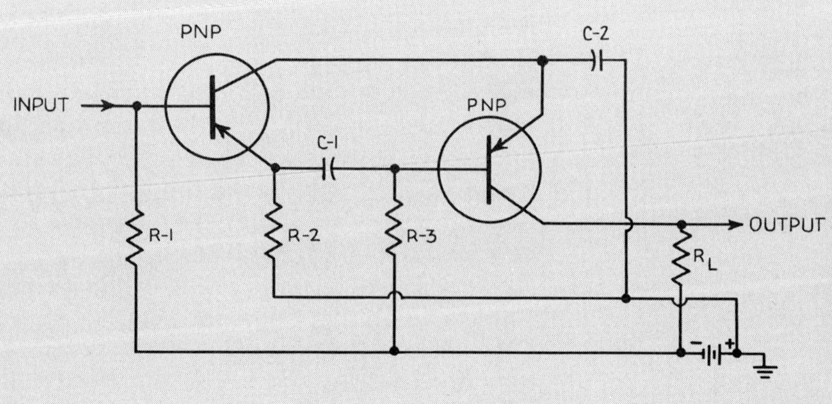

FIG. 20-18. Common emitter-collector current stabilization.

through it rises rapidly while the voltage across it remains constant. This specific reverse bias voltage point is known as the breakdown or Zener voltage. This Zener voltage can be designed to occur at any value from 2 to 60 volts. In the circuit, the breakdown diode is across the load resistor. By maintaining a set voltage, even though the collector current increases as a result of temperature rise, the voltage drop across the load resistor remains constant. As a result, collector voltage remains constant.

Junction diode *CR*-2 is forward biased and is used to compensate for resistance changes in the breakdown diode resulting from temperature changes. The breakdown diode has low a-c impedance and would shunt out the signal from load resistor R_L. To prevent this effect, a high impedance choke coil *L*-1 is placed in series with the diode.

A final stabilization circuit is shown in Fig. 20-20; it is used particularly in transformer-coupled stages. The sudden termination of the input signal or the injection of a strong noise signal may drive the normally forward-biased emitter junction into a state of reverse bias. As a result internal oscillation may occur which could destroy the transistor. This possibility is eliminated by the insertion of a reverse bias junction diode in the base circuit. Normally, the resistance of the diode is so high that it may be considered out of the circuit. When the large voltage appears that would reverse the emitter junction forward bias, the junction diode is driven to forward bias and shunts out this surge voltage. When the diode conducts this surge current, the voltage drop across it is negligible, and the emitter junction is prevented from reverse biasing.

Positive and negative ground. In different receivers, we find that sometimes the positive pole of the battery is grounded and sometimes the negative pole is grounded. Actually, it is immaterial whether battery plus or minus is grounded. Examine the circuits in Fig. 20-21. Both schematic drawings employ

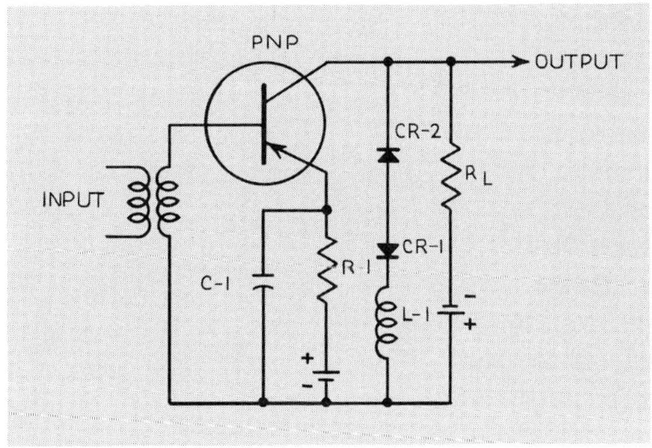

FIG. 20-19. Using diodes for stabilization.

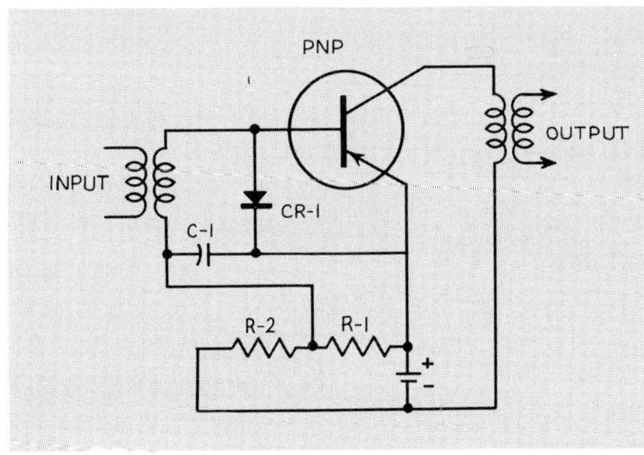

FIG. 20-20. Stabilization by diode in the base circuit.

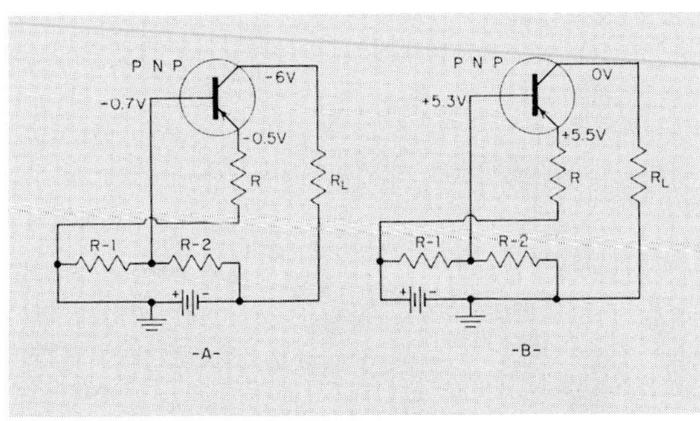

FIG. 20-21. Effect of grounded positive and grounded negative on a PNP transistor. All voltages are measured with respect to ground.

a voltage divider across the battery to develop forward bias for the emitter junction.

In both cases, the voltages shown on the transistor leads have been measured from the lead to ground, as manufacturers do on their schematic drawings. Examine Fig. 20-21*A* with the grounded battery positive. The collector is highly negative with respect to ground and also with respect to the base which is at a low negative potential, 0.7 volt, with respect to ground. Thus, for the collector junction there is the necessary reverse bias. The emitter is even less negative than the base with a potential of minus 0.5 volt with respect to ground. Hence the emitter is 0.2 volt more positive than the base and the emitter junction is properly forward biased. Thus the biasing for this PNP transistor is correct.

Now examine the circuit in Fig. 20-21*B* with the grounded battery minus. The collector is at maximum negative potential which is ground potential. Therefore, there is no voltage between it and ground. The base potential is highly positive with respect to ground; namely +5.3 volts. The collector is therefore 5.3 volts more negative than the base and the collector junction is properly reverse biased. The emitter is at +5.5 volts positive potential above ground. It is therefore 0.2 volt more positive than the base, and the emitter junction is properly forward biased. Again, biasing for this PNP transistor is correct.

So it is immaterial which battery terminal is grounded. A positive ground produces negative voltages on the transistor leads. A negative ground produces positive voltages on the base and emitter leads and zero voltage on the collector lead. A parallel development could also be made for an NPN transistor.

Signal feedback. In the previous section, several circuits were presented that maintained d-c stability. One method was to utilize current or voltage feedback. This section refers to a-c signal feedback for a different purpose. Degenerative or negative signal feedback is used for the following purposes: to fix the amplifier gain, to increase the bandwidth, to reduce distortion, to change the amplifier input and output impedances, and to unilaterize (neutralize) transistor feedback from output back to input.

As with d-c feedback circuits, we may have either signal current feedback or signal voltage feedback. An example of the former is shown in Fig. 20-22. This circuit is like the one employed for d-c stabilization. When only stabilization was desired, the emitter resistor was bypassed by a capacitor. When negative feedback is required, the capacitor is omitted. Such negative feedback has the effect of increasing the input resistance greatly, and is used as the input stage of a voltage amplifier.

An example of signal voltage feedback is shown in Fig. 20-23. Once again, the placement of a resistor between collector and base resembles the circuit described for d-c stabilization. If only the latter were desired, the resistor would be replaced with two equal ohmage resistors each half the value of the single resistor, joined in series, and a capacitor would be connected from their junction to ground. The feedback circuit has the effect of reducing the input resistance of the amplifier and may be used as the input stage of a current amplifier.

In both feedback circuits, the feedback resistor is designated as R_F. In the first circuit, the amount of signal returned to the input is proportional to the current flowing in the output. In the second circuit, the signal fed back to the input is proportional to the output voltage. In both circuits, feedback from output to input is possible in order to achieve negative feedback because in the common emitter circuit, as in a tube, input and output signals are 180 degrees out of phase.

Negative feedback is also sometimes employed to prevent oscillation of a transistor amplifier. A transistor is not a unilateral device in which energy moves forward only in

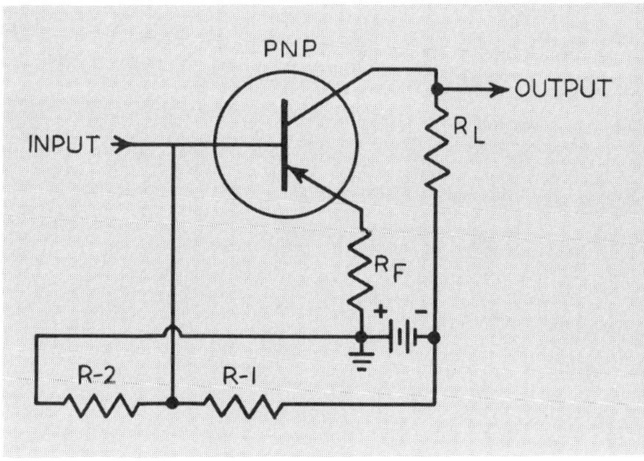

FIG. 20-22. Current feedback circuit.

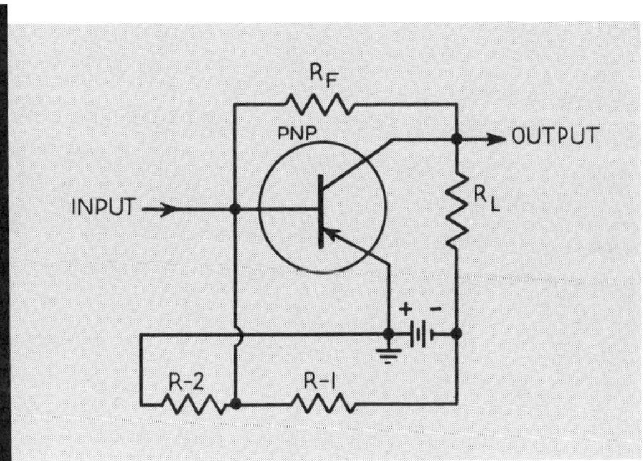

FIG. 20-23. Voltage feedback circuit.

one direction. As in a triode tube, energy tends to be fed back from output to input, across the capacitance existing between the collector and the base, producing regeneration. If this positive feedback is strong enough, the transistor amplifier will tend to oscillate. The higher the frequency of the signal, the greater the feedback and the greater the tendency to oscillate, particularly if there is tuned coupling. Fortunately, modern transistors are designed with small collector-base capacitance, so that amplifiers will not oscillate at broadcast frequencies. But they might do so at higher frequencies. With a triode tube, an external neutralizing circuit was used to prevent the oscillation. With a transistor, an external unilateralizing circuit is employed.

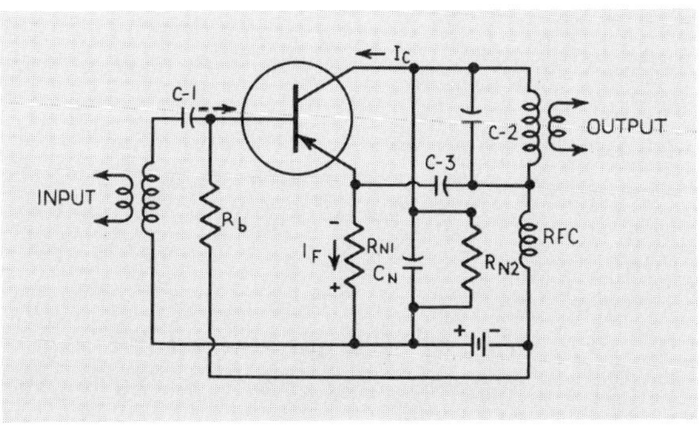

FIG. 20-24. A unilateralizing circuit.

Such a unilateralizing circuit is shown in Fig. 20-24.

When a negative input pulse is fed to the base, forward bias is increased and collector current I_c also increases. A portion of this collector current is coupled back through the collector-base capacitance to the input, producing positive feedback. To offset this, a portion of the signal current is directed through resistor R_{N1} and the parallel R_{N2}/C_N. This feedback unilateralizing current I_F develops a voltage across R_N with the polarity indicated. The voltage developed across this resistor is degenerative and made to equal the positive regenerative collector-base feedback. As a result, the latter is canceled out and oscillation is prevented.

In the circuit shown, resistor R_b forward biases the base-emitter circuit. Capacitor C-1 prevents the shorting of the base voltage. Capacitor C-2 and the output-transformer primary winding form a tuned output circuit. Capacitor C-3 blocks battery negative from the emitter and couples a portion of the collector output current to the emitter.

Interstage coupling. Transistor stages may be coupled in the same manner as tube stages. An RC coupled circuit is shown in Fig. 20-25A. Resistors R-7 and R-4 and capacitor C-1 make up the usual coupling components. The effi-

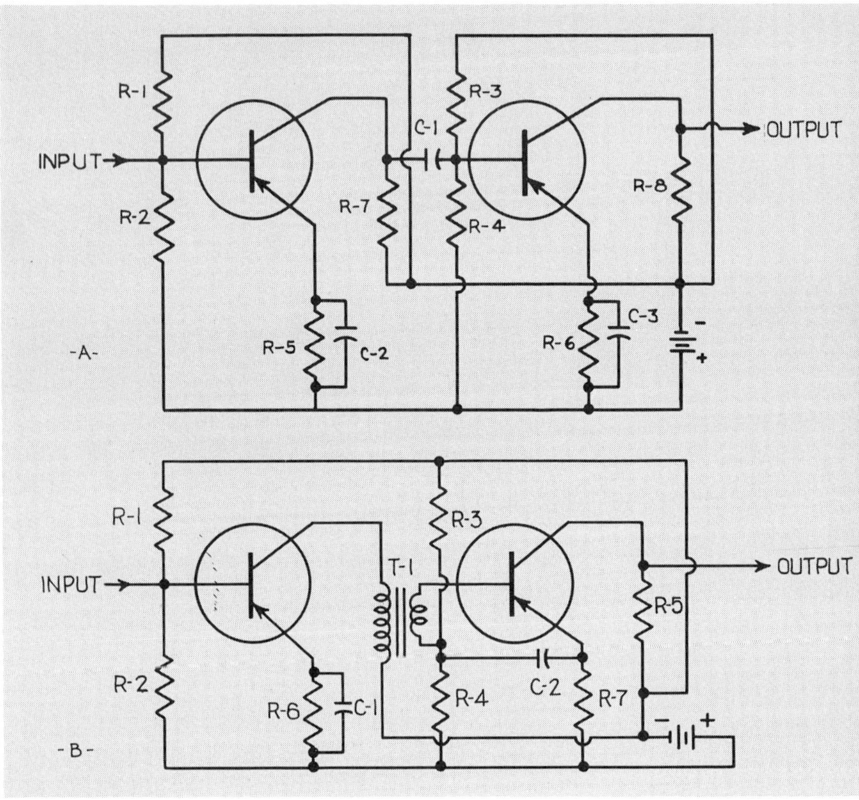

FIG. 20-25. An R-C coupled stage and a transformer-coupled stage.

ciency of such an amplifier is low because of the dissipation of d-c power in the collector load resistor *R*-7. Coupling capacitor *C*-1 blocks the d-c collector voltage of the first stage from appearing on the base of the second stage. To prevent a large signal voltage drop across this capacitor, its reactance must be small compared to the input resistance of the following stage (usually about 800 ohms). Hence its capacitance must be high.

Resistor *R*-4 is usually 7 to 15 times the input resistance of the second stage in order to feed maximum signal to the base. It cannot be too high because it would then adversely affect d-c temperature stabilization of the stage.

A transformer-coupled stage is shown in Fig. 20-25*B*. The primary winding of the transformer is the collector load of the first stage. The secondary winding introduces very little resistance in the base circuit of the second stage, a fact that produces better temperature stabilization. Little d-c power is dissipated in the primary winding; therefore, the power efficiency of this amplifier is high. For this latter reason, transformer coupling is almost universally used in small receivers, even though its frequency response is not as good as that of an RC coupled stage.

Identifying leads in transistors. It is important that transistors be connected with proper battery polarity connected to correct terminals. The base drawings shown in Fig. 20-26 will help in proper identification, E standing for emitter, B for base, and C for collector.

Basic transistor receiver. The signal in a transistor radio is handled basically in the same superheterodyne circuit as in a tube receiver. A block diagram of such a transistor

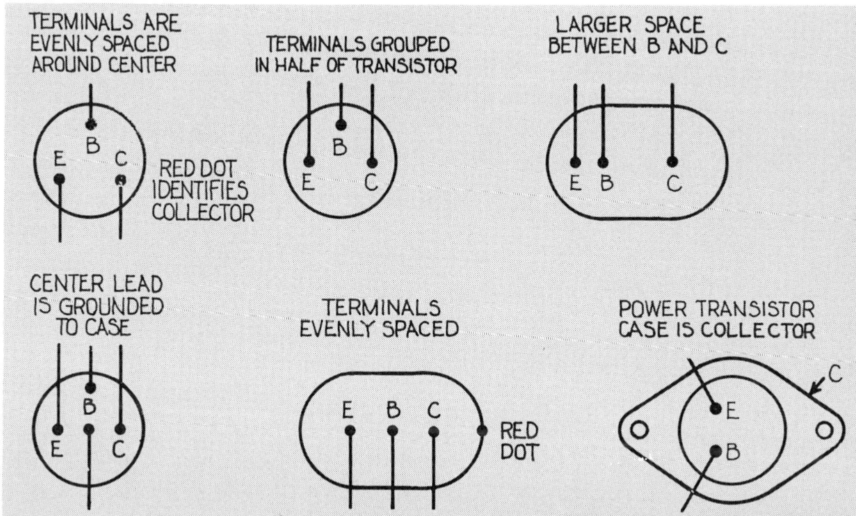

FIG. 20-26. Base diagrams for transistors.

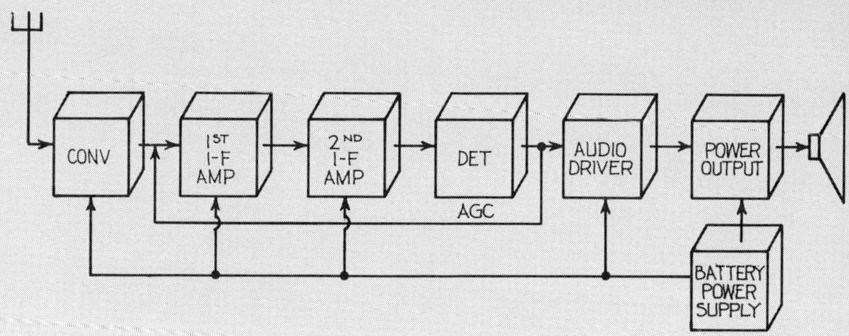

FIG. 20-27. Block diagram for a transistor superheterodyne receiver.

receiver is shown in Fig. 20-27. In such a receiver, one transistor usually serves as a converter. Sometimes, two transistors, a mixer and a local oscillator, perform the function of the converter. There then follow two i-f amplifier stages to give sufficient signal amplification. A junction diode serves as the detector. Sometimes, a transistor performs this function. The detector feeds into an audio amplifier, known as the audio driver, which in turn drives the power output stage. All transistors are powered by a battery.

A typical transistor receiver is shown in schematic form in Fig. 20-28. An analysis will be made, stage by stage. In the discussion that follows, it will be helpful to keep in mind the similar tube circuits with which you are already familiar. The terms emitter and cathode, collector and plate, and base and signal grid are closely allied and will help in understanding the various circuits.

Converter. The converter is shown in Fig. 20-29. Note that ground is battery positive, making the battery minus the hot lead. Component *L*-1 is a ferrite antenna feeding the signal to the base of the transistor. Resistors *R*-1 and *R*-2 comprise a voltage divider across the battery, for emitter-base bias. Resistor *R*-3 in the emitter lead is a d-c stabilizer. It is unbypassed to provide negative feedback for the signal. The oscillator coil is *L*-2 and it feeds its local oscillation to the emitter through capacitor *C*-3. The tuning

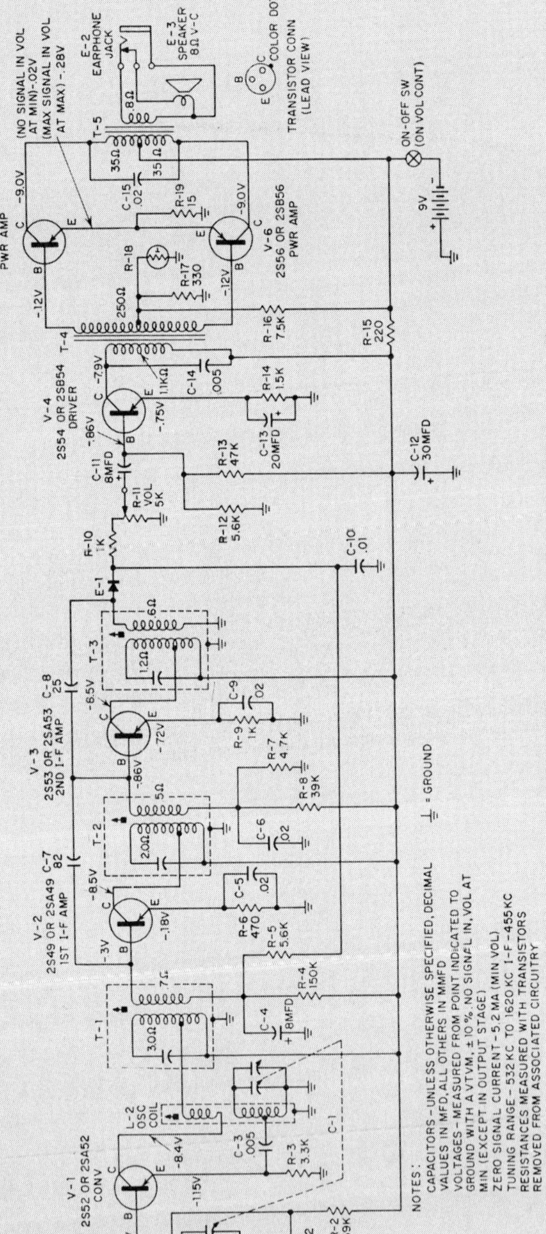

FIG. 20-28. Schematic diagram of a typical transistor receiver.

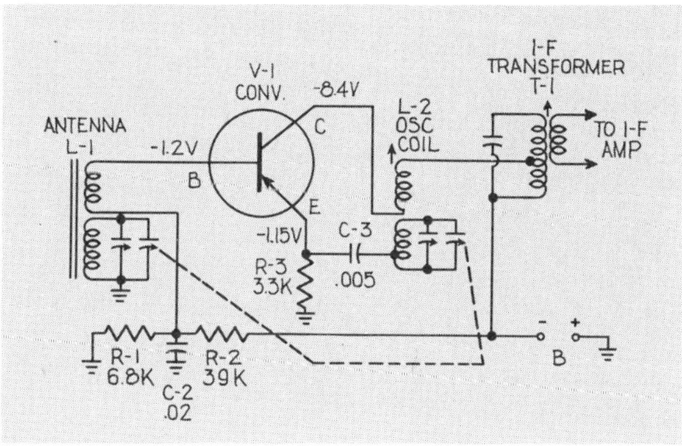

FIG. 20-29. Typical transistor converter circuit.

and oscillator capacitors are ganged, as in a tube receiver. Station signal to the base, and local oscillator signal to the emitter, mix to produce the modulated i-f signal at the collector, which feeds it to the i-f transformer. Capacitor C-2 is an i-f decoupling capacitor to prevent i-f signals from being fed back to the base.

Note that the local oscillator signal is taken from a tap on the oscillator tuning coil. This is done to provide a better impedance match for feeding the oscillator signal to the emitter. Similarly, the collector is connected to a tap on the primary winding of the i-f transformer for better impedance match. Also note that this is a PNP transistor with the emitter 0.05 volt positive with respect to the base, even though both are at a negative potential with respect to ground. The collector is 7.2 volts negative with respect to the base, even though both are at a negative potential with respect to ground.

I-f amplifier. Since a transistor amplifier has less gain than a similar tube amplifier, two stages of i-f transistor amplifiers are used, as shown in Fig. 20-30. The signal from the converter is fed to the i-f input transformer T-1. Transformers T-1, T-2, and T-3 are i-f transformers with slug adjustment.

Note the various transistor element bias voltages for the PNP transistors. Emitter components R-6 and R-9 are the d-c stabilizers and are bypassed by capacitors C-5 and C-9, respectively, to avoid negative feedback. To give uniform volume output, agc voltage (automatic gain control) is fed to the base of the first i-f transistor. The usual practice is to feed agc voltage to only the first of the i-f transistors.

Resistors R-4, R-5, R-7, and R-8 are voltage dividers to provide emitter-base bias. Capacitors C-4 and C-6 are decoupling capacitors. Because the input and output of both transistors are fixed-tuned to the same intermediate frequency and gain is high, the i-f amplifier has a tendency to oscillate. To pre-

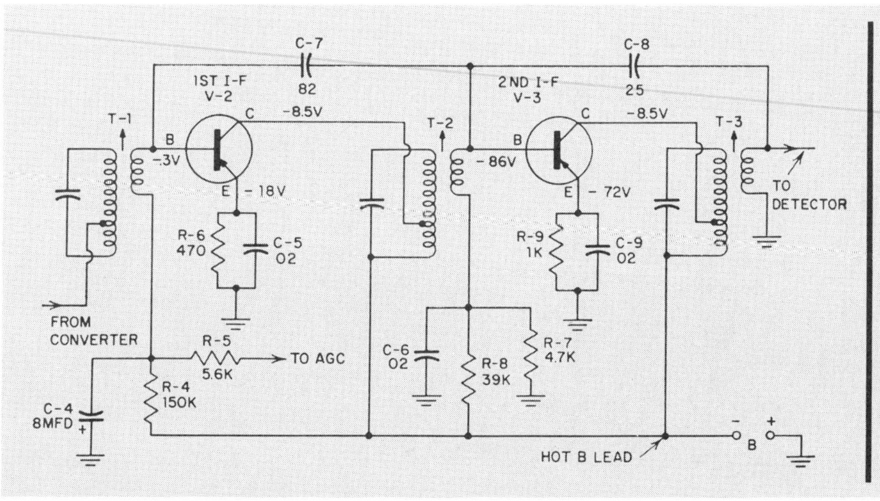

FIG. 20-30. The i-f amplifier in a typical transistor receiver.

vent this effect, neutralizing capacitors are sometimes used. A neutralizing capacitor is a small capacitor which couples signal from the output of a stage back to its input. Note that capacitor *C*-7 is so coupled for transistor *V*-2 and capacitor *C*-8 serves the same function for transistor *V*-3. To get proper phase of neutralizing voltage, signals are drawn from the secondaries of the i-f transformers.

Detector. The detector circuit is shown in Fig. 20-31. In most cases, the detector is a junction diode, with a behavior similar to that of a tube diode detector. It also develops agc voltage for the first i-f amplifier. The diode detector receives its signal from the last i-f stage. The diode load is made up of resistor *R*-10 and volume control *R*-11. The tap between the diode and *R*-10 is the source of the agc voltage. Capacitor *C*-10 filters out any i-f signal in the detector output. Audio components in the agc line are filtered out by *R*-5 and *C*-4.

Audio driver. The remaining stages of the receiver are audio amplifiers. Since the detector output is insufficient to drive the audio power output stage, an audio amplifier called the driver is placed between the two stages. Figure 20-32 shows the driver stage, corresponding to the first a-f tube amplifier.

Audio signal from the detector is fed to the base of the driver via coupling capacitor *C*-11. Note that the latter component is an electrolytic capacitor, made necessary by its required high capacitance. Resistors *R*-12 and *R*-13 make up the voltage divider for base-emitter bias. Resistor *R*-14 is the d-c stabilizing resistor, bypassed by capacitor *C*-13 to prevent signal degeneration. Note that it, too, is of the electrolytic type. Capacitor *C*-14 is an r-f filter which bypasses any r-f signal that is present. Transformer *T*-4 is an audio transformer that couples the signal to the power output stage. Capacitor *C*-12, an electrolytic one, is shunted across the battery supply and decouples signals from all stages.

A-f power output stage. The a-f power output stage in our typical receiver is a push-pull circuit. Of course, it is also possible to have single-ended output. The power output stage and the power supply of our set are shown in Fig. 20-33.

The power supply for the receiver consists of the 9-volt battery, the on-off switch, and

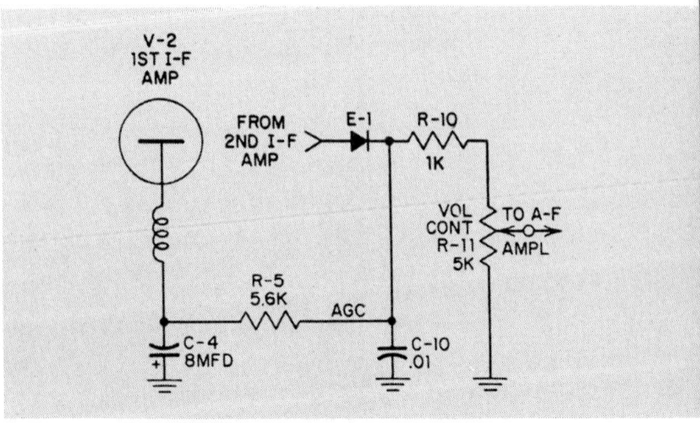

FIG. 20-31. Detector and agc circuit in a typical transistor receiver.

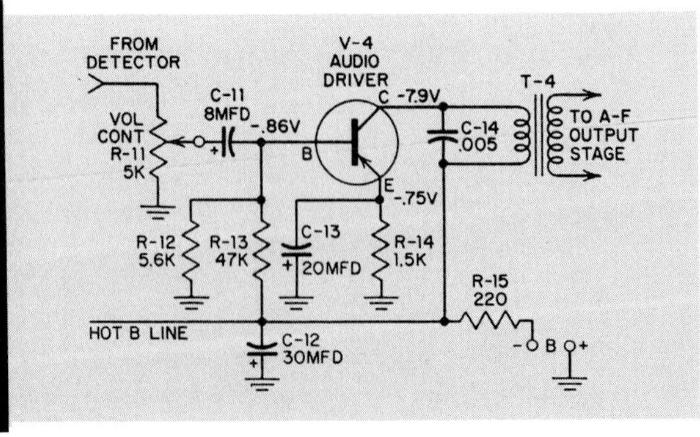

FIG. 20-32. Audio driver in a typical transistor receiver.

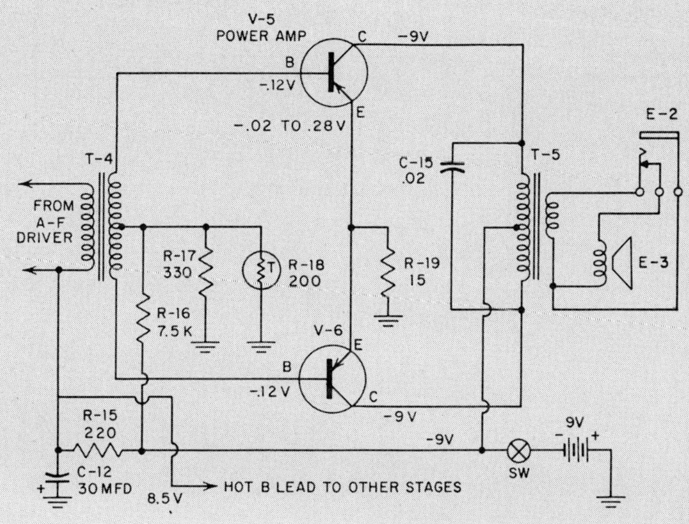

FIG. 20-33. Push-pull power amplifier output stage in a typical transistor receiver.

the decoupling circuit made up of *R*-15 and *C*-12. Observe that the full battery potential of 9 volts is applied to the a-f output transistors, while the potential on the hot *B* lead for the others drops to 8.5 volts. Note this similarity to the *B*++ and *B*+ leads in the tube-type receiver.

Now examine the power output stage. The audio signal is fed to the bases of the output transistors from the driver stage through audio transformer *T*-4. Note that collector voltages are at −9 volts. The range of voltage shown for the emitters is the swing from no signal to that of maximum signal.

As a signal appears across the center-tapped secondary of transformer *T*-4, one transistor base is driven negative and the other positive. The negative base increases forward bias and produces a surge of collector current. The positive base reduces forward bias, causing collector current to decrease or cut off. The situation is reversed when the signal reverses across the secondary winding of transformer *T*-4.

Resistor *R*-19 is the d-c bias stabilizing resistor. It is unbypassed in order to furnish negative signal feedback. Resistors *R*-17 and *R*-16 make up a voltage divider to furnish proper emitter-base bias. Thermistor *R*-18 helps to stabilize the d-c operating point with thermal changes. Resistor *R*-15 serves with capacitor *C*-12 as a decoupling filter. Capacitor *C*-15 is a bypass capacitor across the primary of output-transformer *T*-5. It tends to bypass the higher audio frequencies and to favor the bass audio frequencies. The output transformer feeds the signal to loudspeaker *E*-3. Earphone jack *E*-2 enables the listener to open the loudspeaker voice coil circuit and to listen on an earphone. Finally, battery power is turned on and off by switch *SW*.

General servicing procedure. The general procedure for servicing transistor radio receivers is similar to that for a tube receiver. In fact, the defective section of the receiver is found by exactly the same procedure. Since no receiver will work without proper operating voltages, the first step in all cases is to check the power supply. Here it is a battery, and detailed instructions for checking the power supply will follow.

Having determined that the power supply is good, we next want to isolate the trouble to the audio amplifier, the i-f amplifier, or the converter. The procedure for transistor and tube receivers is exactly the same. Feed an audio signal to the hot end of the volume control. The amplified note out of the loudspeaker shows that the audio section is working. Next, place an injection loop fed by a modulated signal at 455 kc near the ferrite antenna of the receiver. The response or lack of response from the loudspeaker places the trouble in the converter or the i-f sections.

Once we know a section that is at fault, further signal checks localize the trouble to the individual stage. Then it is necessary to find the defective component, lead, or connection in that stage. No possible cause must be overlooked. The voltmeter and ohmmeter are extremely useful in this last localizing procedure.

We shall first describe all these servicing procedures. Then, we will find out how to make replacements and repairs. Because of the nature of the components in a transistor receiver and the miniaturization of parts, special precautions are necessary.

Checking the power supply. In a transistor radio, it is important that the battery voltage be sufficiently high for proper operation. When batteries are weak, the result may be weak or no reception, distortion, or motorboating. Do not be misled by the customer's statement that he did not use the receiver very much. Batteries have a way of deteriorating simply by standing unused.

The batteries may be tested as shown in Fig. 20-34A, with a multimeter. With the radio switch open, measure the voltage across the battery. Convenient test points are ground and battery end of the switch. Unless the battery is almost dead, it will measure very close to the full rated voltage. Then turn the

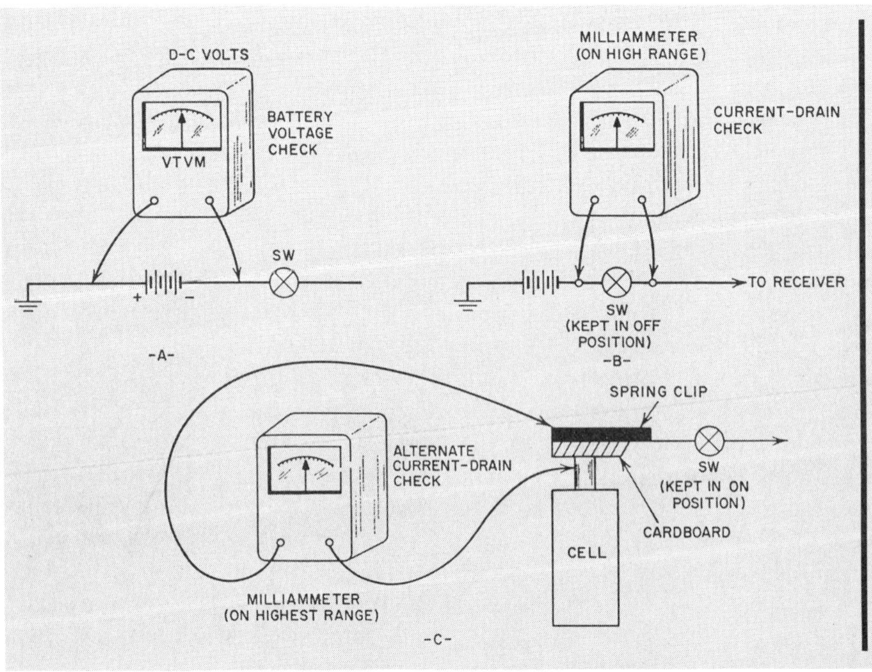

FIG. 20-34. Checking supply voltage and receiver current drain.

set on for a few minutes, while watching the voltmeter. If the voltage drops by 15% or more, replace the battery. If the drop is less than 15%, consider the battery good. About four out of five defective sets need only new batteries. Of course, you may check batteries by substitution with good ones if they are available.

It is important that new batteries be replaced with proper regard for the correct polarity. If zinc-carbon type cells are used, the center button is positive and the zinc can is negative. In mercury cells, the center button is negative. Where a battery pack is used, it is wise to follow the indicated polarity on the pack. Some battery packs are designed for use with snap-on plugs. The male plus is usually for the negative battery terminal and is connected with a black wire; the female plug is usually for the positive terminal and is connected with a red wire. If batteries show any signs of chemical leakage, replace them at once.

When checking the power supply, it is necessary to check that the battery plug makes clean contact with the battery terminals. The same precaution on cleanliness and good contact is essential in the ON-OFF switch. If the switch is defective, replace it and its associated volume control.

Sometimes the battery decoupling capacitor (*C*-12 of Fig. 20-33) opens. When it does, motorboating or oscillation occurs. If *C*-12 shorts, there will be no reception in the receiver. To check this, open one lead and shunt the capacitor with a similar one to see if the defect is removed.

To free the power supply of suspicion definitely and to point up possible trouble sources, test the overall current drain of the receiver. This measurement is made with a d-c milliammeter in either of the ways shown in Fig. 20-34. In Fig. 20-34*B*, the milliammeter, switched to its highest range, is placed across the switch terminals with the switch in the OFF position. If the switch is not readily accessible, place a piece of cardboard between the battery and the battery clip that is connected to the switch. Connect the milliammeter, with due regard to polarity, between the clip and the battery terminal, as in Fig. 20-34*C*. Using the highest range on the milliammeter is a precaution to guard against instrument damage if there is a shorted condition in the set. When the meter barely reads, reduce the current range to get the proper reading. If the cardboard is used, the switch must be turned on. Take the reading with the receiver in a no-station position and the volume control at its minimum position.

Normal current drain for any receiver is given by the manufacturer in service notes. However, where such notes are not available, remember that smaller sets may draw from 4 to 20 ma. Larger sets will draw up to about 30 ma. Where the current drain is excessively high, there may be a shorted transistor, a short of battery decoupling capacitor, or a short in any other component across the battery. Where current drain is extremely low, there may be a defective transistor, weak batteries, poor contacts at the battery or switch, or poor or open components and connections.

Localizing defective stage by signal substitution. Having determined that the defect is not in the power supply, you had best not waste time with a random check. By means of a signal substitution procedure similar to that for the conventional tube receiver, localize the defective stage. The base of a transistor is a test point similar to the control grid of a tube. Similarly, the collector is a test point similar to the plate of a tube. A summary table of signal generator input and test points is presented in Table 1. Note the similarity of the test point in this table to a tube-type superheterodyne receiver. For each step, the signal is fed to the test point through a 0.1-mfd isolation capacitor.

In these stage isolation tests, the dummy antenna is a capacitor in the hot lead of the signal generator. This capacitor serves to

TABLE 1. ISOLATING DEFECTIVE TRANSISTOR RADIO STAGE BY SIGNAL SUBSTITUTION

Signal injected	Test point	Normal result	Analysis
1. A-f	Ungrounded terminal of vol control	Signal from speaker	If OK, audio stages are OK. Go to step 3. If weak or no signal is heard, go to step 2
2. A-f	Driver collector	Signal from speaker	If OK, defect lies in volume control, coupling capacitor, or driver stage. If no signal or weak signal, defect lies in audio transformer, power output stage, output transformer, or speaker
3. Mod i-f	Second i-f base	Modulation signal from speaker	If OK, go to step 4. If weak or no signal, check second i-f amplifier, last i-f transformer, and detector stage
4. Mod i-f	First i-f base	Modulation signal from speaker	If OK, go to step 5. If weak or no signal, check first i-f amplifier and second i-f transformer
5. Mod i-f	Converter base	Modulation signal from speaker	If OK, trouble is in oscillator or antenna. Go to step 6. If no signal, check converter transistor and first i-f transformer
6. Mod 1,600 kc	Converter base. Tune set to 1,600 kc	Modulation signal from speaker	If OK, trouble is in antenna and its input circuit. If not heard, trouble is in oscillator stage

block d-c currents from the test points from shorting through the signal generator. The other lead of the signal generator is clipped to ground. The signal from the signal generator should be at the lowest possible level to make these tests. Naturally, all tests are made with the receiver turned on and in no-station settings; turn the volume control up high.

Voltage analysis. After having localized the defect at one stage, voltage analysis becomes a powerful means for finding the specific defect. Most defects, except open capacitors and misaligned circuits, will upset the normal voltages to be found in the receiver. We thus have an effective means for quickly getting at the heart of trouble.

Most manufacturers show the normal voltages at transistor electrodes. These are measured from electrode to ground. Where these voltages are not found, trouble is indicated.

Let us examine how we may pinpoint areas of defect in a transistor circuit by means of abnormal voltage readings. Take, for example, the circuit in Fig. 20-35*B* which is the second i-f amplifier stage of the basic receiver, showing normal voltage to ground. The circuit defects and their effects in upsetting normal voltages are shown in Table 2. As a rule, misaligned circuits and many capacitor defects have little or no effect on stage voltages. Here other types of test procedures must be used to locate the defect.

TABLE 2. HOW CIRCUIT DEFECTS AFFECT TRANSISTOR STAGE VOLTAGES

Defect	Effect on voltage
Open base circuit: Open resistor *R*-8, open transformer *T*-2, or break in base lead	Base voltage falls to zero. Emitter voltage falls to zero. Collector voltage remains normal. No voltage across *R*-9
Open emitter circuit: Open resistor *R*-9 *or* open in emitter lead	No voltage across *R*-9. Base voltage normal at –0.86 volt. Emitter at same voltage as base, –0.86 volt
Open collector circuit: Open transformer *T*-3 or break in collector lead	Collector voltage drops to same voltage as emitter, –0.72 volt
Open in transistor: Open section of transistor	Base voltage normal at –0.86 volt. Collector voltage normal at –8.5 volts. Emitter voltage zero. High emitter-base voltage
Transistor defective: Leaky transistor	Reversed emitter-base bias
Defective battery decoupler: *C*-12 short or leaky	All voltages lowered. Large battery current drain
Bypass capacitor defect: Bypass capacitor shorted	Abnormal base, emitter, or collector voltage, depending on circuit bypassed

If we analyze the data in Table 2, we discover several general hints which are applicable to any transistor stage. When there is an open in the collector circuit, the voltage on the collector becomes approximately that of the emitter. If there is an open in the emitter circuit, the emitter voltage becomes approximately that of the base. When there is an open in the base circuit, base and emitter voltages fall to zero. A transistor that develops internal leakage will reverse the forward emitter-base bias and should be replaced.

There is another voltage effect not obvious from the circuit just described. This is the case of a shorted coupling capacitor. When there is a shorted or leaky coupling capacitor, the base bias of the following stage will be off and little or no voltage will appear across the capacitor. There is still another voltage effect. When the agc filter capacitor becomes leaky or shorts it would throw off the bias voltages of the controlled transistor.

Bridging techniques. Capacitors may open and have no effect on the test voltages. Electrolytic capacitors, particularly, tend to dry up and lose capacitance or open. If it is a coupling capacitor, it will fail to pass the signal to the following stage. If it is an emitter resistor bypass capacitor, it will cause signal degeneration. If it is an agc filter capacitor, it will cause oscillation or distortion.

Although the overall effect in the receiver may be disastrous, the check for the cause is simple. Bridge the suspected component with a capacitor of the same capacitance. With electrolytic capacitors, be sure to observe proper polarity.

Replacement hints. Most transistor receivers are composed of miniaturized components, compactly linked by printed-circuit boards. In making component replacements, consideration must be given to space requirements and to the possibility of damage to parts through excessive heat from a soldering tool.

Generally, it is advisable to replace defective components with exact replacement parts. Where such replacement is not possible, due consideration must be given to the fact that all characteristics, including size, must be the same for the new and the defective component.

The small components in a transistor receiver are especially subject to damage when overheated. For this reason, you should use a pencil soldering iron with a rating of 25 to

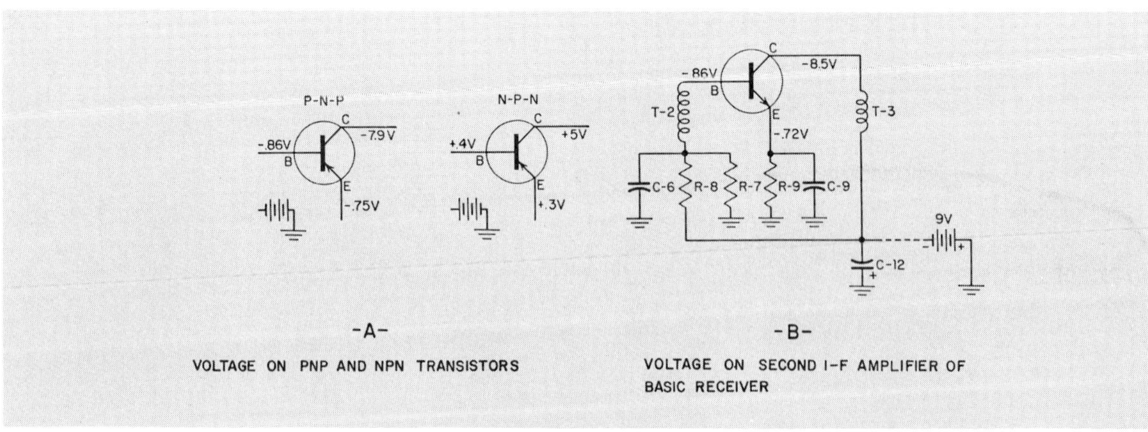

FIG. 20-35. Voltage distribution on transistors and an i-f stage.

35 watts and with a 3/16-inch diameter tip. To prevent overheating because of lengthy application of the iron, keep the tip clean and hot. Use a solder with 60 percent tin and 40 percent lead because it has a low melting point. Tin the leads of new components before connecting them into the circuit.

To prevent burning of the printed board, a good procedure when possible is to clip the defective component leads so that a small portion of the lead remains soldered to the printed wiring through the board hole, as shown in Fig. 20-36*A*. Then solder the tinned leads of the new component to the wire stubs. Where the complete leads of the defective component must be removed, apply the soldering iron to the lug of the printed board where the leads are connected. With a stiff glue brush, brush away the melted solder. It may be necessary to straighten the lead with a long-nose pliers at the same time as the iron is applied, as shown in Fig. 20-36*B*. Then remove the component. Be sure the hole through the board is clean and open. Insert the leads of the new component and solder to the lugs. Never apply the iron too long.

To dissipate heat, it is advisable to hold the new component lead with a long-nose pliers between the component and the point of application of the soldering iron, as shown in Fig. 20-36*C*.

Occasionally, you will find a break in the printed wiring on the board. Improvised repairs and attempts to place a blob of solder across the break can be more damaging than effective. The best procedure is to locate the lugs at the ends of the printed lead where the break occurs and then to solder a wire between these lugs. This procedure is shown in Fig. 20-36*D*.

Although normally rugged in use, the transistor itself is easily damaged by misuse. When making voltage checks, it is most important that you avoid even momentarily shorting of transistor leads. This precaution is particularly important with respect to the power amplifier transistors. Never remove or replace transistors with the receiver turned on. When making continuity checks in a receiver stage, you must not use the low R x 1 range of the ohmmeter. Where ohmmeters are powered by high potential, there is danger of ruining transistors with the high voltage of the test instrument.

Testing transistors and diodes. Junction

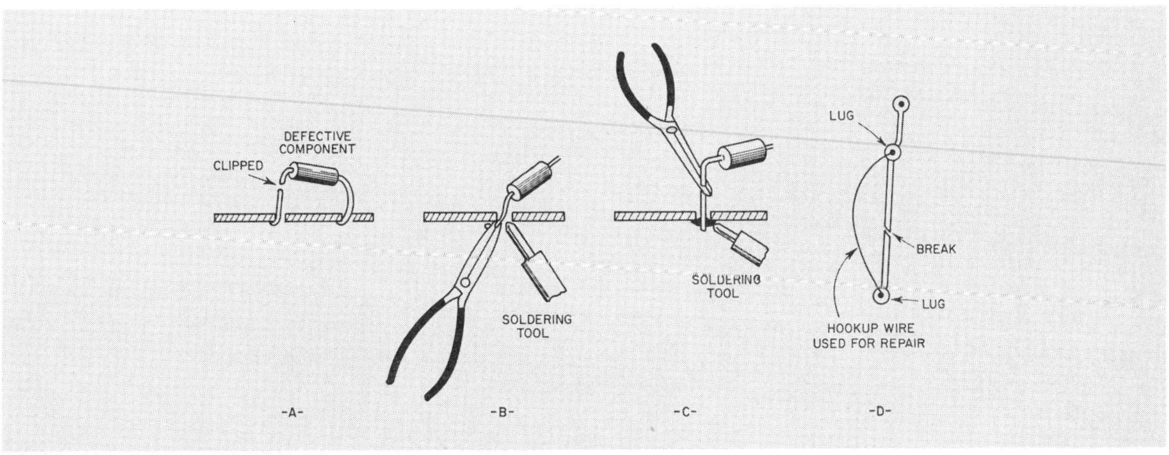

FIG. 20-36. Replacement hints for printed-circuit components.

transistors tend to short more frequently than open. The result of a short is to cause an excess current drain. This would show up in the power-supply check where overall receiver current drain is measured. When a transistor is believed to be defective, the best service procedure is to replace it with an identical one that is good. It is most important that the components of a defective stage be carefully examined before installing a new transistor.

When out of the receiver, transistors may be checked simply by means of an ohmmeter. Recall that a transistor is simply a set of diodes placed back to back. A diode has a lower resistance in the forward direction than in the reverse direction, unless it is open or shorted.

With power transistors, the ohmmeter should be set to the R x 1 range, as in Fig. 20-37. First connect the meter leads to the emitter and the base, and then reverse the leads. In one connection, the forward position, the resistance will be about 2 ohms. In the reverse position, the resistance will be close to infinity. Then connect the ohmmeter from the emitter to the collector; take a reading, and again reverse the leads. In each case, the resistance should read well over 100 ohms. Any marked deviation from the values given above indicate a short or an open in the power transistor.

With low-power transistors, the ohmmeter must be set only to the R x 100 range as in Fig. 20-38. The meter leads once again are connected between the emitter and the base, and then are reversed. In one position, the forward position, the reading will be less than 1,500 ohms. The other reading will be near infinity. Similar readings will be obtained when making the measurement between the emitter and the base. Any marked deviation from these readings indicates a short or an open in the transistor.

To check for leakage in a low-power transistor between the emitter and the collector, connect the ohmmeter between those two pins, using the R x 100 range; then reverse the leads. For r-f and i-f transistors, both readings should be above 5,000 ohms. For a-f driver transistors, both readings should be above 500 ohms. Lower readings indicate leakage or a partial short in the transistor.

Diodes may also be checked in a similar manner. Measure the reverse and forward resistance with an ohmmeter. With a good diode, reverse resistance should be 500,000 ohms or more; forward resistance should be 40 to 100 ohms.

We have presented the transistor and diode test as though they were out of the set. In some cases where they are readily removed, this is the case. But often, they may not be easily removed without damage. Where this is the case, it is often more convenient to make a cut across the printed wiring that leads to

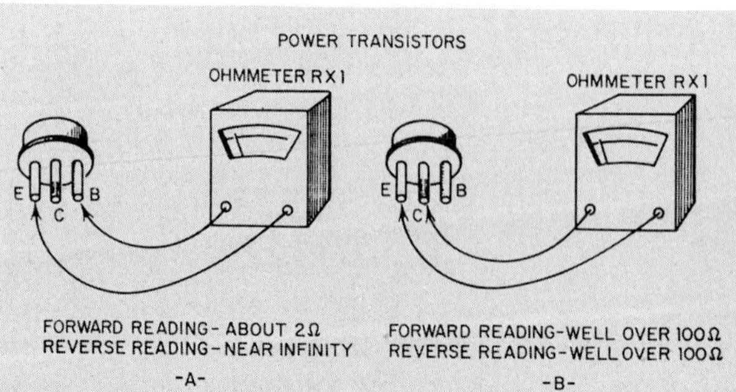

FIG. 20-37. Checking power transistors with an ohmmeter.

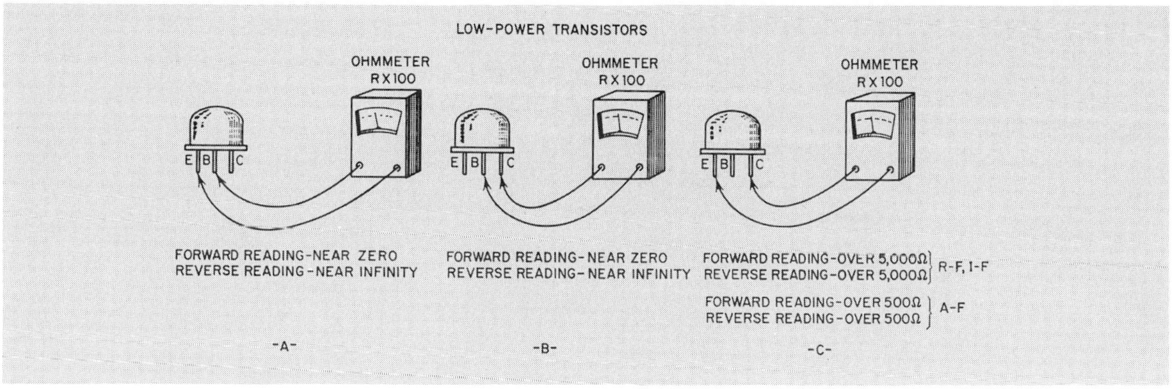

FIG. 20-38. Checking low-power transistors with an ohmmeter.

the transistor or diode terminals. Be sure no other components are connected to these terminals. Now check as though the transistor or diode were out of the set. If they are bad, replace with new ones. Then solder in wire jumpers just as you would with broken printed wiring.

Transistor-receiver alignment. Whenever an i-f transformer, antenna coil, oscillator coil, or even transistors have been replaced, it may become necessary to realign the receiver. Alignment of a transistor radio is no different in procedure than that for an ordinary tube radio. However, in transistor receivers there is more likely to be interaction between the r-f and i-f circuits; so a little more care must be used in setting adjustments.

It is generally advisable to follow the manufacturer's alignment procedure notes. Where these are not available, the procedure listed below will serve as a basic guide. Begin by placing an output meter across the speaker voice coil. Turn on the receiver, and set its volume control to maximum. Connect the signal generator output across a radiation loop made up of five turns of about 5-inch diameter. Place this loop at least 1 foot from the receiver antenna. Turn on the signal generator, and then attenuate its signal as low as possible so as to avoid bringing on agc action. This is most easily achieved by keeping the reading in the output meter below 1 volt. Now follow the steps shown in Table 3.

You now have enough information to service any transistor receiver that you are given. As you can see, these sets present problems very similar to those of tube receivers. There remains only the need to study a few circuit variations.

Transistor circuit variations. An examination of a few receiver schematics will familiarize you with some of the variations to be found. Those that we do not cover will be unusual, but you should be able to reason them out for yourself. The schematic for a Sylvania receiver is shown in Fig. 20-39.

This circuit presents some interesting points. Note that this receiver uses NPN transistors in the audio stages. In these NPN transistors, note that the emitters are negative with respect to their base. In the converter and single i-f stages, emitters and base are at the same potential. Forward bias is furnished by the positive portion of the signal arriving at the base input. These latter two transistors are shielded, and the shield is grounded for signal voltage through capacitors C-13A and C-13B.

TABLE 3. ALIGNMENT PROCEDURE FOR TRANSISTOR RADIOS

Signal generator		Receiver	
Connection to receiver	**Dial setting**	**Dial setting**	**Adjustment**
1. Radiation loop	I-f 455 kc	Tuning gang fully open (highest frequency)	Adjust 3rd i-f, 2nd i-f and 1st i-f transformer for maximum output (in that order) on output meter
2. Radiation loop	600 kc	600 kc	Adjust oscillator coil core for maximum output on output meter
3. Radiation loop	1,620 kc	Tuning gang fully open	Adjust oscillator trimmer for maximum output on output meter
4. Radiation loop	1,400 kc	1,400 kc	Adjust antenna trimmer for maximum output on output meter

5. Repeat steps 2, 3, and 4 until maximum output is obtained. Always end with step 4

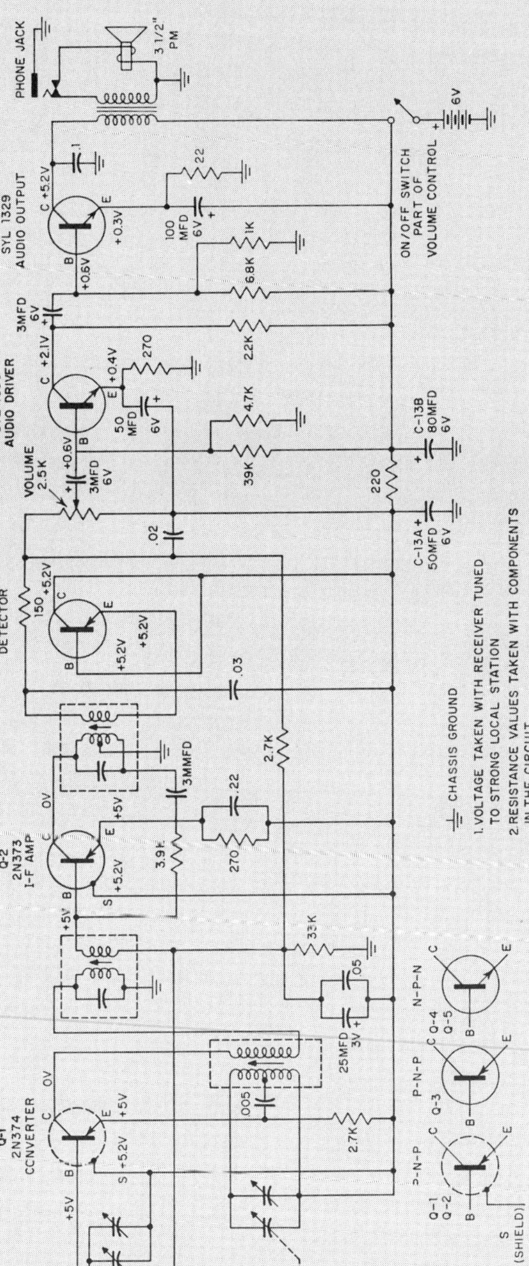

FIG. 20-39. Sylvania transistor receiver.

The detector is a PNP transistor. But note that its collector is directly connected to the base. In effect then this is a crystal diode with emitter as one section and collector-base as the other. Note also that the audio output stage is not push-pull, but single-ended. Finally, note that battery plus is the hot lead and battery minus is ground.

In Fig. 20-40, we see the schematic diagram of the Admiral Model 8T1A transistor receiver. This set has a clock, operating from a 1½-volt cell, with automatic switching. Note that there is an extra stage of r-f amplification before frequency conversion. The converter is composed of separate mixer and oscillator transistors. All transistors are of the PNP type with emitter biased positive with respect to base, and collectors biased negative with respect to base. Note that in this circuit, battery plus is ground and battery minus is the hot lead. So long as proper forward and reverse biasing is present, it does not matter which battery pole is ground. Note that agc voltage is fed to the r-f amplifier, the mixer, and the first i-f amplifier.

An interesting variation is shown in Fig. 20-41, the schematic diagram for an Arvin transistor receiver. This set uses three NPN transistors as converter, i-f amplifier, and reflex i-f amplifier. A crystal diode is used as a detector, and a PNP transistor is used as a single-ended power output stage. Note that battery minus is ground and battery plus is the hot side. For the three NPN transistors, collectors are connected to battery plus; for the PNP transistor, collector return is to ground (battery minus). In all cases, emitters are forward-biased—negative with respect to base for the NPN transistors, and positive with respect to base for the PNP transistor. Capacitor *C*-10 is a block electrolytic capacitor, the two outer sections serving as bypass capacitors for emitter resistors; the center one is the battery line decoupling filter. When one of these becomes defective, the entire block is usually replaced.

The reflex second i-f amplifier is an interesting variation. A reflex amplifier is one that amplifies two signals of widely differing frequencies, in this case the intermediate frequency and the audio frequencies. Special networks prevent interaction of the two signals. Note that the output of transistor *TR*-3 feeds into the tuned primary of the i-f transformer. This signal is then fed to the crystal-diode detector, part of whose load is the volume control *R*-11. Note that the movable arm of the control does not feed the audio signal to the base of the power amplifier transistor, but back to the base of *TR*-3 for amplification. This amplified audio signal is now fed through the primary winding of the output transformer, which offers little impedance to signals at audio frequencies, to the base of the power amplifier transistor *TR*-4. Although a unique design, this receiver is serviced in the same manner as the standard one and presents no special problems.

Figure 20-42 shows a Motorola transistor receiver. The set is designed for two-way operation—either from the 120-volt a-c line or from batteries, the mode of operation controlled by switch *S*-1. The line mode employs a step-down transformer and full-wave rectification furnished by a dual selenium diode rectifier. These are followed by the filter circuit, composed of capacitors *C*-20*A* and *C*-20*B* and resistor *R*-26. Servicing of this power supply is similar to that of the standard tube power supply. Resistor *R*-27 is a bleeder which gives better voltage output regulation.

Diode *E*-1 is known as an overload or clamping diode and serves an interesting function. At the low functioning voltages of transistors, a strong signal could develop enough agc control voltage to block the controlled first i-f amplifier *V*2. The diode tends to overcome this situation. When the set is functioning properly on a moderate or weak signal, collector current through resistor *R*-7 will develop a voltage drop across the resistor. The direction of current will be such as to make

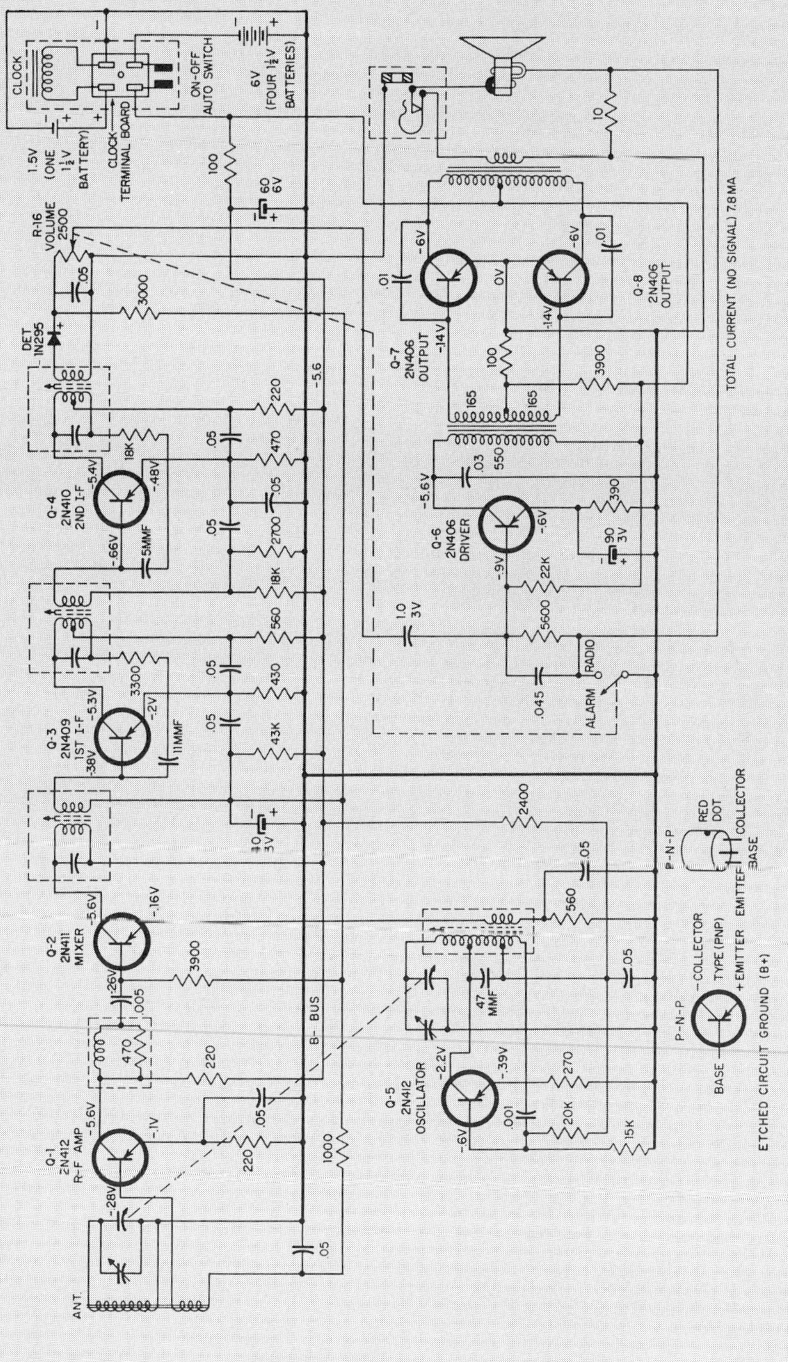

FIG. 20-40. Admiral transistor receiver. Voltages shown are measured with no signal, using fresh batteries, with volume control at minimum, and dial set at low-frequency end.

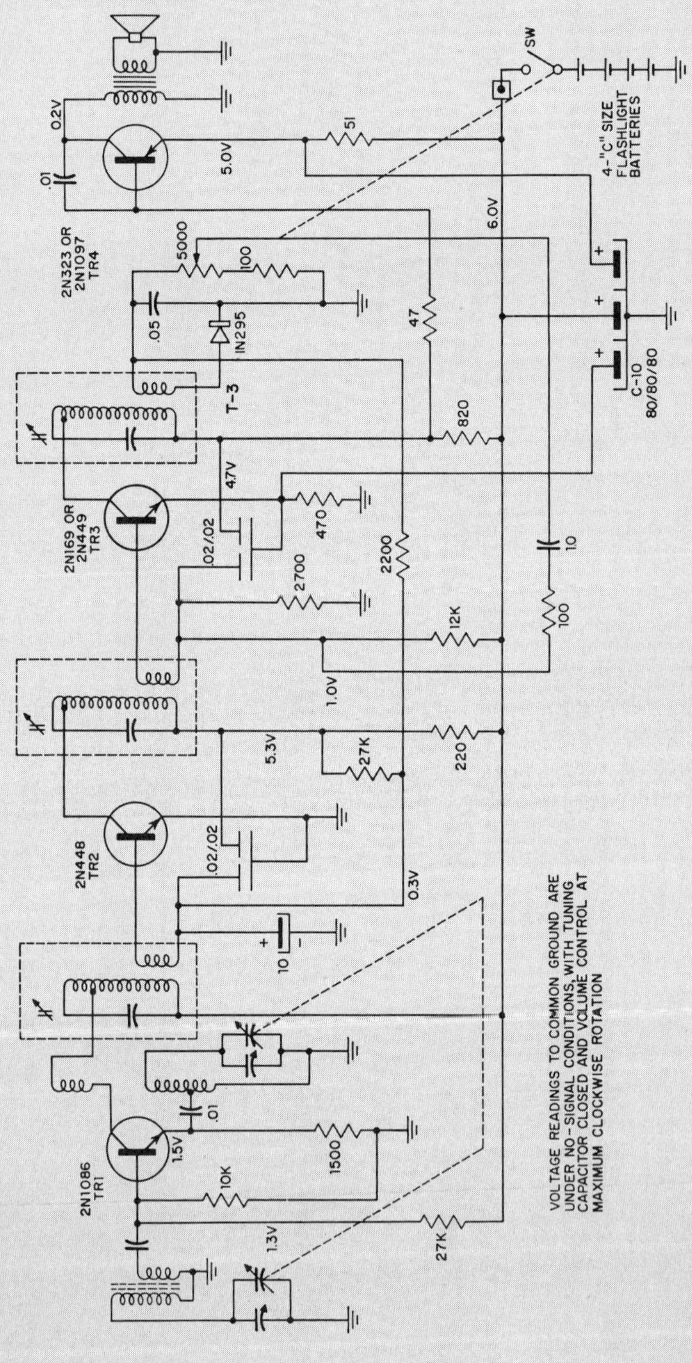

FIG. 20-41. Arvin transistor receiver.

the top end of the resistor, and thereby the cathode of the diode positive with respect to ground. The anode of the diode will be negative with respect to the cathode. The diode is thus reverse-biased and does not conduct, thereby having no effect on the gain of the stage.

Now along comes a strong signal. The agc control voltage rises and cuts off the first i-f transistor. No collector current flows, there is no voltage drop across *R*-7, and the top of the resistor and the diode cathode are at battery minus potential. The incoming signal on the diode anode now makes it conductive, thereby shunting a large portion of the signal out of the i-f amplifier. As a result agc control voltage drops and the stage starts functioning again. The dynamic relation between the stage functioning and the diode conduction gives proper reception.

Thermistor *R*-19 gives transistor operating point stabilization. All transistors are of the PNP type. The receiver operates with positive ground. Therefore, battery minus is hot and all transistor elements have a negative voltage with respect to ground. But remember that nevertheless the emitters must be positive with respect to the base and the collector must be negative with respect to the base. In the circuit diagram, the smaller voltages shown are taken with the battery mode of operation; the larger voltages shown are taken with the line mode of operation.

A final variation is shown in Fig. 20-43, the schematic diagram for an RCA transistor receiver. All transistors are of the PNP type. Note that battery minus is ground and battery plus is hot. Emitters are positive with respect to base, as shown. Collectors are highly negative with respect to base, as indicated by the low voltage with respect to ground (battery minus).

Note that earphone output is obtained in the collector circuit of the audio driver transistor Q-4. To match collector circuit impedance, the earphones must have high impedance, approximately 2,000 ohms.

Overload diode *CR*-2 is a crystal diode that, through agc action, gives signal strength control. If it shorts, then no voltage drop would occur across resistor *R*-4, and the collector current would rise to a point where it could destroy the transistor.

Note the output circuit of the two push-pull output transistors. There is no output transformer. Output from transistor collectors feeds the two high-impedance speakers connected in parallel. Each speaker is center-tapped to present a push-pull circuit. If a speaker should become defective, it is imperative that you obtain an exact duplicate replacement.

You will find other variations from receiver to receiver. But you have enough of the basic information to figure the other ones out for yourself. Just keep in mind the superheterodyne circuit, whether working with a tube or transistor radio. The similarities have been pointed out to you.

When replacing transistors, it is advisable to get a similar replacement. If a similar transistor is not in your stock, you may use a different transistor with similar characteristics. There are many interchangeability charts. One manufacturer, Sylvania, puts out a list of nine transistors that may be used to replace 300 frequently used types. Write to Sylvania and other manufacturers for information.

Transistor radio troubleshooting chart. A troubleshooting chart for transistor radios, arranged by symptoms, is given in Table 4.

Generally speaking, weak reception at either end of the tuning dial, whistles when tuning, reception of one station over the entire tuning range, and cutout of stations at the high-frequency end of the dial can be traced to the antenna circuit and the converter. Because of the greater gain of the i-f stages, we can look there for weak reception, distortion on strong signals, noise, and oscillation squeals. Defects like motorboating and distortion on all stations usually are traced to the audio section. Weak batteries can be the cause of almost any defect.

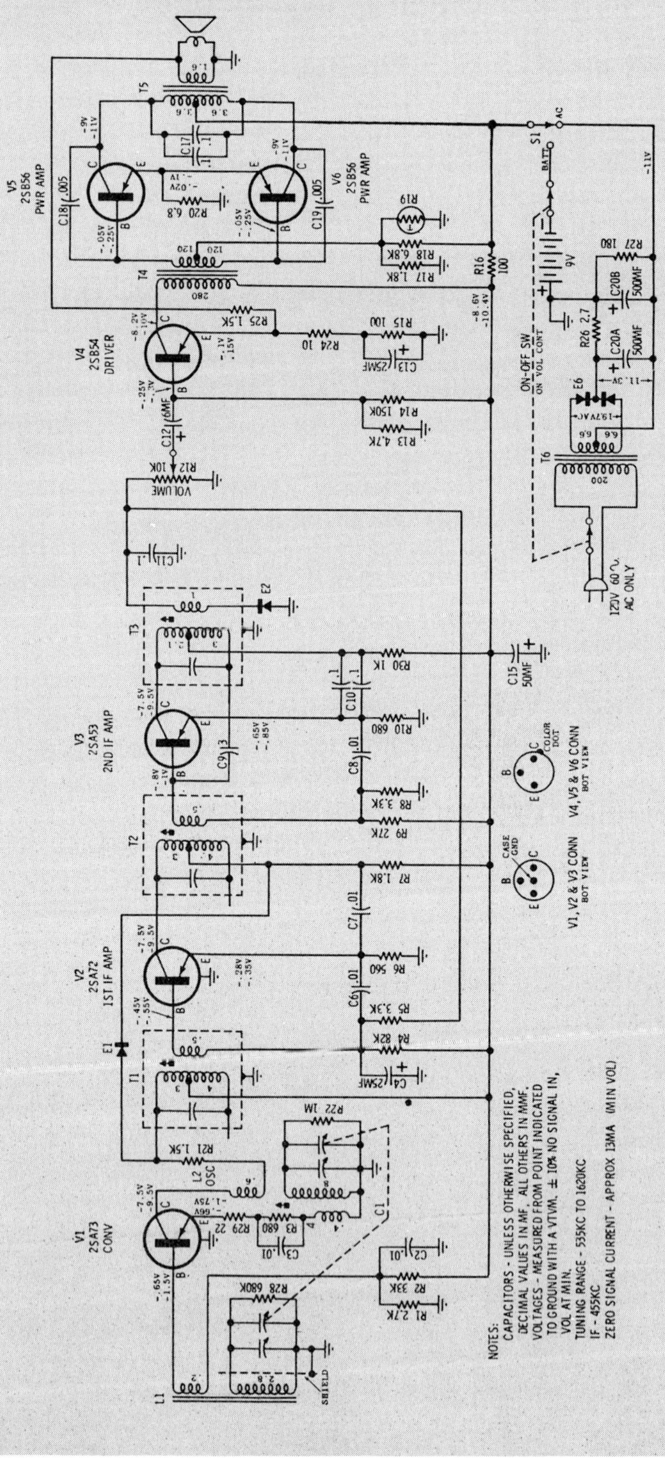

FIG. 20-42. Motorola two-way transistor receiver. Resistance values alongside coils are in ohms.

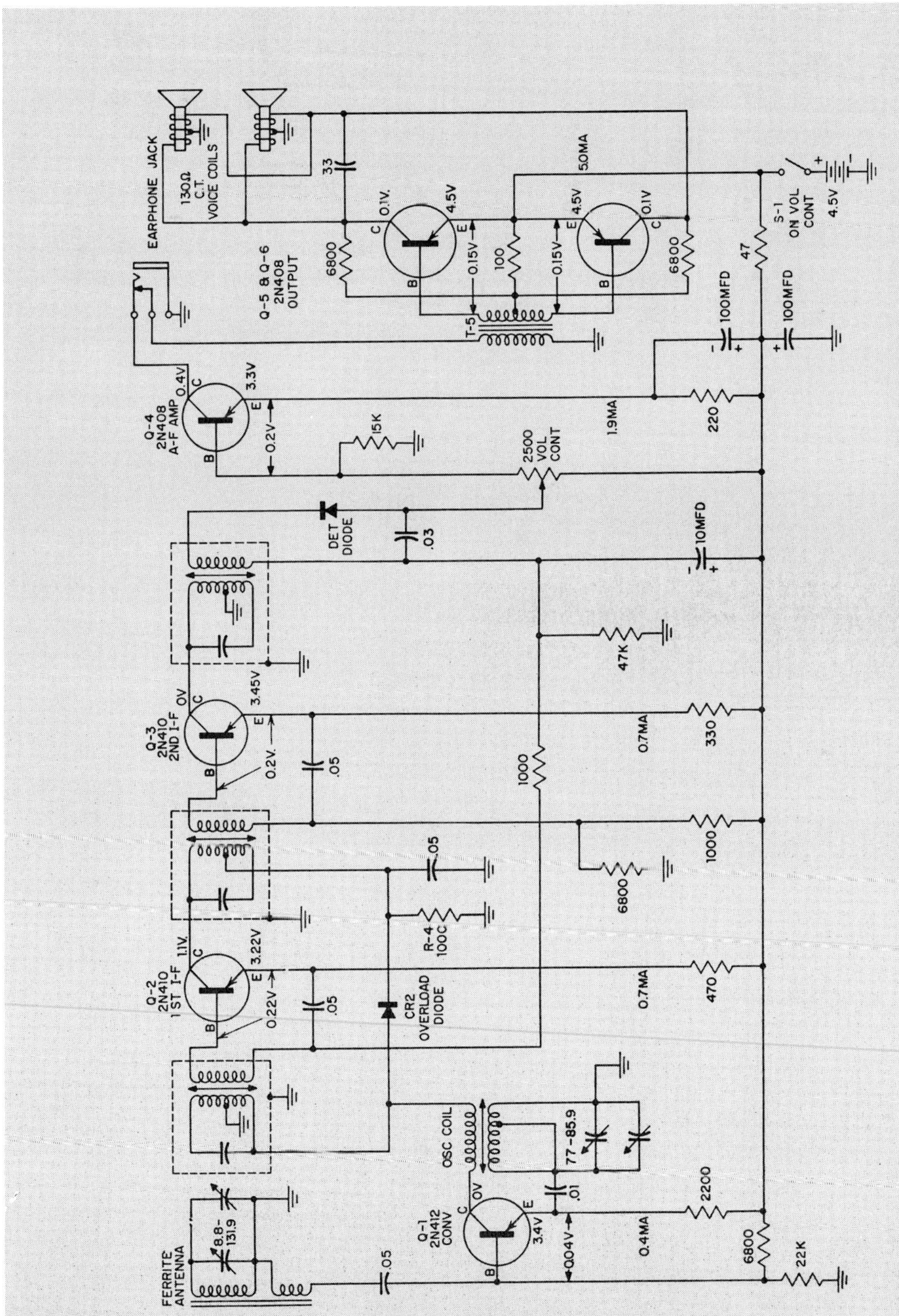

Fig. 20-43. An RCA transistor receiver. Voltages are measured from battery minus with volume control at minimum and no signal. Both speakers are 3-inch P-M units with 130-ohm center-tapped voice coils.

TABLE 4. TRANSISTOR RADIO TROUBLESHOOTING CHART

Symptom	Possible cause
Dead receiver	Batteries dead or weak, reversed batteries, poor battery contacts, broken printed wiring, defective on-off switch, shorted battery decoupling capacitor, shorted detector-driver coupling capacitor, inoperative oscillator, shorted bypass capacitors, open antenna coil primary, open i-f transformer windings, open primary of output transformer, defective speaker, defective earphone jack, defective volume control, defective transistors or diode detector
Weak signal	Weak batteries, open bypass capacitors, open detector-driver coupling capacitor, cracked antenna coil core, misalignment, leaky agc overload diode, defective transistor or diode detector
Weak signal at low-frequency end of the dial	Weak batteries, poor oscillator adjustment, defective converter transistor
Weak signal at high-frequency end of the dial	Poor antenna adjustment, defective converter transistor
Extremely short battery life	Defective on-off switch, shorted or leaky battery decoupling capacitor

Symptom	Possible cause
Motorboating and squeals	Weak batteries, open agc filter capacitor, defective battery decoupling capacitor, open neutralizing capacitors, cracked antenna coil core, open bypass capacitor in i-f stage, poor battery contacts, misalignment
Distortion	Weak batteries, leaky detector-driver coupling capacitor, shorted bypass capacitors, short or leaky agc filter bypass capacitor, defective bias resistors, defective speaker, short in output transformer, defective transistors or diode detector, open overload diode
Fading	Weak batteries
Intermittent operation	Cracked printed wiring, defective volume control, dirty on-off switch, poor battery contacts, poor lead contacts, dirty tuning capacitor, leaky r-f or a-f bypass capacitors, loose i-f transformer slug, defective earphone jack, intermittent open coils
Noisy reception	Weak batteries, leaky capacitors, dirty on-off switch, worn volume control, dirty tuning capacitor, corrosion in transformer windings, defective transistors
Only one station received	Defective oscillator

QUESTIONS

1. Explain how N-type germanium differs from P-type germanium.
2. Draw a diagram showing the biasing of an NPN transistor and of a PNP transistor.
3. Explain how a common emitter circuit transistor resembles a triode tube.
4. Explain several ways by which a transistor circuit may be stabilized against runaway collector current.
5. Explain the biasing voltages of a PNP transistor, first when it is positively grounded and then when it is negatively grounded.
6. How would you test the power supply of a transistor receiver?
7. Describe how you would localize a defective stage in a receiver by signal check.
8. The circuit of Fig. 20-42 does not work. A voltage check of the driver stage shows zero voltage at the base and at the emitter, and normal voltage at the collector. Describe how you would locate the defect.
9. What are the basic voltage conditions in an open in: the collector circuit; the emitter circuit; the base circuit?

Servicing Auto Radios

21

The design of auto radios has undergone considerable change in recent years. Nevertheless, the service technician should have no difficulty in servicing any of them, since they are all superheterodyne circuits. All are powered by the car battery, previously a 6-volt battery but now almost universally a 12-volt one. Signal checking and alignment will follow basically the same procedure that was followed with the standard receiver. This chapter will emphasize the few differences in circuit design to increase efficiency.

Types of auto radios. There are three types of auto radios that are still being used. The oldest variety is an all-tube receiver, operating with high plate voltages. To achieve these voltages, a vibrator-type power supply is used. A later development was the hybrid receiver, employing tubes in the front end that operate with a 12-volt plate supply. The detector is usually a diode crystal, and the audio section employs transistors. Thus, the set can be operated directly from the battery. The latest development, becoming more universal, is the all-transistor receiver, also operating directly from the battery.

We shall use the all-transistor receiver as our starting point and describe the hybrid and older all-tube sets as variations. In the description that follows, keep certain concepts in mind. How is the auto radio like the standard receiver? In what way are transistors similar to tubes? How can the circuits of transistor stages, learned in the previous chapter, help in the understanding of similar transistor circuits in the auto radio?

All-transistor auto radio. The all-transistor auto radio is a superheterodyne circuit, powered by the car battery. The block diagram for such a receiver is shown in Fig. 21-1. All use an r-f stage for the greater selectivity needed under the conditions of use. With the greater overall gain, only one stage of i-f amplification is employed. The agc control voltage is usually drawn at the detector output and is coupled back to one or more previous stages.

The power supply of the receiver is indicated as 14 volts. Most car batteries today are 6-cell storage batteries. Since each cell is rated at about 2 volts, they make up a 12-volt battery. However, in automobile operation with the generator attached, the supply voltage is stepped up to the 14 volts indicated.

Figure 21-2 is the schematic diagram of a Motorola car radio that corresponds to the block diagram of Fig. 22-1. An analysis of this set will help in the understanding and servicing of any similar type of receiver.

Like all American automobile radios, this is a grounded negative receiver, and battery plus is hot. As a result, all voltages shown are positive with respect to ground. Note that, at the bottom, the intermediate frequency is indicated as 262.5 kc. This is a common auto radio intermediate frequency. The set is permeability-tuned, where the control of iron slugs in the tuning coils varies inductance, rather than capacitor-tuned. Note that the slugs of the antenna circuit, r-f amplifier, and the converter oscillator are ganged for tuning. This is shown by the dotted lines. The converter signal grid is untuned.

Let us examine some of the circuits in more detail. Remember, however, that despite the special adaption, the design is that of a superheterodyne receiver and is checked for operation in the usual manner.

Several things should be noted about this schematic diagram. Note the convenient indications of the voltages across the emitter resistors. They indicate to the service technician whether or not proper currents are flowing in the stages. Also note the code using the squares and ells. These code markings are printed alongside similar leads on the printed wiring board. They give convenient points for locating various leads; for example, the 13.8-volt lead, the 12.2-volt lead, the agc lead, and the i-f agc lead. Voltage measurements may be made from these points.

Power supply. The 12-volt storage battery is the source of voltage for the receiver. The complete power supply is shown in the diagram of Fig. 21-3. Recall that all American cars use negative ground. In bench testing, be sure to keep this fact in mind lest you damage receiver components.

It would seem that the power supply is complex to get the 14 volts that are already furnished by the battery and generator. But it should be remembered that high-frequency noises from the generator and from the battery connected as well to the car ignition system might enter the receiver signal chain. To filter out such noise before it reaches the set, choke *L*-5 and capacitor *C*-21 are employed. Capacitor *C*-21 is known as a spark plate. It usually consists of two metal plates separated by a thin dielectric. Often, one plate is the metal chassis itself.

The battery-plus lead is connected to the battery in all cases through a fuse. Many diagrams do not show this component because it is part of the car installation, rather than a component in the receiver. But it must not be overlooked as a cause of trouble.

Note the filter circuit composed of capacitors *C*-20*B* and *C*-20*C* and resistor *R*-24. This filter serves to decouple the battery and so prevent feedback of signals through the common component, the battery.

Power-supply troubles. There are few things that can go wrong in the power supply because

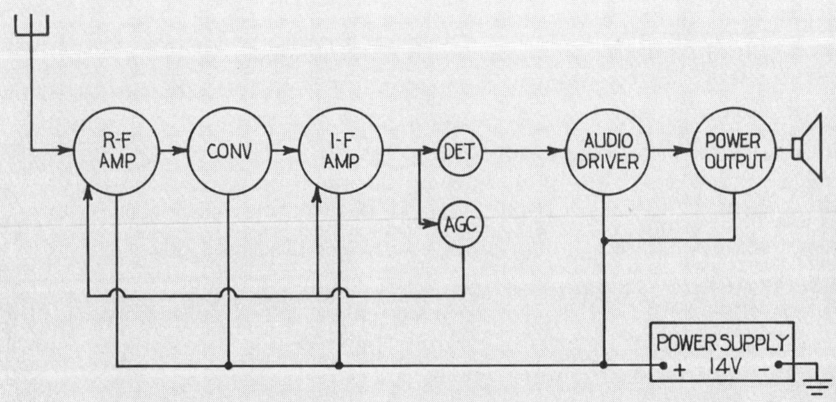

FIG. 21-1. Block diagram of an auto radio.

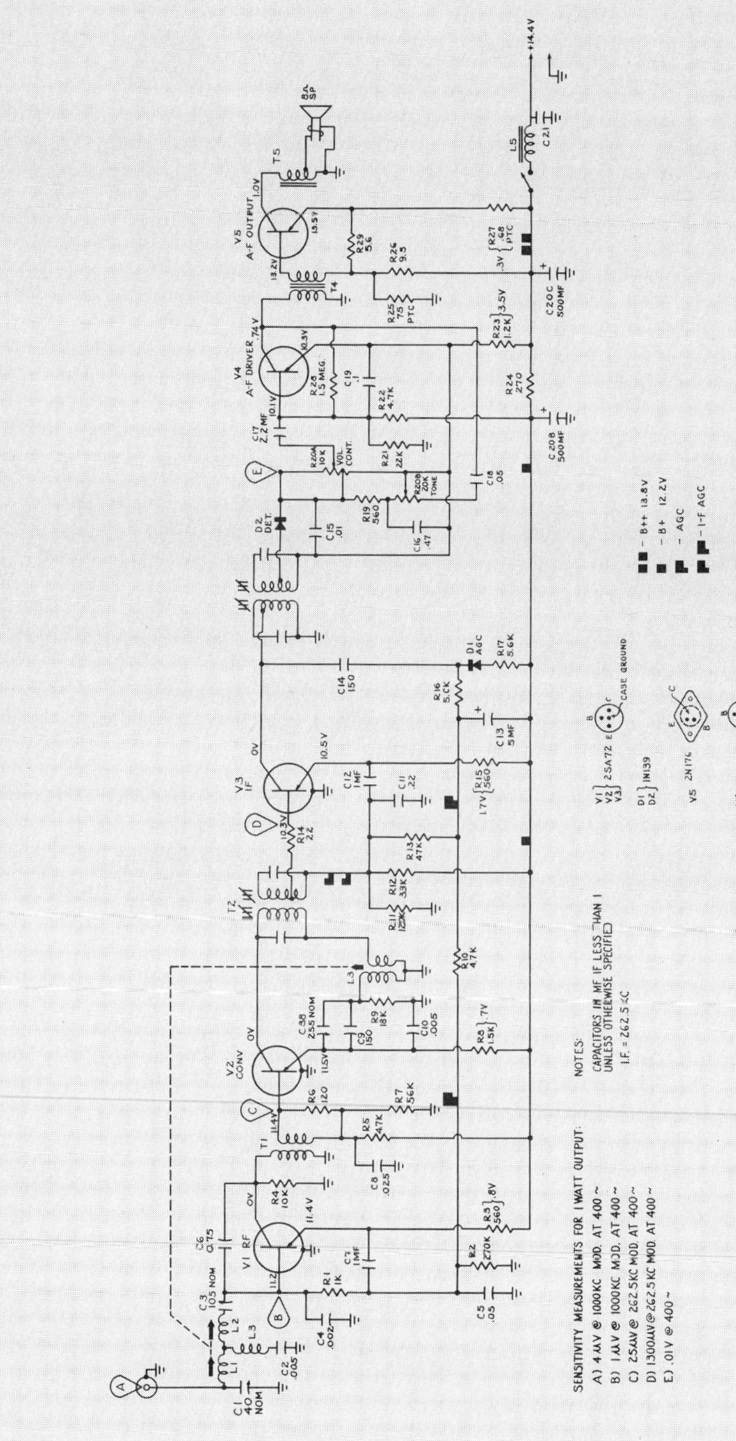

FIG. 21-2. Motorola auto radio.

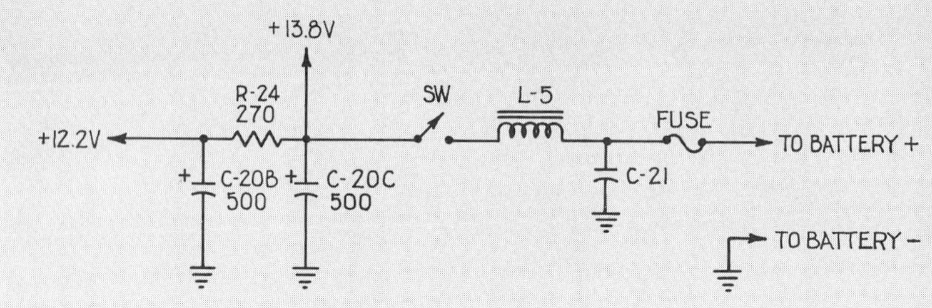

FIG. 21-3. Power supply for the basic auto radio.

it handles low voltages and currents. Yet this circuit is the one to be checked first. Begin with the fuse. Check that it is not burned out and that it sits firmly in its fuse case. If it is blown, check the stage circuits for the cause before replacing with a new one, since a defect in the receiver could be the cause for the overload. Note that no fuse is shown in Fig. 21-2 in the schematic diagram. All cars have an external fuse from the battery to the power input to the receiver, as shown in Fig. 21-3.

Poor contact in the switch or fuse case could produce lowered voltages, giving weak or no reception, noise, intermittent operation, distortion, fading, or oscillation squeals. These are similar to defects in the transistor portable receiver.

Antenna circuit. Several problems are encountered when considering the design of the antenna input circuit to a car radio. First, there is an extremely short antenna for signal pickup. Second, the capacitive loss along the antenna cable can be great for the weak signal coming in. Third, this loss varies with different antenna installations. And last, the input to the r-f amplifier transistor must be low impedance for good impedance match.

The antenna input circuit of the basic receiver has been redrawn in Fig. 21-4*A* to show how the problems are met. Note that the antenna wire feeds through its mounting; the latter is grounded to the car body. The antenna cable capacitance is shown as C_4. It shunts capacitor *C*-1 and is therefore in the tuning circuit. Thus, antenna losses are minimized. To compensate for different antenna installations with different antenna capacitances, capacitor *C*-1 is a variable trimmer. It is usually adjusted with a screwdriver through a hole in the radio cabinet, usually located near the antenna receptacle. In a few sets, it is reached through a hole behind the tuning knob.

Coil *L*-1 is a tunable inductor to tune for the desired station. It is ganged to the other tunable inductor *L*-2. Inductor *L*-2, together with capacitor *C*-3, makes up a series resonant circuit at the same station frequency as that for which *L*-1 is tuned. The series resonant circuit feeds high current to the input of the r-f amplifier transistor, and also provides proper impedance match.

Coil *L*-3 and capacitor *C*-2 make up a low-impedance wave trap. The trap is tuned to the intermediate frequency of the set. Were the trap not there, any station at or near the intermediate frequency of the set would interfere with other stations being received. The trap offers an easy path to ground for the interfering station. Wave traps rarely present any service difficulties.

Antenna input variations. There are many antenna input circuit variations to achieve the same effects as the basic circuit. A common variation is the circuit shown in Fig. 21-4*B*. Coil *L*-1 loads the short antenna. Tunable inductor *L*-2 and capacitor *C*-1

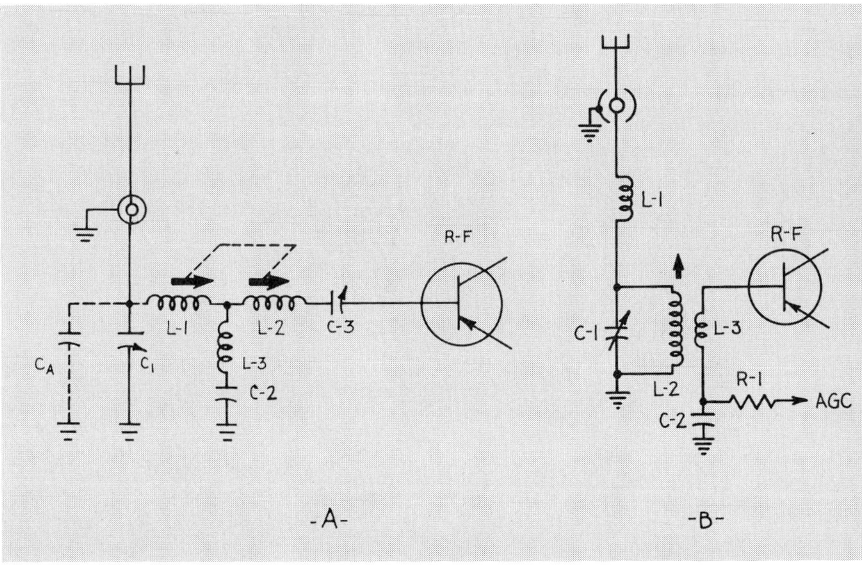

FIG. 21-4. The basic antenna input circuit and a variation.

make up a resonant circuit for tuning to the desired station. Capacitor *C*-1 is also a trimmer for compensating for different antennas. Coil *L*-3 is the secondary winding of the input transformer. It has few windings to match the low input of the r-f amplifier transistor.

Antenna input circuit troubles. The first possibility of trouble is an antenna that becomes grounded to the car body through its mounting. If the grounding is permanent, the set will be dead. If intermittently grounded, reception will be noisy. The defect is most easily discovered when a replacement antenna is connected to the receiver. Sometimes, resetting the antenna mounting is all that is required. Otherwise, the entire antenna assembly must be replaced.

Often, particularly after an antenna replacement, the radio will be weak because the antenna does not match the input to the set. A quick check of the condition is to tune in a station and then to touch the antenna whip. If the sound level increases, there is a mismatch. If the level decreases, the antenna is probably properly matched. To remedy a mismatch, tune in a weak station around the high-frequency end of the tuning dial. Adjust the antenna whip to a height at which the customer normally has it. Then adjust the antenna capacitor, *C*-1 in the basic receiver, for maximum loudness and minimum noise through the hole in the cabinet described previously.

Occasionally, the antenna mounting fails to make good grounding connection to the car body, thereby introducing noise in station reception. Sometimes, simple tightening of the mounting nut will cure the defect. If this does not improve the condition, it becomes necessary to remove the antenna mounting and to scrape the car body directly below it. Then reassemble the mounting.

The r-f amplifier stage. The r-f stage of the basic receiver is shown in Fig. 21-5. Note that the signal is fed from the antenna input to the base of the transistor, which in turn is tied to the agc voltage line. Capacitor *C*-5 and resistor *R*-2 make up the agc decoupling circuit. Capacitor *C*-6 furnishes degenerative signal feedback to prevent oscillation. Capacitors *C*-20*B* and *C*-20*C*, together with resistor *R*-24, make up the battery filter. The stage is coupled to the converter through untuned transformer *T*-1.

Note that the case of the transistor is grounded to shield the component. Such

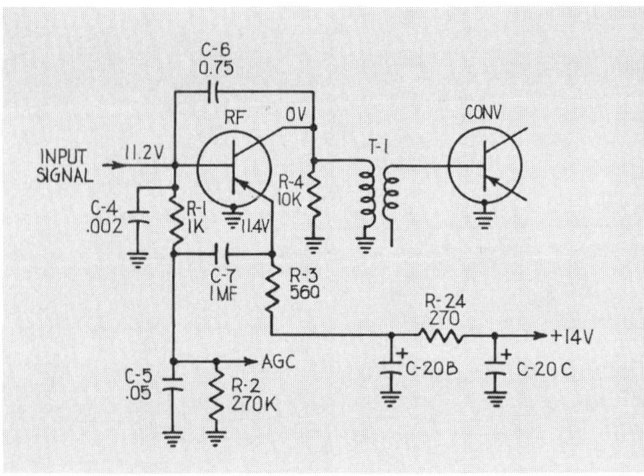

FIG. 21-5. The basic r-f amplifier stage.

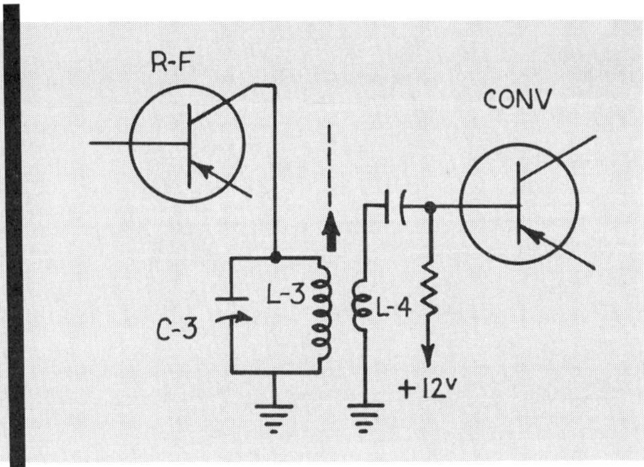

FIG. 21-6. Variation of the r-f amplifier stage.

shielding is not always necessary. It should be generally noted that resistances are smaller and capacitances are larger in transistor circuits than in analogous tube circuits. This is so because of the relatively low input and output impedances of transistors as compared with those of tubes.

The manufacturer of the basic receiver has indicated the voltage drop across emitter resistor *R*-3. This is a convenient measurement for determining that the stage is functioning properly. It eliminates the need for opening circuits to measure emitter and collector current.

Variation of the r-f amplifier stage. The front end of the receiver of Fig. 21-2 uses three tuned circuits: two before the r-f amplifier and one for the oscillator. The coupling between the r-f stage and the converter is untuned. Variations in auto radio receivers often provide tuning for the converter signal grid. Such a circuit is shown in Fig. 21-6.

Coil *L*-3 and capacitor *C*-3 form the tuned circuit. The slug in *L*-3 is operated from the gang tuner. Capacitor *C*-3 is adjustable for alignment purposes. Coil *L*-4 is a stepdown winding to provide proper impedance match to the converter transistor base.

In some cases, the tuning slug of coil *L*-3 is a fourth slug on the gang tuner. In other cases, it is a third slug because only one tuned circuit is used before the r-f amplifier stage.

These variations in circuitry are primarily a matter of design. From the servicing point of view, tests are made in the same way, and the circuit variations cause no complications.

Troubles common to the r-f amplifier. Fading is usually a characteristic defect of the antenna input circuit or the r-f amplifier stage. Open or corroded coils can cause noise and weak reception. A leaky transistor can cause noise, weak reception or a dead receiver. Checks may be made for locating the defective component as described in previous chapters.

The converter. The circuit for the converter of the basic receiver is shown in Fig. 21-7. The input to the converter is through untuned transformer *T*-1. Coil *L*-3 is the oscillator coil, feeding back collector current to the emitter for heterodyning. Base bias is obtained from the tap between resistors *R*-5 and *R*-7. Capacitor *C*-8 is a decoupling capacitor.

Variation of the converter. Some auto radios use a Colpitts oscillator in the converter circuit. The schematic drawing for such a receiver circuit is shown in Fig. 21-8. Feedback of oscillator signal is from the tap between capacitors *C*-10 and *C*-11 to the emitter.

Note that base and emitter voltages are equal; namely, 10 volts. This condition is not

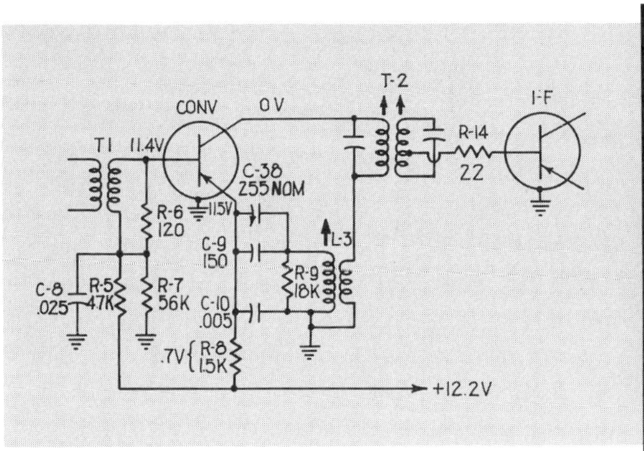

FIG. 21-7. The basic converter.

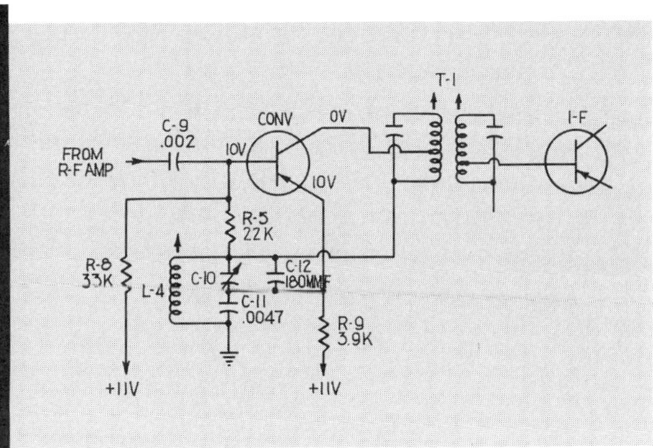

FIG. 21-8. Another converter circuit.

unusual in transistor converters. Some circuits have forward base-emitter bias; some have zero bias, as in our circuit; and some may even have reverse bias. The functioning oscillator and signal voltage will, however, furnish the necessary base-emitter forward bias.

Troubles in the converter. Many of the troubles in the converter are similar to those of a tube receiver. The oscillator may not be working, producing a dead receiver. To check for this condition, connect your voltmeter from emitter to base. Watch the reading as you touch the hot end of the oscillator coil with a screwdriver. Or, swing the tuning dial through its complete range. If the voltage varies when you do so, the oscillator is working. An open oscillator coil or capacitor can stop oscillation.

Sometimes, the signal will fade in and out. This is the result of an oscillator that drifts off frequency. Such an effect may be caused by leaky capacitors or leakage across dirt on the trimmers. Checking for the condition and replacement are handled as previously described.

The i-f amplifier stage. The i-f amplifier stage is shown in Fig. 21-9. Input is from i-f transformer *T*-2 and output is to i-f transformer *T*-3. Resistor *R*-24 reduces loading on the secondary tuned circuit. Again, note the convenient voltage measurement across emitter resistor *R*-15. Base bias is obtained from the tap between resistors *R*-11 and *R*-12. The stage is controlled by agc voltage fed to the base. Resistor *R*-13 and capacitors *C*-11 and *C*-12 make up a decoupling filter for r-f signals and the agc voltage line.

The intermediate frequency of the receiver is 262.5 kc. This is the usual frequency in almost all auto radios, because it allows for design that gives greater selectivity. However, this frequency makes the receiver more susceptible to image-frequency interference. A tuned r-f stage tends to minimize such interference. However, the basic receiver uses an image trap, coil *L*-4 and capacitor *C*-2, to by-

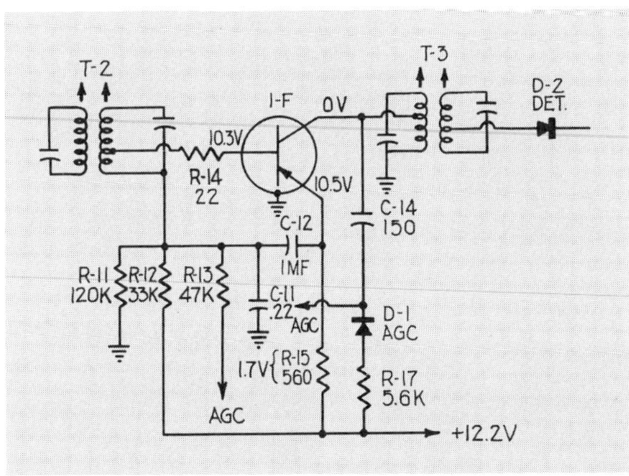

FIG. 21-9. The basic i-f amplifier stage.

pass such interference signals to ground at the antenna input.

Troubles in the i-f amplifier stage. Many of the troubles are similar to those in an analogous tube stage and need not be repeated here in detail. Sometimes the stage will oscillate because of a defective transistor, an open agc bypass capacitor, or an open (though rare) battery decoupling capacitor. A voltage check would disclose the condition.

The i-f transformers also give troubles similar to those described in previous chapters. If trying to peak them results in failure, it is almost certain that the transformer or its associated parts are defective. A leaky or shorted i-f transformer will upset the transistor bias and increase emitter current. Normally, emitter current should range between about 0.5 and 3 ma for an i-f stage. Multiply the range of current by the resistance of the emitter resistor and you will get the normal range of voltage across this resistor. Then measure it and see if the i-f transformer assembly shows leakage or a short. The defect will produce scratchy noise in the receiver. Of course, an open i-f transformer will produce a dead set, but bias voltage measurements will quickly show the condition.

Detector stage and agc circuit. The detector stage and agc circuit employ two separate diode crystals, as shown in Fig. 21-10. Crystal *D*2 is the detector and crystal *D*1 supplies the agc voltage. Unlike a tube detector which furnishes both the detected signal as well as the agc voltage, the two separate diodes in the transistor set are necessary. Most car radios use PNP transistors and negative ground. The agc voltage fed to the base of a controlled transistor must be a positive signal to reduce its base-emitter bias voltage. The detector diode will furnish an incorrect negative agc voltage. So a separate agc diode becomes necessary.

The detector signal is taken from the i-f transformer and fed to the volume control. Capacitor *C*-15 serves to bypass to ground any i-f signal at the output of the detector.

The agc voltage is obtained from the collector of the i-f stage. Capacitor *C*-14 blocks the positive voltage from the collector which is at zero potential. Trace the current path from +12.2 volts, through resistor *R*-17, through diode *D*1, and in turn through re-

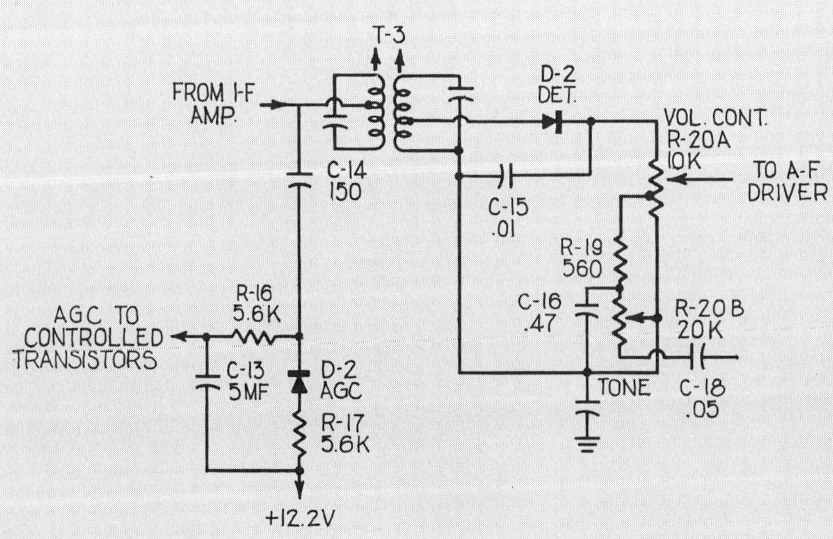

FIG. 21-10. The basic detector stage and agc circuit.

sistors *R*-16, *R*-13, and *R*-11 to ground. Normally, with no signal, the base bias of the i-f amplifier is +10.3 volts. With applied signal, the diode is biased even more so that it conducts more current and sends a positive voltage to the base of the i-f amplifier. As a result, the forward bias of the base-emitter junction of the i-f transistor is reduced as is the gain of the stage. The stronger the signal, the greater the reduction of the forward bias of the transistor. Note that the agc control voltage is also fed to the r-f amplifier base.

Variation in the detector-agc circuit. The circuit shown in Fig. 21-11 is from a Delco auto radio. Two agc crystal diodes are used as agc rectifiers in a voltage doubling circuit. Otherwise, the agc supply is fairly standard.

Note that in this circuit the detector output is directly coupled to the audio driver stage. This is achieved by returning the detector *D*3 to the +11-volt line, rather than to ground. Without affecting detector operation, it provides for the +9.3 volts bias required by the audio driver transistor base. The i-f filter is composed of resistor *R*-5 and capacitors *C*-3 and *C*-4.

Troubles in the detector-agc circuit. Once again, we may be guided to a great extent by analogous tube radio troubles. Diodes occasionally open, leak excessively, or short. These defects for the detector crystal will give little or no audio output. The technician can check the crystal diodes in circuit with an ohmmeter. Normally, the reverse reading should be ten or more times greater than the forward reading. Any variation from this ratio points up a defective crystal diode.

An open agc diode will produce distortion, particularly on strong stations. Where the base return of a controlled transistor is through the agc line, the defect will be discovered when checking the bias voltages of that transistor.

The audio-driver stage. The driver stage is similar to the first a-f stage in the tube receiver. The driver circuit used in the basic receiver is shown in Fig. 21-12. The input is fed from the detector to the base of the driver transistor through coupling capacitor *C*-17. Degenerative feedback is furnished from collector to base through resistor *R*-28. Base bias is obtained from the tap between resistors *R*-21 and *R*-22.

A bass compensation circuit is tied to the tap on the volume control *R*-20A. Its purpose is to bypass some of the treble notes and thereby give a better bass response, particu-

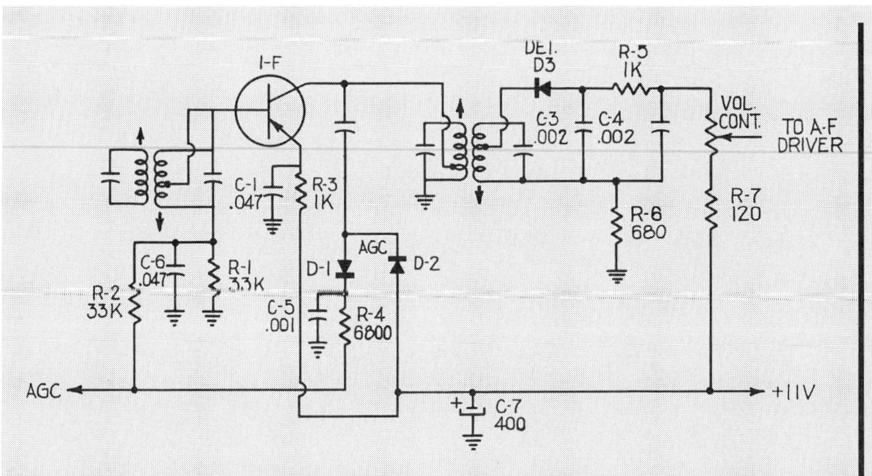

FIG. 21-11. Variation of the detector-agc circuit.

larly at low volume level. Potentiometer *R*-20*B* is the continuous tone control that bypasses more treble notes to ground at the top position and fewer treble notes at the bottom position.

Troubles in the driver stage. Distortion may occur in the receiver from the driver stage as a result of a leaky coupling capacitor *C*-17, a defective transistor, or defective biasing resistors. Each of these defects would affect the bias voltages on the transistor and may be checked by such measurements. In other respects, troubles are similar to analogous tube circuits.

The output stage. The output stage of the basic receiver is a single-ended one, shown in Fig. 21-13. Input from the driver stage is through audio transformer *T*-4 to the base. Output is through the autotransformer *T*-5 to the speaker. The transistor is a power transistor whose base wiring is shown at the bottom of Fig. 21-2. Note that biasing voltages are higher than those for the other transistors, since the emitter and base returns are to the +13.8-volt side of resistor *R*-24 in the power line. Base and emitter bias is obtained from taps on the voltage divider string made up of resistors *R*-27, *R*-29, and *R*-25. To prevent runaway collector current, use is made of a fusible resistor *R*-27 in the emitter circuit. Should runaway occur, the resistor opens and stops the stage from functioning. There is another control. Resistor *R*-25 is one with a positive temperature coefficient (PTC). When it heats up with runaway current, its resistance increases. As a result, the voltage drop across this resistor increases, making base bias more positive and reducing forward base-emitter bias. This effect reduces the collector current.

Variations of the output stage. Another single-ended output stage is shown in Fig. 21-14. Here, base-emitter bias is obtained from the voltage drop along the voltage divider string made up of resistors *R*-23 and *R*-24 in parallel, both in series with

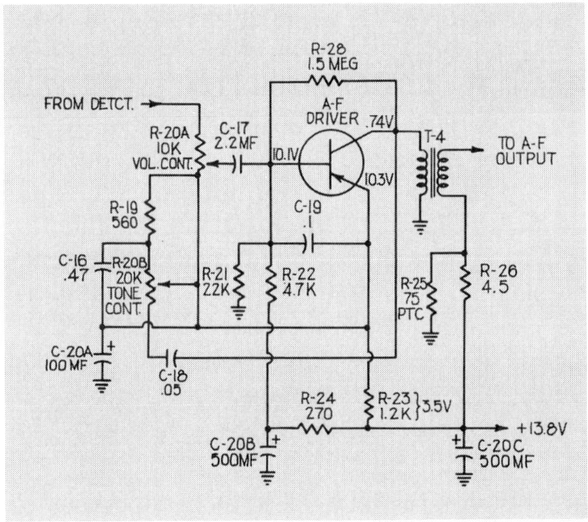

FIG. 21-12. The basic audio-driver stage.

rheostat *R*-22. Resistor *R*-24 is a thermistor with a negative temperature coefficient (NTC). With runaway collector current, it heats up and its resistance decreases. As a result, the total resistance of the parallel pair *R*-23 and *R*-24 decreases, with a resulting smaller voltage drop across them. This effect drives the base more positive, reducing forward bias and reducing collector current. Resistor *R*-25 is a fusible resistor in the emitter circuit.

Note rheostat *R*-22. Being in the base-emitter bias string, its adjustment determines the magnitude of collector current.

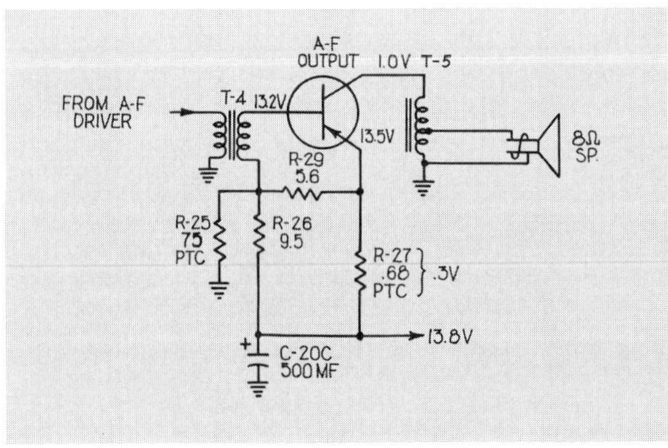

FIG. 21-13. The basic output stage.

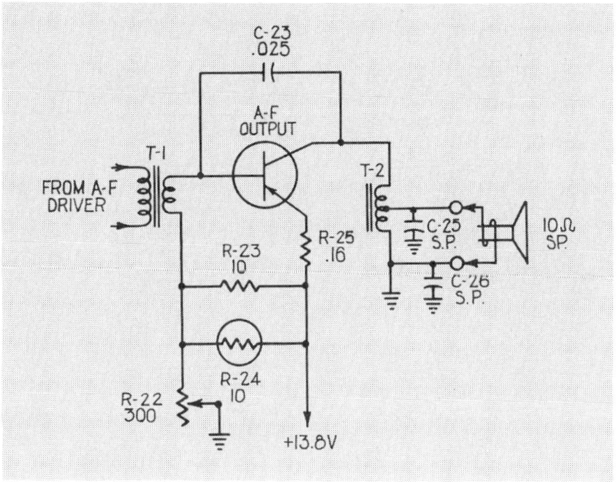

FIG. 21-14. Variation of the output stage.

Too little current will reduce output and may cause distortion; too much current might destroy the transistor. Each manufacturer indicates to what value of collector current to set it. For this circuit, the adjustment recommended is for 400 ma of collector current. In some schematics, where the power output collector voltage is given but the collector current is not, connect a vacuum-tube voltmeter between collector and ground; then adjust the bias rheostat until you get the specified collector voltage.

Note the two capacitors, *C*-25 and *C*-26, from each end of the voice coil to ground. They are spark plates to bypass any r-f noise picked up by the external speaker leads.

Some power output stages do not use an output transformer in a single-ended output. Such a circuit is shown in the diagram of Fig. 21-15. Impedance match between the transistor and the output is made by a parallel circuit of a choke and a 10-ohm voice coil. It is important that a speaker with a similar impedance be used when replacing a defective one.

Note the range of voltages given for the emitter and base biases. The lower limits are for use when the car battery is the sole source of voltage. The upper limits are for use when the car is running and the generator with its somewhat higher voltage is the voltage source.

Frequently, the power output stage is a push-pull stage rather than single-ended. Such an output stage is shown in Fig. 21-16. The input is to the bases of the output transistors through a push-pull input transformer. Rheostat *R*-18 is the control adjustment for the collector currents. Base-emitter bias is obtained from the voltage divider made up of resistors *R*-14, *R*-17, and *R*-18.

Note the controls for runaway collector current. Resistor *R*-14 has a negative temperature coefficient and resistor *R*-17 has a positive temperature coefficient. Together, they serve to reduce emitter-base forward bias when they heat up as a result of runaway collector current, thereby reducing that current.

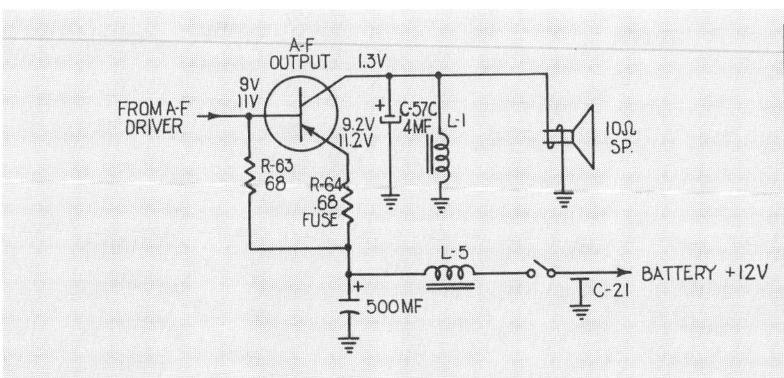

FIG. 21-15. Output stage without an output transformer.

Because of the relatively low output impedance of the transistors, there is no output transformer. Instead, the transistors feed into a 20-ohm center-tapped voice coil. If it should become necessary to replace the speaker, it is important to use a similar one for proper impedance match, and push-pull operation.

Rheostat *R*-18 is used to set the level of collector current as follows. Connect the ammeter in the center-tap lead, as shown in Fig. 21-17*A*. Turn on the radio for about 15 minutes. The current that you should read is that recommended by the receiver manufacturer, and is here between 150 ma and 340 ma with a 14-volt input to the radio.

After the reading is taken, base return resistor *R*-18 must be changed to give the desired collector current. This value is given for actual operation in the automobile with the charger across the battery, giving about 14 volts. The manufacturer also specifies a collector current range of 120 ma to 260 ma with a 12.6-volt input, which is from a storage battery on your shop bench.

Where you find difficulty in connecting the ammeter into the circuit with the speaker, you may substitute a dummy load, as shown in Fig. 21-17*B*. The ammeter connection is then made from the junction of the two resistors to ground and the reading is taken as usual.

With some receivers, collector voltage is given on the schematic diagrams. Where such is the case, you can adjust collector current by connecting a d-c vacuum-tube voltmeter from each collector to ground and adjusting the base bias resistor until you read the specified collector voltage on the instrument, as shown in Fig. 21-17*C*. Be sure to use the proper speaker and supply voltage when using this technique.

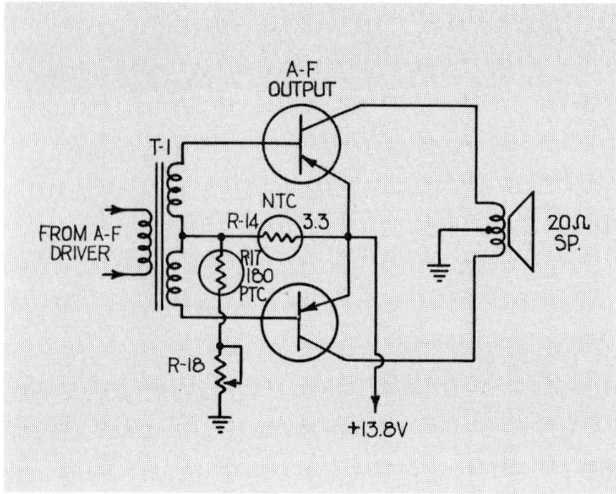

FIG. 21-16. Push-pull output stage.

Troubles in the output stage. When a receiver is dead and the trouble is traced to the output stage, a likely cause is an open fusible resistor. This must be replaced with a resistor of the same type and resistance. But it would be wise to see if there is another de-

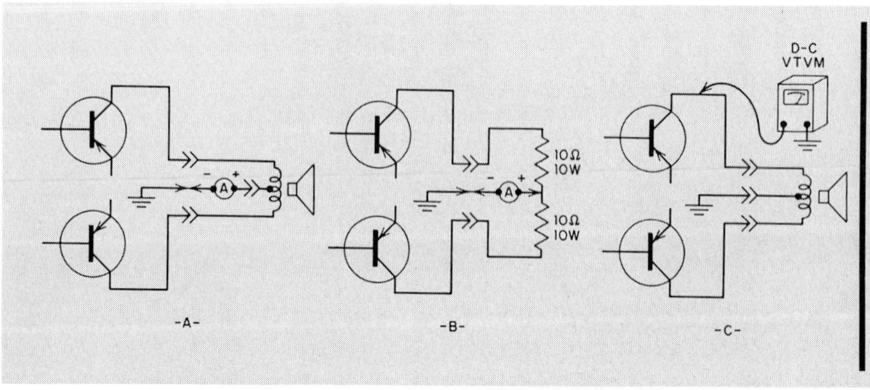

FIG. 21-17. Collector current adjustment for push-pull output.

fective condition that caused the resistor to open, before replacing the open resistor and checking for the restoration of normal operation.

Resistance checks on all components in the output stage is the first step. If these are normal, a dynamic check comes next. This should be done at a lowered power supply voltage to avoid burning out the new fusible resistor. An input of 10 to 12 volts should be safe and, at the same time, provide readable emitter, base, and collector voltages a little below normal. Any marked deviation from these lowered norms will give a good indication as to the location of the defect. When the defective condition is cleared and the voltage readings are close to the lowered norms, it is safe to try operation of the set under standard conditions. If an adjustment for output collector current is provided in the receiver, it should be checked as the last step in the repair procedure.

Push-button tuning. Our typical receiver, in addition to normal manual tuning, has push-button tuning for immediate selection of several desired stations. It employs a mechanical system for moving the powdered iron cores or slugs of the antenna, r-f, and oscillator tuning coils to the exact position required for the station.

This push-button mechanism rarely gives any difficulty. However, you may be called upon to reset the buttons for other stations. The simple procedure for doing this is listed in the steps below:

---

1. Keep antenna in its lowest position.
2. Turn on the receiver, and let it operate for 15 minutes.
3. Unlock the push buttons by pulling them forward about ½ inch beyond the normal position with your fingers.
4. Accurately, tune the receiver manually to the desired station.
5. Lock one of the push buttons to this station by pushing it in very firmly.
6. Repeat steps 4 and 5 for each other desired station.

Preliminary service procedure. Before you remove a radio from the automobile, a preliminary service check should be made. Inspect the fuse in the hot battery line. If the fuse is good, turn on the receiver. Then make a careful inspection of the antenna. Wiggle it to be sure it is not grounding. If it appears suspicious, try a replacement antenna.

When all seems well but the set is still defective, you will have to remove it from the automobile for systematic servicing. On the bench, you should use a 6-cell storage battery as your power source. Do not use a battery eliminator unless it is well filtered and has good voltage regulation. Be sure that you use the correct polarity for your hot power lead, and be sure that the other pole of the battery is clipped to the receiver chassis. If connected incorrectly, the set will not work and damage to components may result.

After the chassis is out on the bench, check that the speaker is connected to the receiver and turn on the set. You are now ready for a systematic procedure to locate the defective stage.

Localizing defective stage. The automobile radio receiver is a superheterodyne circuit and may be analyzed in the usual manner by signal substitution to locate a defective stage. Again, the ground lead of the signal generator is clipped to the chassis. Connect a 0.1-mfd capacitor in series with the hot lead. The signal from the generator should be at the lowest possible level to produce a tone in the speaker. Tune the receiver to a non-station position and turn it on. Set the volume control up high. A summary of the signal injection test is given in Table 1, based on the typical circuit of Fig. 21-2. Note the similarity to previous signal-check procedures.

TABLE 1. ISOLATING DEFECTIVE STAGE OF AUTO RADIO BY SIGNAL INJECTION

Signal injected	To test point	Normal result	Analysis
1. A-f (400 cps)	Arm of volume control	400-cycle note from speaker	If heard, audio stages are OK. Go to step 5. If not heard, go to step 2
2. A-f (400 cps)	Base of output transistor	400-cycle note from speaker	If heard, go to step 3. If not heard, defect is in transistor, bias resistors of output stage, or speaker
3. A-f (400 cps)	Base of a-f driver	400-cycle note from speaker	If heard, go to step 4. If not heard, defect is in driver stage or a-f coupling transformer
4. Modulated i-f	Base of i-f amplifier	Modulated note	If heard, go to step 5. If not heard, defect lies in i-f amplifier, output i-f transformer, detector, volume control, or tone control
5. Modulated i-f	Base of converter	Modulated note	If heard, go to step 6. If not heard, defect is in mixer stage of converter or in input i-f transformer

Signal injected	To test point	Normal result	Analysis
6. Modulated 1,600 kc (receiver tuned to 1,600 kc)	Base of converter	Modulated note	If heard, go to step 7. If not heard, defect is in oscillator stage of converter. In this test, wobble signal generator around 1,600 kc to get response
7. Modulated 1,600 kc (receiver tuned to 1,600 kc)	Base of r-f amplifier	Modulated note	If heard, defect is in antenna circuit. If not heard, defect is in preamp circuit or r-f amplifier stage

For checking a weak receiver, sensitivity measurements are given with Fig. 21-2 from various stage points indicated by capital letters. These measurements help to locate the weak stage. Although the service technician may not be able to so accurately adjust his signal generator, he can calibrate his instrument at the various similar test points of several good receivers. Then he can use his settings as a guide for checking other sets.

Component analysis. After you have located a defective stage, voltage checking with a 20,000-ohms-per-volt multimeter or a vacuum-tube voltmeter and resistance checking with an ohmmeter will enable you to locate the defective component. Remember that all voltages are measured to ground. The defects in an all-transistor auto radio are similar to those described for the tube radio and the portable transistor receiver. Most of the component defects have been described in the previous stage analysis.

When all voltages are low or at zero, the hot battery line is either open or has developed a resistive connection. Check for a defective ON-OFF switch, an open filter resistor or choke, a rosin joint, or a poor connection.

If the set is weak and distorted, check the agc voltages with a vacuum-tube voltmeter. A defective agc filter capacitor may be the cause. Check the audio stages. For the transistors, check base, emitter, and collector voltages to ground with the receiver turned on. Check that base-emitter voltage is properly forward biased. If voltages are incorrect, shut off the set and make your ohmmeter measurements. Be sure to look for poor connections and high-resistance connections in the defective stage. Transistors are checked in the manner described in the previous chapter.

Replacement notes. Replacement of the power output transistors requires special treatment. The collector current should be adjusted after replacement to ensure proper level. Connect a 0–1 or 0–2 ampere d-c ammeter with a low internal resistance, 0.05

ohm or less in the collector circuit. Turn the receiver on for about 15 minutes. The current that is read should be that recommended by the receiver manufacturer. If the reading is incorrect, proper adjustment can be made with the rheostat in the base circuit. Remember that if collector current is not given, you can adjust the rheostat control to correct collector voltage. By careful design, some manufacturers avoid the use of this rheostat control. This is the case in the basic receiver of Fig. 21-2.

A further word must be said about the power output transistor. Only the base and emitter leads come out of the component. The collector is the shell of the transistor itself. To avoid grounding of the collector to the chassis, an insulating washer is placed between the shell and the chassis. To utilize the full heat-dissipating capability of the chassis as a heat sink, the insulating washer is coated on both sides with a special heat-dissipating but current-insulating grease, like the Motorola DC-4 grease, when replacing a power output transistor.

Auto radio alignment. The alignment of a transistor auto radio is essentially the same as for any other type of superheterodyne receiver It is generally advisable to follow the alignment notes of the manufacturer. However, the procedure given in Table 2 will serve as a basic guide, referring to Fig. 21-2.

When performing the alignment procedure, some preliminary steps are necessary. Set the volume control to maximum and the tone control to treble. Place an output meter across the speaker voice coil and turn on the receiver. The signal from the test generator should be attenuated to keep the output meter reading below 0.5 volt at all times to minimize agc interference with the alignment procedure.

The general procedure is to begin with the i-f circuit alignment and to work back to the antenna. When performing the r-f alignment, a dummy antenna is substituted for the actual one. The manufacturer usually suggests how to make one up. The dummy antenna for the basic receiver is shown in Fig. 21-18.

A pictorial diagram of the front and rear view of the basic receiver is also shown in the same figure. The location of all the adjustment controls to which reference is made will be found there. Similar alignment adjustments will be found in other auto radios.

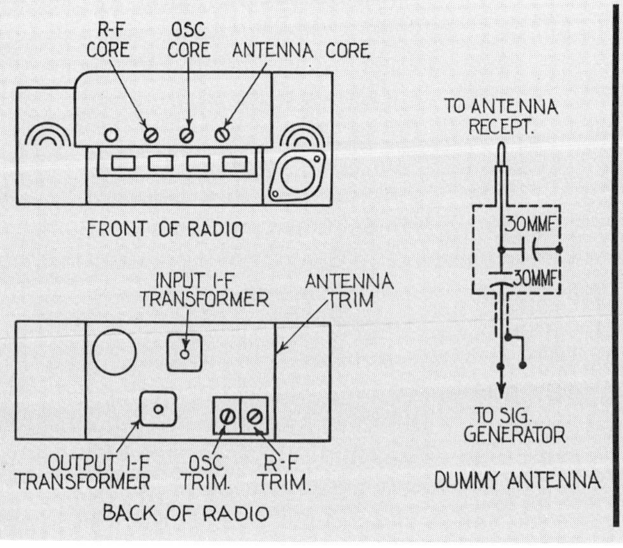

FIG. 21-18. Location of alignment controls. Dummy antenna is at the right.

TABLE 2. AUTO RADIO ALIGNMENT PROCEDURE

Step	Signal generator connection	Input signal	Receiver dial setting	Adjustment
		I-F ALIGNMENT		
1.	To collector of r-f amplifier through 0.1-mfd capacitor	262.5 kc mod. at 400~	High-frequency end	In sequence, adjust transformer core of secondary of output i-f transformer; then the primary for maximum on the output meter. Then peak the secondary and the primary of the input i-f transformer, in the same manner
		R-F ALIGNMENT		
2.	To antenna receptacle through dummy antenna	1,610 kc mod. at 400~	High-frequency end	In sequence, adjust oscillator trimmer *C*-38, r-f trimmer *C*-1 for maximum on the output meter
	Do not perform steps 3 and 4 unless tuner has been tampered with or associated components have been replaced. Before performing steps 3 and 4, back the tuning cores as far as possible out of the coils to eliminate their effect on trimmer adjustments. Repeat step 2			
3.	To antenna receptable through dummy antenna	1,200 kc mod. at 400~	1,200 kc	In sequence, adjust oscillator core in *L*-3, r-f core in *L*-2, and antenna core in *L*-1 for maximum on the output meter

Step	Signal generator connection	Input signal	Receiver dial setting	Adjustment
4.	To antenna receptacle through dummy antenna	1,610 kc mod. at 400~	High-frequency end	Repeat adjustments made in step 2
Repeat steps 3 and 4 until no further increase, making step 4 the last adjustment. Cement core screws in place				
			ANTENNA	
			Weak station around 1,400 kc, radio in car	With antenna set in usual height, adjust antenna trimmer *C*-1 for maximum loudness

Eliminating noise in auto radio. The auto radio is particularly vulnerable to electrical noise pickup because of its location near electric sparking devices. An otherwise fine receiver may annoy its owner no end because of the static-like noise accompanying all reception. Proper installation can reduce this noise to a tolerable level.

There are currently two main ways by which the original installation, aside from devices used within the receiver, provides for noise pickup reduction. Many manufacturers use a special ignition wire known as resistance ignition wiring. It is identified by having RADIO WIRE printed on the outside insulation. If an auto mechanic has replaced this with regular ignition wire, advise your customer to get a replacement with resistance ignition wire.

The other noise-suppression device is the use of special spark plugs with internal spark-suppressor resistors. An auto mechanic may in time replace the plugs with corresponding ones without the internal resistors. Advise your customer to have his mechanic replace spark plugs with exact duplicates.

When noise stubbornly persists, there are several things that you can do to help remove it. Several devices are illustrated in Fig. 21-19. Here we have special filters, obtainable from auto or radio parts dealers, which are placed across the generator, ignition coil, and the voltage regulator. The use of all three filters may not be necessary. When connecting these filters, be sure that all connections are clean and firm. Make certain that there is no paint, dirt, or grease at the mounting points. Be sure to tighten all connections

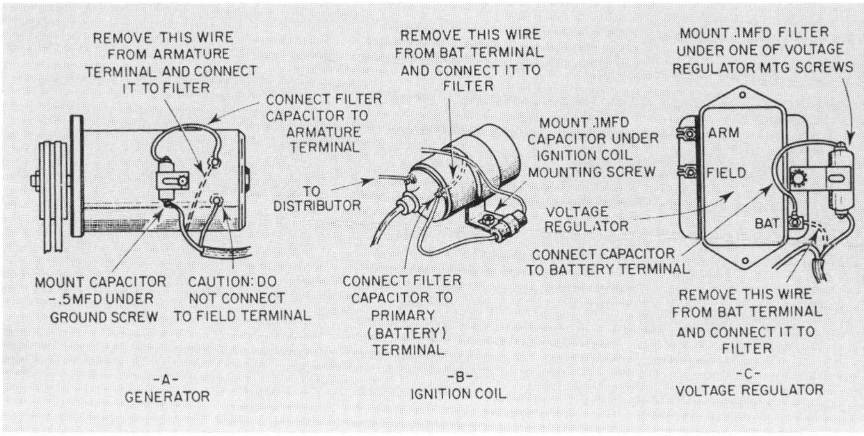

FIG. 21-19. Methods of mounting and connecting filter capacitors to suppress ignition interference.

securely. Cars that use alternators, rather than generators, do not require a filter across the alternator.

Installation of ground straps to ensure good electrical bonding between various metal parts of the car reduces noise pickup. The best points for bonding such ground straps are the engine block to the firewall, from the block to the automobile frame at one of the points where the engine is shock-mounted, from the grounded terminal of the battery to the automobile body, and from the grounded terminal of the battery to one of the voltage regulator mounting screws. In some cars, a support for the dash is bolted to the firewall. Connecting a bond strap from the engine block to the firewall at such a point is very effective.

Sometimes it becomes necessary to bond other points. When the hood is well grounded, a shield is interposed between the motor compartment and the antenna. But if the hood hinges and its holding clamp fails to make good contact to the automobile frame, this shielding effect is lost. A hood bond is then mounted as shown in Fig. 21-20A. The bond is a piece of brass, one surface of which has been serrated like the teeth of a rasp.

Another source of noise is wheel static. The wheels generate static electricity when they turn. This static electricity sparks across to the axle, thereby producing noise signals. The front wheels are greater offenders because of their weighting.

Wheel-static interference usually starts when the automobile is traveling at a fair

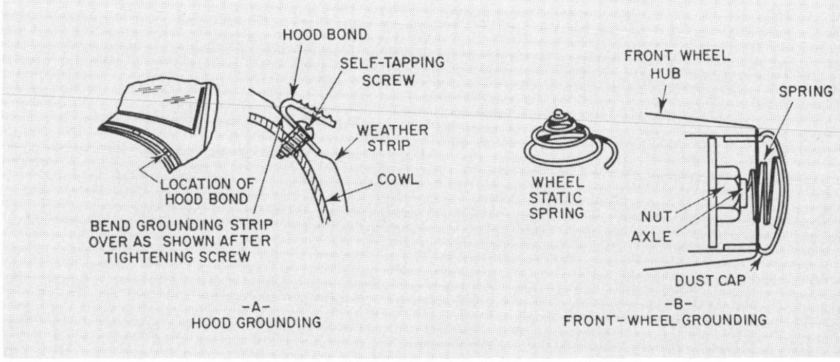

FIG. 21-20. Method of grounding hood and front wheels to reduce noise.

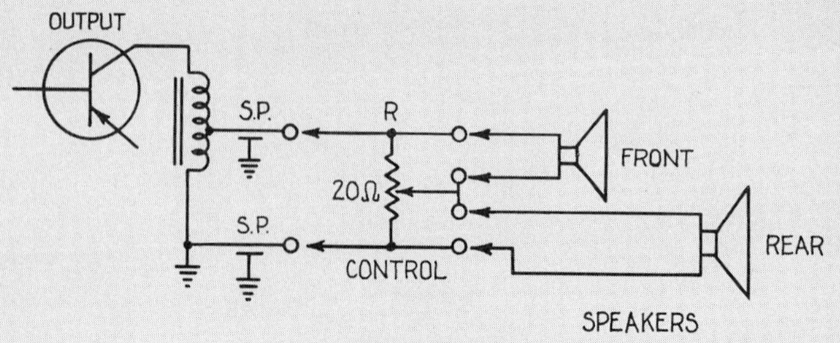

FIG. 21-21. Installation of a two-speaker setup.

speed above 20 miles per hour on a smooth, dry road, and sounds like atmospheric static. To identify it, drive the automobile until the interference starts; then turn off the ignition key with the set still on and coast. If you still hear noise and it gradually disappears as the automobile slows down, the cause is wheel static. Wheel-static interference is reduced by installing collector springs (spring contacts that ground the wheel to the axle), as shown in Fig. 21-20*B*.

Two-speaker arrangement. Many automobiles with factory installation of the radio have a second speaker mounted in the rear of the car. When a car is not so equipped, a service technician is often asked to install a second speaker in the rear.

Rear-seat kits are available which include the speaker, grill, control unit, and wiring. If the front speaker is not in the radio cabinet, but mounted separately and connected by a cable, the installation is easy. A typical circuit is shown in Fig. 21-21.

The control is of the fader type and plugs into the speaker jacks on the radio. It is generally mounted on the bottom lip of the instrument panel on the car. Both speakers plug into the control. When the control is at mid-position, as shown, both speakers are working. If the control arm is moved toward the position marked R, the rear speaker gradually becomes louder, and the front speaker weaker. Reverse rotation strengthens the front speaker at the expense of the rear. If the original front speaker is an 8-ohm one, an 8-ohm speaker is chosen for the rear. With a fader resistance of 15 to 20 ohms, the impedance match is not disturbed. Also, since the speaker leads in the radio are equipped with spark plates, any noise picked up by the additional wiring is filtered out.

If the radio and speaker are in one cabinet, the job becomes more complicated. The same circuit and components may be used, but it will be necessary to remove the receiver, install speaker jacks with spark plates, and to extend the speaker leads to reach the control unit. As the speaker leads leave the radio cabinet, spark plate capacitors should be attached to filter out any stray r-f pickup by the wiring.

The hybrid auto radio. The hybrid auto radio, like all other car radios, is a superheterodyne circuit. The schematic diagram of Fig. 21-22 is that of a Delco hybrid auto radio. In many ways, it combines features of all-tube and all-transistor sets. Examine some of the differences. The front end through the first audio amplifier is made up of tube stages. The rear end is made up of an audio driver transistor stage and a transistor output stage. The tubes are of the low-voltage variety which require plate and screen voltages of 12 volts. Therefore both the tubes and the transistors may be powered directly by the car battery.

Notice the power supply of the receiver in Fig. 21-23. The heaters of the tubes require 12 volts. So they are connected in parallel

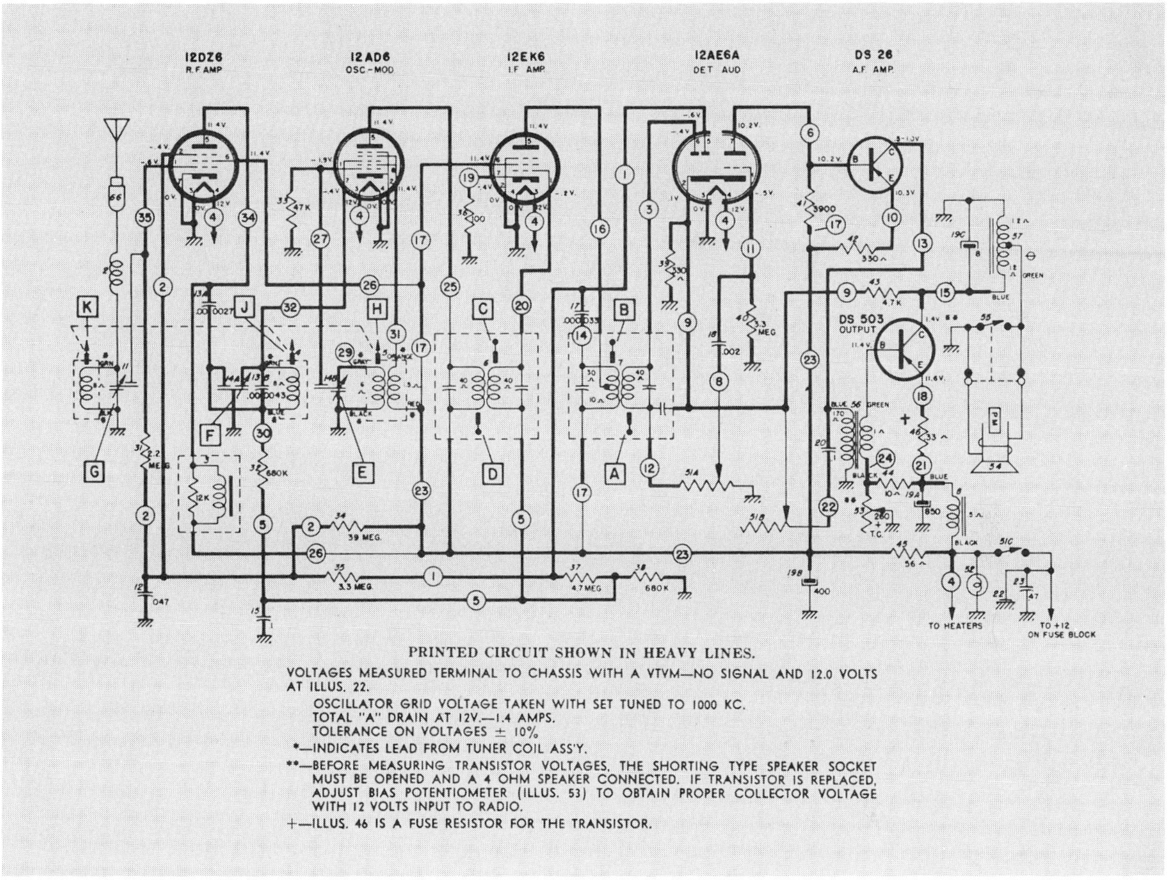

FIG. 21-22. A Delco hybrid car radio.

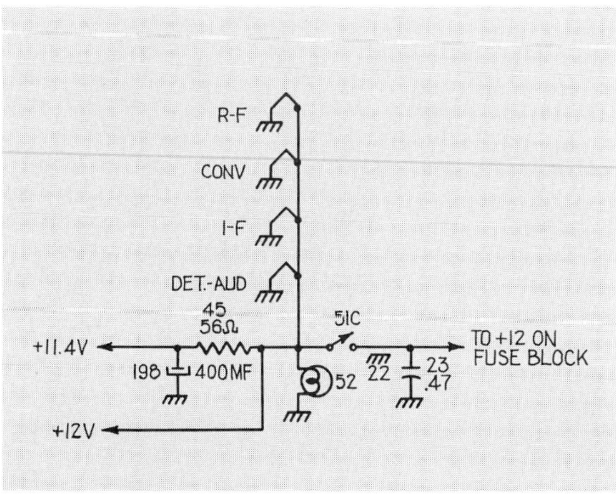

FIG. 21-23. Power supply and heater circuit of the Delco car radio.

across the battery. As a result, it is possible for one of the heaters to burn out while the other tubes light. Keep this fact in mind when making a first inspection of a set brought in for servicing.

The detector is one diode of the 12AE6A tube. The other diode is the agc rectifier. The agc control voltage is fed to the grids of the r-f and i-f amplifiers to control their gain.

Notice items 11 and 14A. They are combination trimmers and fixed capacitors, using a common plate. Item 22 is a spark plate capacitor to bypass noise signals along the battery line.

The letters in the squares identify adjustment components used in aligning the receiver. The numbers in the circles identify the heavy lines which are the printed wiring.

Observe the transistor section. Item 53 is the collector current adjustment. Item 51B and item 20 make up a tone control. Item 55 is a safety switch that closes when the speaker is removed. Item 46 is a fusible resistor in the emitter circuit of the output stage.

Servicing the hybrid auto radio. Service analysis of a hybrid receiver is not very different from service notes given for the tube receiver for the front end and for the transistor receiver for the audio end. To isolate a defective stage, reference should be made to Table 3. Note its similarity to Table 1 for the transistor radio.

TABLE 3. ISOLATING DEFECTIVE STAGE OF HYBRID AUTO RADIO BY SIGNAL INJECTION

Signal injected	To test point	Normal result	Analysis
1. A-f (400 cps)	Arm of volume control	400-cycle note from speaker	If heard, audio stages are OK. Go to step 5. If not heard, go to step 2
2. A-f (400 cps)	Base of output transistor	400-cycle note from speaker	If heard, go to step 3. If not heard, defect is in transistor, bias resistors of output stage, or speaker
3. A-f (400 cps)	Base of a-f driver	400-cycle note from speaker	If heard, go to step 4. If not heard, defect is in driver stage or a-f coupling transformer
4. A-f (400 cps)	Signal grid of a-f amplifier	400-cycle note from speaker	If heard, go to step 5. If not heard, defect is in a-f amplifier stage or decoupling circuit

Signal injected	To test point	Normal result	Analysis
5. Modulated i-f	Signal grid of i-f amplifier	Modulated note	If heard, go to step 6. If not heard, defect lies in i-f amplifier, output i-f transformer, detector, volume control, or tone control
6. Modulated i-f	Converter signal grid	Modulated note	If heard, go to step 7. If not heard, defect is in mixer stage of converter or in input i-f transformer
7. Modulated 1,600 kc (receiver tuned to 1,600 kc)	Converter signal grid	Modulated note	If heard, go to step 8. If not heard, defect is in oscillator stage of converter (see note below). In this test, wobble signal generator around 1,600 kc to get response
8. Modulated 1,600 kc (receiver tuned to 1,600 kc)	Signal grid of r-f amplifier	Modulated note	If heard, defect is in antenna circuit. If not heard, defect is in preamp circuit or r-f amplifier stage

Note: To determine whether oscillator is working, check for negative voltage at oscillator grid. If customer's complaint is that he receives stations at one end of dial but not at other, check oscillator grid voltage as you swing tuning dial across band. If grid voltage drops to zero or goes slightly positive somewhere along dial, you have a critical oscillator. Usually, a new converter tube remedies the situation. It is perfectly normal to read a higher negative grid voltage at high-frequency end of band than at low-frequency end.

The general procedure, after isolating a defective stage, is to check the bias voltages. If these are absolved of blame, then check the signal paths which often do not affect the voltages. Keep in mind the fact that open bypass capacitors in transistor circuits cause greater loss of signal than similar capacitors in tube circuits because of the lower impedances involved in the former case.

The alignment procedure for a hybrid receiver is basically the same as for the transistor receiver. Refer to Table 2 for the steps to be followed.

Tube-and-vibrator auto radio. The last variation is the oldtimer, using higher-voltage tubes and vibrator power supply. An auto receiver of this type is shown in Fig. 21-24. These receivers were designed to operate from the 3-cell, 6-volt storage battery of the car of its day. They are superheterodyne receivers incorporating an r-f stage, a converter, an i-f amplifier, a detector, two a-f amplifier stages, and a power supply. This chain makes this type of receiver very similar to the standard receiver we studied in the early chapters of this book. The same servicing procedures which locate defective stages and components in the latter set will locate similar faults in this type of receiver. The primary difference lies in the vibrator power supply and the special problem it presents.

Note the high voltages on the plates of the signal chain tubes. The function of the power supply is to take the 6 volts direct current of the battery and to step it up to the high voltage required by the tubes. This job is essentially performed by a vibrator, step-up transformer, rectifier, and filter system. Let us examine this in more detail.

How a vibrator works. To understand vibrator operation, consider the circuit of Fig. 21-25*A*. The battery sets up a steady unvarying magnetic field in the core of the transformer as a result of the steady direct current driven through the primary winding. No voltage is induced in the secondary winding because the magnetic field is unvarying. The magnetic field can be made to vary by using a single-pole double-throw switch, as shown in Fig. 21-25*B*. When the switch is thrown to contact *A*, the magnetic field builds up as direct current flows down the top half of the transformer primary winding. When the switch is opened, the magnetic field collapses. When the switch is thrown to position *B*, the magnetic field builds up again as direct current flows up the bottom half of the transformer. When the switch is opened again, the magnetic field collapses again, and does so repeatedly. The constantly changing magnetic field induces an alternating current voltage in the secondary winding. When the transformer is a step-up transformer, the peak voltages across the secondary winding are higher than the battery voltage across the primary winding.

Vibrator circuit. When the hand-operated switch is replaced by an automatically operated switch, such as a magnetic vibrating reed between two contacts, we have the basic circuit of the vibrator-type power supply. Such a vibrator is shown in Fig. 21-26. At the starting position, current flows through the top half of the transformer primary winding and the vibrator coil, which are in series. The latter becomes an electromagnet which attracts the steel reed upward, touching contact *A*. This shorts out the vibrator coil and a heavy current flows through the top half of the primary winding. The flexible reed, now freed from the vibrator coil, springs back and touches contact *B*. This sends a heavy current through the bottom half of the transformer primary winding. The vibrator coil is now no longer shorted out and is energized again, attracting the reed upward. Thus the cycle repeats itself. The effect is the same as with the manual switch in Fig. 21-25*B*, and an alternating voltage with a frequency of about 115 cycles per second results.

Spark plates and hash filters. It would seem

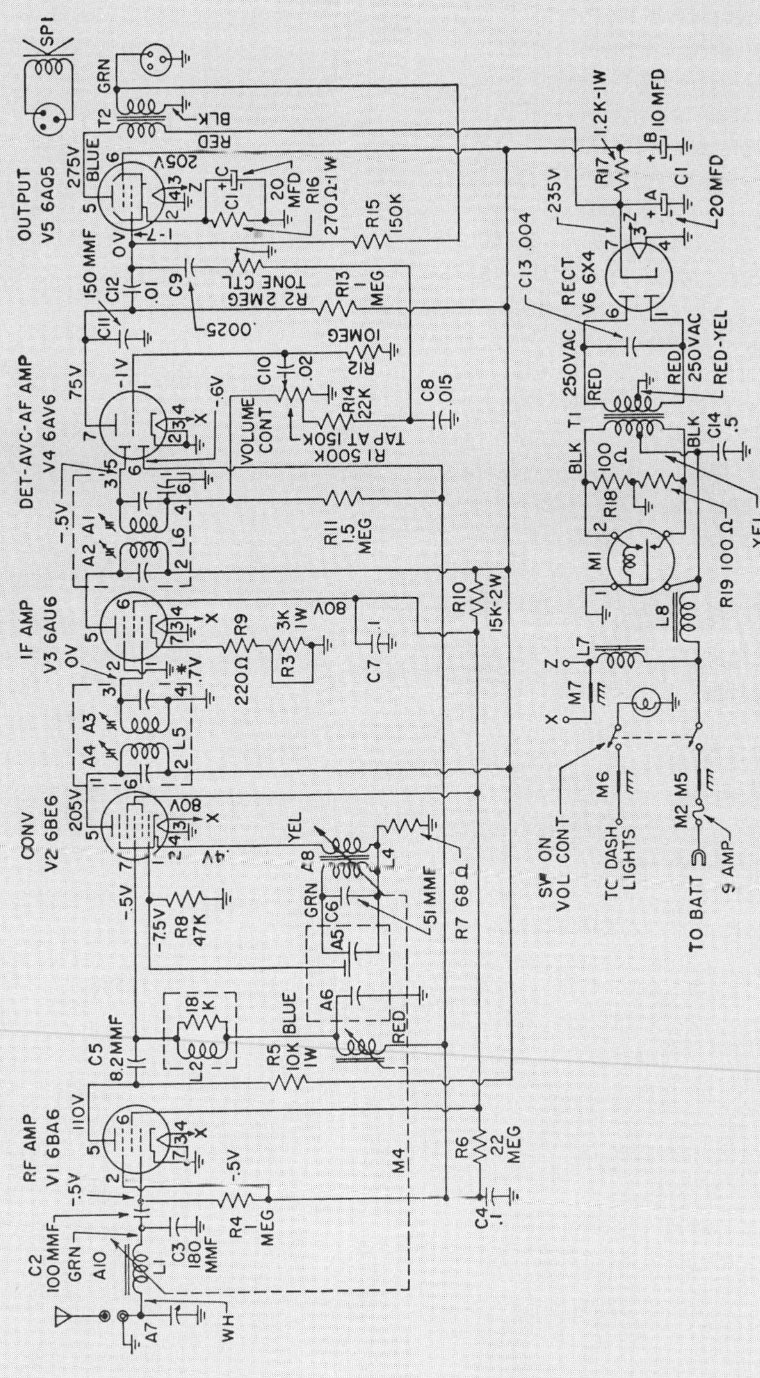

FIG. 21-24. Schematic diagram of Ford auto radio.

that all we need do now is to rectify and filter. However, the make and break of current at the vibrator contacts is accompanied by sparking which causes r-f interference, known as "hash." Practical circuits include provision for spark suppression and hash filters. Let us examine this circuitry further. In Fig. 21-27, we have the power supply of the receiver of Fig. 21-24, through the transformer.

The 6-volt supply comes from the car battery. A 9-ampere fuse, *M*-2, is in the hot feed line to protect the battery from shorts in the radio. The next component, *M*-5, is a spark-plate capacitor to bypass to ground any engine noise signals picked up by the hot lead. Next in line is the ON-OFF switch. Component *L*-8 is known as the vibrator choke. Together with capacitor *C*-14, it acts to keep vibrator r-f hash from being fed to the tube heater leads. Further hash filtering is performed by choke *L*-7 and spark plate *M*-7. The hot battery lead now feeds on to the center tap of the primary winding of transformer *T*-1.

Buffer capacitor. Capacitor *C*-13, known as a buffer capacitor, surge capacitor, or timing capacitor, is an important component. It takes up the high-voltage surges that would otherwise result from the rapid make and break of the vibrator reed. It is also effective in reducing sparking at the contacts. Its capacitance is important in design and its value should not be changed in service work. Common values are from 0.005 to 0.03 mfd. They are of the oil-filled type rated at 1,600 volts.

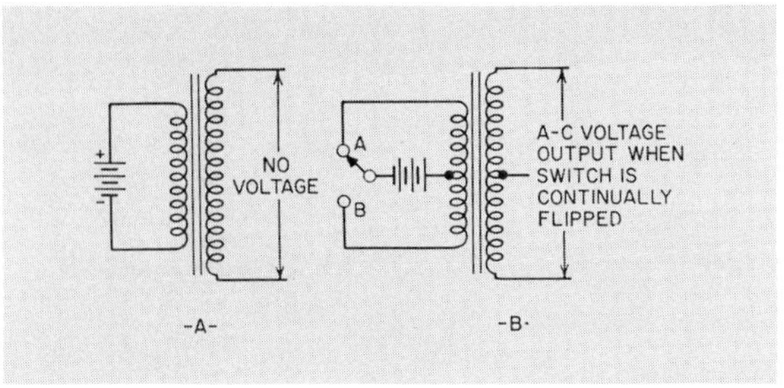

FIG. 21-25. Transformer fed with pure direct current (A) and interrupted direct current (B).

FIG. 21-26. Basic vibrator circuit.

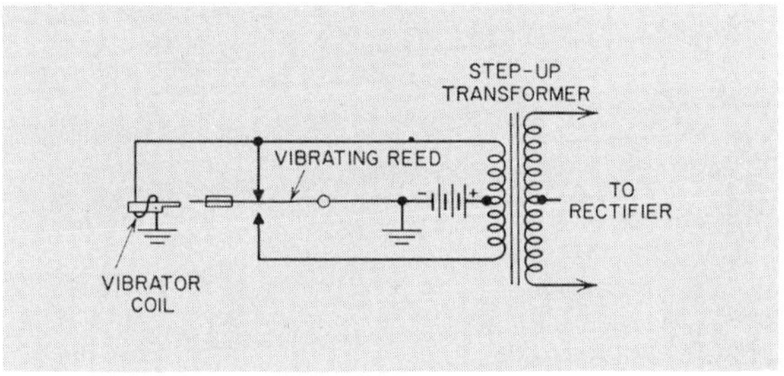

Resistors *R*-18 and *R*-19 are connected across the vibrator contact points and are also effective in reducing sparking and hash. They form a discharge path for back electromotive force (emf) in the primary winding which would otherwise cause a heavier spark at the contact break. These resistors rarely cause any service difficulty. Sometimes,

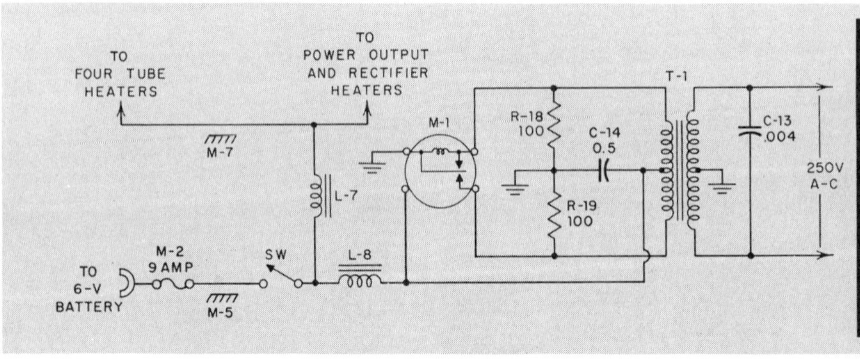

FIG. 21-27. Front end of vibrator power supply.

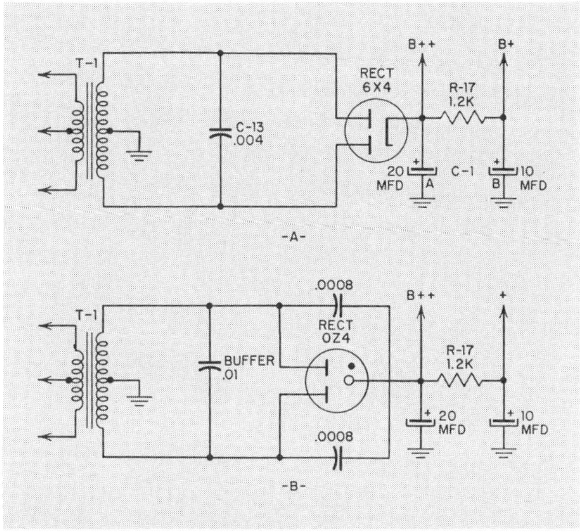

FIG. 21-28. Two types of rectifier tubes used with vibrator power supply.

capacitors are used in a similar function in place of these resistors.

Rectifying the vibrator output. Component *M*-1 is the vibrator unit itself. Note the vibrator coil, the reed, and the two contact points. Now all that is left to do is to rectify and filter the high alternating voltage. The circuit for this is drawn in Fig. 21-28*A*.

The transformer secondary winding feeds the 6X4 twin-diode tube in a full-wave rectifier circuit. Each plate, in turn, draws current through the tube as its end of the transformer becomes positive with the alternating voltage across the transformer. Sometimes a 6X5-GT heater-cathode type of tube is used, or sometimes an 0Z4 gas rectifier tube, in a similar circuit, shown in Fig. 21-28*B*. The final components make up the filter circuit composed of electrolytic capacitors *C*-1*A* and *C*-1*B* and resistor *R*-17.

To prevent hash from being fed to the receiver, the vibrator unit is shielded by being enclosed in a metal can. The entire power supply is often shielded by being enclosed in a metal compartment of the receiver assembly, as shown in Fig. 21-29.

Vibrator checking. We can quickly check whether the vibrator power supply is working normally. Turn on the set. If there is a muffled buzz, hash and hum level are normal, all tubes in the receiver light, and *B*+ voltage measures from 150 to 250 volts, the power supply is all right. Absence of vibrator buzz leads you to suspect the vibrator unit, the fuse, or the ON-OFF switch. Too high a hum or hash level indicates trouble in the filter or hash-suppression circuits. The current drain of the set is

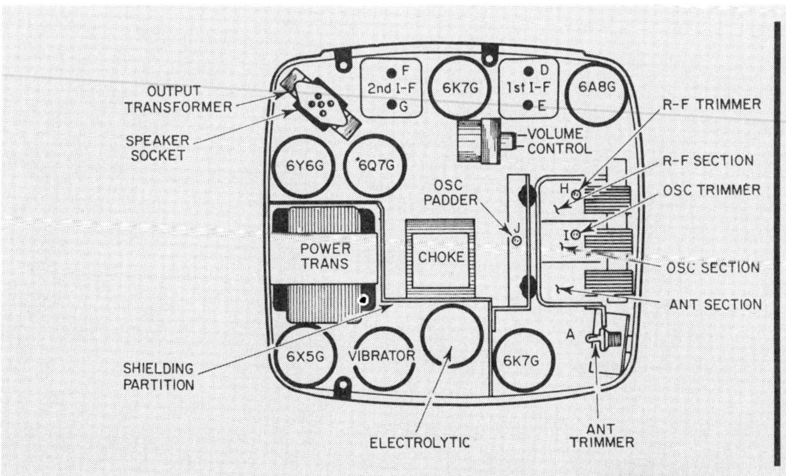

FIG. 21-29. Older car radio, showing shielding of vibrator power supply.

also a check for normal operation. A d-c ammeter across the ON-OFF switch, in its open position, normally reads 5 to 8 amperes. Shorts in the *B* circuits or in the buffer capacitor or stuck vibrator points would give higher current drain and probably blow the fuse.

Normal resistor data for such a power supply is given below:

Chassis to rectifier plate	150–300 ohms
Rectifier plate to plate	300–600 ohms
Chassis to *B*+	Capacitor action
Transformer primary winding	Less than 1 ohm

Vibrator troubleshooting. The troubles that commonly develop in the vibrator power supply are in the vibrator, rectifier tube, buffer capacitor, filter capacitors, and the switch. Other components rarely break down. In addition, vibration in a car sometimes loosens a grounding lug or screw. When hash is too loud, check for loose shielding and loose grounding screws.

The vibrator sometimes fails to operate. Because of loss of reed tension or pitted contacts, the reed sticks to one contact. Occasionally, a sharp thump starts it up for a while. Continued blowing of fuses is a good indication of the condition. If the vibrator fails to work on a 4-volt input, then it is definitely defective. When such is the case, replace it with an exact duplicate of the original or one specified by the manufacturer as being correct.

The buffer capacitor sometimes shorts. When it does so, there will be no *B*+ voltage and no a-c plate voltage fed to the rectifier plates. Receiver current drain will be larger than normal and the fuse will burn out. This capacitor should then be replaced with an oil-filled capacitor of the same capacitance as the original and the same or higher voltage rating.

The rectifier tube has the usual ills of any rectifier tube—weak or no electron emission and cathode-heater leakage. Checking the tube in a tube checker will point up the former condition, and in some cases cathode-heater leakage as well. The latter defect is definitely indicated when there is no *B* voltage, the vibrator is working, there is a heavy receiver current drain, but there is some a-c voltage at the rectifier plates. Before a poor tube is replaced, the *B*+ circuit of the set should be checked for a short, since such a condition may harm the new tube.

The filter capacitors are the main offenders in the filter circuit. When the input filter capacitor opens, the receiver hums. Hum in an auto radio is higher-pitched than hum in the a-c/d-c receiver because the input frequency in auto radios is about 115 cycles per second. The condition can be checked by bridging the capacitor with a similar one and observing the improvement.

When the output filter capacitor opens, the receiver will hum and oscillate. Again, bridge with a similar test capacitor. If either of these capacitors become shorted, there would be no *B* voltage, the fuse would blow and the set would be dead. An ohmmeter check would disclose the condition.

ON-OFF switches in vibrator-type power supplies give more trouble than in hybrid sets because they must break 5 to 8 amperes of current, as compared with the smaller currents of the latter type. In the vibrator type of receiver, a poor contact will cause considerable loss of power, great heating, and eventual burning up of the switch. A similar condition would occur with a poor fuse contact. If the switch (turned on) or the fuse shows about 1 volt across its terminal points, undesirable contact resistance is indicated and should be remedied. When replacing a defective switch in this type of car radio, use a switch that has been designed for heavy-duty use. It is a high-current, low-voltage type, marked "10A-12V."

This description completes the special problems with respect to the vibrator-type power supply. For the rest of the car receiver, servicing and alignment procedures are similar to those for the standard tube receiver.

Search tuners. Some higher-priced auto radios have a unique system of tuning. To avoid the dangerous practice of tuning while driving or the searching for stations in unfamiliar areas, the user pushes in a special button or steps on a similar button. When he does so, a search tuner is activated, scans the tuning range, and comes to a halt whenever it locates a station. If the user wants another station, he presses the special button and the search tuner starts up again. In some search tuners, the tuner slowly scans from one end of the dial to the other, and then quickly springs back to start the search in the original direction. In others, the tuner scans first in one direction, then reverses and scans back in the other direction.

Superficially, it would seem to be an easy matter of design. However, there are several requirements for a search tuner. First, the tuner must stop right on the carrier of a received station and not off on a sideband. Second, it must be able to stop on weak stations as well as on strong stations. Third, the search tuner must not be activated by noise signals.

Yet, the principle by which the device works is relatively simple. Basically, there is a motor which causes the tuner to sweep across its tuning range. The search can only be stopped when a voltage pulse of a definite magnitude from the trigger circuit cuts out the motor. The trigger circuit responds to a received station in the following manner. The trigger tube is biased to near cutoff, so that it acts like a detector. When a station is tuned in on the dial, the developed negative agc voltage fed to the trigger tube grid makes it even more negative. At the same time, a strong positive voltage pulse from the i-f transformer is fed to the same grid. It overrides the negative agc voltage and the negative bias voltage, causing a pulse of plate current. This pulse stops the motor and the tuner is now on station.

The manner in which the search tuner responds to both strong and weak stations is indicated in Fig. 21-30. The strength of the i-f signal, for both a strong and a weak station, is shown by the amplitude of the i-f resonance curves in the diagram. The magnitudes of the canceling negative agc voltages also vary with strong and weak stations, as shown. Note that in both cases, even though the negative agc voltage and positive i-f voltage are greater for a strong station than for a weak station, the magnitude of the posi-

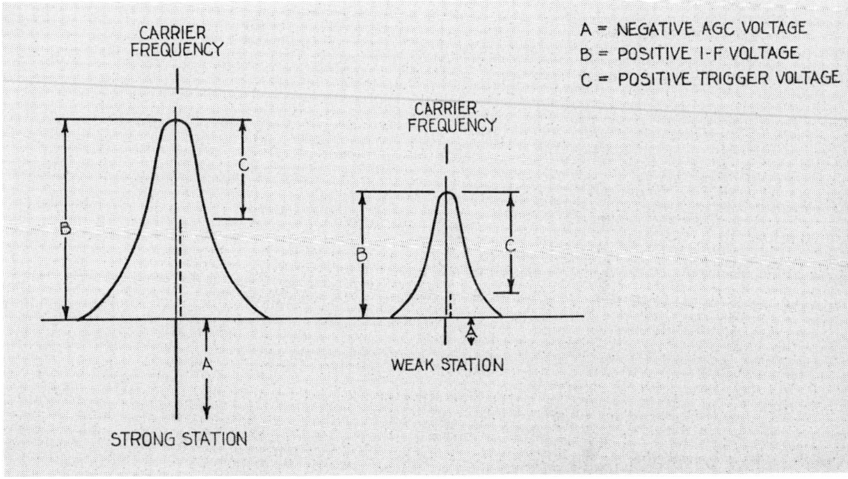

FIG. 21-30. Search tuner characteristics.

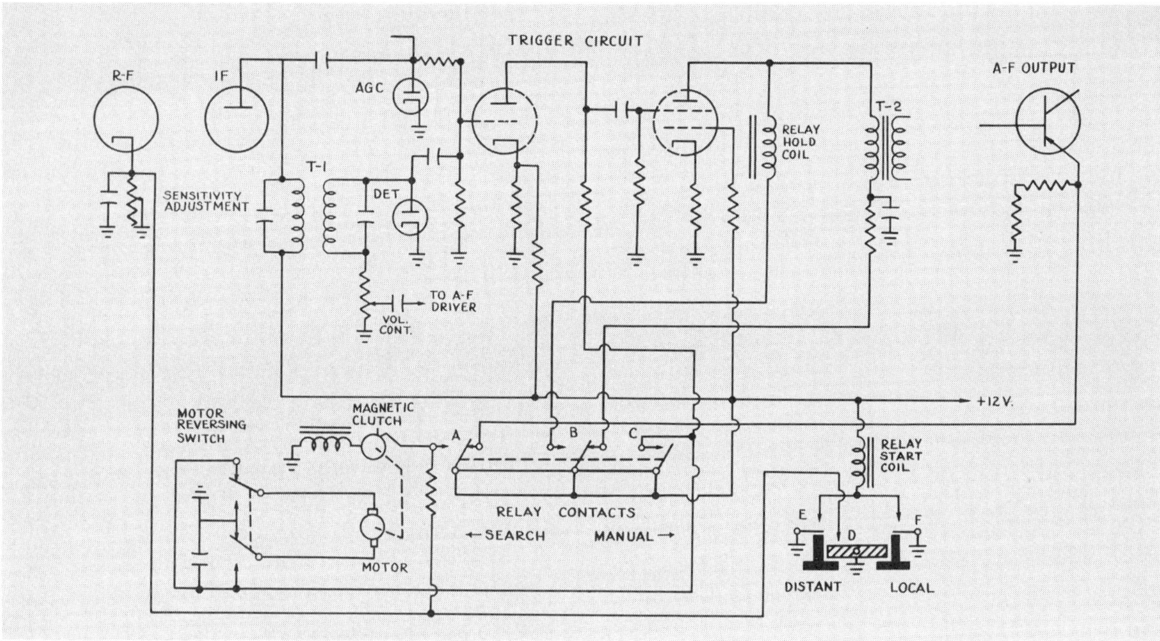

FIG. 21-31. Schematic diagram of a search tuner.

tive pulse to the trigger tube grid is about equal.

Now examine a typical circuit, shown in Fig. 21-31, used in a hybrid auto receiver. The set is turned on and all switches in the diagram are in the normal position so that the user may tune manually or by means of push buttons; the search tuner is not functioning. There are two buttons used to start the search mechanism. One is the LOCAL button for local areas with strong station signals; the other is the DISTANT button for distant areas with weak station signals. Assume that the LOCAL button is pushed in. The pivoted grounded bar tilts up at the local end and remains that way with no effect. At the same time, switch *F* is closed. Current now runs from the +12-volt point, through the relay start coil to ground, thereby energizing the relay. The relay armature swings switches *A*, *B*, and *C* to their opposite positions.

Switch *A* opens the voltage supply to the emitter of the audio output transistor and mutes the output signal during the search. It also feeds voltage to the magnetic clutch. Switch *B* opens the voltage supply to the primary winding of transformer *T*-2 and connects the relay hold coil from plate to +12 volts, as the plate load. It is the plate current through this hold coil that energizes the relay even after the search button is released. Switch *C* applies voltage to the motor and to the plate of the triode trigger tube. The motor is ganged to the clutch which in turn drives the tuner. When the tuner reaches one end of the dial, it flips the motor reversing switch and the search continues over the dial in reverse.

Note the connections to the tetrode tube. The first grid next to the cathode has a positive potential and acts as an accelerator for electron current in the tube. The second grid receives the signal from the triode trigger tube. The latter is biased to near cutoff by its cathode resistors, so that it acts like a detector and will only pass plate current when a strong positive pulse of a certain magnitude is applied to its grid.

As the tuner approaches the carrier frequency of a station, an i-f voltage builds up and develops a negative agc voltage through the agc diode and feeds this negative voltage to the grid of the trigger triode. At the same time, the greater i-f signal from the secondary winding of i-f transformer *T*-1 feeds positive voltage on its positive phase to the same grid. When the difference between the negative agc voltage and the greater positive i-f voltage reaches trigger pulse magnitude, the triode trigger tube passes plate current. This plate current increases the voltage drop across its plate resistor, sending a negative voltage pulse through the coupling capacitor to the second grid of the tetrode tube. This negative grid pulse reduces the tetrode plate current and the relay is deenergized. Switches *A*, *B*, and *C* flip back to their original positions. The deenergized magnetic clutch cuts off further tuning as it disengages from the motor which coasts to a stop. The receiver is now receiving a station. To start the search again, the search button must be pushed in again.

The rheostat in the cathode circuit of the r-f amplifier is a manual sensitivity control to set the desired magnitude of the trigger pulse. It is used to compensate for variations in output of the various tubes of the receiver. Now look at the DISTANT search button. When pushed in, it does everything that the LOCAL button did. But in addition, it closes switch *D* which remains closed because the pivoted grounded bar tilts up and remains that way even after the button is released. Switch *D*, when closed, grounds out the cathode of the r-f tube, thereby increasing its sensitivity. This is desirable for weak distant stations. Switch *D* will open only when the LOCAL button is pushed in and the pivoted bar tilts in the other direction.

Troubles in the search tuner. If the tuner fails to search when a search button is pushed in, but the set works when operated manually, check switches *E* and *F*. Depress them by hand and see whether searching action begins. If this brings no response, the relay may be open. Flip the contacts to the search position and see if the search begins now. Check for good connections to the relay contacts. Check the motor and its associated clutch.

If the tuner searches but fails to bring in any station, yet the receiver works in the manual position, the defect probably lies in the trigger circuit. Check the triode trigger tube for correct voltages given by the manufacturer. Check the associated components.

If the tuner has poor stopping action, passing over some stations without stopping, check the trigger circuit. Also check the antenna for good signal pickup. Check the i-f transformer, since a defect here could decrease its positive pulse. The defect, however, would probably show up in manual operation.

If the search tuner stops but no station is heard, check the relay holding coil for a defect. Also check switch *A* to see that it restores the muted a-f power transistor. If the search tuner runs to the end of the dial and stops, check the motor reversing switch. If the search tuner fails to stop on station frequency but stops off on a sideband, check the agc circuit. Also try to improve action by adjusting the sensitivity control.

QUESTIONS

1. What are the first steps in checking a transistor car radio that does not work?
2. Explain how to check for a defective car antenna and list the necessary adjustments after installation of a new one.
3. The receiver of Fig. 21-2 has a fading signal. What is the probable cause and how would you check for the defect?

4. How would you check that the oscillator is working in the receiver of Fig. 21-2?
5. The receiver of Fig. 21-2 oscillates. What are the possible causes and how would you check for each possibility?
6. The receiver of Fig. 21-2 distorts. What are the possible causes?
7. When a transistor car radio is dead and the trouble is traced to the output stage, explain your first check.
8. Explain how you would install a second car radio speaker.
9. The vibrator in an inoperative auto radio fails to start. List the possible causes and explain how you would check for each.
10. A vibrator-type auto radio blows fuses repeatedly. What are the possible causes and how would you check each one?
11. List the possible causes for excessive hum in a vibrator-type car radio and indicate how you would check for each.
12. Outline a procedure for removing excessive hash in a vibrator-type car radio.
13. What troubles can occur in a search tuner? Tell how to check for each.

22 Servicing F-M Radios

The popularity of frequency-modulation receivers, as well as the use of frequency modulation in the sound system of television receivers, makes the understanding of the f-m signal necessary for any book on radio servicing. In this chapter, you will find that an f-m receiver is a superheterodyne set, in many ways similar to a standard a-m receiver. At the start, you will have a stage analysis of an f-m set. This will be followed by a servicing procedure that will enable you to repair such a receiver.

The a-m signal. We begin with a comparison of the a-m and f-m signals as received at the antenna, disregarding how the transmitters achieve their purposes. What is implied in an a-m signal? First, every station is assigned a specific carrier frequency. Thus, in New York City, station WABC has a carrier frequency of 770 kc, and station WNBC has a carrier frequency of 660 kc.

Second, the a-m carrier is made to carry intelligence (speech or music) by varying the carrier amplitude. The rate at which the amplitude is made to vary is the pitch of the modulating note. For example, examine the a-m signal shown in Fig. 22-1A. The pitch of the modulating note is 400 cycles per second. The carrier is 1,000 kc. The a-m signal is shown at the right of the diagram, where the amplitude of the original carrier varies up and down 400 times a second.

Third, the a-m carrier must also carry the loudness of the original modulating signals. Loudness in an a-m signal is indicated by the extent to which the amplitude of the carrier is varied from the unmodulated carrier. The two waveforms in Fig. 22-1*B* make this point clear. The greatest permissible loudness is that in which the amplitude is reduced to zero (100 percent modulation). To go beyond this point would result in a distorted signal. Hence, the modulating percentage is carefully controlled at the broadcast station.

The fourth consideration with respect to the a-m signal is that known as sidebands. Modulation of a carrier frequency produces a mixture of frequencies, representing the mixing of the carrier and modulating frequencies. The spread of this mixture is dependent on the frequency of the audio modulating signal. The spread on each side of the carrier is known as the upper and the lower sideband. Thus, if the carrier frequency were at 710 kc and the modulation note were at 1,000 cycles per second, the lower sideband frequency would be 710 minus 1, or 709 kc, and the upper sideband frequency would be at 710 plus 1, or 711 kc. This relationship of upper and lower sideband frequency is shown in Fig. 22-1*C*.

With a few exceptions, a-m broadcast stations are restricted to a bandwidth of 10 kc, that is, 5 kc on either side of the carrier. This means that for the most part no audio note can go above 5 kc. The station, therefore, suppresses any note above that frequency at the transmitter, thereby failing to give true fidelity. Fortunately, the quality of radio reception is generally satisfactory even though some of the higher notes are missing.

The f-m signal. Now compare the f-m signal with the a-m signal described above. The same considerations enter the picture, but in a different manner. Again, every f-m station is assigned a carrier frequency. In New York City, WCBS-FM operates at 101.1 mc, and WABC-FM operates at 95.5 mc. The complete f-m band is from 88 to 108 mc.

The second consideration is that of how the carrier is modulated to carry voice or music. In the f-m system, the mixing of the audio modulation note and the carrier results primarily in a frequency variation above and below the carrier frequency at a rate equal to the frequency of the modulating note. Inspection of Fig. 22-2*A* will make this point clearer.

Recall that in the a-m system, the highest permissible audio note was 5 kc because of sideband limitations. In the f-m system, audio modulation does not affect the sideband considerations. Hence, higher-frequency audio signals are permitted and greater fidelity results.

The third consideration is the loudness of the modulating sound. In the f-m signal, the louder the modulating sound, the greater the extent to which the frequency of the carrier swings above or below the carrier center frequency. Inspection of Fig. 22-2*B* will make this point clearer. Note that the rate at which the frequency swings around the center carrier frequency is the same for both a loud or a soft sound. However, the amount of swing or deviation is different for the two signals.

Since the amount of deviation of carrier frequency is of little concern in the f-m system, the FCC has given 75 kc on each side of carrier frequency to each station. This permits the receiver to give truer response with respect to the loudness of the modulat-

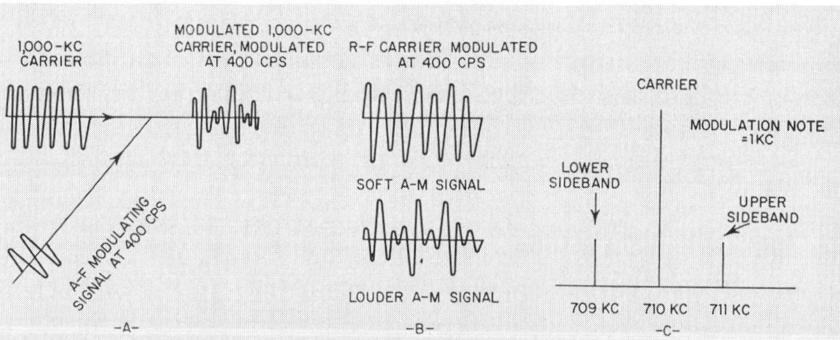

FIG. 22-1. Characteristics of an a-m signal.

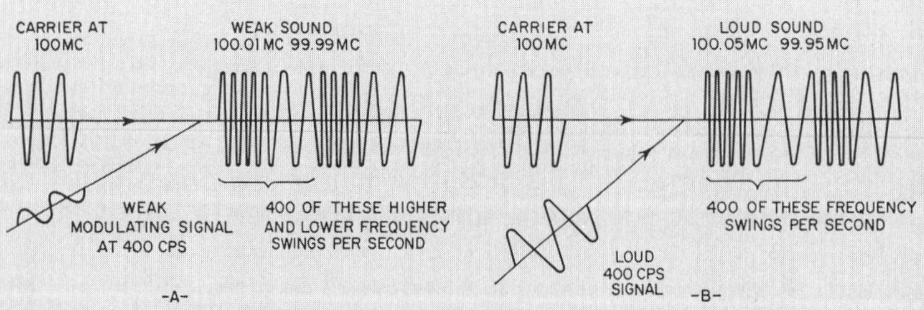

FIG. 22-2. Soft and loud f-m signals.

ing sounds. To prevent overlapping of stations, the FCC further allots 25 kc at each end beyond the 75-kc deviation range. The two extreme protective bands are known as guard-bands.

Advantages of the f-m receiver. There are two main advantages that the f-m receiver has over the a-m receiver. The first is the matter of noise rejection. The f-m listener will report that he can listen to music or speech without annoying crackling sounds or whistles or hum, even during lightning storms. This he cannot do when listening to an a-m radio. This characteristic is the primary advantage of f-m reception. It has been estimated that a noise signal 1/100 or more times the intensity of the desired station signal will mar good reception with an a-m receiver. With f-m reception, a noise signal even one-half as intense as the desired station signal will not interfere with proper reception.

The second advantage exists because of the legal allocations of the f-m band, rather than in f-m design. From the previous description, we saw that the f-m system can handle higher frequency sounds better; the loudness of sounds also is handled better. Of course, if the audio amplifier of the f-m receiver is poorly designed, these advantages will be lost to some degree.

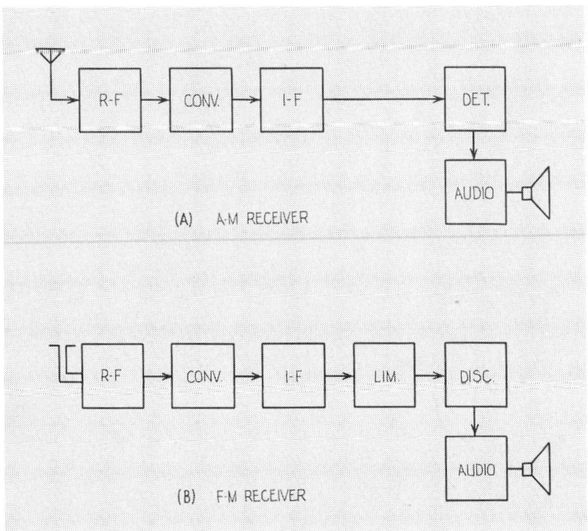

FIG. 22-3. Comparison of a-m and f-m receivers.

Block diagrams of a-m and f-m receivers. In many respects, the f-m receiver is similar to the standard a-m superheterodyne receiver. The primary difference lies in the second detector stage before the audio amplifier input. In the f-m receiver, this second detector stage is most frequently of the discriminator or ratio detector type. Examine Fig. 22-3 for block diagrams of an a-m and an f-m receiver. Note the similarity of stages used. In the case of the f-m receiver with its discriminator, a limiter stage is used. It is a specialized i-f stage whose function will be described in detail later.

Let us now analyze the various stages of an f-m receiver as compared with the familiar a-m superheterodyne receiver stages. In most cases, the function is similar, although design will differ considerably.

The current trend, particularly with less expensive equipment, is to combine an a-m and f-m receiver into one unit. However, for better understanding, the f-m receiver will be treated as a separate unit. The combination of the two will be taken up later in the chapter.

F-m receiver antennas. Most f-m receivers make provision for an external antenna installation which gives maximum signal pickup. This external antenna is similar to a TV antenna installation, except that the former is somewhat larger because of the higher frequencies involved in frequency modulation. The simplest version is a plain or folded dipole, as shown in Fig. 22-4A. The dipole is resonant to a signal whose wavelength is twice the length of the dipole from tip to tip. Since a dipole must respond to all frequencies from 88 to 108 mc, how long must the dipole be? The usual practice is to cut it to a length such that it is resonant at the middle of the band, at 98 mc. The total dipole length is figured by dividing 5,904 by the frequency in megacycles. Thus, for 98 mc the length should

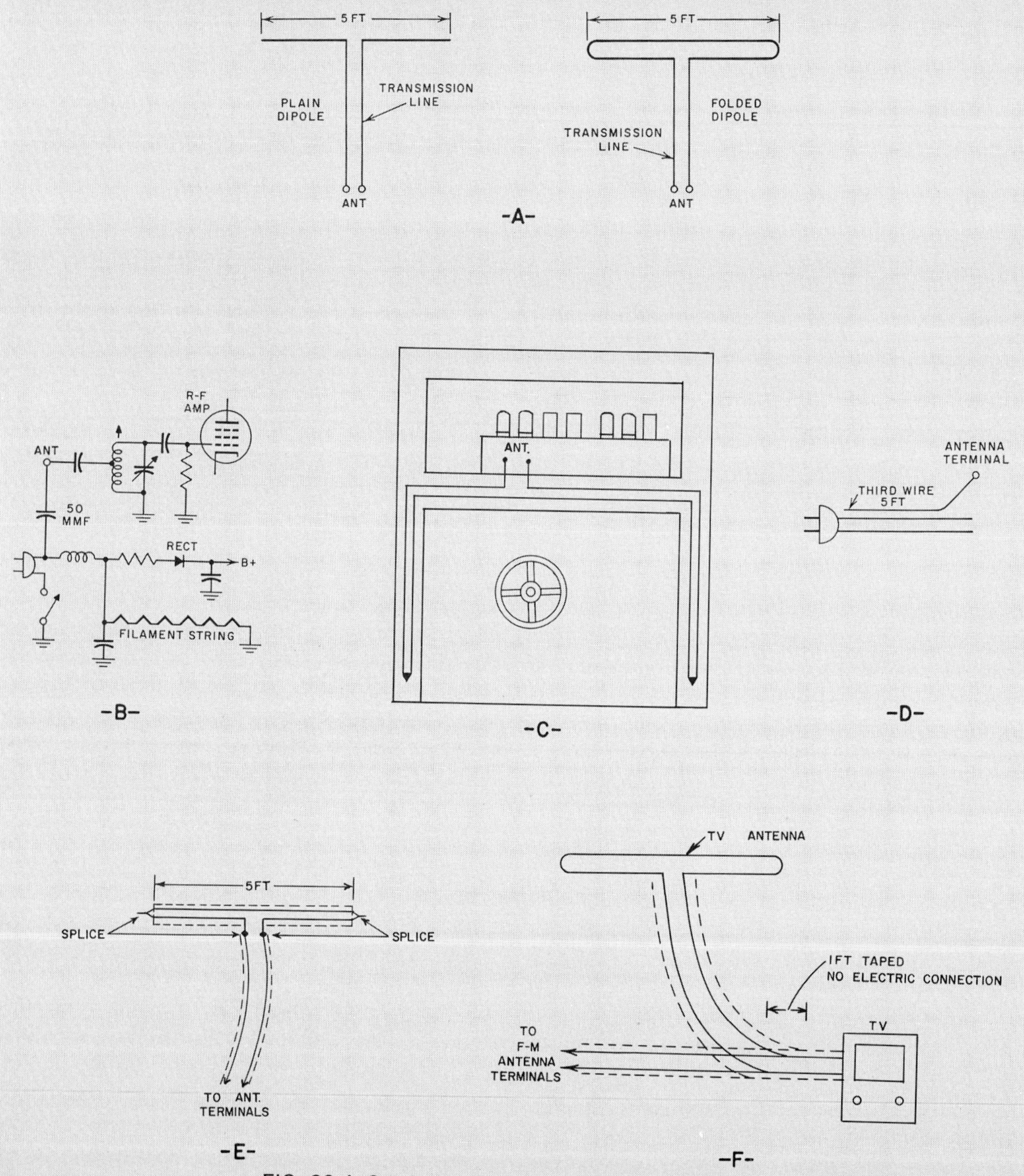

FIG. 22-4. Several common f-m antenna systems.

be 5,904 ÷ 98, or about 60 inches from tip to tip.

For best efficiency, a dipole must be matched to the input of the receiver. The impedance of a plain dipole is about 73 ohms. It must be matched with a transmission line with an impedance of 73 ohms. This transmission line in turn is connected to the receiver with an input impedance of 73 ohms.

Most f-m receivers have an input impedance of 300 ohms. To meet this condition, use is made of the folded dipole shown in Fig. 22-4*A*, which is a variation of the simple dipole. It consists of an aluminum tube, folded like a trombone slide into two parallel sections. Its overall length is the same as the plain dipole. The two parallel sections are separated by a few inches and the transmission line is connected to the two ends, located at the center. Such an antenna has an impedance of about 300 ohms. The transmission line is usually a twin-lead line with a 300-ohm impedance. This is suitable for connection to a receiver with a 300-ohm input impedance.

All dipole antennas receive most efficiently when the broad side faces toward the transmitting station. It is usual practice to rotate the antenna so as to get best results from most f-m stations. At best, this will be a compromise condition that serves no one station best. For better results, the antenna is sometimes mounted on a motor-driven rotator to give best efficiency for each station.

Smaller f-m receivers that will probably not have an external antenna, even though provision in the set may be made for one, provide for an alternate antenna pickup system. The antenna installation of Fig. 22-4*B* is frequently used. It consists simply of a 50-mmfd capacitor connected from the power line cord to the antenna input post. The power lines serve as the antenna.

The type of antenna shown in Fig. 22-4*C* is also frequently used in larger console-type receivers. A length of twin-line used in TV is shorted at both ends and tacked around the cabinet. One wire in the twin-line is cut at the center of the total length, and the cut ends are then connected to the antenna input terminal. This antenna simulates a folded dipole.

Another type of antenna is the line cord, shown in Fig. 22-4*D*. Here, a third unconnected wire is included in the power line cord. One end of this wire is connected to the antenna input post. This wire is about 5 feet in length. The capacitance between this wire and the power line makes this type of antenna similar to the one in Fig. 22-4*B*.

Frequently, a compromise antenna between the outdoor installation and the indoor installation is used. In one case, a 5-foot length of 300-ohm twin lead is spliced at both ends. Then one of the leads is cut at the center and spliced to another length of similar twin lead. The ends of this latter twin lead are then connected to the antenna terminals of the receiver. This antenna simulates a standard folded dipole, and is shown in Fig. 22-4*E*. The dipole end may then be placed on the floor or under a rug in such position as to give the best directional pickup for most f-m stations.

Another compromise antenna takes advantage of an existing outdoor TV antenna. As shown in Fig. 22-4*F*, a length of 300-ohm twin lead is connected at one end to the antenna terminals of the f-m receiver. The other end is taped without making any electrical contact for a distance of about one foot to the lead-in wire to the TV set. The pickup between the two twin leads is capacitive.

In a-c/d-c f-m receivers, safety precautions are taken so that the user of the equipment is isolated from common negative. Built-in antennas require no precautions. However, when a pair of antenna terminals are provided for connection to an external dipole antenna, isolating precautions must be taken. The circuit generally used is shown in Fig. 22-5. Capacitors *C*-1 and *C*-2 are virtually open circuit for the 60-cycle line frequency, and excellent conducting paths for the 1-

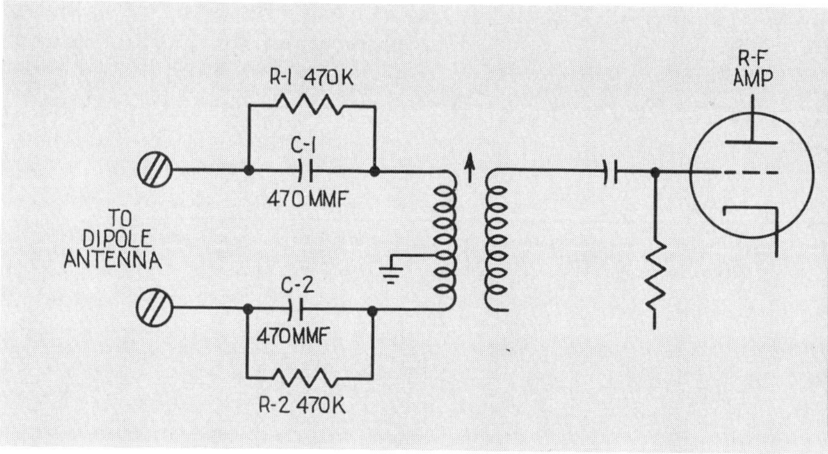

FIG. 22-5. Antenna input circuit for f-m receiver using an a-c/d-c power supply.

megacycle f-m signal. Resistors *R*-1 and *R*-2 prevent any charge from building up across the capacitors. These resistors and capacitors rarely present any servicing difficulty. This protection is unnecessary and is not used in receivers operated by a transformer-type a-c power supply.

F-m r-f subchassis. The r-f amplifier and converter in an f-m receiver serve the same functions as the similar stages in an a-m receiver. The stages are different because of the problems encountered at the high frequencies of the f-m band. Feedback effects from stray wiring are of major importance.

Most of the problems are solved by mounting and wiring the r-f amplifier and converter separate from the rest of the receiver, on a small subchassis. This subchassis is then mounted in a cutout section of the main chassis. Manufacturers generally make one such front end subchassis which is then used for all their f-m receivers, including the a-m/f-m models.

The schematic diagram of the r-f and converter stages of a typical subassembly of this type is shown in Fig. 22-6. The tube used is a twin triode 6AQ8. One triode is the r-f amplifier, the other is the converter. In a-c/d-c receivers, a twin triode 12DT8 is used in the same circuit with isolation capacitors added in the dipole leads.

The dotted line surrounding the unit indicates that it is a subchassis assembly, in addition to its usual significance as shielding. The subchassis is grounded to the main chassis, as indicated by the ground lead connected to the bottom dotted line. The power feed lines are also brought in through the bottom dotted line.

Note capacitors *F*-1, *F*-2, and *F*-3 also located in the bottom dotted line around the power feed lines. These are known as feed-through capacitors. They are mounted on the subchassis, and filter the power leads as they enter, very much like the spark-plate capacitors used in auto radios. Note the additional capacitor *C*-11 on the heater lead. This one is located right at the tube socket.

The input lead from the dipole antenna and the output lead to the i-f grid go right through the shielding without feed-through capacitors because they carry high-frequency signals which would be bypassed to ground by capacitors.

The antenna feeds signal through transformer *T*-1 to the cathode, not the grid of the r-f tube. The circuit is known as a grounded-grid amplifier. The grid is grounded and acts like a shield between the tube cathode and plate. The grid is at zero potential, and the cathode potential changes with the input signal, thereby changing plate current and

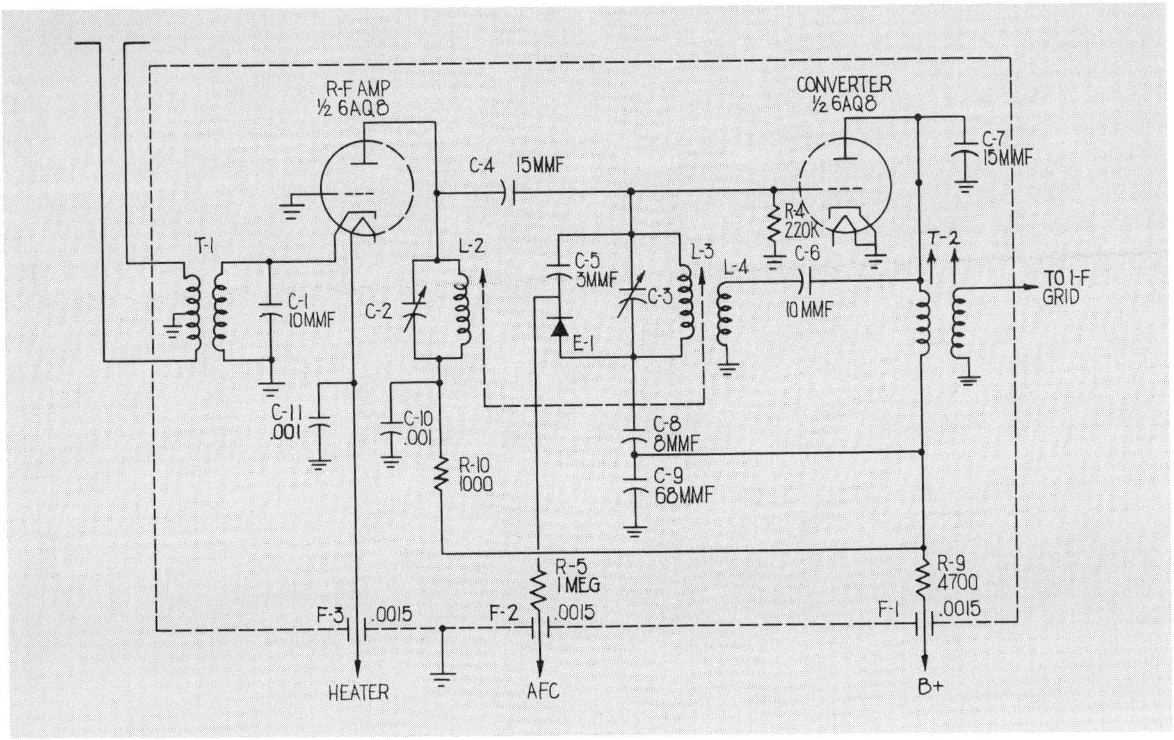

FIG. 22-6. Schematic diagram of an f-m r-f subchassis.

providing amplification. The grounded grid between the cathode and plate prevents feedback from the output to the input circuit, and so makes for a stable amplifier. Input transformer *T*-1 is untuned and, with its ferrite core, is designed to feed broadband coverage of 88 to 108 mc signals to the r-f tube cathode. The grounded-grid triode is generally preferred for the high-frequency r-f amplifier because of the low noise characteristics of triode tubes.

The r-f plate circuit is tuned to station frequency by the *L*-2/*C*-2 combination. Capacitor *C*-2 is the alignment trimmer and coil *L*-2 is slug-tuned from the tuning dial. Coupling capacitor *C*-4 feeds the amplified station signal to the mixer grid.

The converter circuit is known as a self-oscillating mixer. Plate signal is fed through capacitor *C*-6 to coil *L*-4. The signal is then inductively coupled to coil *L*-3 in the grid circuit, completing the feedback chain. Coil *L*-3 and capacitor *C*-3 form the oscillator tank. The slug in *L*-3 is gang-driven from the tuning dial to keep the oscillator circuit 10.7 mc ahead of the station frequency. Capacitor *C* 3 is a trimmer and acts as the high-frequency oscillator aligner. Tuning coils *L*-2 and *L*-3 generally consist of 3 or 4 turns of heavy wire wound on opposite ends of a coil form. The tuning slugs are then pulled through the coils by a simple mechanical arrangement.

Automatic frequency control. Another high-frequency problem is the tendency for the oscillator frequency to drift, primarily because of temperature changes in the oscillator tank circuit. The drift in oscillator frequency causes loss of part of the signal sideband, with distortion as a result. In some cases, the drift is large enough to lose the signal entirely. This effect can be corrected by manually retuning the receiver. In our basic f-m receiver, the retuning is done automatically by a circuit known as automatic frequency control, abbreviated afc.

The *L*-3/*C*-3 oscillator tank curcuit is repeated in Fig. 22-7, together with the additional components needed for the afc circuit. Note that capacitor *C*-5 and afc crystal diode *E*-1 are also across the oscillator tank. A voltage applied to the crystal causes it to change the capacitance between its two sections. Since this capacitance is across the oscillator tank, the voltage can be used as a control of the oscillator frequency.

The automatic control voltage is obtained from the f-m detector or discriminator, very much in the same way as automatic gain control voltage is obtained from an a-m detector. In the f-m receiver, when the oscillator is tuned to the correct frequency, average voltage from ground at point *A* is zero, point *A* being the output of the discriminator. Normal frequency deviations caused by signal modulation produce alternating voltages at point *A*. This is the audio output; but the average voltage is still zero. Resistor *R*-13 and capacitor *C*-13 form the detector r-f bypass circuit. Resistor *R*-12 and capacitor *C*-12 filter the audio signal so that the voltage on the afc line is zero.

If the oscillator should go slightly off the correct frequency, the signal voltages fed to the two halves of the discriminator transformer would become unbalanced. The a-f output at point *A* would still be present, but its average value would go positive or negative, depending on whether the frequency drift was an increase or a decrease. This condition places a positive or negative voltage on the afc line which, when applied to crystal *E*-1, changes its capacitance in the proper direction to correct the original frequency drift.

The afc circuit includes an ON-OFF switch. In the ON position, the control voltage is present and tends to hold the oscillator on station. If the receiver is then tuned to another station, the control voltage would still try to hold the oscillator on the first station. As a result, tuning would be very broad and adjacent stations might be missed. Correct operating procedure for tuning is to first turn the afc switch off. This shorts out the control voltage and allows for sharp manual tuning. When the desired station is found, the afc switch is returned to the ON position, and

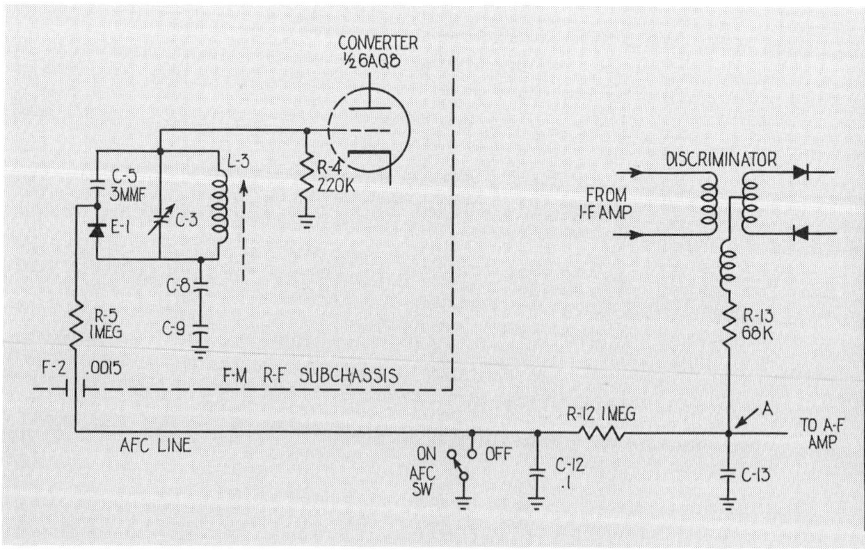

FIG. 22-7. Automatic frequency control circuit (AFC).

the station is held against oscillator drift.

Variations in the front end r-f section. In some f-m tuners, the station signal is fed to signal grid of the r-f amplifier tube, instead of the use of a grounded-grid triode. Although the tube now used is a high-frequency pentode, it may also be a high-frequency triode tube. Figure 22-8 shows such a circuit using the pentode half of a 6AQ8 tube. To prevent the possibility of oscillation, a neutralizing circuit is used.

Signal from the plate circuit is brought back to the grid through neutralizing capacitor *C*-3 in such phase as to oppose the normal oscillatory feedback through the grid-plate capacitance of the tube. When capacitor *C*-3 is adjusted to the correct capacitance, its feedback will exactly balance out the oscillation-producing feedback effect and so prevent oscillation of the r-f amplifier. In practice, the adjustment of capacitor *C*-3 is not critical, and alignment notes for receivers using this circuit do not mention the adjustment.

If the adjustment has been tampered with and the receiver oscillates as a result, the following readjustment procedure may be followed. Turn the adjustment for the capacitor back and forth through its full range, leaving it at approximately mid-position. Then tune the receiver from the low-frequency end of the band toward the high-frequency end, stopping at the first station that causes oscillation. Readjust the neutralizing capacitor. A small turn in one direction or the other should stop the oscillation. Continue to tune the receiver toward the high-frequency end of the band, readjusting the neutralizing capacitor if necessary.

Note that in this circuit the antenna input circuit is also untuned. The slug in transformer *T*-1 is not controlled by the gang tuner. It is a permeability adjustment on transformer *T*-1 and is meant to give optimum performance at the low-frequency end of the tuning band.

Another variation that is often found is the use of separate triodes for the mixer and for the oscillator in the converter. Such a circuit is shown in Fig. 22-9. Coupling between the oscillator stage and the mixer grid is through capacitor *C*-4. The use of a separate oscillator and mixer gives greater stability than the use of a single converter. A separate subchassis is not used.

Note that the circuit is permeability tuned with the slugs being ganged. Sometimes, the coils are fixed and there are ganged tuning ca-

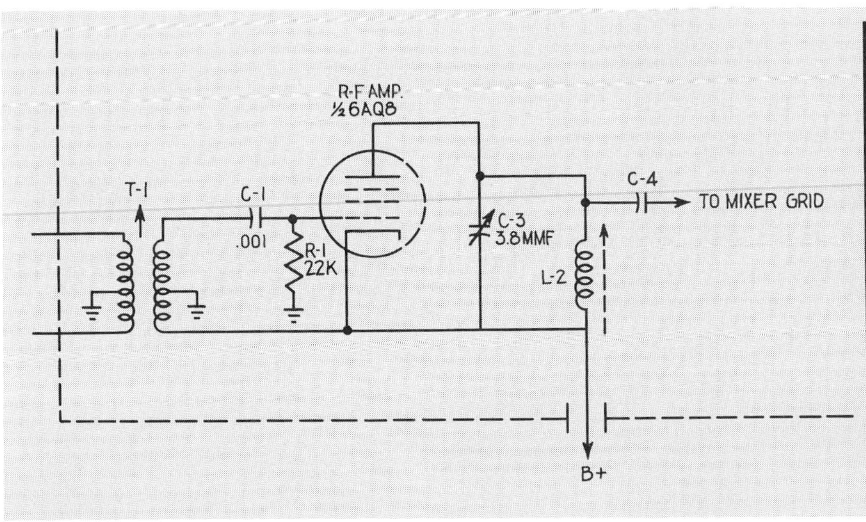

FIG. 22-8. R-f stage using a neutralizing capacitor.

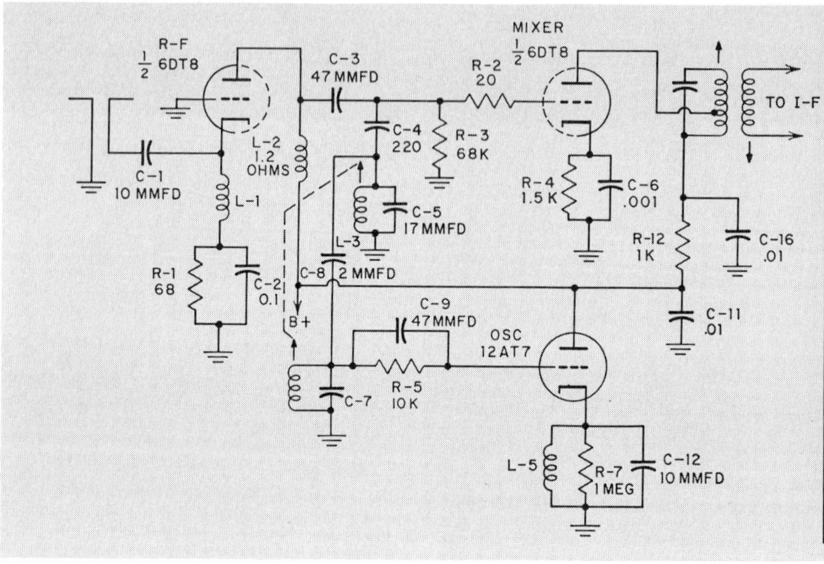

FIG. 22-9. An f-m frequency converter with separate oscillator and mixer tubes.

pacitors instead. When this is the case, the tuning capacitors are much smaller than corresponding tuning capacitors found in a-m tuners. Sometimes, special ganged tuning capacitors are used. They consist of a metal sleeve on a glass tubing. A metal cylinder is drawn through the inside of the glass tubing to give various desired capacitances.

I-f amplifier. Again, the i-f amplifiers of f-m and a-m receivers serve similar functions. A few differences should be noted. Many f-m sets use two or three stages of i-f amplifiers, the last of them known as the limiter, having a special function to be described later.

In Fig. 22-10, we have the schematic diagram of a typical i-f amplifier. In many ways, it resembles the i-f amplifier of an a-m receiver. Capacitor *C*-1 and resistor *R*-2 make up a plate and screen decoupling filter to prevent coupling of signal among the various stages. These should be carefully noted, since they frequently present service problems.

Note also the omission of the cathode bypass capacitor across cathode resistor *R*-1. This omission introduces degeneration into the i-f amplifier and tends to prevent the tube from oscillating. The high gain of an i-f amplifier more than compensates for this degenerative loss.

The trend is to use permeability tuning for the i-f transformers. In this way, it is possible to use a higher inductance and thereby to cover a wider tuning range than that of a small variable capacitor. Use of a larger inductance also gives greater gain than use of a larger capacitor.

The common intermediate frequency for all f-m receivers is standardized at 10.7 mc. The use of this frequency tends to eliminate image-frequency interference. Image-frequency interference is introduced by a station operating at twice the intermediate frequency

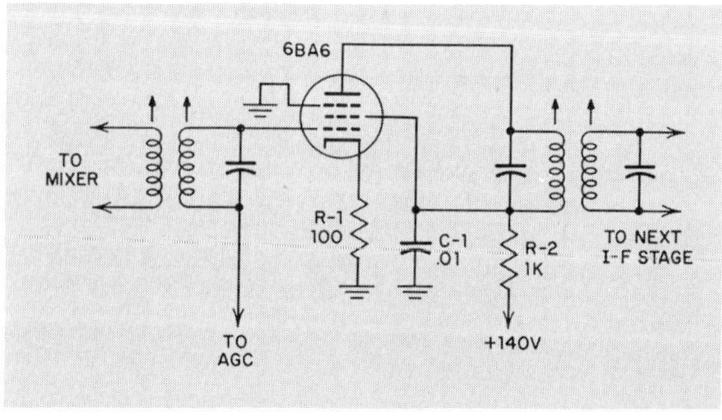

FIG. 22-10. A typical f-m i-f amplifier.

of the receiver away from that of a desired station that has been tuned in. Thus, if you take twice the intermediate frequency (21.4 mc) and add it to the lowest possible f-m frequency (88 mc), you get 109.4 mc, which is outside the f-m band where no station is found.

Limiter. Here is the first point at which we seem to depart from the standard a-m superheterodyne receiver in function. But this is not so. The limiter stage is simply a specialized i-f amplifier. Its action is indicated in Fig. 22-11A.

Why is the limiter necessary? The f-m signal, as it travels through the receiver signal chain, is given amplitude modulations of various sorts because of air static, nearby spark discharges, tube noise, etc. To eliminate these undesired effects from the loudspeaker, the amplitude variations must be eliminated before they get to the discriminator-detector. The limiter stage performs this job.

The limiter tube is a sharp-cutoff tube with grid-leak bias, operating with reduced plate and screen voltages so that it is readily driven to saturation. It is this saturation point which removes the amplitude modulations that produce noise in the receiver.

A strong input signal is required at the input of the limiter stage. If the signal is weak, the amplitude variations ride through and produce noise, as shown in Fig. 22-11*B*. For this reason, it is desirable to have a good antenna installation, an r-f stage, and two i-f stages ahead of the limiter to get sufficient gain of the signal and eliminate the noise.

The limiter is used to develop automatic gain control voltage (agc) in addition to its limiting action. Here, use is made of the fact that grid current flows, and the grid circuit acts as a diode rectifier. A typical circuit for this limiter function is shown in Fig. 22-12. The low plate and screen potentials are obtained by using a large dropping resistor in the supply circuit. This also acts as the decoupling filter for the stage. The amount of agc voltage that is fed back depends on the strength of the signal fed to the limiter grid. A filter will be placed in the agc line to eliminate audio variations and to feed unvarying d-c voltage to the controlled tubes.

Demodulation. It is at the demodulator or second detector that we encounter the first major difference between a-m and f-m sets. In a-m receivers, amplitude variations result in

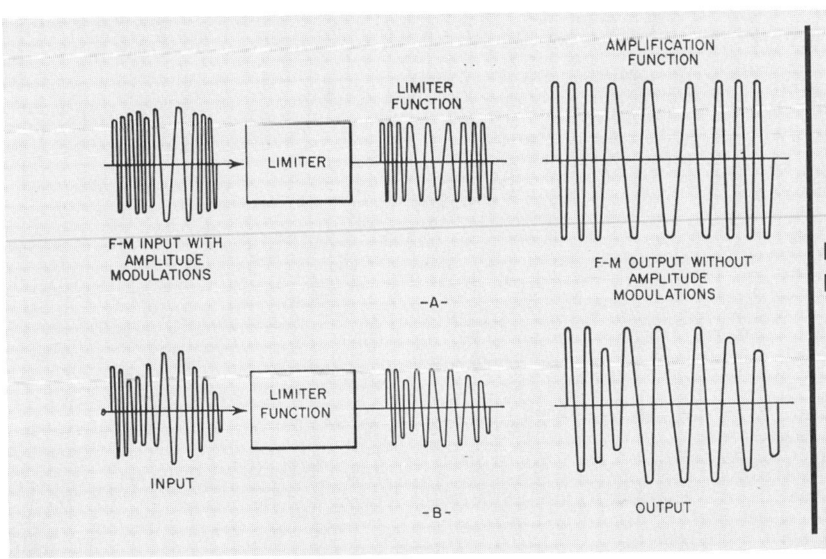

FIG. 22-11. Action of a limiter stage.

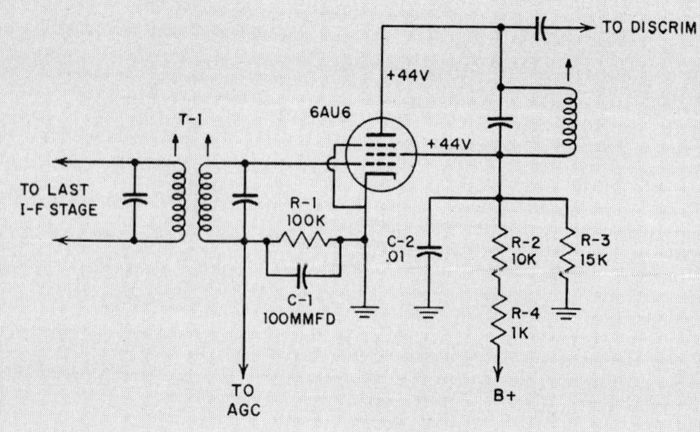

FIG. 22-12. Typical f-m limiter circuit.

voltage variations fed to the grid of the first a-f amplifier. In f-m sets, there are no amplitude variations after the limiter. It becomes necessary to change the rate of frequency variations into voltage variations fed to the first a-f grid. That function is performed by the f-m demodulator.

There are two main types of demodulators used in f-m receivers. They are the discriminator type and the ratio detector type. The latter circuit is normally used without a preceding limiter stage.

A typical discriminator is shown in Fig. 22-13. There are other variations of this circuit in use, but they operate in basically the same manner. The plates of two diodes are connected to the secondary winding of the discriminator transformer *T*-1 and receive the frequency-varying i-f signal inductively through transformer action between the primary and secondary windings of *T*-1. The diodes also receive the same i-f signal capacitively from the limiter stage via capacitor *C*-2, which is connected to the center tap of the secondary winding of the discriminator transformer. The phase relationships of these two inputs combine to produce the usual a-f voltage variations at the a-f output. Thus, frequency variations are changed to voltage variations.

Note resistors *R* and capacitors *C* in the discriminator output circuit. They make up what is known as a deemphasis network. At the transmitter, the high-frequency notes are amplified more in order to get a better signal-to-noise ratio for those frequencies. To restore the natural relationship of all frequencies, the

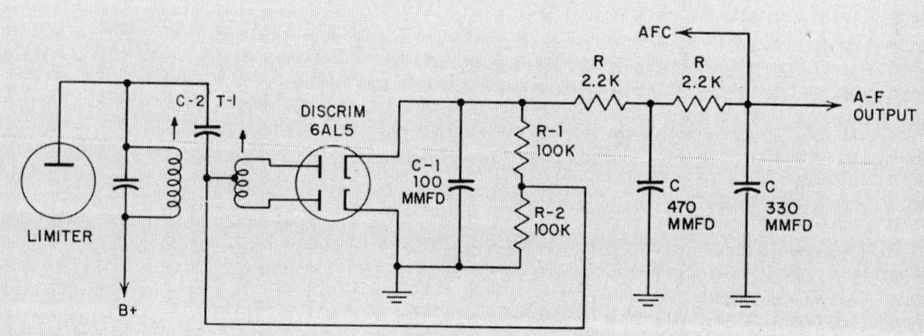

FIG. 22-13. Typical f-m discriminator circuit.

high-frequency notes are reduced or deemphasized at the receiver by the deemphasis network.

Ratio detector. A type of detector also in common use is the f-m ratio detector. Figure 22-14 shows the circuit of a typical ratio detector. Its function is the same as that of the discriminator—to change the rate of frequency variations into voltage variations fed to the a-f section. Note, however, that the detector is fed directly by an i-f amplifier, not a limiter. It is a characteristic of the ratio detector that it does its own limiting. Such a setup saves the manufacturer one stage.

Note also the agc voltage tap in the detector. As explained previously, the agc voltage was obtained from the limiter when the discriminator was employed. Since there is no limiter here, the agc voltage must be found in the detector itself.

And finally, note the deemphasis network, made up of resistor *R*-6 and capacitor *C*-5. It serves the same function as that described in the section on the discriminator.

Sometimes, even in tube f-m receivers, the ratio detector is a circuit built around two crystal diodes. Such a circuit is shown in Fig. 22-15. Its operation is similar to the circuit previously described.

Tuning guide in f-m receiver. The f-m re-

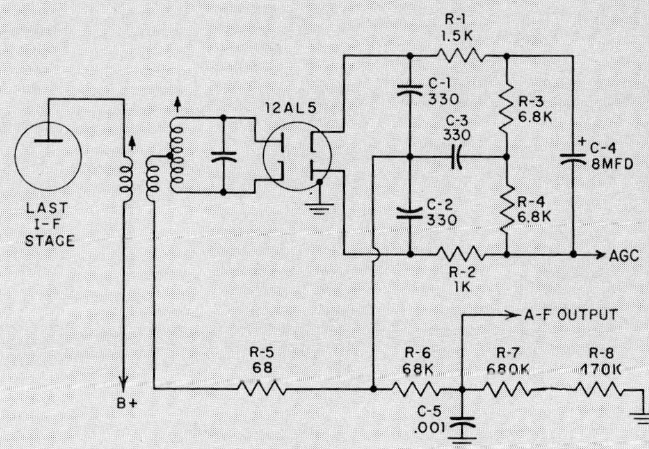

FIG. 22-14. Typical f-m ratio detector circuit.

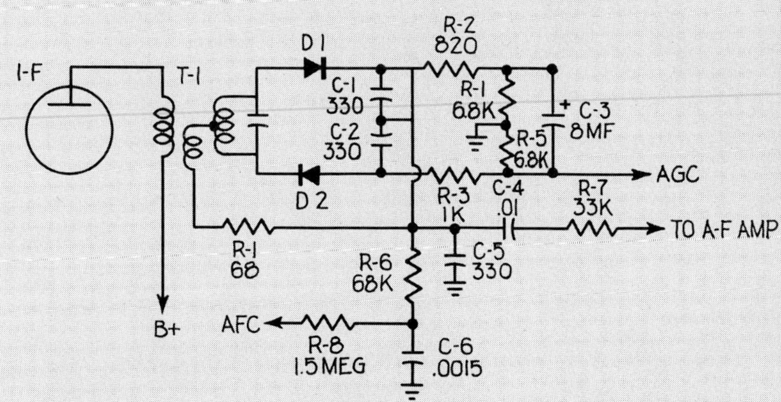

FIG. 22-15. Ratio detector using diode crystals.

ceiver gives best response when the incoming signal is exactly tuned in on the center of the carrier frequency. To aid the ear in this sharp tuning, many sets incorporate a visual tuning indicator. A common device is the 6E5 electron-ray indicator tube. This is a type of cathode-ray tube that shows a wide-angle shadow when a low voltage is applied to its grid, as shown in Fig. 22-16*A*. The deflection narrows as the applied grid voltage is increased, as shown in Fig. 22-16*B*.

The grid of the indicator tube is connected to a point in the f-m set where the signal, at varying strengths, may be tapped off. This is usually the grid of last limiter tube or at the f-m detector—the same point from which agc voltage is tapped. When the set is tuned in sharply, signal voltage is at a maximum, maximum high voltage is fed to the grid of the indicator tube, and the shadow is narrowest.

In Fig. 22-16*C*, we see the indicator tube connected in the grid circuit of a limiter tube. The negative voltage across the 350K resistor is proportional to the strength of the station signal. Resistors *R*-4 and *R*-5 and capacitor *C*-2 filter the audio components before feeding the voltage to the indicator tube grid.

In Fig. 22-16*D*, a 6E5 indicator tube is connected across the output of a ratio detector. Potentiometer *R*-12 is adjusted until the eye just closes when the receiver is in a no-station setting. Detuning of any desired station causes overlap of the bright area or increasing of the angle of the shadow. Sharp tuning just closes the eye or makes the shadow as small as possible. Again, *R*-7 and *C*-5 filter the input to the grid of the indicator tube.

There are other types of tuning indicators, using other types of tubes and shadows. However, the principles are the same. In some

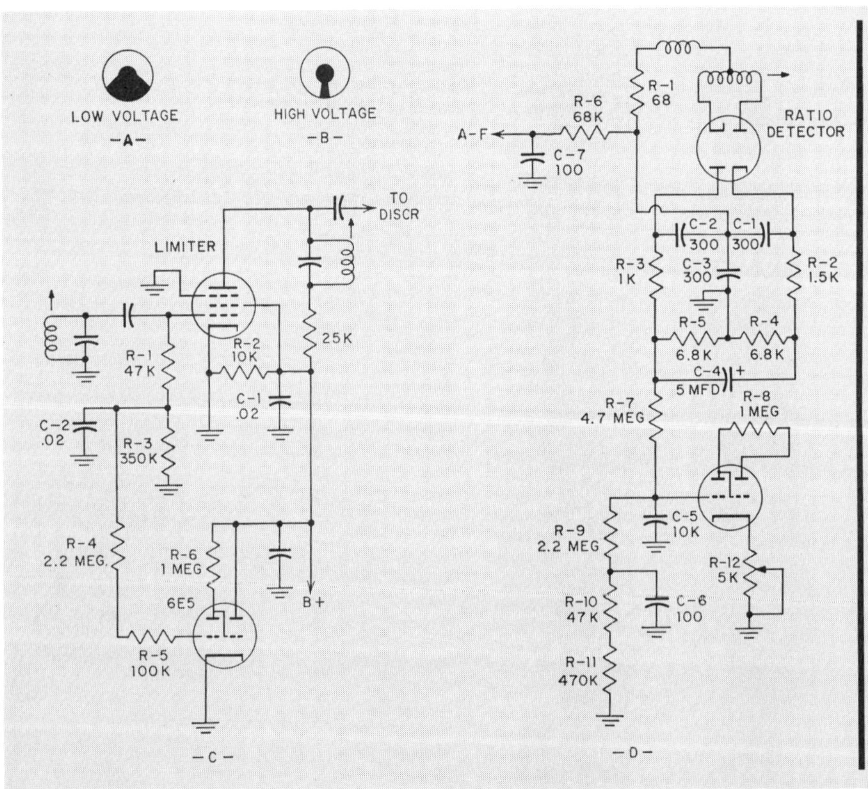

FIG. 22-16. Tuning guides in an f-m tuner.

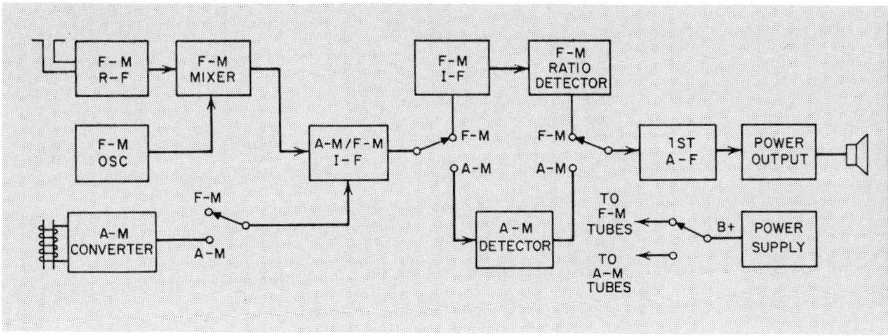

FIG. 22-17. Block diagram of a typical a-m/f-m receiver. The four switch sections are ganged together.

cases, direct-reading meters replace the tube for sharp tuning indication.

A-f section. The audio system of an f-m receiver is the same as that of an a-m set. In some cases, it is designed more carefully to handle the f-m system. Especially in high-fidelity f-m sets is this true, with push-pull output feeding treble and bass loudspeakers. Tone controls for treble and bass control are present also.

Combined a-m/f-m receiver. Higher-priced equipment utilizes separate a-m and f-m tuners, a separate audio amplifier, and a separate power supply. The units are interconnected and switching chooses the various modes of operation: a-m, f-m, f-m with afc, and phono. Smaller equipment combines a-m and f-m in a single unit. In order to keep down cost, some stages are made to serve more than one function. Here, too, switching controls the mode of operation.

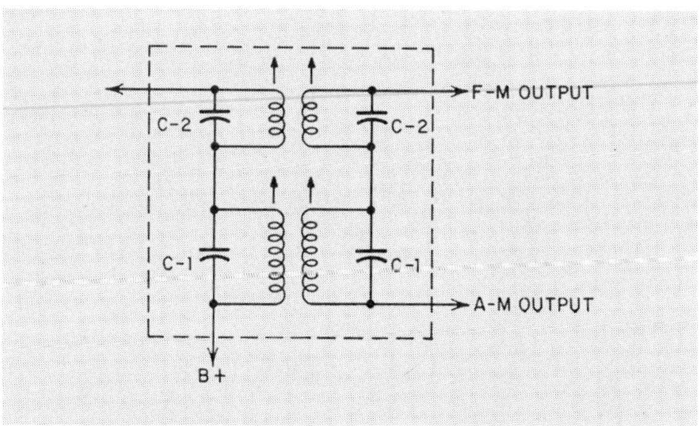

FIG. 22-18. An a-m/f-m i-f transformer assembly.

The block diagram of such a setup is shown in Fig. 22-17. There are many other variations. Sometimes a separate a-m converter tube is used and the f-m section uses separate oscillator and mixer tubes. Sometimes separate i-f tubes are used for the a-m and f-m functions. In other sets, one tube serves as an a-m converter and an f-m oscillator with a separate f-m mixer tube. The block diagram has been chosen because it shows some interesting points.

Note that separate antennas are used. The f-m antenna is either a dipole or line-cord antenna, while the a-m antenna is usually the standard ferrite loop. The functions are selected by a ganged set of function switches which select the antenna input, the required oscillator, and the a-m or f-m demodulator. Note also that the first i-f stage is common to both a-m and f-m operation. This point requires a little further investigation.

The i-f transformers for both functions are often connected in series and placed within one shield can, as shown in Fig. 22-18. Because the a-m coils operate at 455 kc and the f-m coils at 10.7 mc, the former coils have many more turns and higher inductance than the latter. As a result, when operating on the a-m function, the a-m signal at 455 kc meets very little impedance from the small f-m coil. The signal behaves as though the f-m coil is shorted out. When operating on the f-m function, capacitor C-1 across the a-m coils offers almost no impedance to the f-m signal at f-m frequencies, and serves to short out the a-m coils. Thus, the i-f transformers do not affect each other and may be connected as a single unit.

Figure 22-19 is the schematic diagram of an a-m/f-m receiver. Many of the circuits previously described are illustrated in the diagram. Note the grounded-grid f-m r-f amplifier. The f-m converter is an autodyne circuit. Note that tube *V*-3 and its circuitry double as the a-m i-f stage and as the second f-m i-f stage. Because of such double function, the secondary windings of i-f transformers *T*-3 and *T*-5 are tied together. The f-m demodulator uses a twin diode 12AL5 tube. Note the grounded screen which shields the two diodes from each other. Resistor *R*-13 in the heater line is a thermistor that helps to maintain a constant voltage across the heaters. As the filaments heat up and increase their resistance, the thermistor heats up and reduces its resistance, thereby keeping the overall resistance of the heater string constant. Finally, note the code symbols that identify the various wires: the various levels of *B*+ voltage, agc, afc, and heater leads.

The same a-m/f-m combination can be achieved with transistors. The schematic drawing of such a circuit is shown in Fig. 22-20. There are several things to be observed in this receiver. Note that the f-m r-f and the f-m converter transistors are operated in the common base configuration. They simulate a triode tube with a grounded grid. Also notice how some of the transistors function in both the a-m and f-m mode of operation: a-m converter and first f-m i-f amplifier, first a-m i-f amplifier and second f-m i-f amplifier, and second a-m i-f amplifier and third f-m i-f amplifier. Of course, the audio section is common to both modes.

Switch *S*-1 selects the a-m or f-m mode of operation. It switches in the circuits required for either mode of operation.

Note the diode crystal *E*-1 that taps the output of the second f-m i-f amplifier and develops f-m agc voltage to control the r-f amplifier. The a-m agc voltage is developed in the usual manner and is fed to the base of the first a-m i-f amplifier to control its gain.

Component *E*-2 is a crystal diode used as an overload control in the a-m signal chain. The diode is connected to the resonant load of the a-m converter and acts like a variable r-f load on that resonant circuit. Normally, it is reverse biased so that it will not conduct when a weak signal is received; the loading on the resonant circuit is negligible. When a very strong signal comes in, agc voltage to the controlled first a-m i-f stage increases, reducing its gain and collector current. Collector voltage rises and forward biases the crystal diode. It begins to conduct and load down the converter resonant load, lowering the gain of that circuit.

Finally, note the power supply. There are two methods of operation. There is a line power supply with a full-wave rectifier. The set may also be operated from a 9-volt battery. Switch *S*-2 selects the power supply that is to be used. Note that battery plus is grounded, making all transistor voltages negative.

Power supply. The power supply for an a-m/f-m receiver has been fully described in Chap. 10 and need not be repeated here. The basic principles of operation and the common defects are fairly similar for all receivers. In transistorized sets, like the one shown in Fig. 22-20, the battery and its associated decoupling circuits are similar to those of transistorized receivers previously described. If a line power supply is used, a special voltage dropping transformer is used to develop the required transistor bias voltages. Otherwise, the circuit is fairly standard.

Servicing a-m/f-m receivers. The servicing of a combined a-m/f-m receiver brings in more complications than that of a simple f-m receiver. This is why the combined receiver will be taken up first. Then, this servicing procedure may be readily amplified and adapted to fit the receiver made only for f-m or for an f-m tuner.

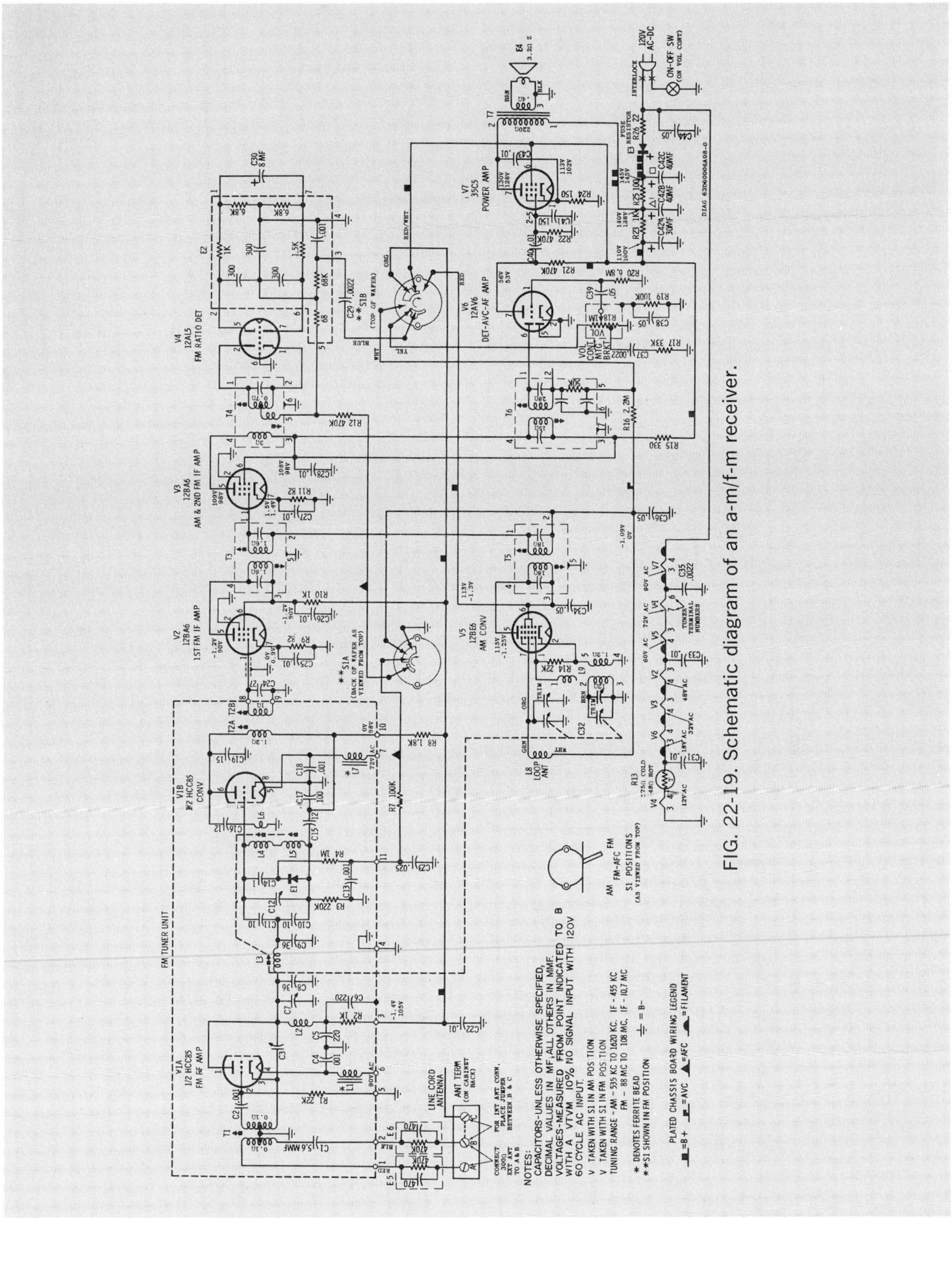

FIG. 22-19. Schematic diagram of an a-m/f-m receiver.

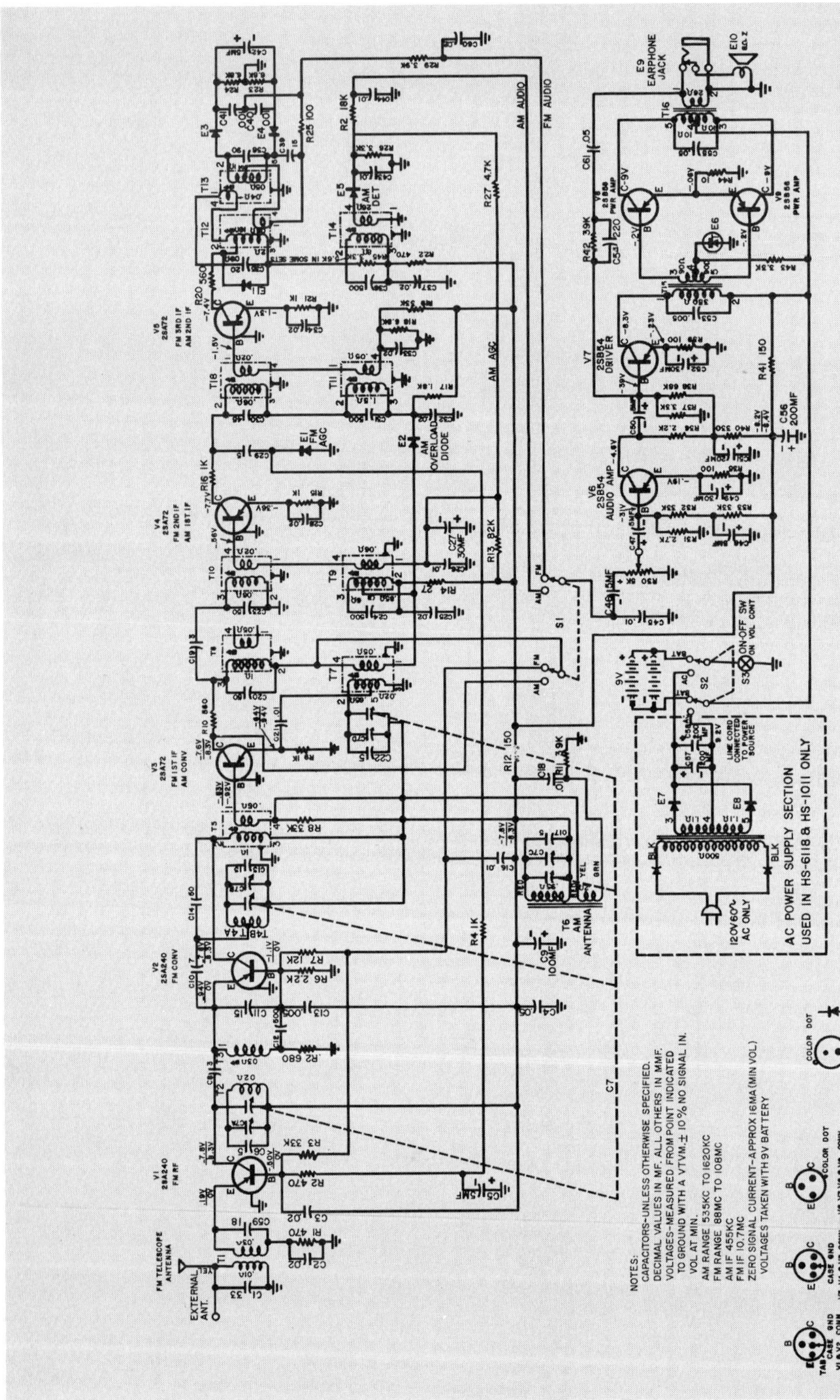

FIG. 22-20. A transistorized a-m/f-m receiver.

The block diagram of Fig. 22-17 has been chosen as being representative of most a-m/f-m receivers. Keep this diagram in mind as we present servicing procedures. Although many other possible circuit arrangements may be found, this one is basic enough to give an understanding of the others.

The servicing of a-m/f-m receivers need not be a frightening task. It is in many respects so similar to the servicing of the familiar a-m set that an over-all approach will well serve the purpose.

The mode of operation of an a-m/f-m receiver can be most useful in localizing where a defect may exist. Assume that we have a set designed with an f-m tuner, an a-m tuner, and a phono input. All feed into an audio amplifier. Switch to the phono mode of operation. If the output is normal, we may conclude that the audio amplifier and the power supply are good.

Now switch to the a-m mode of operation. If we do not have normal reception, we can assume that the defect lies in the a-m tuner section. If reception is normal, then the a-m tuner is good. Now switch to the f-m mode of operation. If reception is not normal, we may assume that the defect lies in the f-m tuner. If reception on none of the modes of operation are normal, then we must look for the defect in those circuits that are common to all three. These are the audio amplifier, the power supply, and the mode switch.

We have previously taken up the servicing of the a-m tuner, the power supply, phonograph operation, and switches. Therefore, we shall now turn to the servicing of the f-m section of the receiver.

Quick f-m stage check. If the a-m section is found to be good, then the f-m section should be checked. The block diagrams for the two basic f-m sections are shown in Fig. 22-21. The one in Fig. 22-21*A* utilizes the discriminator. The one in Fig. 22-21*B* utilizes the ratio detector. The same check works for both types. The overall procedure for checking the f-m

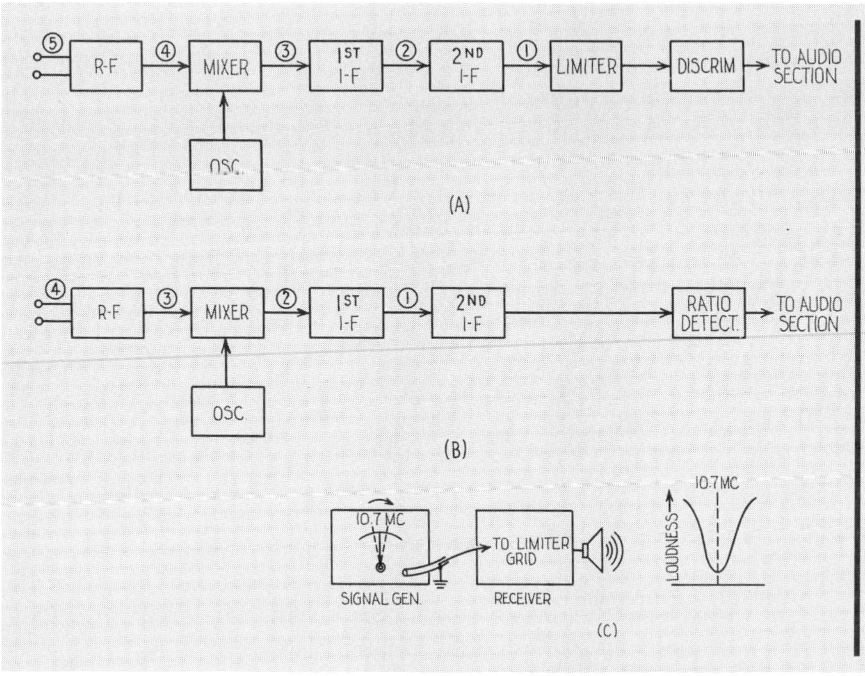

FIG. 22-21. Quick check for an f-m tuner.

section is performed with a standard a-m signal generator whose frequency range goes up to about 15 mc. Set the receiver for f-m operation.

Feed a modulated signal at 10.7 mc (the intermediate frequency for f-m) to the last i-f grid. Remember that the limiter is considered to be the last i-f stage. Adjust for full output, and wobble the frequency control to each side of 10.7 mc. If the demodulators (discriminator or ratio detector) are operating properly, the modulation note will be heard in a specific way. A little off frequency, the modulation note will be fairly loud. As you come onto frequency, the modulation note dips sharply in loudness. As you move off frequency on the other side, the modulation note rises back to its previous loudness. The wobble procedure should be performed slowly to hear the sharp dip in loudness. The procedure is illustrated in Fig. 22-21*C*. This is the quick check for normal operation of the f-m demodulator.

Now feed the same signal to the preceding i-f tube grid. If that i-f stage is not functioning, the note will not be heard. If the note is heard, feed the same signal to the first i-f tube grid. Failure to hear the note indicates that the second i-f stage is not functioning. To simplify this quick check from the second i-f tube grid on, the hot clip lead from the signal generator may be latched onto an insulated section of the grid wire or to the insulated body of the grid resistor. This loose coupling may be even more desirable than direct contact if the signal-generator test signal is strong. It is also advisable to reduce the test signal by means of the attenuator to as low a level as possible. If the note from the first i-f grid is heard, feed the same signal to the mixer grid. Failure to hear the note indicates a defect in the first i-f stage or in the mixer. If the note is heard, those two stages are working.

The next check is that of the converter and r-f stages. Use is made of an a-m signal generator like the Precision E-200-C, described in Chap. 6. Tune the generator so that it delivers a modulated output at 100 mc. Feed this signal to the mixer grid. The modulation note will be heard if the oscillator is good, when the receiver is tuned to 100 mc. Next, feed the same signal to the antenna terminal, and again tune the receiver to 100 mc. Failure to hear the modulation note indicates trouble in the antenna input or r-f stage.

If your signal generator is an older unit, it may not be able to deliver a signal as high as 100 mc. But it is certain to be able to deliver a signal of 9 mc. This signal may be used to perform the same tests as have just been described, if we take advantage of the harmonic output of the generator. The tenth and twelfth harmonics of the 9-mc signal will be at 90 mc and at 108 mc. Then, by tuning the receiver to those two high frequencies, we should be able to hear the harmonics of the 9-mc test signal.

F-m servicing procedure. The quick checks just given located the defective stage in the f-m section. The next job is to localize the defective part. This technique is essentially the same as for an a-m radio. Check tubes first, then make voltage and resistance checks.

The basic a-m/f-m tuner circuit shown in Fig. 22-22 will be used to illustrate the test procedure for locating the defective part. This tuner is the heart of the a-m/f-m receiver. The tuner has a phonograph input and is designed to be used with a conventional hi-fi two-channel audio amplifier.

Let us examine this basic circuit in detail for servicing procedures. The power supply has already been covered and will not be repeated here. For the remainder, emphasis will be on the f-m section. The a-m tuner will not be covered, since previous chapters described the checking of those stages. Where the a-m tuner ties in with the f-m tuner, both will be analyzed.

A-m/f-m receiver audio section. The audio section of the a-m/f-m receiver presents no new servicing problems. The procedures you learned previously also apply here.

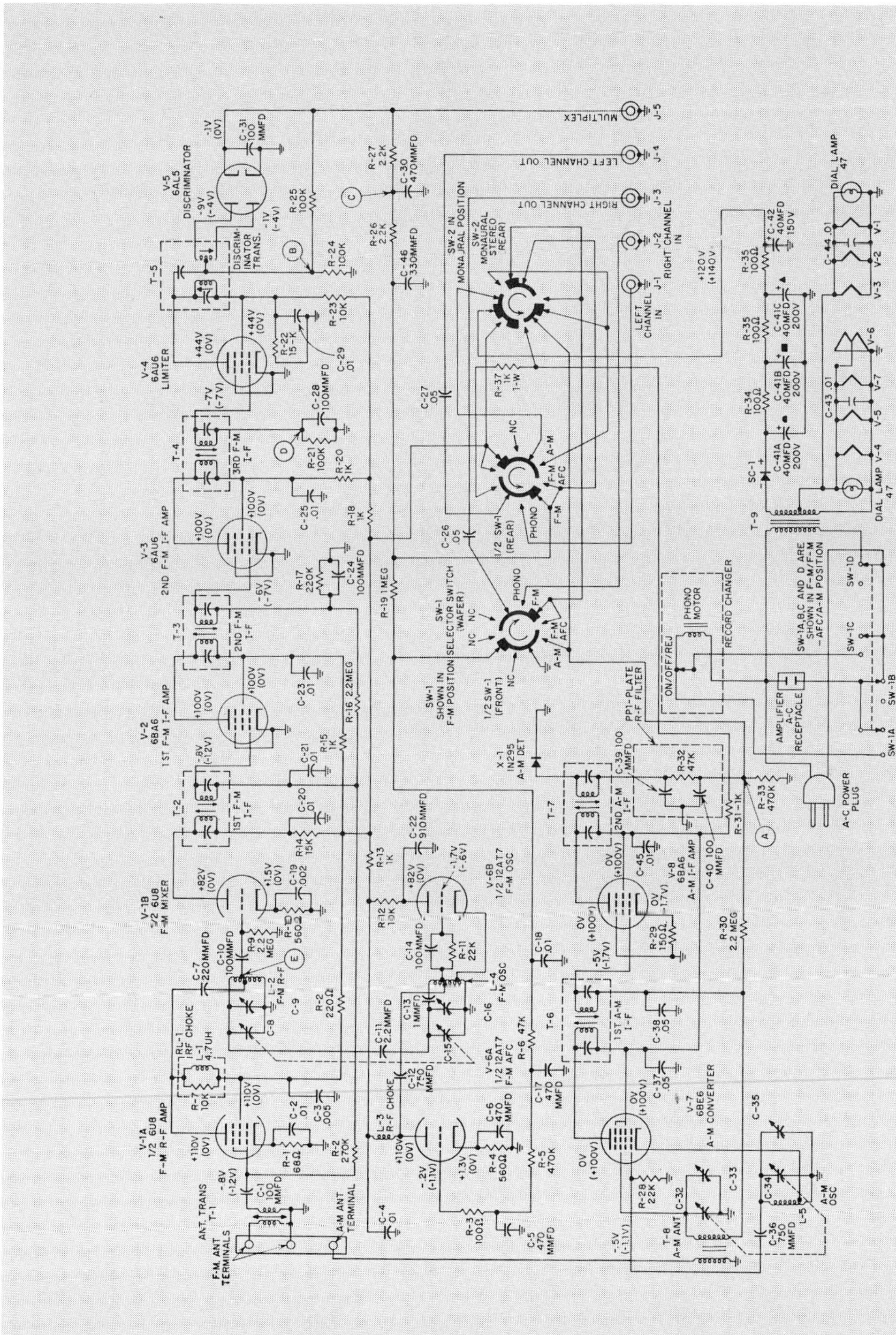

FIG. 22-22. Complete circuit of a basic a-m/f-m tuner. Voltage readings in parentheses are for a-m position of selector switch; other voltages are for f-m position.

As with phonographs, there may be many different designs in the audio section. The a-m/f-m tuner of Fig. 22-22 is designed to feed into a two-channel audio amplifier with a right and left channel. Input jacks *J*-1 and *J*-2 are for stereo record input. The output to the audio amplifier channels is through jacks *J*-3 and *J*-4.

The type of output fed to the audio section is controlled by selector switches *SW*-1 and *SW*-2. In the F-M position, the grid of the afc tube is grounded out, making it inoperative. In the F-M/AFC position, the afc tube is functioning with the f-m oscillator.

F-m demodulator. The detector of our basic set is of the discriminator type. It is redrawn in Fig. 22-23A. The tube used is a 6AL5 twin diode. Note that it is permeability-tuned. Capacitor *C* which couples signal from the limiter stage to the discriminator tube is usually part of the discriminator-transformer assembly, shown enclosed by dotted lines. Resistors *R*-24 and *R*-25 make up the diode loads across which the audio output voltage appears. Capacitors *C*-46 and *C*-30 and resistors *R*-26 and *R*-27 make up the deemphasis circuit. Troubles in the discriminator are due to defective tubes and misalignment. Tubes are checked in the usual manner by substitution of new ones for those suspected as poor.

Misalignment causes poor tone quality and a weak signal. Alignment may be checked by the following procedure. Connect a d-c voltmeter probe to the junction of the diode load resistors (point *B* in Fig. 22-23A). Connect the common voltmeter lead to ground; then feed an unmodulated signal at 10.7 mc to any convenient i-f tube grid. Wobble the signal generator around 10.7 mc for a peak indication on the meter; then shift the voltmeter prod to point *F*. The reading should be zero. Any volt-

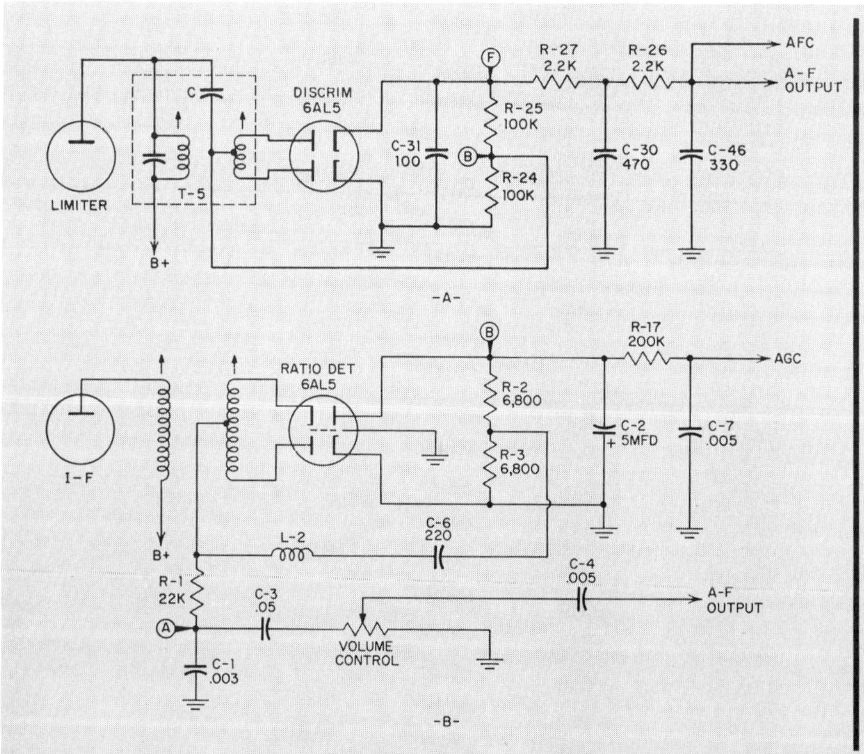

FIG. 22-23. Test points in f-m demodulators.

age at this point—plus or minus—indicates misalignment. A complete alignment procedure is given later in the chapter.

The circuit of a ratio detector is shown in Fig. 22-23*B*. The tube used is a 6AL5 twin diode with a grounded internal shield to prevent coupling between the two halves. Note that the stage is permeability-tuned. Also note that in the discriminator, the secondary of the discriminator transformer is connected to the plates of the twin diode; in the ratio detector it is connected to the plate of one diode and the cathode of the other diode.

Capacitor *C*-1 and resistor *R*-1 make up the deemphasis network. Resistors *R*-2 and *R*-3 make up the ratio detector load. Capacitor *C*-2 is known as the stabilizing capacitor and serves to eliminate amplitude variations from the output.

An overall check of the ratio detector may be made with an a-m signal generator and a vacuum-tube voltmeter or a 20,000-ohms-per-volt voltmeter. The check points are shown in Fig. 22-23*B*. The meter is connected from the audio input (point *A*) to the chassis. Now feed an unmodulated r-f signal at 10.7 mc into the second i-f tube signal grid. The meter should read zero unless alignment is bad or a defect exists in the detector circuit. Later in the chapter, we shall see how to align the set. Now connect the meter from the plate end of resistor *R*-2 (point *B*) to the chassis. The voltage at this point depends on the strength of the input signal. The size of this voltage is unimportant for the signal check, but its absence indicates trouble in the stage.

The tubes are the most likely source of trouble in the ratio-detector circuit. Loss of emission will cause a weak signal and distortion. The condition is cleared when the tube is replaced with a new one. If replacing tubes does not help the situation, then a resistance check of individual components is necessary. In this regard, remember that capacitor *C*-2 is an electrolytic capacitor and may in time lose capacitance. When that happens, the circuit loses the ability to remove amplitude variations from the signal, and the set becomes noisy like any a-m set. A short in capacitor *C*-2 would make the set completely inoperative. This capacitor is checked in the usual manner; one end should be disconnected when checking, to separate it from the shunt resistive load.

Multiplex operation. Refer back to our basic circuit shown in Fig. 22-22. Note the output jack *J*-5 marked MULTIPLEX, connected to the discriminator. This jack is for a special adapter used in some communities. Its development arose from the desire of many listeners for high-fidelity music listening.

An early solution was to have two stations, one on amplitude modulation and the other on frequency modulation, broadcast the same music program simultaneously. A listener with an a-m receiver and an f-m receiver at different parts of the room would get a stereo effect. In other communities, two independent f-m stations would agree to transmit simultaneously the same music program. The listener with two f-m sets would now also get a stereo effect. The currently approved multiplex system enables a listener with a two-channel audio system and a special multiplex adapter in his f-m receiver to get the stereo effect from the transmission of a single f-m station.

At present, such multiplex programs are transmitted by only a relatively few f-m stations. To avoid obsolescence of f-m receivers, many manufacturers include the multiplex jack for use with an external adapter. Some manufacturers are including the adapters in the f-m receiver itself. The multiplex f-m signal, however, is such that the listener will hear a monaural standard f-m signal without the multiplex unit.

F-m multiplex signals. A quick look at how the multiplex system works would be helpful. Examine Fig. 22-24 which shows the block diagram of the transmitting f-m station. Two microphones, *L* and *R*, are placed for stereo

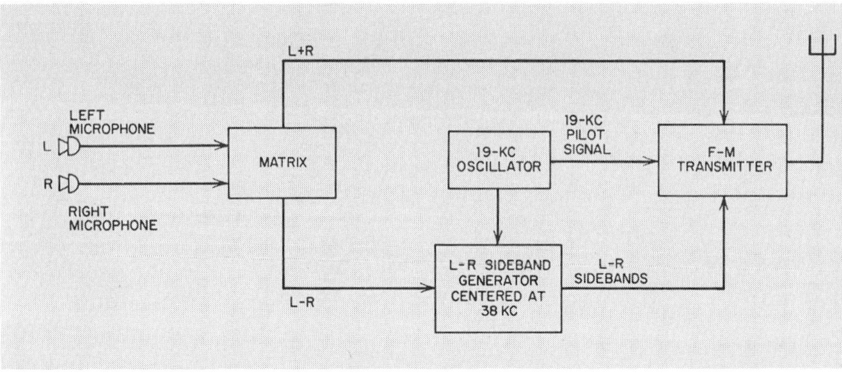

FIG. 22-24. An f-m multiplex transmitter.

pickup. Their outputs are fed into a special matrix mixing circuit from which there are two outputs. One is a blending of the two audio signals, producing an $L + R$ output. This signal is fed directly to the f-m transmitter. The other matrix output, obtained by means of phase shifting, is the difference between the two microphone outputs and is called the $L - R$ signal. Beyond the matrix circuit is a 19-kc oscillator. This oscillator feeds its signal, known as the 19-kc pilot signal, to the f-m transmitter. The oscillator also feeds its signal to a frequency-doubling circuit to produce an output at 38 kc, known as the subcarrier. This 38-kc subcarrier then receives the $L - R$ signal and is amplitude-modulated by it, producing sidebands above and below the 38-kc subcarrier. The subcarrier is then filtered out and the remaining $L - R$ sidebands are fed to the f-m transmitter. The f-m carrier is thus frequency-modulated by the $L + R$ signal, the $L - R$ sidebands, and the 19-kc pilot signal.

Let us see then how the 75 kc of frequency allotted to an f-m station is utilized. The frequency spectrum of Fig. 22-25 will illustrate the information just given. The $L + R$ sidebands occupy 15 kc of the channel, at the lower end. The upper and lower $L - R$ sidebands occupy 30 kc of the channel, centered on the suppressed 38-kc subcarrier. This leaves the 19-kc pilot signal neatly placed away from the $L + R$ and $L - R$ information to avoid interference. The upper portion of the frequency range is often used for private f-m storecasting (SCA) programs. This latter program is also placed so as to avoid interference with the $L - R$ information; it is centered on 67 kc.

Although there are several ways by which L and R signals are mixed at the transmitter, the final signal is the one just described. It should be remembered that this signal makes it possible to receive a station on a standard monaural f-m set or stereophonically with a multiplexing unit.

How the multiplex decoder works. The multiplex decoder in an f-m receiver is connected to the f-m tuner at the output of the demodulator, before the deemphasis circuit. Its function is to restore the left and right

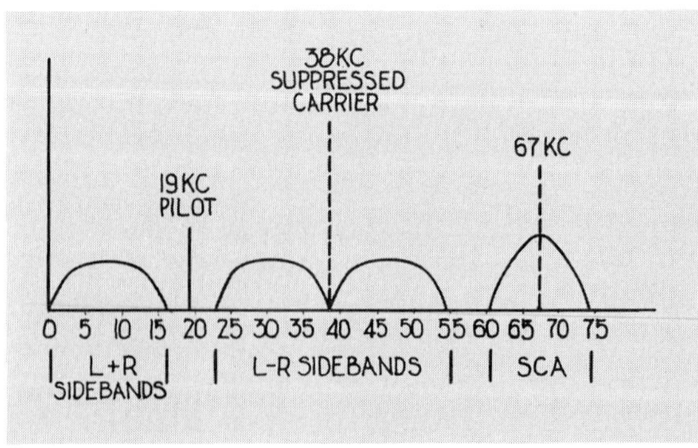

FIG. 22-25. F-m frequency spectrum for stereo transmission.

audio signals produced at the transmitting station. It must give good balance between the levels of the left and right audio signals; it must provide good separation between the two signals; and it must do the two previous jobs without distorting the signals.

There are many variations of the multiplex decoder circuits in use today. However, they boil down to three basic methods of performing the job: the matrix type, the switching or time-division type, and the envelope detection type. Sometimes, there is a combination of types. In all types of multiplex decoders, the 19-kc pilot signal must regenerate the 38-kc subcarrier in correct phase with the subcarrier that was suppressed at the transmitting station. The 19-kc pilot signal synchronizes the restored carrier. The further handling of the composite stereo signal is then handled differently in the different types of decoders.

In Fig. 22-26, we have the block diagram of a matrix type decoder, the earliest type. The f-m signal passes through the f-m tuner to the detector output. After passing through an amplifier, the signal is passed through a low-pass filter with a cutoff between 15 and 19 kc. As a result, its output is only the $L+R$ signal.

The remaining $L-R$ sideband signal is passed through a 23- to 53-kc bandpass filter. This filter permits audio signals originally up to 15 kc ($38-15=23$ and $38+15=53$) to pass. Meanwhile, the 19-kc signal, which has also been present, is fed to a frequency doubler which produces a 38-kc output. This latter output is mixed with the $L-R$ sideband output, producing an $L-R$ amplitude-modulated subcarrier of 38 kc. These are then fed to two detectors, from which emerge a positive $L-R$ signal and a negative $L-R$ signal. The three signals ($L+R$, positive $L-R$, and negative $L-R$ signals) are then fed to a matrix mixing circuit. The output from the matrix is the original L signal and the original R signal, available for feeding separately to the two channels of the audio amplifier.

If the f-m receiver has no decoder, the 19-kc pilot signal and the $L-R$ sidebands in the transmitted f-m signal are filtered out by the regular f-m circuits because they are way below the lowest f-m station frequency, leaving only the $L+R$ signal for normal monaural reproduction.

An actual matrix decoder is shown in Fig. 22-27*B*. Tube *V*-1*A* amplifies the composite stereo signal; tube *V*-1*B* is an impedance matching cathode follower. Tube *V*-2*A* amplifies the 19-kc pilot signal; tube *V*-2*B* is the synchronized oscillator with a 38-kc output. Tubes *V*-3*A* and *V*-3*B* are cathode followers

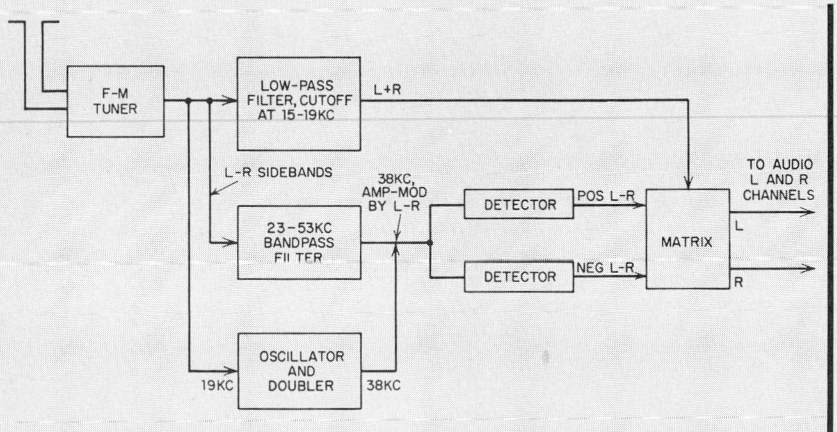

FIG. 22-26. One system of multiplex f-m reception.

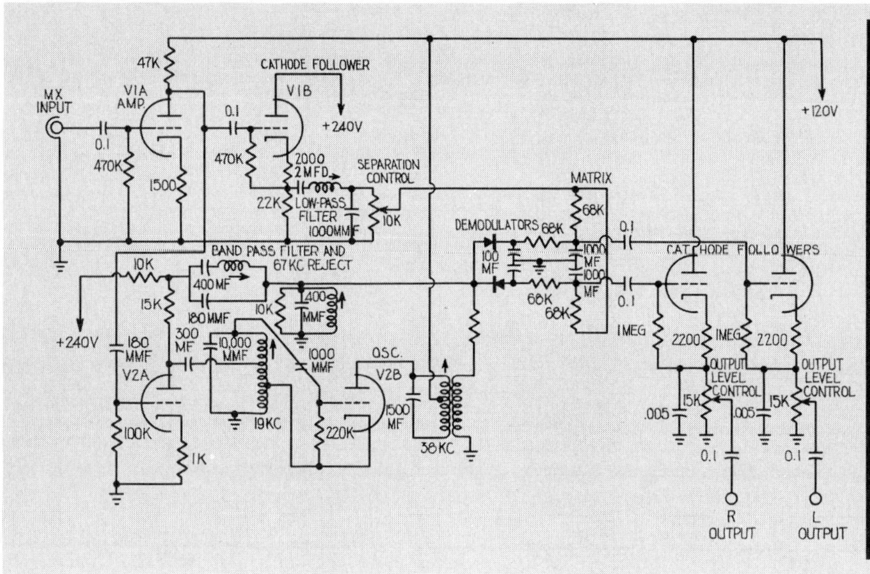

FIG. 22-27. A stereo matrix decoder.

that provide a low output impedance to prevent signal loss and hum pickup in the cables that connect to the stereo amplifiers. Note the separation control. Its purpose is to control the $L + R$ signal so that the ratio of its amplitude to that of the $L - R$ sidebands is the same as it was at the transmitter. This condition is necessary to get back the original L and R signals with proper separation in the stereo audio amplifier. Also note the individual output level controls for the R and L output.

The block diagram of a switching multiplex decoder is shown in Fig. 22-28*A*. As with the matrix type, the composite stereo signal is first amplified. Also, a 19-kc amplifier, oscillator and 38-kc doubler are used. But the 38-kc signal is used, not to reinsert the suppressed carrier, but to synchronize or switch the detectors so that they pass signal at the right instant. The bandpass filter which is difficult to design is not used.

It is observed that the switching diodes are fed both the composite stereo signal and the synchronized 38-kc signal. The latter signal will place a positive voltage on the L diode when the L portion of the composite signal is fed to the two diodes, causing it to detect the L signal. The 38-kc signal will also place a negative voltage on the next alternation on the R diode when the R portion of the composite signal is fed to the diodes. This will cause the R diode to detect the R signal. Thus each diode in turn will conduct only for an L or an R signal. The reject filter removes any 38-kc signal present in the output, leaving only L and R signals which are then fed to the stereo audio amplifier. No matrixing is necessary.

The actual schematic diagram of a switching multiplex decoder is shown in Fig. 22-28*B*. Tube *V*-1*A* is the wideband amplifier for the composite stereo signal. Its output feeds into a 67-kc SCA trap to remove its interference. Tube *V*-1*B* is an amplifier for the 19-kc pilot signal. Tube *V*-1*C* is a locked oscillator with a tuned 38-kc output. Low-pass filters in the output of each diode remove any high-frequency signals that are present.

There are several variations of the switching multiplex decoder. Figure 22-29*A* shows the diagram of one variation that uses a beam deflection tube, like the 6AR8, which does the switching instead of a pair of diodes. The composite stereo signal is applied to the

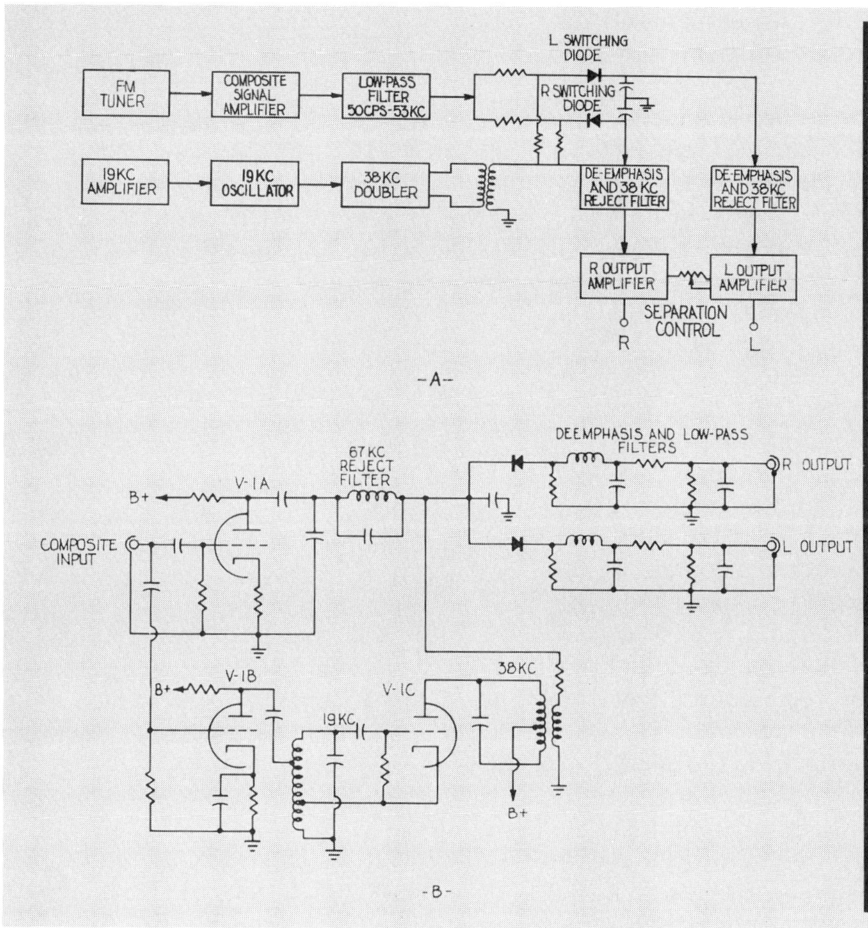

FIG. 22-28. The switching-type stereo decoder.

signal grid. The 38-kc switching voltage is applied to the two deflection plates. The polarity of these latter two plates drives the *R* signal to one of the tube anodes and the *L* signal to the other anode.

Another variation is shown in Fig. 22-29*B*. Here, four diodes are shown in a bridge circuit for switching. The composite signal is fed to the tap between the two 33K-ohms resistors, while the 38-kc switching voltage is fed to the taps between the pairs of 10K-ohms resistors.

The third type of multiplex decoder employs envelope detection. When the 38-kc subcarrier is properly added to the composite stereo signal, it alters the latter signal so that the upper edge of the envelope of the composite signal is the *L* waveform and the lower edge is the *R* waveform. Then the top and bottom of the envelope are detected separately, using no filters.

There are many variations and combinations of the decoders just described. It would be fruitless to analyze all of them. The basic principles have been presented so that understanding of most of the circuits is possible. The following section will present common troubles in f-m stereo receivers. Later a general description for alignment of the decoder will be described.

Common troubles in the multiplex decoder. The problem of developing a procedure for multiplex decoders is a complex one. First, there are a tremendous variety using different systems, and second, there are relatively few stations that broadcast f-m stereo programs. The number of calls that a service technician would have would not warrant the large expenditure of money for specialized test equipment at this time. Therefore, a generalized analysis will be given as an approach to current servicing.

Probably the greatest complaint is that of improper separation of *R* and *L* signals in the receiver. The service technician should first check this condition by verifying that the customer has full knowledge of how to use his equipment. Be sure that all controls have been properly set and that the f-m tuner has been sharply tuned. This defect is readily corrected. Sometimes, however, the condition is caused by insufficient signal pickup. Often this is accompanied by distorted output. The remedy may be the installation of a good outdoor antenna. It should be remembered that the effective fringe area of f-m stereo is closer to the transmitting station than for f-m monaural reception. A simple method to check the need for an outdoor antenna installation is to connect a TV rabbit-ears antenna to the f-m receiver and to vary its direction. If reception improves in some positions, an outdoor antenna is indicated.

A related defect may exist even with a good outdoor antenna installation. Reflections from tall buildings or hills create multipath signals which cause regular switchovers of *R* and *L* signals from their audio channels. Orienting the direction of the antenna, or better, installing a rotatable antenna will improve the reception.

The installation of a good antenna will also improve another defect. The signal-to-noise ratio for stereo reception is higher than that for monaural reception. The outdoor antenna will therefore give less noise when receiving stereo.

Within the receiver, with a matrix type decoder, lack of proper separation may be caused by insufficient or no $L + R$ or $L - R$ signal, or

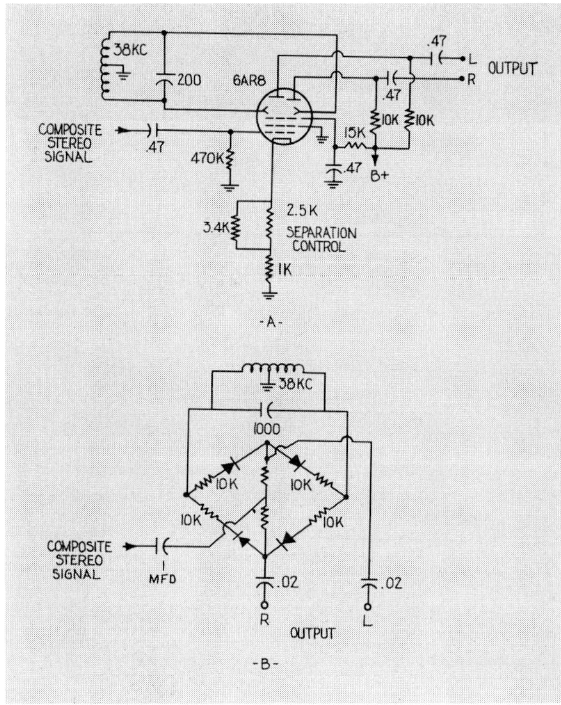

FIG. 22-29. Variations of switching-type stereo decoders.

by a defect in the matrix circuit. The respective signal paths for these signals should be examined with an ohmmeter for defective components. The matrix circuit should be similarly checked for defective components.

In less expensive receivers, separation of *R* and *L* signals becomes poor after a short time of operation. This is due to oscillator drift in the decoder. This condition is remedied by retuning slightly when separation becomes poor.

Sometimes a motorboating sound is produced in the receiver. This is caused by the local oscillator of the decoder failing to lock with the 19-kc pilot signal. The extreme of a distorted wavering sound would indicate the absence of the 19-kc pilot signal. With these conditions, the signal chain for the pilot signal should be inspected for defective components.

Distorted output in a matrix type decoder may be due to loss of the 38-kc subcarrier.

Check the oscillator and its components. In the switching type decoder, this defective condition would give only monaural reception. Distortion in both types of decoders could be caused by a defective demodulator too. Check the components carefully.

If the receiver gives monaural reception and no stereo reception, check the diode demodulators. One may be defective.

Sometimes a whistling or swishing sound accompanies reception. This is caused by SCA interference. In this situation, try readjusting the 67-kc rejection trap.

In any type of defect, it is advisable to check the filter circuits for shorts or opens. Such conditions can result in loss of any transmission or the feeding of signals that should have been filtered out.

Generally, check components after you have decided where the defect probably exists as you did in other servicing jobs. Check the tubes first. Look for overheated components, bad connections, defective diodes, open coils, etc. Many of the defects will show a remarkable improvement by simple realignment of the decoder and the f-m tuner. More will be stated about realignment later in the chapter.

F-m limiter. The limiter of our basic tuner is shown in Fig. 22-30. The limiter is simply the last i-f stage, utilizing a sharp-cutoff tube and operating at lowered plate and screen voltages. Bias for the tube is obtained by means of resistor *R*-21 in the grid-return lead. Since this voltage is developed by signal strength and varies with signal strength, it is tapped off from point *D* in the diagram to be used in some cases for agc voltage where desired. In our basic circuit, agc voltage is tapped from the previous i-f stage. The voltage at point *D* is of value in service work for showing the presence and amount of signal.

The lowered plate and screen voltages for the limiter are obtained from the voltage divider, consisting of *R*-23 and *R*-22. Capacitor *C*-29 is the bypass to ground.

Troubles in the limiter are due to defective tubes, shorts or leakage in bypass capacitor *C*-29, and misalignment. Voltage analysis would disclose a defect in *C*-29. The other defects would produce results similar to those described for the discriminator and are checked in the same manner.

F-m i-f amplifier. The f-m i-f amplifier, consisting of the first and second amplifier stages of the basic receiver, is shown in Fig. 22-31. Note the absence of the self-bias resistor in the first i-f amplifier stage. Grid bias for this stage is obtained from the agc voltage line. The second i-f stage develops grid leak bias. Resistors *R*-15 and *R*-20, acting with capacitors *C*-23 and *C*-25, make up the plate decoupling circuit. Decoupling capacitors are usually of the ceramic disk or button variety. On occasion, they grow leaky or short. When this

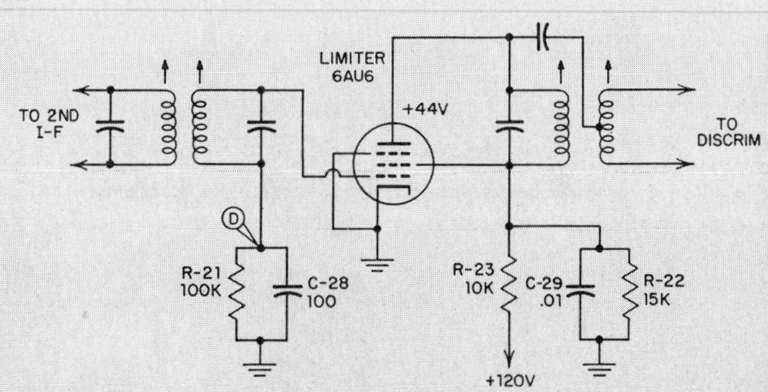

FIG. 22-30. Checking an f-m limiter stage.

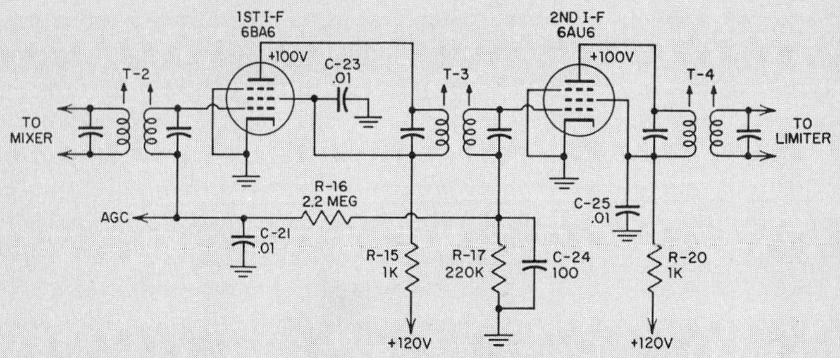

FIG. 22-31. Checking an f-m i-f amplifier.

happens, a large current flows through the decoupling resistor and tends to burn it up. When this is the case, *B*+ voltages on the plates of the i-f tubes drop very low. An ohmmeter check from plate to chassis would show a short.

When such a condition is found, be sure to check the decoupling capacitor before replacing the resistor, since a shorted capacitor will only burn out the replacement resistor. Be sure to replace the capacitor with a similar ceramic one, not with a higher-voltage paper capacitor. The long leads of the latter may introduce unwanted coupling and produce oscillation. Try to put the replacement component in the exact position of the old one.

Note the higher plate and screen voltages on the first and second i-f amplifiers, as against those for the limiter. For the former, they run about 100 volts; for the limiter, we have about 44 volts.

A-m/f-m i-f amplifier. Our basic tuner employs separate i-f tubes and i-f coils. With small, low-cost a-m/f-m receivers, manufacturers sometimes make one tube serve as both an a-m i-f amplifier and an f-m i-f amplifier.

Although tuned to 455 kc and 10.7 mc, respectively, the i-f transformers are often housed in one shield can. Such a circuit is shown in Fig. 22-32. Note the series connections of the a-m and f-m intermediate-fre-

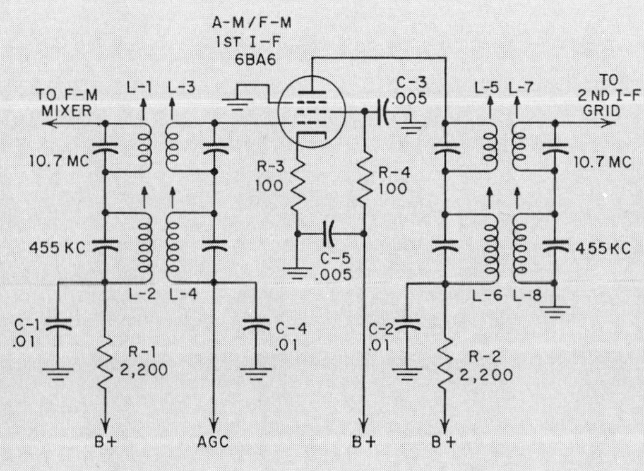

FIG. 22-32. An a-m/f-m i-f amplifier.

quency coils L-1, L-2, L-3, L-4, L-5, L-6, L-7, and L-8. The operation of this circuit was previously explained.

Resistors R-1 and R-2 and capacitors C-1 and C-2 make up plate decoupling circuits. They present the same difficulties as similar circuits described for the limiter and should be handled in the same manner. The other components present problems similar to those described for a-m radio and are handled in the same way.

F-m frequency-conversion system. Frequency conversion for f-m receivers is achieved in many ways. In Fig. 22-33, we see the system used in our basic a-m/f-m tuner. The f-m signal string employs a separate mixer and oscillator tube, the latter kept from drifting by afc. The a-m signal string employs a separate standard converter. The tuning controls of each string are separately ganged. The function switch selects the functioning string. In many a-m/f-m sets, there is only one tuning control for both a-m and f-m, with a different dial scale for each. Here, all tuners are ganged together.

In Fig. 22-34, we see another frequency conversion system, using an f-m mixer and f-m oscillator/a-m converter tube. Coils L-14 and L-15 make up the a-m oscillator coil, functioning when the function switch is in the a-m position. Coil L-16, the f-m oscillator coil, is connected to the oscillator grid when the function switch is in the f-m position. Output from the f-m oscillator is coupled through C-29 to the f-m mixer grid. The mixer feeds the i-f signal to the primary winding of the f-m i-f transformer L-10. The 6BE6 tube, when acting as an a-m converter, feeds the a-m i-f signal to L-11, the primary winding of the i-f transformer.

The f-m frequency-conversion system presents, for the most part, the usual servicing difficulties found in a-m receivers. A signal check will determine if the stages are functioning. Voltage and resistance checks will help to locate the defective components.

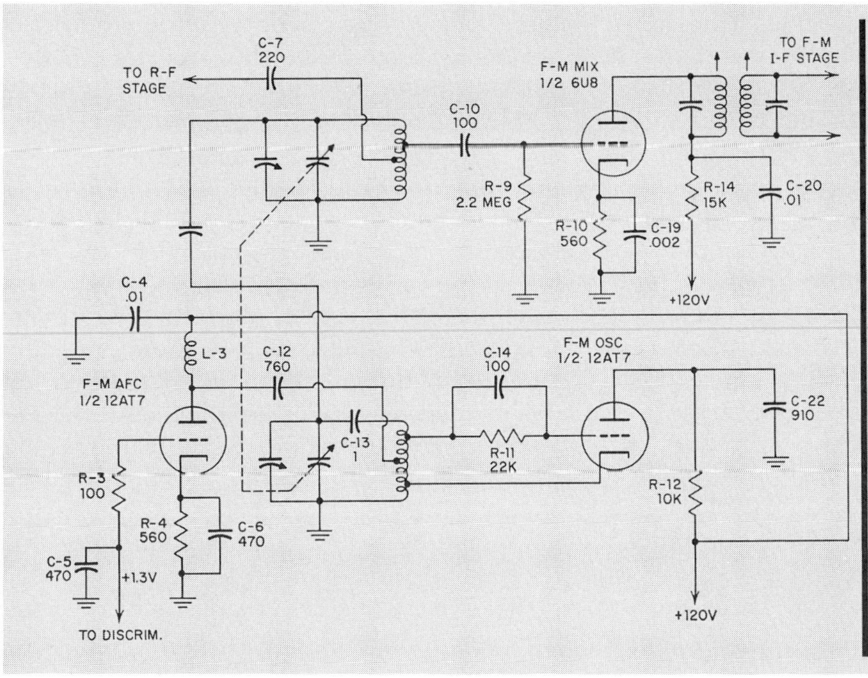

FIG. 22-33. Converter for a basic f-m tuner.

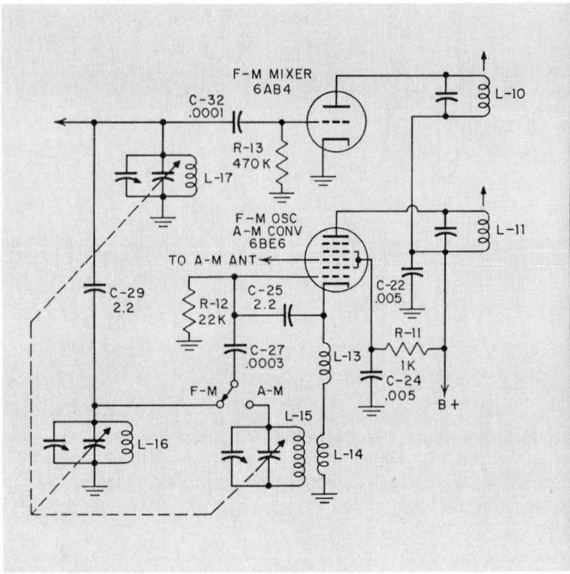

FIG. 22-34. Converter for an a-m/f-m tuner.

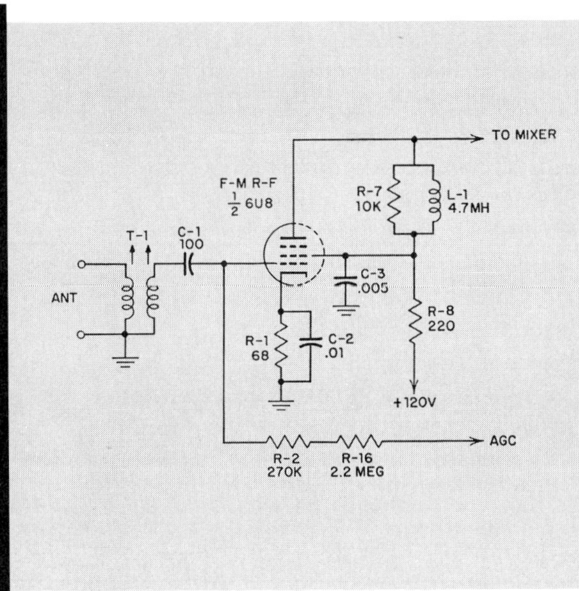

FIG. 22-35. R-f amplifier in the basic f-m tuner.

Defective conditions are often disclosed by means of the function switch. If the set works in the a-m position, but not in the f-m position, we can almost immediately turn to the f-m mixer and oscillator for locating a defective component. However, the function switch itself must first be carefully inspected. There may be a dirty or broken contact. A cleaning or replacement job may be indicated for the function switch.

F-m r-f amplifier. The f-m signal string of an a-m/f-m receiver or tuner usually employs an r-f amplifier stage. The r-f stage of our basic tuner is shown in Fig. 22-35. It employs a high-frequency pentode tube in a conventional circuit. When replacing any coil, an exact replacement should be used.

Common troubles in the r-f amplifier are the usual ones—tubes go bad or decoupling capacitors short. These are found by the standard procedure. Remember that when the set employs a grounded-grid triode, you must feed test signals to the cathode, rather than to the grid.

Servicing a front-end subchassis. When signal checks place the trouble in the r-f amplifier or converter section of a front-end subchassis type receiver, servicing calls for the same procedures already described for the analogous sections of a-m receivers. Tubes are substituted; visual examination is made for charred or broken components; and voltage and resistance measurements are made. This procedure may call for the lifting of the entire subchassis because the underside is generally inaccessible. If this becomes necessary, look for and remove the mounting screws. Then carefully lift the end opposite to the i-f transformer. The antenna dipole leads generally move with the subchassis. If necessary, the power feed leads may be removed from the feed-through capacitors. This exposes the underside for inspection and resistance checks. If voltage checking is considered necessary, the power feed leads may be lengthened and reconnected.

In service work, precautions must be taken because of the problems involved with high-frequency circuits. For example, if a component must be moved slightly to insert a test prod, the technician should be careful to return the component to its original position after the check has been made. Also, when the defective component has been isolated, it should be replaced with one as close to an exact duplicate as is possible. This is especially true for capacitors which often have specifications as to temperature coefficients, as well as to capacitance and tolerance. Before replacing the components, note their physical location and soldering points. Then replace the new components in the same position, with the same lead length and lead orientation, and solder them to the original connection points.

The precautions given for replacement of components in the front-end subchassis apply equally for receivers that do not have subchassis construction. They should be followed when servicing any f-m receiver.

F-m antenna. For the f-m tuner function, a-m/f-m sets commonly use a line-cord antenna or built-in dipole antenna for indoor installations. The outdoor dipole is found less often.

The complaint of weak operation of a set may sometimes be made for an indoor installation. If a signal check indicates that the receiver is good, try improvising a simple outdoor dipole antenna with several lengths of wire. If improvement is evident, proceed with the installation of the more permanent outdoor dipole.

Advantage may be taken of an existing television antenna installation, if present. Just connect the television antenna temporarily to the f-m tuner, to see if reception improves. Sometimes both the f-m tuner and the television set can be connected permanently to the same television antenna. Check to see that f-m operation does not affect television reception.

Checking voltages in an a-m/f-m set. The signal check is effective in locating a defective stage. After that, voltage checking in the defective stage becomes a powerful weapon in locating the specific defective components. The value of this technique is indicated in the skeleton plate-supply circuit shown in Fig. 22-36. The signal circuits have been deliberately omitted to simplify observations. Note that the a-m and f-m detectors have been omitted because they are not in the plate circuit. It is good practice to make or visualize similar skeleton diagrams for other sets to be serviced.

Normal voltages at various test points are included in the diagram. Zero voltage for any plate or screen, with normal voltages for the other tubes, indicates an open in the plate or screen supply for that tube. Zero voltage for any tube, with low voltages for the other tubes, indicates a short in the plate-supply line to that tube.

For example, assume that decoupling capacitor *C*-23 in the f-m i-f amplifier stage shorts. We can readily see that plate current would flow through the shorted capacitor rather than through the tube. Voltage at the tube plate would be zero. Resistor *R*-15, carrying more than its normal current load, would probably overheat. The set would be dead, and plate voltages at the other tubes would be low because of the overload. The overheated resistor *R*-15 would be a clue to the situation.

Do not neglect to check the function switch when hunting trouble in the plate voltage line. Note that no plate voltage is supplied to the a-m stages when the switch is in the F-M position, and vice versa. In the PHONO position, no plate voltage is fed to any of the tuner tubes. A defect in the switch can block plate voltage to any of the tubes.

The skeleton diagram just described is a useful device for service analysis. It is applicable to any circuit, an a-m only set, an f-m only receiver, or an a-m/f-m set.

Tracing unwanted oscillations in an f-m set. In the a-m mode of operation of an a-m/f-m set, oscillation may be checked by the usual

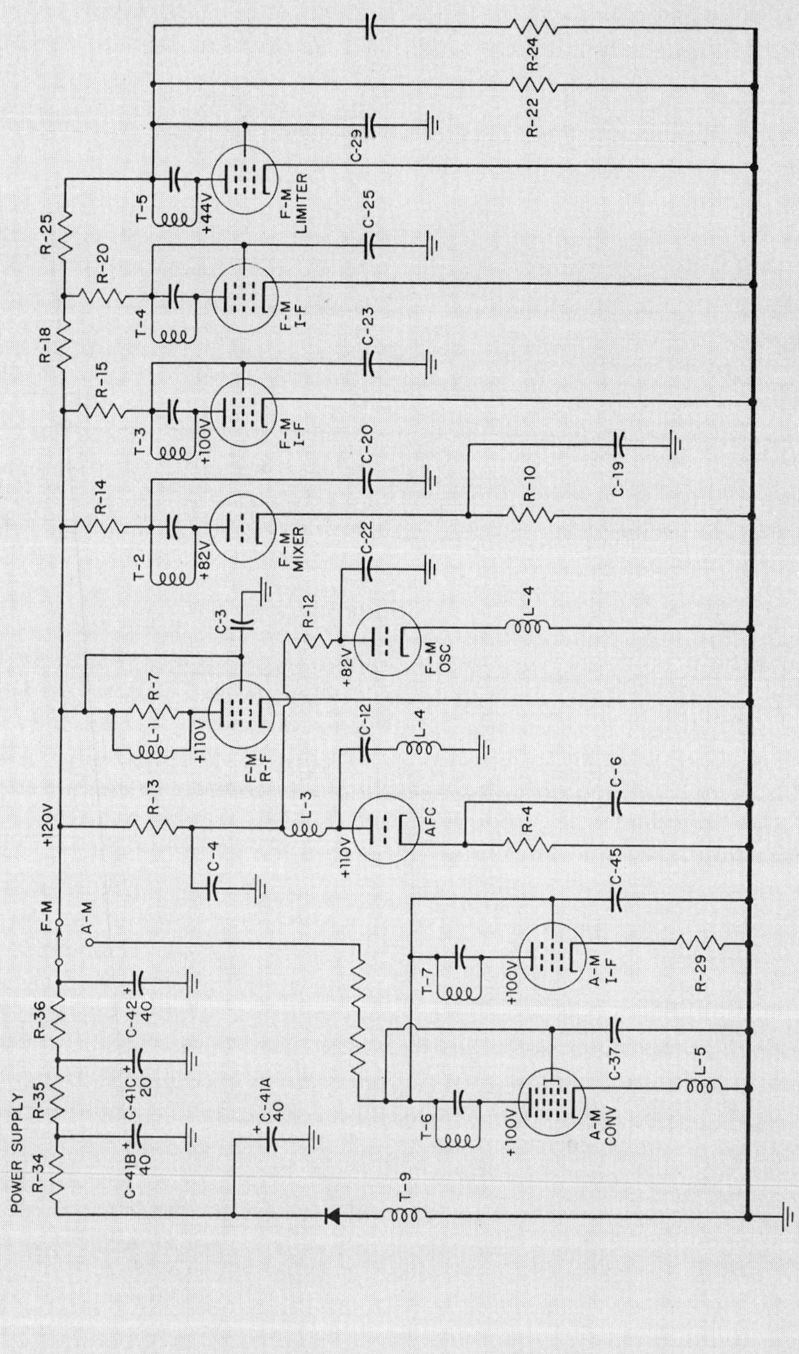

FIG. 22-36. Skeleton plate circuit diagram for the basic a-m/f-m tuner.

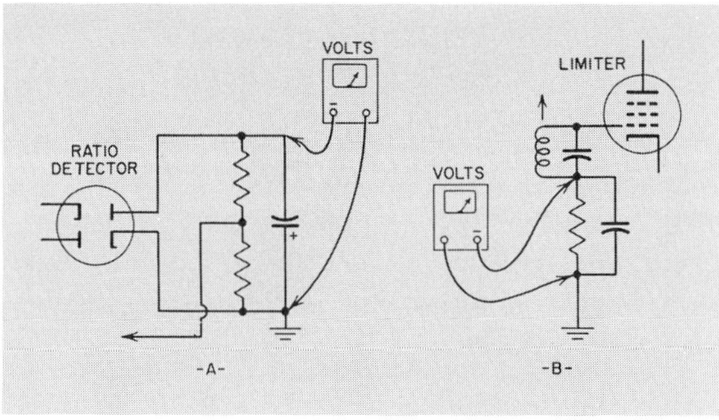

FIG. 22-37. Checking for oscillation in an f-m receiver.

procedure. However, in the f-m mode of operation, a different technique is employed. The procedure depends on the fact that, when no signal is riding through the receiver, no signal voltage should appear at the ratio detector or at the limiter grid in the discriminator type of detector. If it does, then one of the amplifier stages is oscillating.

Set the receiver tuning dial at a non-station position. Now place a voltmeter across the stabilizer capacitor of the ratio detector or across the limiter-grid resistor, as shown in Fig. 22-37. Any voltage indication is due to oscillation. Try replacing the r-f and i-f tubes. If this fails to stop oscillation, try placing a test capacitor across all bypass capacitors. If oscillation still continues, you must realign the set, as explained later.

Switch troubles. Function switches are usually wafer types. There is such great variation in the use and wiring of these switches that a study of one in a receiver would be of little aid in understanding that of another.

Trouble in the plate-circuit section of the switch would show up during a voltage check. Troubles in the signal-circuit sections would appear during a signal check.

The usual trouble in the switch consists of dirty contacts or contact wipers. The application of a cleaning agent, followed by snapping the switch through its positions, takes care of dirty contacts. Sprung wipers are found by visual inspection. Pressure applied with long-nose pliers is usually effective in reestablishing good contact.

Replacement precautions. Because of the high frequencies involved in f-m sets and because of the greater possibilities of unwanted coupling of circuits, special precautions must be taken when replacing parts. Lead dress is extremely important. Route wires in the same positions they had originally. To do this, try to use exact replacement parts. Do not make ground connections where most convenient, but connect parts to their original grounding points. Failure to do so may produce undesirable coupling and unwanted oscillations.

Common troubles in f-m sets. A service chart of common troubles in f-m receivers and tuners is given in Table 1.

There are other defects in f-m receivers and tuners, but these are resolved in a manner similar to that for an a-m set. Such defects are hum, signal over only a portion of the tuning dial, hum only when station is received, etc.

TABLE 1. F-M RECEIVER TROUBLESHOOTING CHART

Defect	Possible cause	Test procedure
Set is dead	Bad tube Set fuse burned out Defect in power supply	Check power-supply voltage and fuse, if present Check tubes Signal check the receiver

Defect	Possible cause	Test procedure
Weak sound on all stations	Defect in power supply Weak audio tubes Defective volume control Defective speaker Defect in i-f stage Dirty function switch	Check power-supply voltage Check volume control Check function switch Check speaker Check voltages on i-f tubes Check tubes
Distortion	Defective tubes Low *B* voltage Open filter capacitor Misaligned f-m detector Leaky stage coupling capacitor	Check power-supply voltage Check tubes Check detector alignment Check coupling capacitors
Noise	Misalignment Loose connections Defective tubes Leaky bypass capacitors Dirty tuning capacitor Dirty function switch	Check tubes Check tuning capacitor Check function switch Check bypass capacitors Check alignment
Motorboating	Open filter capacitor Open bypass capacitor Open grid resistor Poor ground connections Poor lead dress	Check power supply Check lead dress Check other possible components
Intermittent operation	Defective tube Loose connections Dirt in tuning capacitor Dirt in function switch Poorly soldered connections Intermittent open capacitors Intermittent open coils	Check components indicated
Signal drifts	Defective afc circuit Defective oscillator circuit Open or shorted components in the afc line	Check oscillator and afc stages

General service notes. The service information just given was centered around tube receivers. But most of it is equally applicable to transistorized sets. The input to transistorized stages should be considered to be analogous to the input to tube stages for signal check. When checking stages, normal transistor voltages should be substituted for normal tube voltages. Similar components go defective in both cases. Draw upon service knowledge given in the chapters on the portable transistor and transistor car radios for defects and relate them to the information given for the f-m tube radio.

ALIGNMENT OF F-M RECEIVERS

The aligning of a-m/f-m receivers takes on the nature of a twofold job, because there is an a-m function and an f-m function. Because of the different frequencies involved, there is no interaction between the two modes of operation. As a result, the a-m section and f-m section may be aligned separately.

The procedure for aligning the a-m section is the same as that already given for a-m sets. Aligning of the f-m section is performed with the function switch in the f-m mode of operation. Although it is advisable to follow the manufacturer's specific alignment notes, the generalized procedure given will serve when such notes are not available.

Circuits involved in alignment are the discriminator transformer, i-f transformers, oscillator coil, and r-f coil. The discriminator and i-f coils are usually permeability-tuned; this means that tuning-slug adjustment is involved. The oscillator and r-f coils are usually capacitor-tuned and are adjusted by means of trimmer capacitors.

When to realign. The technician, when deciding whether or not to realign the f-m section of the receiver, is guided by the symptoms of trouble and by his own purposes. The receiver should not be realigned unless, by a process of elimination of causes for poor operation, that procedure is shown to be necessary.

What symptoms indicate the need for realignment? The signal check can point up several conditions. If the receiver gives a weak response to a 10.7 mc signal and temporary changing of the alignment adjustments gives some improvement, realignment is indicated. If the modulated test signal sounds distorted, realignment is again indicated.

Whenever a component of a tuned circuit is replaced, it is usually routine procedure for the technician to realign the set. Such realignment assures optimum performance.

Finally, the set may be realigned to remove undesired oscillation. However, before proceeding, be sure to substitute new tubes, check for open bypass capacitors, and check lead dress carefully for unwanted coupling of circuits.

Location of alignment adjustments. Before attempting alignment, the technician must be able to locate the adjustment screws and must know which circuits are being adjusted. In this regard, the manufacturer's service notes serve best. However, the following hints will be helpful where such notes are not available.

The adjustment screws involved in alignment are those for the discriminator transformer, the i-f transformers, the oscillator coil, and the r-f tuner. The discriminator transformer is the one that feeds the second detector, whether of the Foster-Seeley or ratio-detector type.

The discriminator and i-f transformers are normally located in metal cans on the top side of the chassis. Most frequently, their slug adjustment screws are located at the top and bottom of the can. This arrangement necessitates setting the receiver on its side position so as to make both adjustments accessible.

The question arises as to which screw, top or bottom, aligns the primary or secondary winding of the transformers. This is of no great importance except for the discriminator transformer. Usually the primary coil adjustment will be found at the top of the can. In other cases, the primary adjustment screw will be found by trial and error during the alignment procedure.

If manufacturer's service notes are not available, it would seem that confusion might arise in determining which are the f-m i-f cans and which are the a-m i-f cans. In some cases, the cans have identifying labels. But where such markings are not found, a simple experiment will disclose which are which. As part of the alignment procedure, feed a 10.7-mc test signal to the stage, and adjust the aligning screw involved. If no effect appears on the output indicator, then the i-f can belongs to the a-m section. The discriminator transformer can is usually a little larger than the other i-f cans and nearest to the second detector.

The r-f and oscillator trimmer adjustment screws are usually located on the ganged tuning capacitor, in their appropriate sections of the gang. Sometimes, the f-m oscillator trimmer capacitor is found on the underside of the chassis.

The top view of the chassis of our basic a-m/f-m tuner is shown in Fig. 22-38. Learning tube types and their functions aids in the job of identification. Note the discriminator transformer nearest to the 6AL5 discriminator tube. Note the third f-m i-f transformer to the right of the 6AU6 second f-m i-f amplifier tube. The second f-m i-f transformer is to the right of the 6BA6 first f-m i-f amplifier tube. The first f-m i-f transformer is closest to the 6U8 f-m mixer tube. The f-m oscillator trimmer and f-m r-f trimmer are located near the tuning capacitors. The antenna transformer is found near the 6U8 f-m r-f amplifier tube.

Alignment methods. There are two basic techniques for aligning the f-m section of an a-m/f-m receiver. Generally, manufacturers list both methods in their service notes. One procedure utilizes an a-m signal generator and an output voltmeter. It is known as the *meter method* and is somewhat like the standard a-m alignment procedure. The other

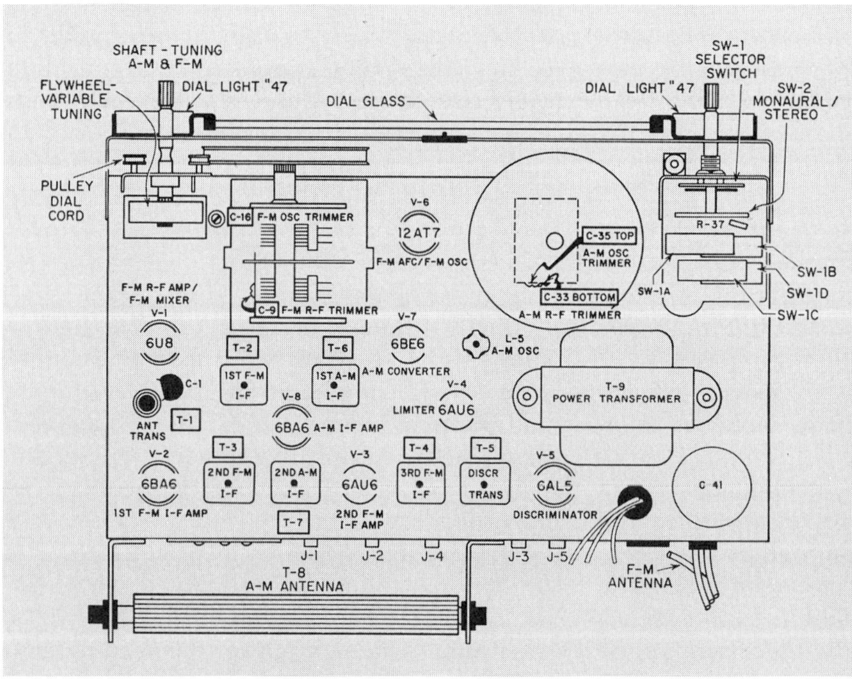

FIG. 22-38. Location of alignment adjustments in the basic a-m/f-m tuner.

method, known as a visual alignment procedure, utilizes a special type of signal generator, known as a sweep-signal generator, with a cathode-ray oscilloscope as an output indicator.

Both methods will do the job. However, the visual alignment method is generally considered to be a more refined technique. The procedure that is followed by the technician will depend on the instruments available to him. Both will be described in this chapter.

Equipment needed for meter method. The first requirement in the meter method is an a-m signal generator with a frequency range up to about 15 mc. Such a test instrument takes care of the intermediate frequency of 10.7 mc. Its harmonic content takes care of any other frequency in the 88- to 108-mc range. The usual test oscillator for servicing a-m receivers may be used, although more elaborate instruments like the Precision E-200-C generator give better results.

The second requirement in the meter method is a vacuum-tube voltmeter or a 20,000-ohms-per-volt voltmeter. The meter leads should terminate at one end with alligator clips for ease in connecting to the various test points.

Another alignment instrument recommended for both methods of alignment is a recessed-nib screwdriver, such as the one shown in Fig. 22-39A. The hollow screwdriver fits over the adjustment screw which is then engaged by the recessed nib, as shown. This tool is extremely useful when aligning permeability-tuned i-f transformers. Parts suppliers list many other similar alignment tools.

Setting up for meter alignment. When preparing to align the f-m section, place the receiver on its side, properly supported. This position enables you to work from the top and bottom of the receiver without moving it. For a-c receivers, connect a piece of bonding between the signal generator and the receiver chassis; then run a connection from the bonding to a good ground, like a grounded water pipe. This setup is shown in Fig. 22-39*B*.

When performing the alignment procedure, it is most desirable that you connect the set to the power line through an isolation transformer, particularly with an a-c/d-c receiver. However, if this test equipment is not available, you will have to connect the ground clip of the signal generator to the chassis of the a-c/d-c set through a 0.1-mfd, 400-volt paper capacitor.

When using a 20,000-ohms-per-volt meter as the output indicator, an isolating resistor of about 50,000 ohms should be connected in series with the hot lead. This resistor prevents the meter leads from affecting the circuit under test. An isolating-resistor clip lead like the one shown in Fig. 22-39C is convenient.

The resistor clip lead is unnecessary when using a vacuum-tube voltmeter, since these

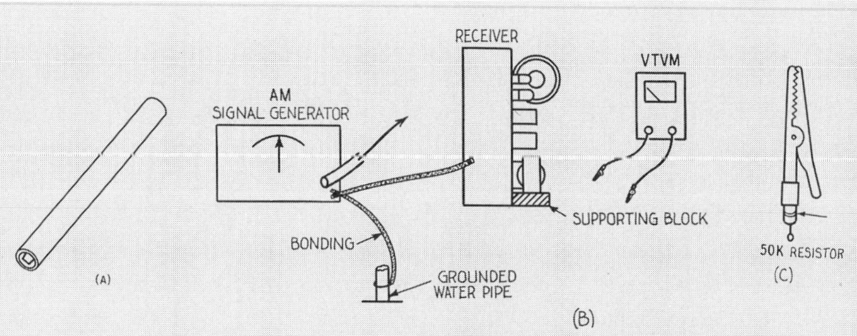

FIG. 22-39. Meter setup for f-m alignment.

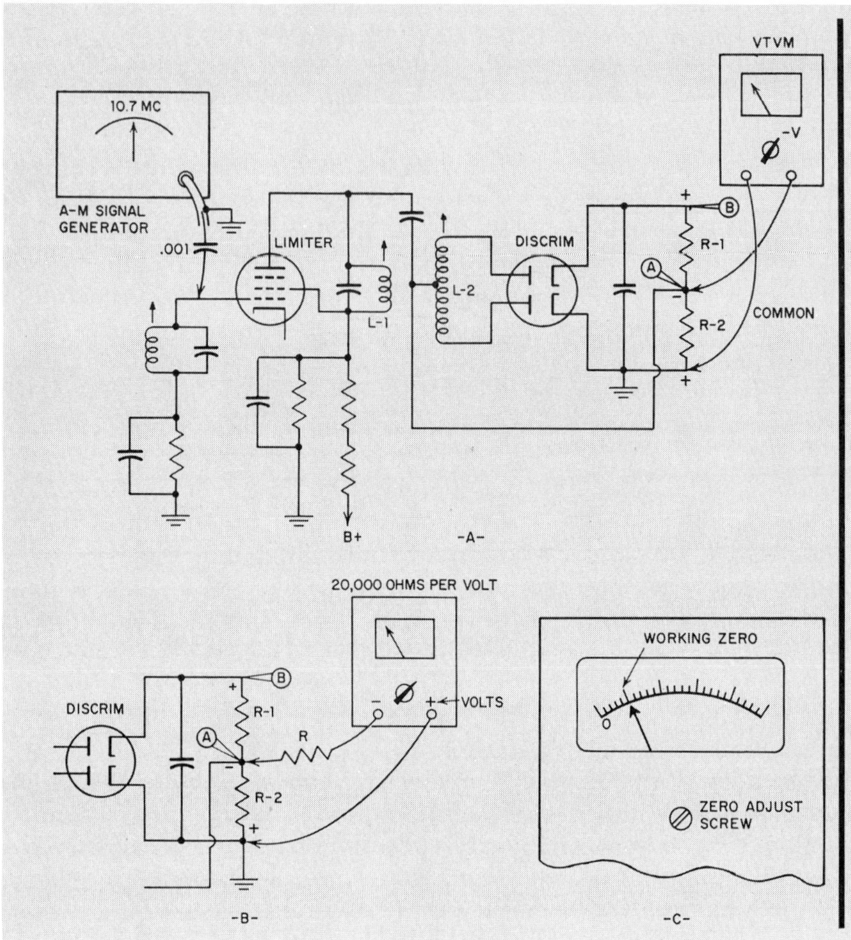

FIG. 22-40. Output meter connections for f-m discriminator alignment.

instruments are already equipped with an isolating resistor located in the test prod of the hot lead. Connections for both types of instruments are shown in Fig. 22-40.

The hot lead of the signal generator must always be connected in series with a dummy antenna to the receiver. For i-f alignment (i-f transformers and discriminator transformer), such an antenna is a 0.001-mfd, 600-volt capacitor. For r-f alignment, the dummy antenna is a 270-ohm, 1-watt resistor.

Aligning the basic tuner. The following generalized f-m alignment procedure for our basic a-m/f-m tuner may be adapted for use with any other similar unit. Of course, where a manufacturer's specific notes are available, follow those directions. Where both a-m and f-m alignment are required, align the a-m section first, as described earlier in the book.

Allow the set and the signal generator about 15 minutes to warm up with the selector switch in the F-M position; do not use the F-M/AFC position.

Limiter-discriminator receiver: i-f alignment. To align the i-f stages of an f-m receiver with a limiter and discriminator, align the discriminator transformer first, then the limiter, and then each successive i-f stage working back. Finally, repeat the complete procedure. This will now be taken up, stage by stage.

Turn the tuning control of the tuner or receiver to a non-station position at the upper end of the dial. Before proceeding, consider the circuit diagram of the limiter and discriminator in Fig. 23-40. Resistors *R*-1 and *R*-2 are the diode load resistors across which voltages are developed. Note the polarities of the voltage drops across these resistors. This condition exists because currents through them from their junction point are in opposite directions. When the primary coil of the discriminator is properly aligned, the voltage across either *R*-1 or *R*-2 is as large as possible when an unmodulated 10.7-mc signal is fed to the limiter grid.

When the secondary winding of the discriminator transformer is properly aligned, the voltage drop across the two load resistors is equal and opposite for the same signal as above. Thus, if a voltmeter were connected from point *B* to the ground, with the unmodulated 10.7-mc signal fed to the limiter grid, the reading would be zero. This measurement and the one given just previously serve as the basis for the alignment procedure for the discriminator transformer.

Feed a 10.7-mc unmodulated signal from the generator to the limiter grid, as shown in Fig. 22-40*A*. The diagram shows the vacuum-tube voltmeter connected across load resistor *R*-2 for alignment of primary winding *L*-1. The common lead of the meter, as usual, is connected to ground. The hot lead goes to negative point *A*, and the voltmeter function switch is set to minus volts. Figure 22-40*B* shows the same setup when using a 20,000-ohms-per-volt voltmeter. Here, the negative lead goes to point *A* and positive lead goes to the ground. The resistor clip is in the negative lead.

Now proceed to adjust the tuning-slug screw of primary winding *L*-1 of the discriminator transformer so as to get a peak reading on the voltmeter. Assume that the top adjustment screw of the discriminator can is the primary. If a normal peak cannot be obtained, try the other adjustment screw on the bottom of the can. It is good practice to rock the control screw through the peak in ever-smaller swings until the true peak is achieved.

Next, align the discriminator transformer secondary winding. Shift the hot voltmeter lead from point *A* to point *B*. Adjust the tuning slug screw of secondary winding *L*-2 until the voltmeter reads zero. To be sure that this is the true zero, rock the slug screw to either side of the zero. On one side, the voltmeter swings sharply up the scale with a negative voltage. On the other side, the voltmeter needle swings sharply down the scale with a positive voltage. Having established the two meter swings, adjust the slug screw for zero between the two swing points.

In this step, it is desirable to have a voltmeter with a zero in the center of the scale. If such an instrument is not available, a little tricky maneuver will suffice. Simply throw the pointer of the unconnected voltmeter to a convenient scale division line by means of the manual zero adjustment screw. This new scale line may now be used as your working zero. Adjust the secondary winding alignment screw so as to bring the meter pointer to the working zero. This indication is shown in Fig. 22-40*C*.

The third step is alignment of the limiter stage. The setup is shown in Fig. 22-41*A*. The 10.7-mc test signal is fed to the grid of the i-f amplifier tube before the limiter. The hot voltmeter lead is connected to point *C*, the negative end of the limiter grid resistor. Remember to couple the signal generator loosely to the grid and to turn down the attenuator. Then, by means of the adjustment rocking procedure, align i-f transformer coils *L*-3 and *L*-4 by adjusting each for a peak reading on the voltmeter. It is usual procedure to align the secondary coil before aligning the primary coil. However, no great harm results if the order is reversed when identification is difficult.

The remainder of the procedure is the alignment of the i-f stages, as shown in Fig. 22-41*B*. Throughout alignment of the i-f stages, the voltmeter is left across the limiter grid resistor,

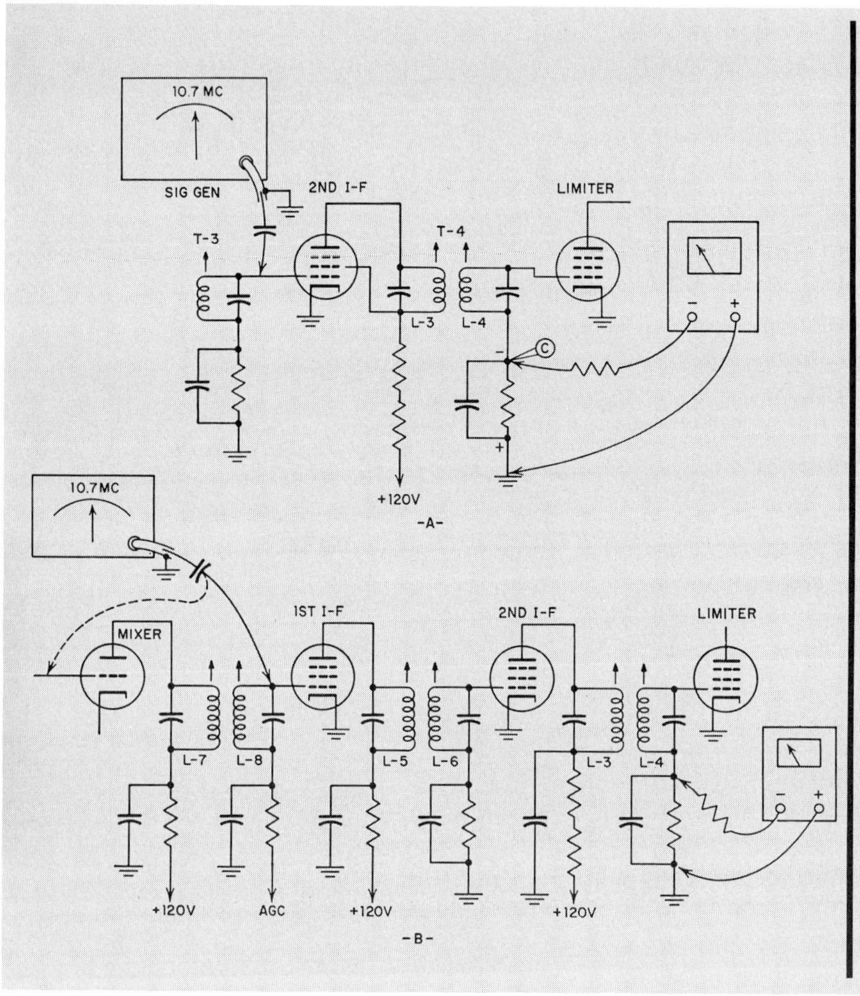

FIG. 22-41. Alignment of the i-f amplifier.

as was done in the previous step. Remember to couple the signal generator loosely to the set and to turn the attenuator down.

Feed the 10.7-mc unmodulated signal to the first i-f grid. Adjust the alignment slug screws of i-f transformers *L*-5 and *L*-6 in the usual manner to get peak readings on the voltmeter. Again, no great harm is done if the primary coil is aligned before the secondary coil.

Finally, shift the signal-generator hot lead to the f-m mixer grid and peak i-f transformer coils *L*-7 and *L*-8 for maximum reading on the voltmeter in the same manner as was done for the previous step. Repeat the entire procedure for final touchup alignment. As a last step, check to see that the discriminator audio output is zero.

Limiter-discriminator-type receiver: oscillator and r-f alignment. The f-m oscillator is aligned in a manner similar to that of the a-m radio. Set the tuning control of the receiver to about 108 mc. Leave the output voltmeter across the limiter grid resistor, as in the previous step. Feed an unmodulated signal at 108 mc to the antenna input, as shown in Fig. 22-42. Note the dummy antenna resistor *R*-1 of 270 ohms in the hot lead of the signal gen-

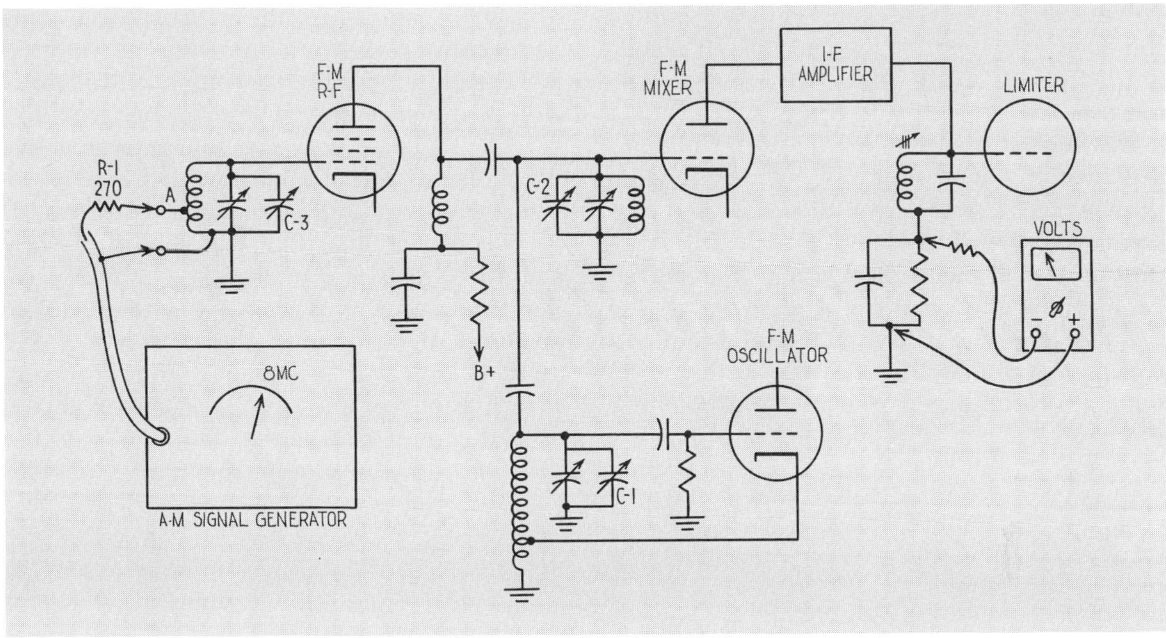

FIG. 22-42. Oscillator and r-f alignment.

erator. Keep the generator output at a very low level. Adjust the oscillator trimmer to give peak output on the output voltmeter.

If your signal generator does not go up to 108 mc, you can make the same adjustment by relying on the harmonic output of the generator. Adjust the tuning control of the tuner to 104 mc, then adjust the signal generator for an 8-mc output, whose thirteenth harmonic is 104 mc. Now adjust the oscillator trimmer as before for peak output on the voltmeter.

There remain now the alignment of the r-f trimmer and the f-m antenna transformer. With the signal generator and the tuner set to 104 mc and the output meter in the same positions as before, adjust the r-f trimmer (*C*-9 in Fig. 22-38) for maximum output meter reading. Make this adjustment while slowly rocking the receiver tuning control.

Finally, set the tuner to 96 mc. Tune the signal generator for an unmodulated output at the same frequency. Now adjust the f-m antenna transformer (*T*-1 in Fig. 22-38) for maximum output meter reading. Remember that if your signal generator does not run up to 96 mc, you can use the twelfth harmonic of an 8-mc signal. The tuner or receiver is now completely realigned. A second quick alignment will make for greater accuracy.

Ratio detector-type receiver: i-f alignment. The procedure for realigning a ratio-detector type of f-m receiver is essentially the same as the basic one just presented. However, because of circuit variations, the output voltmeter must be connected at a different point for readings.

In the ratio detector, the diodes are connected in series through the load resistors *R*-1 and *R*-2, as in Fig. 22-43. Current flows in the same direction through these resistors. As a result, the voltage drops across them do not buck each other, but are additive.

The voltage across these resistors is at a maximum when the i-f amplifier and the primary winding of the discriminator transformer are tuned to 10.7 mc. Therefore, the voltage at point *A*, the ungrounded end of the load resistors, may be used as the test point for alignment of the i-f trimmers and the discriminator primary coil.

In the ratio detector, the discriminator secondary coil is properly balanced when the voltages in the two diode circuits are equal for an unmodulated signal. This condition may be measured by zero voltage between the tap for the audio output of the detector and the midpoint of the stabilizer capacitor terminals. In Fig. 22-43, the audio output tap is labeled point *B*, and the midpoint or balance reference point is labeled *C*. These points *A*, *B*, and *C* are important alignment points in the ratio-detector type of receiver.

There are several variations of the ratio detector, especially where some manufacturers use point *A* as the source of negative agc voltage. These are shown in Fig. 22-44, with points *A*, *B*, and *C* indicated on each.

In some circuits, like that in Fig. 22-44*B*, the load resistors across the stabilizer capacitor are not two resistors with equal resistance. Instead, a single resistor is used. In that case, to get the balance reference point, temporarily solder two resistors of about 100,000 ohms each across stabilizer capacitor *C*-1, and use the middle junction. The resis-

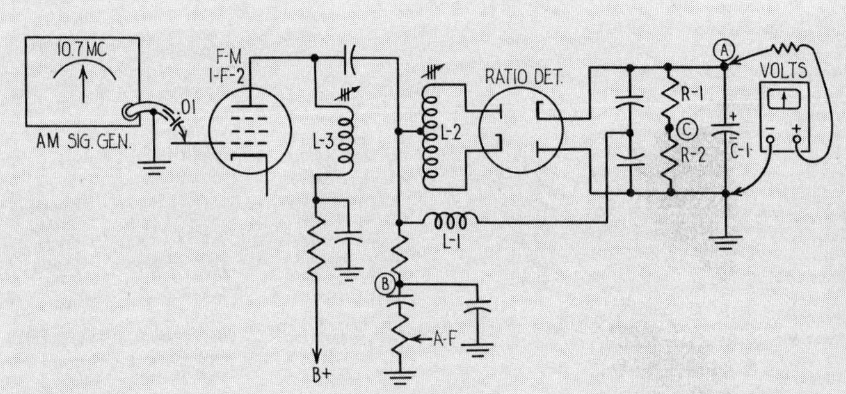

FIG. 22-43. Output meter connections for a ratio detector.

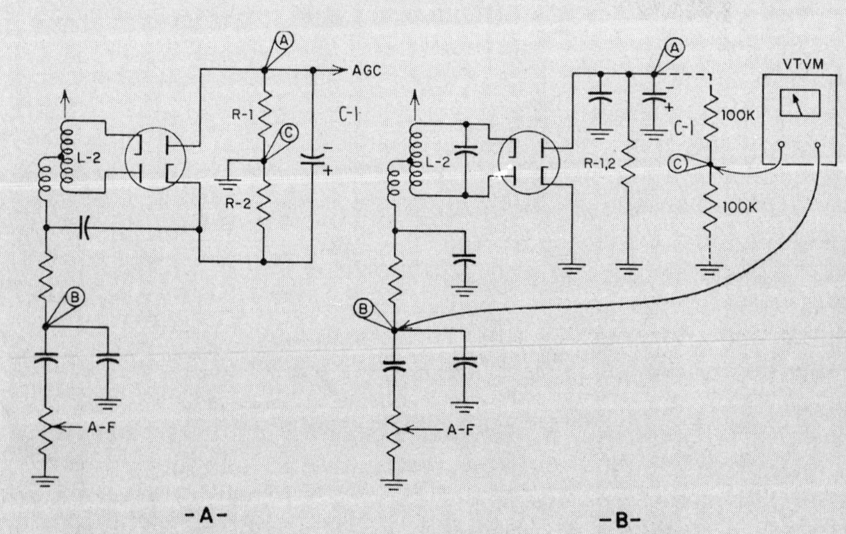

FIG. 22-44. Output meter connections for variations of the ratio detector circuit.

tor connections are indicated by the dotted lines in the diagram. Remove the resistors after the discriminator alignment.

Begin by connecting the voltmeter with the hot lead to point *A* and the other to the ground. Observe proper polarity for that point. Such a connection is shown in the diagram of Fig. 22-43. Turn the receiver tuning capacitor to the fully OPEN position and turn the afc switch, if present, to the OFF position. Loosely couple the signal generator to the second f-m i-f tube grid, and feed in an unmodulated signal at 10.7 mc. Adjust the tuning slug alignment screw of discriminator transformer primary coil *L*-3 so as to get a peak reading on the voltmeter.

Now shift the voltmeter, so that one lead goes to the balance reference point *C* and the other to a-f output point *B*. Such a connection is shown in Fig. 22-44*B*. Adjust the tuning slug alignment screw of secondary winding *L*-2 so as to get a zero reading on the voltmeter. The discriminator transformer is now properly aligned.

The remainder of the procedure is alignment of the i-f stages. Throughout the alignment of these stages, the voltmeter is left connected from point *A* to the ground. Refer to Fig. 22-45. Feed a 10.7-mc unmodulated test signal to the first a-m/f-m intermediate-frequency amplifier tube grid, loosely coupled with the attenuator down. Adjust the tuning slug alignment screws of i-f transformer windings *L*-4 and *L*-5 so as to give peak readings on the voltmeter.

Next, feed the signal to the f-m mixer grid, with loose coupling and with the attenuator down. Adjust the tuning slug alignment screws of i-f transformer windings *L*-6 and *L*-7 so as to give peak readings on the voltmeter. Repeat the entire procedure for final touchup alignment.

Ratio-detector-type receiver: oscillator and r-f alignment. The alignment of the oscillator and r-f stages is identical with that of the limiter-discriminator type of receiver. The only difference is that the voltmeter is kept connected from point *A* to the ground. The oscillator trimmer and mixer trimmer are peaked as usual, with the unmodulated test signal fed to the antenna input. The receiver is now completely realigned.

Equipment for visual alignment. In visual alignment, the sweep generator replaces the a-m signal generator as the source of test signal and the oscilloscope replaces the voltmeter as the output indicator. Essentially,

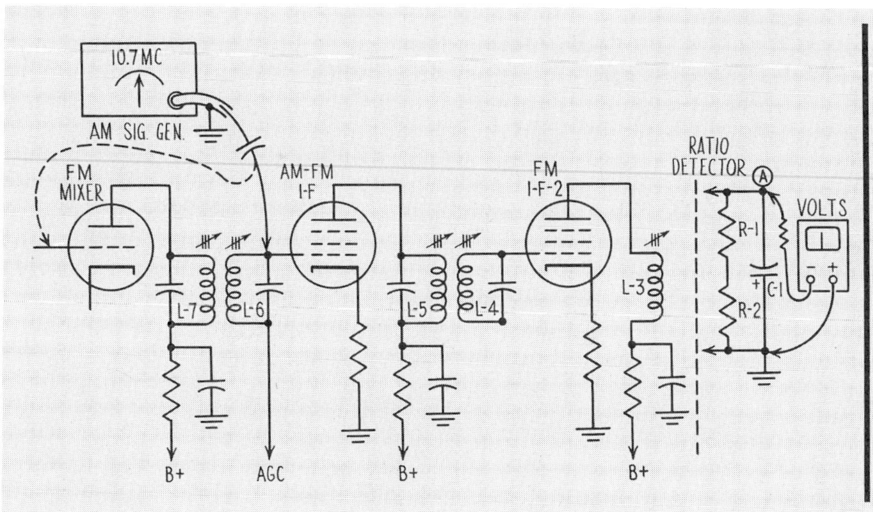

FIG. 22-45. Alignment of the i-f amplifier in a receiver using a ratio detector.

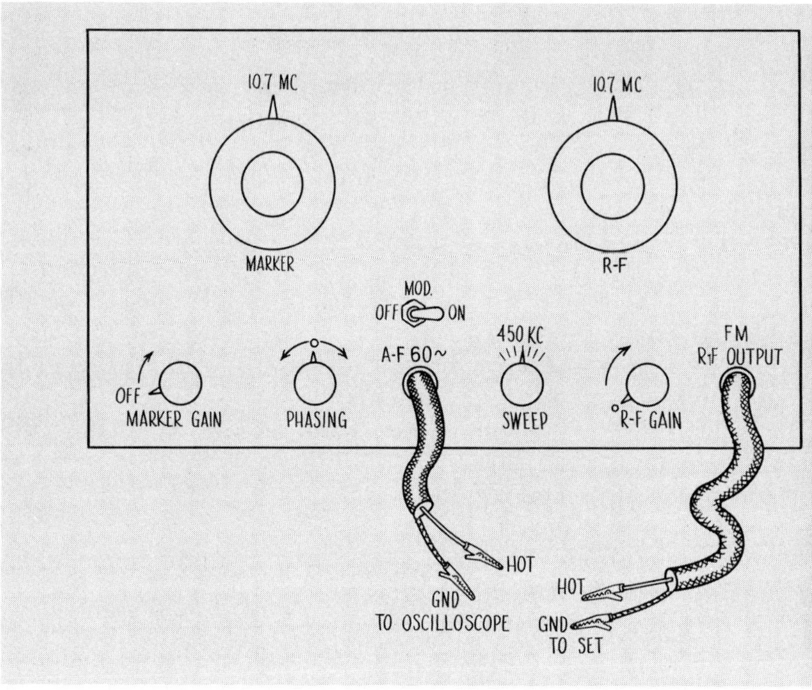

FIG. 22-46. Typical controls on a sweep generator.

the same test points are used for alignment in both methods.

Most manufacturers, in their service notes, confine the visual-alignment procedure to that for the discriminator and i-f stages. For the oscillator and r-f stages, the standard meter method is followed. That will be the plan in this book, even though the oscilloscope method could be used for the latter stages. Now, let us familiarize ourselves with the visual-alignment test instruments.

Sweep-frequency signal generator. This instrument is often called an f-m signal generator. There are many varieties of sweep generator to serve many purposes. It is advisable to read the manufacturer's service notes to learn how to use a particular instrument.

To explain f-m alignment, we shall assume that the sweep generator has the controls shown in Fig. 22-46. At the upper right is the r-f control which selects the center frequency of the f-m output. This frequency-modulated r-f output is obtained at the leads of the shield cable connected to the R-F OUTPUT jack. The lead connected to the shield is the ground lead. The magnitude of output voltage is controlled by the R-F GAIN control—our familiar attenuator.

The modulation, when the MOD switch is in the ON position, is an internal 60-cycle voltage. It causes the f-m carrier (shown at 10.7 mc) to deviate above and below the center frequency 60 times a second. This modulating voltage is tapped at the A-F 60 CPS jack by means of a shielded cable for connection to the oscilloscope. Again, the shield lead is the ground lead. The frequency deviation of the f-m carrier is determined by the amplitude of the 60-cycle modulating voltage. This amplitude is controlled by the SWEEP control, which is calibrated in frequency deviation. For f-m alignment, a frequency sweep or deviation will be 225 kc on each side of the carrier.

The PHASING control is a built-in phase-shifting network to avoid double images on the oscilloscope. By adjusting this control, double images tend to be superimposed upon each other to produce a single image.

The MARKER control (upper left) is used to locate the frequency of the center of a trace on the oscilloscope. It is a separate r-f oscillator whose frequency is selected by the control knob. Its output is through the F-M R-F OUTPUT cable. Below the MARKER control is the MARKER GAIN control. In its OFF position, it removes marker voltage from the output. As the knob is turned clockwise, the voltage of the marker signal is increased.

To summarize, the sweep generator, as used for f-m alignment, produces an r-f carrier voltage at 10.7 mc, deviating 225 kc on each side of 10.7 mc. This deviation swing is repeated 60 times a second. The 60-cycle modulation voltage is brought out through a separate cable for connection to an oscilloscope.

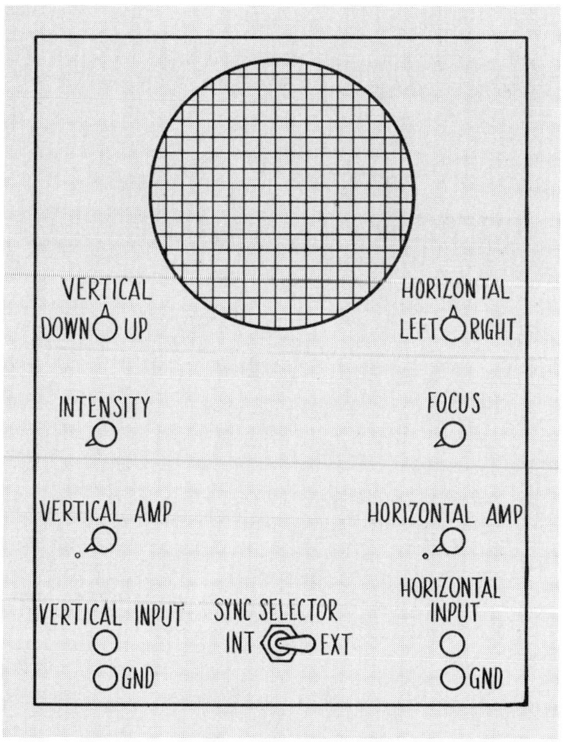

FIG. 22-47. Typical controls on a cathode-ray oscilloscope.

Cathode-ray oscilloscope. The oscilloscope, like the sweep generator, exists in a variety of forms. For any specific instrument, the manufacturer's service notes should be read carefully. An oscilloscope will have the controls shown in Fig. 22-47 in one form or another.

The heart of the oscilloscope is the cathode-ray tube. In this tube, a stream of electrons from a cathode is made to focus to a fine dot on the face of the tube. The face is coated on the inner side with a fluorescent phosphor which glows when hit by the electron stream.

The electron stream, on its way to the face, passes through two sets of plates known as deflection plates. A front view of these plates is shown in Fig. 22-48*A*. The pair above and below the stream are known as the vertical deflecting plates. The pair on each side of the stream are known as the horizontal deflecting plates. By applying a voltage to the vertical deflection plates, the electron stream is deflected up or down. As a result, the dot is similarly deflected on the screen. If rapid pulses of voltage are applied to the vertical plates, the dot will move up or down rapidly. When this movement is rapid enough, persistence of vision will make the dot appear to be a continuous vertical line, as shown in Fig. 22-48*B*.

Similarly, applying a voltage to the horizontal plates moves the dot to the right or left on the screen. If very rapid pulses of voltage are applied to these plates, the dot appears to be a continuous horizontal line.

If voltages are applied to the horizontal plates at the same time as voltages are applied to the vertical plates, the dot takes positions up and down and to the right and left. If the pulse voltages applied to the plates are rapid and repeat themselves, a figure appears on the screen. That is what happens in alignment work.

The face or screen of the cathode-ray tube is on the front panel of the oscilloscope. The face is often covered with a cross-section screen, as an aid in judging the trace image.

At the top right corner of the oscilloscope is the horizontal centering control, used to move the image to the right or left of the screen. Opposite it is the vertical centering control, used to move the image up or down on the screen.

The INTENSITY control makes the image brighter or fainter. Usually, it is toned down to increase the life of the cathode-ray tube. The FOCUS control is used to make the image sharper, if it is blurred.

At the bottom on the left are two input terminals, marked VERT INPUT. Signal voltages are fed through these terminals to the vertical deflection plates of the cathode-ray tube. One of these terminals is a ground terminal. Since signal voltages fed to them are usually not strong enough to give enough vertical deflection, an internal vertical amplifier is included in the instrument. The knob marked VERT AMP controls the gain of this amplifier.

Similarly, signal voltages are fed to the horizontal deflection plates through the HORIZ INPUT terminals. Again, an internal horizontal amplifier is included to give greater horizontal deflection. Its gain is controlled by means of the HORIZ AMP knob.

There is an internal synchronizing oscillator in the oscilloscope which feeds pulse voltage to the horizontal plates to produce a regular and repeated horizontal deflection or sweep of the fluorescent dot. When the SYNC SELECTOR switch is in its EXT position, it becomes possible to use an external pulse voltage to give the repeated horizontal sweep. Thus, we can use the 60-cycle modulating voltage from a sweep generator to give the synchronizing sweep to the oscilloscope. The external pulse voltage is fed to the horizontal plates of the instrument.

Setting up for visual alignment. With the brief description of the test instruments just given, we are now prepared to set up the equipment for visual alignment of the i-f section of the f-m receiver. Refer to Fig. 22-49.

Place the receiver on its side, properly supported, so that the top and bottom of the chassis are readily accessible for connection and adjustment. Then, ground both test instruments by running a bond jumper from a metal screw on the test instrument chassis to a ground, like a cold-water pipe. Turn on the receiver, sweep generator, and oscilloscope for a 15-minute warmup before beginning the alignment procedure. If an afc control is present, turn it off.

Connect the 60-cycle modulation output

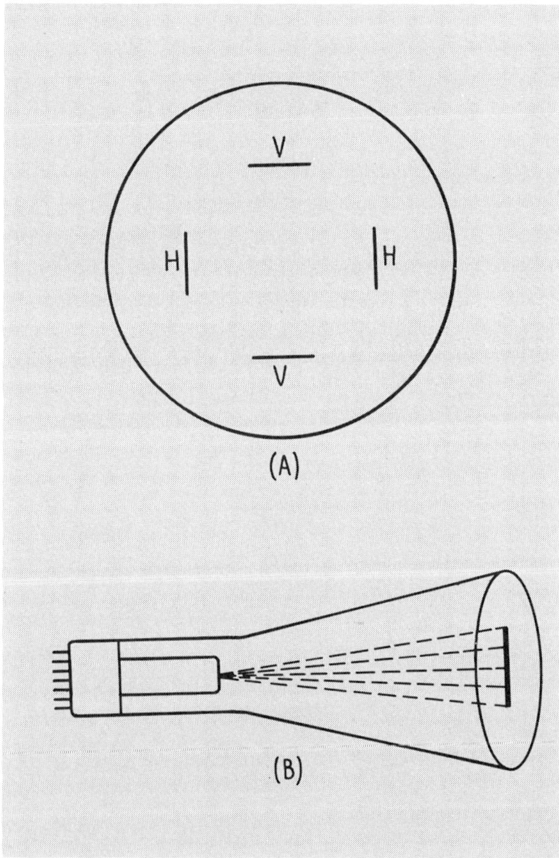

FIG. 22-48. How the beam is deflected in a cathode-ray tube.

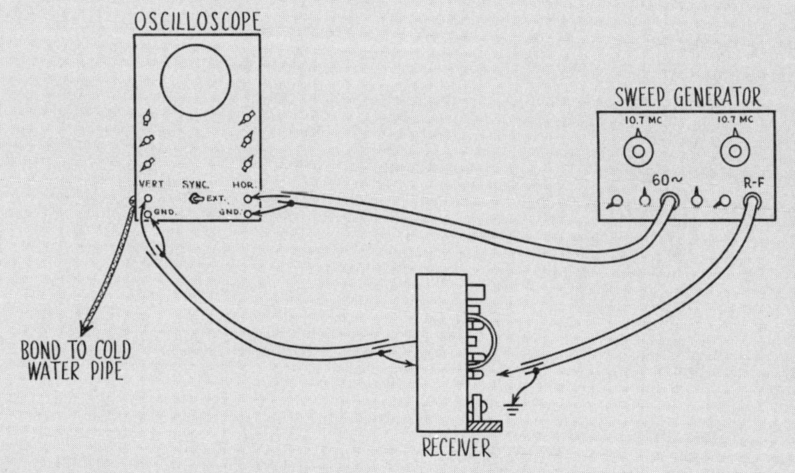

FIG. 22-49. Setup for visual alignment.

from the sweep generator to the horizontal input terminals of the oscilloscope by means of the generator cable. Be sure to connect the ground terminal of the cable to the ground terminal of the oscilloscope. Set the frequency control of the generator to 10.7 mc, set the modulation switch to its ON position, and turn the sweep control to a frequency deviation of 450 kc. Adjust the marker dial to 10.7 mc and turn its gain control down to zero.

On the oscilloscope, set the SYNC SELECTOR switch to its EXT position. A horizontal trace line now appears on the screen. If the line is displaced from center, adjust it by means of the centering controls. Adjust the FOCUS control so as to get a sharp unblurred line. Turn the INTENSITY control down so as to get a trace that is bright enough to see but not so bright as to damage the screen. And finally, if the horizontal line is too narrow, spread it horizontally by adjusting the HORIZ AMP control. We are now ready for the alignment.

Limiter-discriminator type receiver: i-f alignment. Connect the ground leads of the sweep generator and the oscilloscope to the receiver chassis. Feed the 10.7-mc signal to the second i-f grid through a 0.001-mfd capacitor serving as a dummy antenna. Connect the vertical hot lead of the oscilloscope to the hot end of the limiter grid resistor through a 22,000-ohm resistor. The setup is illustrated in Fig. 22-50A.

If the i-f transformers are considerably out of alignment, there may be no indication other than that of the horizontal trace line. Turn up the R-F GAIN control of the generator and the VERT AMP control of the oscilloscope until some sort of small pip appears on the screen. To determine if this pip is at 10.7 mc, set the marker frequency control of the generator at 10.7, and turn up the MARKER GAIN control. A hash marker response will appear on the screen at the 10.7-mc position. If alignment is very poor, the marker hash and the response curve peak will not coincide, as shown in Fig. 22-50*B*.

Adjust the r-f frequency control of the generator to bring the marker hash to the center of the screen. Now adjust the alignment slug screws for i-f transformer coils *L*-3 and *L*-4 to increase the peaking of the response image and to more nearly center it on the 10.7-mc marker hash, as shown in Fig. 22-50*C*.

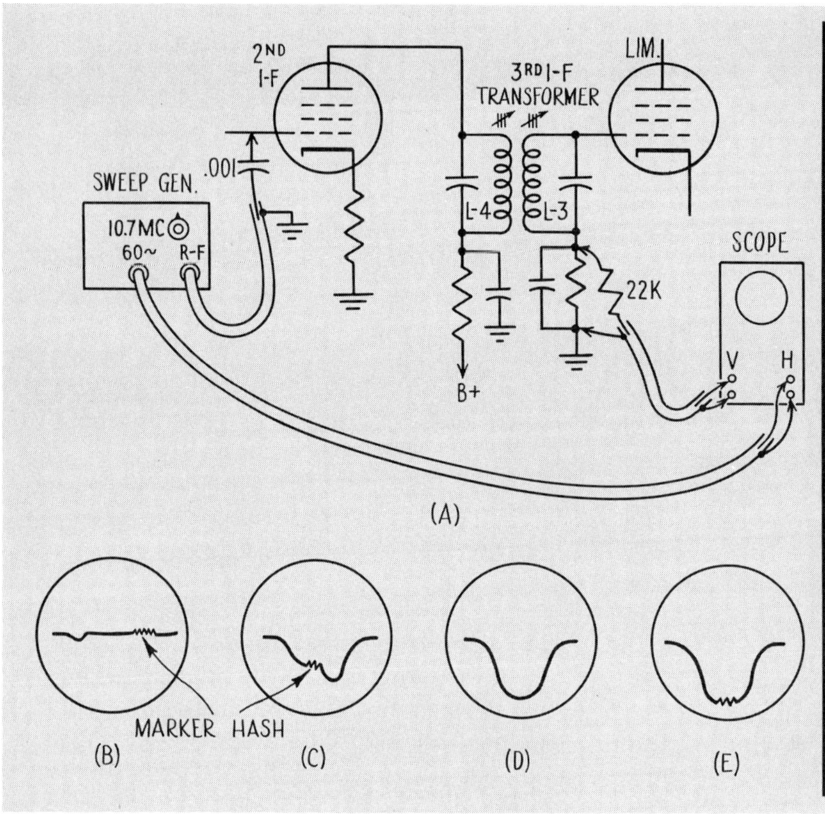

FIG. 22-50. Visual alignment of the limiter stage.

If a double response curve appears on the screen, adjust the PHASING control of the signal generator to get a single image.

The lopsided curve of Fig. 22-50*C* indicates a rough alignment where the alignment screws are close to their correct settings at 10.7 mc. Turn down the marker gain, and slightly retouch each screw so as to get the symmetrical curve of Fig. 22-50*D*. To make sure that the response curve centers at 10.7 mc, turn up the generator MARKER GAIN control, and see if the hash is at the center of the response curve, as shown in Fig. 22-50*E*.

Now move the hot lead of the signal generator to the grid of the first i-f tube, as shown in Fig. 22-51*A*. If coils *L*-5 and *L*-6 are way out of alignment, the trace might be shorter and broader than when last viewed, and the symmetry could be poor, as seen in Fig. 22-51*B*. Adjust the tuning slug screws for i-f transformer coils *L*-5 and *L*-6 for peak height and symmetry, as shown in Fig. 22-51*C*. When the trace gets too large, reduce the R-F GAIN control of the generator to keep it on the screen. The final trace should not be as broad as the peak trace (Fig. 22-50*D*) for the second i-f stage. Finally, check centering by turning up the MARKER GAIN control. The marker hash should appear as shown in Fig. 22-51*D*.

Next, shift the hot lead of the signal generator to the grid of the f-m mixer tube, as shown in Fig. 22-52. The vertical input lead of the oscilloscope is still at the limiter grid resistor. Adjust the alignment screws of input i-f transformer coils *L*-7 and *L*-8, by the same procedure, for maximum height and symmetry. The final response curve will be even less broad than that of Fig. 22-51*C*.

The discriminator stage is now ready for alignment. Shift the hot probe of the vertical

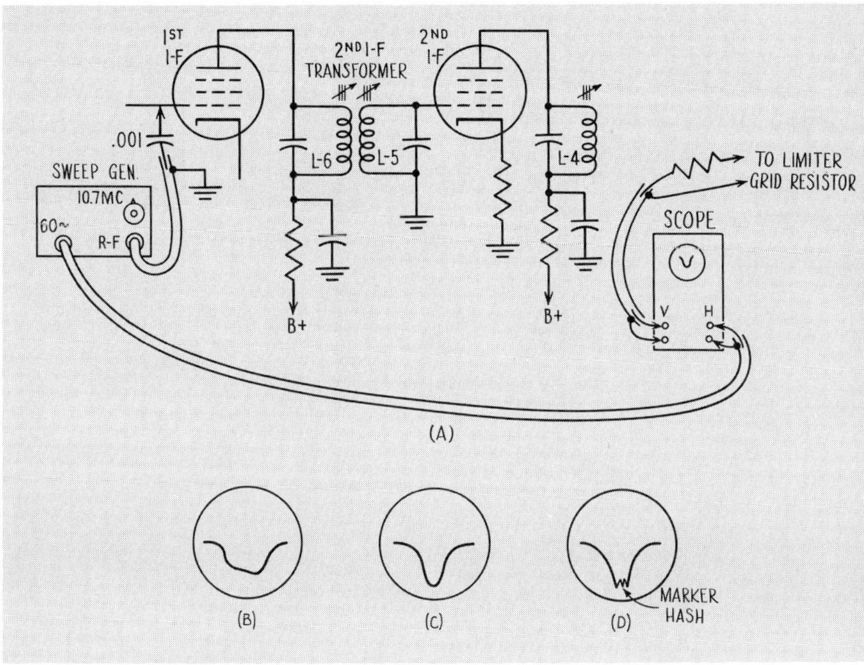

FIG. 22-51. Visual alignment of the second i-f stage.

input of the oscilloscope to audio output point *B* of the discriminator, as shown in Fig. 22-53*A*. Adjust the alignment tuning slug screw for the secondary of discriminator transformer *L*-2 until you get a trace like that shown in Fig. 22-53*B*; then adjust the primary screw of coil *L*-1 for peak amplitude. Retouch both adjustment screws so as to get symmetry of the curve as shown in Fig. 22-53*B*. The center slant trace should be straight, and the upper and lower peaks should be equal in height. Finally, check that 10.7 mc is in the center of the trace, by turning up the MARKER GAIN control. The marker hash should appear as shown in Fig. 22-53*C*.

The i-f section of the f-m receiver is now completely aligned. As previously stated, the r-f alignment is preferably performed by the output meter method.

Ratio-detector type receiver: i-f alignment.

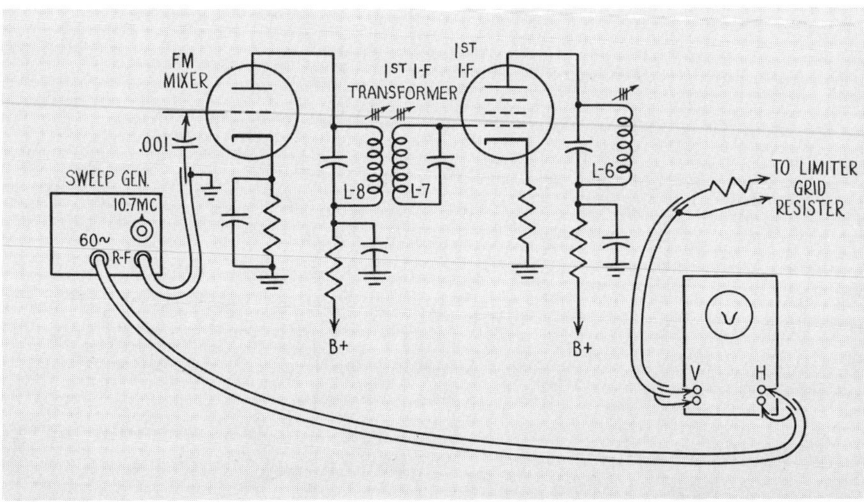

FIG. 22-52. Visual alignment of the first i-f stage.

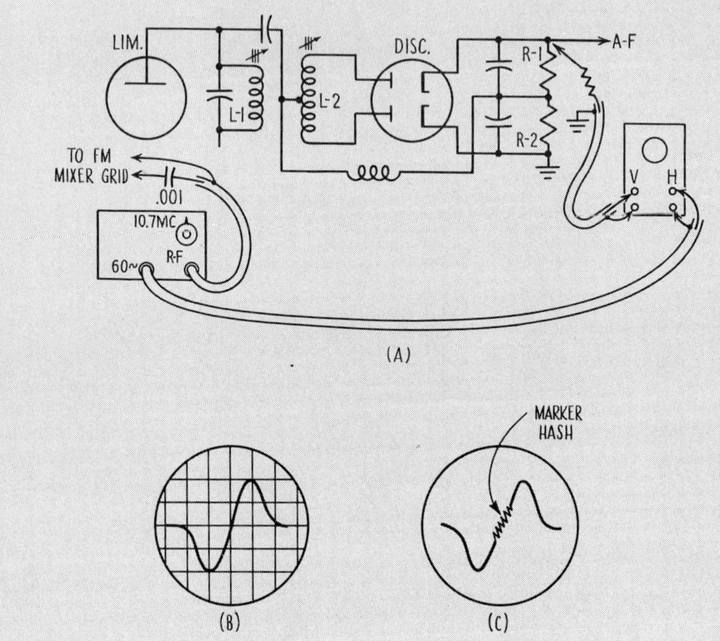

FIG. 22-53. Visual alignment of a discriminator.

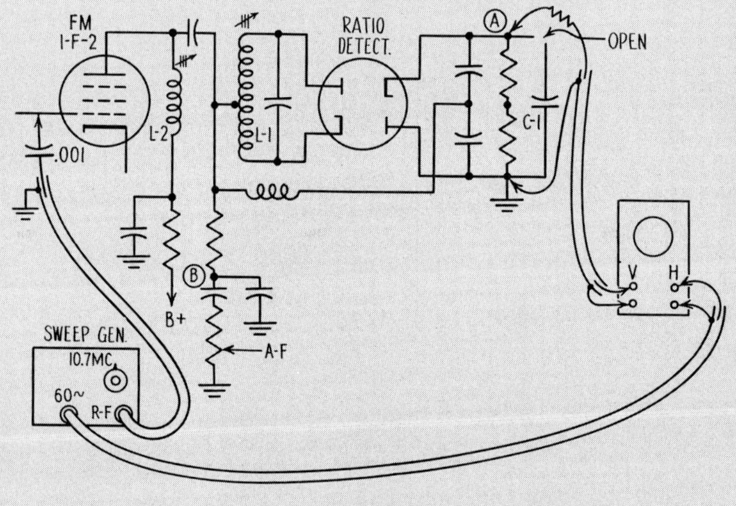

FIG. 22-54. Connections for visual alignment of a ratio detector.

The visual alignment of the i-f and detector section of a ratio-detector type of f-m receiver is similar to the procedure described above. The difference lies primarily in the points to which we connect the oscilloscope. Refer to Fig. 22-54.

Connect the generator and oscilloscope in the same manner as was done in the previous procedure. Disconnect stabilizer capacitor *C*-1. Connect the hot lead from the oscilloscope to point *A*, the ungrounded end of the load resistors. Feed a 10.7-mc signal from the sweep generator (60-cycle modulation and 450-kc frequency deviation) to the second i-f tube grid through a 0.001-mfd capacitor-type dummy antenna. Turn up the marker gain until the marker trace is visible. Adjust the alignment screw of primary winding *L*-2 of the discriminator transformer to center the response curve on the marker hash; then turn off the marker signal and retouch the primary adjustment for height and symmetry. The trace should appear as in Fig. 22-50*D*.

Shift the hot lead of the generator to the grid of the a-m/f-m i-f tube and adjust the alignment screws of the i-f transformer between that tube and the second f-m i-f tube to get a peak response like that of Fig. 22-51*C*. Again check center frequency by means of the marker signal.

Now shift the generator hot lead to the f-m mixer grid. Adjust the alignment screws of the first i-f transformer (between the mixer and the a-m/f-m i-f tube), to get a peaked and symmetrical trace. Check again with the marker signal.

Reconnect stabilizer capacitor *C*-1. Connect the hot lead of the vertical input of the oscilloscope to point *B*, the audio output point. Feed the 10.7-mc signal to the mixer grid. Adjust the alignment screw of secondary winding *L*-1 of the discriminator transformer so as to get the *S* curve of Fig. 22-53*B*. By means of the marker, check to see that the center of this trace is at 10.7 mc. Shut off the marker signal and slightly retouch the aligning screw of primary winding *L*-2 to get a peak trace with a straight center-slant line. The alignment of the i-f section is now complete.

Aligning the multiplex decoder. The lack of standardization of design of multiplex decoders makes it difficult to present a single alignment procedure. At this stage of development, it is best to refer to the manufacturer's notes on the step-by-step alignment of a stereo set. However, there are some general procedures in any multiplex decoder alignment. First, in every case, the maximum strength of the 19-kc pilot signal is desirable. Second, the maximum output of the 38-kc subcarrier for reinsertion is desired. Third, we want good separation of *L* and *R* signals. And last, we want no interference from SCA signals.

Although these tests would best be made with a stereo signal generator that could duplicate a composite stereo signal, current cost of that test instrument would not warrant its purchase for a limited service market. Therefore, a simpler procedure will be presented. Reference will be made to the matrix type decoder shown in Fig. 22-55.

Turn on the set for about 15 minutes for warm-up time. Then connect a vacuum-tube voltmeter to either the *L* or *R* output terminal and ground. With an audio oscillator, feed a 19-kc signal to the input of the decoder (point *A*) of such magnitude as to give a 0.1-volt reading on the voltmeter. Adjust the core of coil *T*-3, the 19-kc tuned input of the 19-kc oscillator until a slight flicker appears on the voltmeter, thereby indicating that synchronization with the 38-kc output is correct. Then adjust the core of *T*-5, the coil of the tuned 38-kc output, for maximum reading on the voltmeter.

We now turn to adjustment of the *L*-*R* circuit and the 67-kc SCA reject filter. Place a jumper from the grid end of coil *T*-3 (point *B*) to ground to disable the oscillator. Readjust the audio oscillator to a frequency of 27 kc and peak coil *T*-4, the bandpass filter, for maximum reading on the voltmeter. Then readjust the audio oscillator for a frequency of 67 kc and adjust coil *T*-2 of the 67-kc reject filter for a minimum voltmeter reading. Remove the jumper from the oscillator.

Now tune in for a stereo station broadcast. It is advisable, if the station does not announce that the broadcast is a stereo one, to call the station to verify that it is. If a beat note is heard steadily, it is an indication that the oscillator is still not locked in with the transmitted 19-kc pilot signal. Touch up the core of coil *T*-3 once again until the beat note disappears.

The last steps are concerned with proper

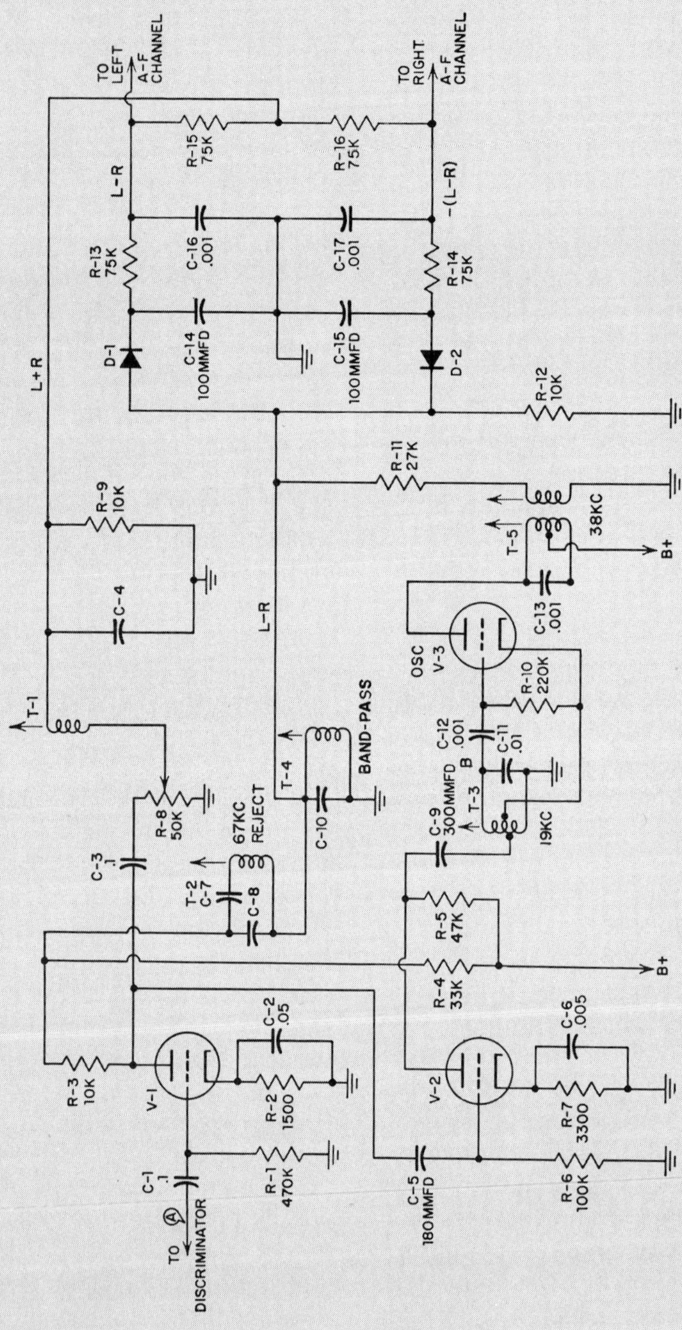

FIG. 22-55. Adjustment of a multiplex unit.

separation of L and R signals. Some stations transmit test signals in which only L, and then only R, signals are broadcast. When this is so, adjust low-pass filter coil T-1 and separation control R-8 for desired maximum separation. If the station does not transmit the test signals, proper separation will have to be judged by ear.

The alignment procedure described above may be adapted to any matrix decoder. With the switching type decoder, the alignment is quite similar. Of course, in this latter type, the composite stereo signal is handled as a unity without separation of the $L+R$ and $L-R$ components. Since the $L+R$ is unaffected by the switching voltage, it is always fully recovered. But any misalignment of any 19- or 38-kc tuned circuits reduces the amount of recovered $L-R$. Therefore, alignment consists in adjusting those tuned circuits.

The following procedure is one for aligning the switching type decoder. Turn on the receiver and allow it to warm up for about 15 minutes; tune in to an f-m stereo program. Then connect a vacuum-tube voltmeter to the output of the 19-kc amplifier. Adjust the input and output coils of the amplifier for maximum output on the voltmeter. Now move the voltmeter to the L or R output terminal and adjust the 38-kc output for maximum reading on the voltmeter. Connect the voltmeter back to the output of the 19-kc amplifier and adjust the input to this amplifier for maximum separation of R and L signals.

The station test signal is once again useful in this adjustment. When this adjustment is complete, the voltmeter should be near peak reading. If a beat note is heard, touch up the tuning circuits of the 19-kc amplifier slightly. Finally, adjust the separation control for maximum separation. The 67-kc SCA trap is adjusted as with the matrix decoder.

QUESTIONS

1. An a-m/f-m receiver operates on a-m, but not on f-m. Plate voltages are normal. What is the likely source of trouble?
2. Describe how you would tune in a station on an f-m receiver when you desire afc operation.
3. What precautions with regard to lead dress should be taken when replacing a component in an f-m receiver? What defects might result from improper lead dress?
4. Explain the action of a limiter stage.
5. How would you quickly locate a defective stage in an f-m receiver?
6. An f-m stereo receiver has poor channel separation. How would you proceed to remedy this defect?
7. State the general procedure for aligning a multiplex decoder.
8. A receiver, in the f-m position, squeals when stations are tuned in. A voltmeter check discloses a voltage at the f-m limiter grid with no signal input. What causes the defect? How would you remedy it?

Survey of the Servicing Procedure

23

The preceding chapters of this book have analyzed each stage of the receiver and discussed troubles that might arise from defective components within each stage. The emphasis has been primarily on defects that produce no reception or weak reception.

However, when a defective receiver is brought in for servicing, the defective stage is not usually self-evident. It is therefore necessary to present an overall servicing procedure for tracking down troubles. In addition, other defects, like hum, distortion, motorboating, modulation hum, noise, and intermittent operation, which have been treated incidentally, require an overall approach. It is the purpose of this chapter to present such an inclusive procedure for all the defects listed.

SERVICING PROCEDURE FOR NO RECEPTION

When the complaint is "no reception," the trouble may be caused by breakdown of almost any component throughout the receiver signal chain. For the beginner or servicing apprentice, a routine check of tubes, followed by a routine voltage check, is a good approach. But it is too time-consuming for the more experienced technician, who will begin with a routine signal check. The following steps represent the more experienced approach:

1. Check the power supply. The technician will ask himself various questions with respect to the inoperative receiver. Do all the tubes in the receiver light or warm up? Is there any sign of unusual overheating? Is the hum excessive? Does *B*+ measure its normal 200 to 300 volts for an a-c receiver, or its normal 100 volts for an a-c/d-c receiver? If the answers to these questions are those applicable to a normal receiver, he proceeds to the next stage. If not, there is trouble in the power supply and it must be found.

The causes for lack of receiver reception originating in the power supply are listed as follows:

Open fusible resistor
Open line fuse
Defective line switch
Defective line cord
Open power transformer primary
Dead rectifier
Open filter choke or filter resistor
Filter choke winding that shorts to chassis
Shorted filter capacitors
Short in the *B*+ line
Open voltage divider resistor
Open heater in an a-c/d-c receiver

---

2. *Check the speaker.* If the power supply checks perfect, the speaker comes up for inspection. To check its normal operation, momentarily unseat the second a-f tube. If a loud click is heard in the speaker, the latter is not the cause of inoperation, and the technician goes on to the next check.

If the click is not heard, the speaker may be defective in some respect. Possible causes for inoperation originating in the speaker or associated circuits are:

Open speaker voice coil
Open speaker voice-coil leads
Open output-transformer primary
Dead second a-f tube

---

3. *Check the second a-f stage.* If the speaker is working, the technician proceeds to check the second a-f stage. A plugged-in soldering tool is applied to the signal grid pin of the second a-f tube. If a low growl is heard in the speaker, the stage is all right, and the next stage is checked.

If the growl is not heard, the trouble is in the second a-f stage, which is then subjected to a voltage and resistance check to localize the cause of the trouble. Causes of lack of receiver reception originating in the second a-f stage are as follows:

Dead second a-f tube
Open output-transformer primary
Shorted plate bypass capacitor
Open cathode self-bias resistor

---

4. *Check the first a-f stage.* With all previous checks showing normal conditions, the technician proceeds to check the first a-f stage. When a plugged-in soldering tool is touched to the ungrounded end of the volume control, a very strong growl should normally be heard in the speaker. If it is heard, the technician may go on to check the next stage in the signal chain.

If the growl is not heard, the cause of no reception is in the first a-f stage and its associated parts. Such possible causes are:

Dead first a-f tube
Open coupling capacitor in the grid or plate circuit
Open volume control
Volume-control lug shorting to chassis
Short in grid wiring (shielding)
Open cathode self-bias resistor

---

5. *Check the detector stage.* The detector stage is the next check when all previous check results are normal. A modulated signal at the intermediate frequency of the receiver is fed to the grid of the i-f tube. If the signal-generator modulation note is heard in the speaker as the generator frequency control is wobbled around the intermediate frequency, the detector is all right. The technician then goes on to check the i-f amplifier stage. With due regard for the different intermediate frequency involved, the same signal check will work for f-m receivers.

If the modulation note is not heard, the trouble is in the detector stage or the i-f tube. Possible causes of receiver inoperation here are:

Dead i-f amplifier tube
Shorted grid circuit in the i-f tube
Open or shorted plate, screen, or cathode in the i-f tube circuit
Defective output i-f transformer:
  a. Open windings
  b. Shorted trimmers
  c. Leads shorting to the shield can
Defective detector tube
Open volume control
Misalignment of the i-f transformer

---

6. *Check the i-f stage.* When the modulated signal-generator output is fed to the control grid of the i-f tube and its note is heard, indicating normal detector operation, the hot lead is shifted to the mixer grid of the con-

verter. If the note is now heard at greatly increased volume, the mixer and i-f amplifier are functioning. The technician then proceeds to check the oscillator of the converter.

If the signal-generator note is not heard when the "hot" lead is applied to the mixer grid, the following factors may be defective:

Dead mixer (converter) tube
Shorted mixer grid circuit (tuning capacitor)
Shorted or open plate, screen, or cathode circuits in the mixer circuit
Defective input i-f transformer:
- *a.* Open windings
- *b.* Shorted trimmers
- *c.* Leads shorting to the shield can

Misalignment

---

7. *Check the oscillator circuit of the converter.* After the signal-generator output is fed at the intermediate frequency of the receiver to the mixer grid and its note is heard, the receiver dial is set near the low-frequency end of its tuning range. The signal-generator frequency control is then wobbled back and forth around this frequency. If the note is now heard at about the same volume as the former modulated i-f signal, the oscillator circuit is functioning, and the technician proceeds to the next check.

If the modulation note from the generator is not heard, the oscillator circuit is inoperative. Possible causes are:

Defective oscillator (converter) tube
Open oscillator coil
Short or resistance in the oscillator section of the gang tuning capacitors
Defective oscillator padder capacitor
Defective oscillator grid capacitor
Defective oscillator grid resistor

---

8. *Check the mixer circuit of the converter.* If the oscillator is functioning normally, the "hot" lead of the signal generator is shifted to the control grid of the r-f tube, or to the antenna if there is no r-f stage. The receiver dial is set near the high-frequency end of its tuning range, and the signal generator is wobbled around this frequency. If the modulation note is heard at increased volume, the mixer circuit is functioning.

If the note is not heard, the trouble lies in a component between the r-f grid (or antenna) and the mixer grid, and these might be:

Dead r-f tube
Shorted r-f control grid circuit (tuning capacitor)
Open or shorted plate, screen, or cathode circuits in the r-f stage

---

9. *Check the r-f input circuit.* If the signal-generator modulation note is heard when the hot lead is connected to the r-f grid but there is no reception from the antenna, the trouble must be in the antenna coil or leads. Possible causes in this regard are:

Antenna lead shorting to chassis
Open connection between antenna and antenna coil
Open or burned antenna coil primary

A 2-point check of a superheterodyne receiver. As the technician gains in experience, he develops shorter methods of procedure which reduce the time consumed. Such a short cut is the 2-point servicing procedure for checking an inoperative superheterodyne receiver.

If visual inspection does not disclose the source of the trouble, the tip of a plugged-in soldering tool is applied to the ungrounded end of the volume control. This is the beginning of the a-f signal chain. Normally, a strong growl from the speaker should be heard. If it is not heard, the trouble is in the audio amplifier (first a-f, second a-f, and speaker) or the power supply, and they are checked stage by stage for the specific defect. If the strong growl is heard, this one check clears the first a-f stage, the second a-f stage, the speaker, and the power supply of blame for the receiver inoperation.

The technician then moves on to the second check point. This is the mixer grid of the converter. A modulated test signal at the intermediate frequency of the receiver is fed to this mixer grid. The normal response is the modulation note of the signal generator in the speaker. If this note is not heard, the trouble is in the i-f amplifier or the detector stage. The signal-generator output is increased and the frequency control is wobbled around the intermediate frequency to see if the receiver is misaligned. If the response is still not heard, the test signal is fed to an i-f grid to localize the defect further.

If the normal response is heard when the modulated i-f test signal is fed to the mixer grid, the i-f amplifier and detector stages may be presumed to be functioning. The signal-generator frequency control and receiver dials are set near the low-frequency end of the tuning range in order to check the oscillator circuit of the converter.

The normal response in this latter check is the signal-generator modulation note from the speaker. If it is not heard, the oscillator circuit is not functioning. A voltage check of the converter stage is then made.

If the normal response is heard, the defect must be before the converter. A check of the r-f amplifier stage and the antenna circuits is now in order.

By this short 2-point check, the signal channel may be quickly analyzed into three blocks, which are checked over all before resorting, if necessary, to stage-by-stage checking.

SERVICING PROCEDURE FOR WEAK SIGNALS

The defects that cause weak reception are different from those which result in no reception. However, the servicing procedure that localizes the stage in which the defect lies is the same signal check just outlined for the complaint of no reception. The main difference in the two checks is the receiver response to the generator signal.

For a dead receiver, all signal checks result in a normal speaker response until the defective stage is reached. At that point, the receiver will give no response. For a weak receiver, all signal checks give a normal response until the defective stage is reached. At that point, the receiver will give a weak response, as shown by a loss or no gain over the last normal check before this check.

Many factors within the receiver may result in weak response. These are tabulated below:

- Weak tube in any stage
- Short in the power transformer winding
- Short in the filament wiring
- Jammed voice coil in the speaker
- Shorted turns in the output transformer
- Open cathode bypass capacitor in the second a-f, i-f, converter, and r-f stages
- Open agc bypass capacitor
- Receiver misalignment
- Open plate bypass capacitor in the i-f, converter, and r-f stages
- Open antenna coil
- Resistance in the gang tuning capacitor
- Poor wiper contact in the gang tuning capacitor

SERVICING PROCEDURE FOR HUM

A common receiver defect is a hum level that is so high that it mars normal receiver reception. This section will describe the type of hum that appears all over the receiver dial.

Checking the power supply. When a receiver is being serviced for the complaint of an abnormally high hum level, the most common defect that causes this condition is the breakdown of the power-supply filter capacitors.

This is usually due to the aging of these capacitors.

The first step, therefore, in trouble shooting for hum is to connect a substitute capacitor across each of the power-supply filter capacitors in turn. If the hum level is reduced as a result, the defective filter capacitor is replaced. The filter choke of the power supply must also be checked for a short that results in inadequate filtering.

Tubes as a source of hum. If the filter capacitors check perfect, a good second step is to replace the tubes with new ones, one at a time. Tubes often introduce hum, especially the a-f tubes. Such hum results from heater-cathode leakage through their insulation, capacitive coupling between the heater and other electrodes, and emission from the heater to other electrodes or vice versa. Although elimination of hum from these sources is primarily a design problem, replacement of tubes with new ones may reduce the hum level.

Grid circuits as a source of hum. Another possibility for hum is an open grid circuit in any stage of the receiver. This type of hum results from a build-up and discharge of signal at the open grid at a rate that may be close to the power-supply hum, and is mistaken for it.

The next check in hum elimination, therefore, is a continuity check made with an ohmmeter of all grid circuits.

Previous service work as a source of hum. If the cause of hum still proves elusive, the next check is to see if previous service work may not have introduced the trouble. For instance, replacement of part of a speaker may have resulted in the reversal of polarity of the hum-bucking coil. Or the wiring may have been disturbed with resulting poor lead dress, particularly in the region of the detector and first a-f tube. The diode-plate leads, volume-control leads, and first a-f grid leads must all be short. They should be dressed close to the chassis and away from the filament or other wiring that carries 60-cycle current.

Tracking down elusive hum. The suggestions made above should locate most of the common causes of hum. Occasionally, an elusive cause will escape the normal procedure that has been suggested. In such a case, the receiver must be examined stage by stage.

To do this, remove all the receiver tubes, except the rectifier. Since it is dangerous to operate a power supply without any load, a heavy-duty 5,000- to 10,000-ohm/25- to 50-watt resistor should be connected as a load from *B*+ to ground. Then turn the receiver on and listen for hum. If hum is present, it is due to some factor that was overlooked in the power supply, and it must be carefully sought for.

If the hum level is normal, insert the second a-f tube and remove the power-supply resistor load. If the hum now is heard, it is due to some defect in the second a-f stage. If the hum level is normal, insert the first a-f tube and listen for hum. In this way, the tubes are reinserted, one stage at a time, until the offending stage is reached. Then the components of only one stage need be carefully checked to find the defect.

In the case of an a-c/d-c receiver, where tube heaters are in series, tubes may not be removed, as above. The stages must be made inoperative in another manner. A short from the second a-f grid to ground makes everything before this point inoperative, so far as their effect on the speaker is concerned. Any hum present limits the defect to the power supply or second a-f plate circuit. If the hum level is normal, the short is shifted to the first a-f grid and ground. This adds the first a-f plate circuit and the second a-f grid circuit to the part of the receiver being checked. Thus, in shifting the short to ground from grid to grid of the various tubes, more and more parts of the receiver that will affect the speaker are brought in for check until the point of hum is located.

Once the tube-removal procedure or grid-grounding procedure has localized the stage in which hum originates, nothing in this stage should be overlooked in the careful recheck. On some infrequent occasions, a new tube that replaces a bad one may have a similar defect that still results in hum. If all else in the hum-producing stage has been found to be good it may be necessary to replace the old tube with several new ones before the hum disappears.

Another possibility in the careful recheck of a stage is the possibility of leakage between sections of a bypass capacitor block. For example, a line filter capacitor or a capacitor connected to a heater lead will carry alternating current. If they are part of a block, leakage to other capacitors in the same block may introduce hum.

Normally, capacitors are checked for opens, shorts, leakage, and intermittent opens. None of these checks requires the removal of the capacitor from the circuit. However, in making a careful recheck of a stage, the capacitor lead must be opened, and a substitute capacitor connected in its place. This procedure will take care of leakage between sections of a block.

Summary of the causes of hum in receivers. The causes of hum in a receiver may now be summarized for quicker use.

Open power-supply filter capacitors
Defective tubes (cathode-heater leakage)
Open grid circuit
Reversed speaker hum-bucking coil
Closeness of audio grid leads to wiring carrying 60-cycle current
Leakage between sections of a bypass or filter capacitor block
Shorted filter choke

SERVICING PROCEDURE FOR NOISY OPERATION

When the consumer complains of noisy reception, he is speaking of hissing and crackling sounds that are extraneous to the desired station signal. Noise may result from one of a great number of causes including noise picked up by the antenna or the power line, or by defective units within the receiver itself.

The first job of the technician is to determine which of the above conditions is the cause. A good check is to try operation with another receiver at the same location. Traded-in receivers have little resale value. One of these kept in repair is ideal for this kind of work. If this radio shows the same noise symptoms, the trouble is in the power line or the location of the receiver in a noise area. Usually moving the receiver to a different outlet in another part of the room will provide a location where reception conditions are improved. If the old trade-in receiver provides good reception, the cause of the noise is within the consumer's set. Tracking it down to the source may require considerable time, so it is best to remove the set for bench repairs in the service shop. It might be a good idea to leave the old set at the customer's home so that he has its use until his own receiver is repaired. This makes for good customer relations.

Causes of noise within a receiver. The components in a receiver that usually cause noisy operation are as follows:

---

1. Noisy tubes (loose elements)
2. Corrosion in coil windings:
   a. R-f transformers
   b. I-f transformers
   c. Audio transformers
3. Speaker defects:
   a. Rubbing voice coil
   b. Torn paper cone
   c. Loose rim
4. Poor connections
5. Noisy volume control
6. Swinging shorts in i-f transformers
7. Conductive dirt in vital spots (like sockets)

8. Tuning capacitors (shorts and poor wiper contacts)

Locating the source of noise in a receiver. Several of the causes of noise within the receiver have already been presented in previous chapters. Chapter 11 gives the checks for the speaker defects that cause noise, and these checks may be used in noise analysis. Probably, replacement with the bench test speaker will disclose this source. In Chap. 16, the noisy tuning capacitor is described. In Chap. 13, defective volume controls are described. Both of them are common sources of noise and are easily identified as the sources, since the noise comes on when the controls are adjusted. A procedure for cleaning tuning capacitors is given in Chap. 16. A noisy volume control should be replaced as described in Chap. 13.

A good procedure to follow when hunting for the source of noise is similar to that used in checking for hum. Remove all the tubes, except the second a-f tube and rectifier tube. In the case of the a-c/d-c receiver, connect a short between the second a-f grid and ground. Tap the tube and other components in the second a-f stage and listen for noise. If the noise is heard, all connections and components in the second a-f stage and power-supply stage are checked until a poor joint or defective component is found. If the noise is not heard, the second a-f and power-supply stages are probably in good condition. The first a-f tube is then added, or the ground connection is moved to the first a-f grid. Then components in this stage are checked as before. If they prove satisfactory, the procedure is repeated for each stage in the receiver until the troublesome stage is found. The search within the defective stage must be thorough and not overlook any odd and unusual condition of a component.

As each tube is replaced or made operative, it should be slightly jarred by tapping the radio sharply. When this is done, a noisy tube will become more noisy. Then replacement with a new tube followed by the jarring test will tell if the tube was at fault. In some cases, simply replacing one tube at a time with new ones and tapping the receiver sharply will locate a noisy tube.

The stage-by-stage analysis may show a noisy stage. If the defect is due to corrosion in coil windings, jarring the radio will not affect the noise. In this case, the defective winding can be found by an ohmmeter check. Good windings in r-f and i-f transformers normally measure less than 100 ohms; a corroded winding usually measures several hundred ohms. The resistance measurement of a-f transformers and chokes also increases when they are corroded. In the case of f-m receivers, good r-f and i-f transformer windings measure less than 1 ohm. They rarely corrode.

Often, unsoldered or poorly soldered connections, or bits of solder or other conducting dirt, may be the cause of noise in a stage. They may be difficult to locate because they may be in an out-of-the-way place. Such defects may cause intermittent noise or intermittent operation. Jarring the receiver usually increases the noise when those defects are the cause. An extremely careful search must be made for them.

If the procedure localizes the i-f amplifier as the source of noise, remove the i-f transformers from their shield cans for a careful inspection. Even though an ohmmeter check shows freedom from corrosion, the leads from the coils to the trimmers, which lie along the shield can, may vibrate into contact with the can, producing noise. Inspect the leads and route them so that they cannot possibly touch the shield can.

SERVICING PROCEDURE FOR INTERMITTENT OPERATION

Intermittent reception can be divided into two main groups. In one, the radio suddenly clicks off and remains inoperative for a while; then, just as mysteriously, it resumes normal operation. In the other type, the volume decreases and then returns to normal a little later. Sometimes, these changes are gradual rather than sudden. This condition is often called "fading."

Causes of intermittent operation. Intermittent operation, when accompanied by noise, may be due to intermittent breaks or other defects in the receiver, which should be carefully checked by the procedure just given for noise.

When the set just dies out, the technician might replace all components likely to cause the trouble, hoping by elimination to remove the cause. Or, he might track down and repair the causative factor. The first procedure is expensive; the second is time-consuming.

The wholesale replacement of suspected receiver components includes:

1. All tubes
2. All bypass and coupling capacitors
3. Any resistors that dissipate heat and may change in ohmic value as a result, like voltage-divider resistors
4. The volume control

In addition, the capacitor gang should be cleaned and overhauled. Finally, a thorough search should be made for poorly soldered connections.

Tracking down intermittent reception. In tracking down the cause of intermittent reception, the receiver is allowed to play on the service bench until the fading out occurs. This condition may be hastened by jarring the receiver; or, the receiver may be made to operate inside a packing box to cut off ventilation and produce overheating; or, it might be operated through an autotransformer connected to the power lines, so that it operates under the condition of an abnormally high line voltage.

Then when the receiver fades out, any accompanying symptom, like noise or squeal, will be of aid in locating the trouble. If the receiver stays out, it is serviced as though it were a dead or weak one. If reception is resumed before any conclusive evidence has been reached, the technician waits for the next fade-out, or attempts once again to induce it.

SERVICING PROCEDURE FOR MODULATION HUM

"Modulation" hum is the name applied to a hum that is heard together with the station voice or music only when a station is tuned in. The hum level is normal at an off-station position on the tuning range. This type of defect in receivers is also called "tunable" hum.

Causes of modulation hum. The most common cause for this condition is an open line filter capacitor, or inefficient grounding of the receiver. When a receiver is checked at the service bench for a complaint of tunable hum, the first step is to operate the receiver on a station where the hum is very noticeable. No ground lead should be connected to the receiver, since the home installation may not use one. Then connect a capacitor of similar capacitance across the line filter capacitor and listen for a reduction in the hum. If this step is not effective, try connecting the capacitor from the other side of the line to the chassis, as shown in Fig. 23-1.

If the modulation hum still persists, the next likely cause is leakage or capacitance effects from the heater to other elements in the r-f or converter tubes. The next step, therefore, is the substitution of tubes, known to be good, for the r-f and converter tubes.

SERVICING PROCEDURE FOR SIGNAL DISTORTION

Distortion in a receiver results in poor tone quality from the loudspeaker. It is usually due to the overloading of some stage in the receiver by a signal that is too large for the stage to handle.

Causes of receiver distortion. It is unusual

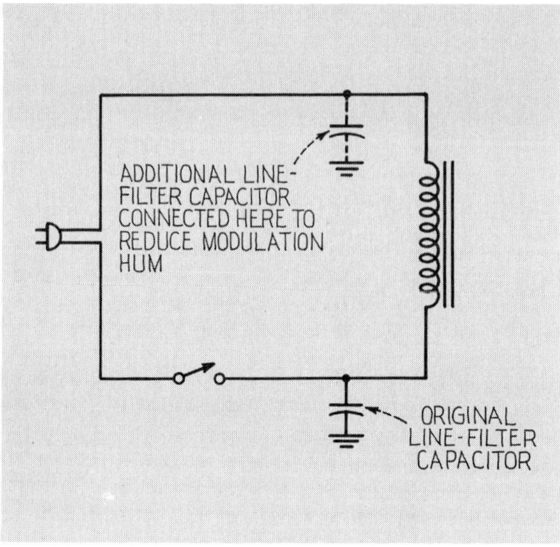

FIG. 23-1. Procedure for reducing hum modulation.

for the signal to be too large. The common difficulty is that the stage operation has deteriorated to a point where it cannot handle a signal of normal strength.

The usual difficulty is trouble in the grid-bias circuits, which is found by voltage analysis. The speaker, of course, is another possible cause of poor tone. This condition is checked by substituting the bench test speaker for the receiver speaker. A more complete list of possible causes of receiver distortion is given below.

---

1. Rubbing speaker voice coil due to
   a. Off-center voice coil
   b. Warped speaker cone
   c. Off-center speaker field gap
2. Shorted cathode bypass capacitor in the second a-f stage
3. Changed value of second a-f bias resistor
4. Open grid leak in the first or second a-f stage
5. Open volume control
6. Defective tubes
7. Shorted or leaking audio coupling capacitors
8. Shorted or leaking agc bypass capacitors
9. Misalignment, especially in f-m receivers

Less frequent causes of poor tone quality are those resulting from previous replacement of defective parts. These include a mismatch resulting from the replacement of a speaker or output transformer; a replacement plate circuit bypass capacitor in the second a-f stage that is too high or low in capacity, resulting in too high or low a response; rarely, sideband cutting resulting from the use of an i-f replacement transformer with extreme selective characteristics. In the latter case, the sideband cutting may be reduced by slightly mistuning each i-f trimmer, broadening its response characteristic.

SERVICING PROCEDURE FOR MOTORBOATING

Motorboating is a defect in a receiver that results in a put-put noise similar to the exhaust of a motorboat. The most common cause for motorboating is an open output filter capacitor in the power supply. The only other common cause is an open grid circuit in any of the stages of the receiver.

Removing motorboating in a receiver. A servicing procedure for this defect in a receiver is to bridge the output filter capacitor in the power supply with a test capacitor of similar capacitance and to see if the trouble is eliminated. If this proves ineffective, the technician proceeds to make an ohmmeter check of all grid circuits, looking for an open. In this regard, it should be remembered that agc decoupling filter resistors are part of their grid circuits. They must not be overlooked, even though they rarely open. The most common opens occur in the grid-load resistors of the first and second a-f stages.

SERVICING PROCEDURE FOR SQUEALS AND OSCILLATIONS

There are many types of squeals and howls in a receiver, all of which are classified under the general term of "oscillation." Their causes are many and varied and call for different servicing procedures.

Chirps or birdies. First, there is the type of squeal or birdie that appears to spoil reception from only one or two stations. This is probably image-frequency interference. A procedure for handling these is given in Chap. 17.

Microphonic noise. There is another type of howl known as "microphonic" noise. It usually starts on a loud signal, or when the radio is jarred, and builds up to a strong howl that drowns out all reception. It can be caused by loose elements in a tube or by vibrating tuning capacitor plates. The howl is started by either the jarring of the receiver or the vibration resulting from a loud signal from the speaker. The loose elements begin to vibrate rapidly and introduce sustained high-pitched a-f notes into the tube.

When a receiver with microphonic howl is serviced, the receiver is operated at low volume. Each tube in turn is gently tapped. When the offender is reached, a "bong" is started which soon dies out, since the speaker volume is too low to sustain the vibration. Any tube in the receiver may be the cause of the microphonic howl. However, the detector first a-f tube is the most common offender.

If a check of the tubes discloses no defect, the tuning capacitor should be investigated. Microphonics due to the tuning capacitor are usually found in small receivers, where the speaker and tuning gang assembly are in close proximity, or in large receivers designed for and operated at high-volume levels. In both cases, original design takes care of the condition by mounting the tuning capacitors or the chassis, or both, on a rubber suspension. Sometimes, even the speaker is mounted on rubber to dampen vibrations. It is only necessary thereafter to check the mounting provisions to see that the rubber has not become old and cracked, or that the suspended mounts are still floating freely.

Squeals over the major part of the receiver tuning dial. Another type of squeal is the one that occurs over the entire tuning range of the receiver or a large part of it. If this squeal is affected somewhat by tuning, the defective component is usually in the r-f or i-f portions of the receiver. If the squeal is unaffected by tuning but is affected by the operation of the tone control, the defective component is probably in the audio end of the receiver. However, these considerations are not of too great consequence, since the servicing procedure is the same for both conditions.

Squeals of either type are usually caused by regenerative coupling. The latter is usually caused by poor contact between a shield and the chassis or by the opening of a bypass capacitor. The service procedure is suggested by the cause. Shields are checked for their contact to the chassis. Ohmmeter checking is inadequate, since even a small resistance contact (too small to be read on the ohmmeter) may still cause inadequate shielding. The best procedure is to clean and tighten all shield-grounded contacts. Where a tube shield has been inadvertently discarded, it should be replaced by the technician. This shielding is especially important in the case of a high-gain tube like the i-f amplifier.

Open bypass capacitors are checked by bridging a test capacitor across each bypass capacitor in the receiver. It is important when making these checks to short the terminals of the test capacitor after each capacitor is checked.

A test capacitor of about 0.1 mfd can be used for all low-frequency r-f bypass capacitors, even though the capacitor being tested differs considerably from that capacitance. The audio bypass and the power-supply filter capacitors should not be neglected in this test. In high-frequency circuits, like those in f-m receivers, the best procedure is temporarily to solder a 0.01-mfd test capacitor across the suspected component. Use a ceramic disk type with short leads.

Sometimes, the broad squeal is due to an error that crept in during previous service work. Disarranged or poorly dressed leads may come about in the replacement of an i-f or r-f transformer. Or, an inverse feedback winding from the secondary of an output transformer may have been reversed during the replacement of the transformer. In the former case, the leads may couple with other parts of the receiver and deliver regenerative feedback. In the latter case, a reversed inverse feedback winding may deliver regenerative feedback, rather than degenerative feedback. As a result, an audio oscillation is set up.

Poorly dressed wiring may be checked by moving the suspected wires with a bakelite rod, while the receiver is oscillating. A change in the squeal indicates that the wire is at fault. Generally, the grid and plate leads are the "hot" leads and should be routed close to the chassis and direct to their connection points without crossing each other or coming close to other wiring.

The reversed inverse feedback winding may be checked for by reversing the primary or secondary wires of the output transformer and by observing if there is any improvement.

A summary listing of factors that might cause broad squeals and oscillations follows:

1. Open power-supply output filter capacitor
2. Open second a-f plate bypass capacitor
3. Reversed feedback winding (after output transformer has been replaced)
4. Open shielding
5. Incorrect lead dress
6. Open agc bypass capacitor
7. Open screen bypass capacitor in the r-f, i-f, or converter stage

AIR CHECK OF A RECEIVER

The final step in servicing a receiver is first to check that the original complaint has been removed, and then to check the receiver in all respects for normal operation. This final check is known as the "air check."

To make the air check, the receiver is connected to an antenna and turned on. The tuning dial is rotated to a non-station position, and the hum level is noted for normal operation.

At the same dial position, the volume control is rotated from minimum to maximum, in order to determine if it is noisy. The same is then done for the tone control, if present.

Then the dial is rotated to the low-frequency (550-kc) end, the volume control is set for a moderate volume level, and the dial is rotated toward the high-frequency (1,500-kc) end. The stations are checked off as they appear. This procedure checks the dial calibration and the sensitivity of the receiver. All stations that the technician knows are normally picked up in his locality should be picked up by the receiver being checked. Failure to receive any of them indicates a weak receiver. Good judgment should be used by the technician in this test. Obviously, a sensitive superheterodyne receiver with an r-f stage and two i-f stages should pick up more stations than a receiver with no r-f stage and only one i-f stage.

As the stations are picked up, the selectivity of the receiver may be determined by the dial space that each station covers, especially the strong local stations. If a strong local station stretches over 30 kc of the dial, the receiver selectivity is poor and should be checked. Misalignment is indicated.

Tone quality is most easily checked by listening to speech rather than to music. Clear, crisp, intelligible speech is a sign of good tone quality, especially for the high audio frequencies. Then turn to some symphonic music program, and listen for the response to the low frequencies.

The next check is for the power handling of the receiver—from whispers to the limit that the speaker will take without rattling. This will not be much for a small speaker. But a large speaker system in a high-fidelity receiver ought to be capable of roaring with good tone quality.

The course over the tuning dial will disclose whistles, birdies, and squeals. The technician must be on the alert for these effects. A sharp slap on the receiver will show up any noisy conditions.

The final step is the check of the operation of any other controls on the receiver, like fidelity controls, function switches, push buttons, short-wave operation, and f-m operation.

This survey has been concerned primarily with tube receivers. A similar survey was presented in the single chapter on transistor receivers.

Appendix

E.I.A. COLOR CODE FOR RESISTORS (OHMS)

Basic Reference Chart

COLOR	1st FIGURE (A)	2nd FIGURE (B)	MULTIPLIER (C)	TOLERANCE (D)
Silver			0.01	10%
Gold			0.1	5%
Black		0	1.0	
Brown	1	1	10	1%
Red	2	2	100	2%
Orange	3	3	1000	3%
Yellow	4	4	10000	4%
Green	5	5	100000	
Blue	6	6	1000000	
Purple	7	7	10000000	
Gray	8	8	100000000	
White	9	9	1000000000	
No color				20%

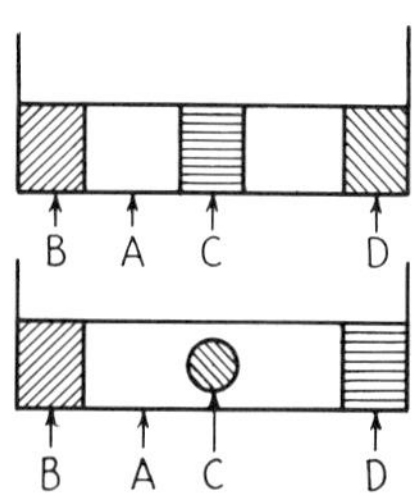

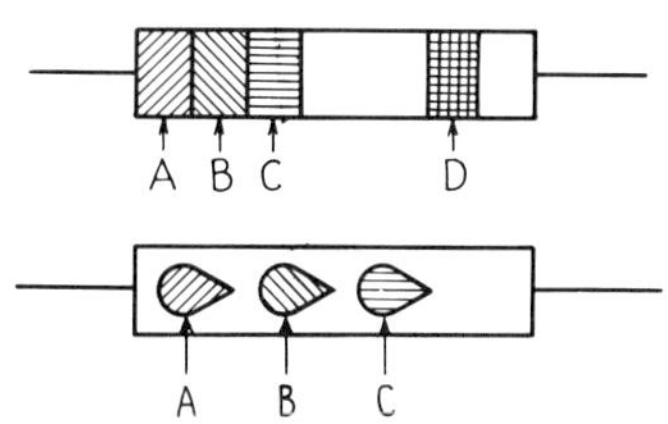

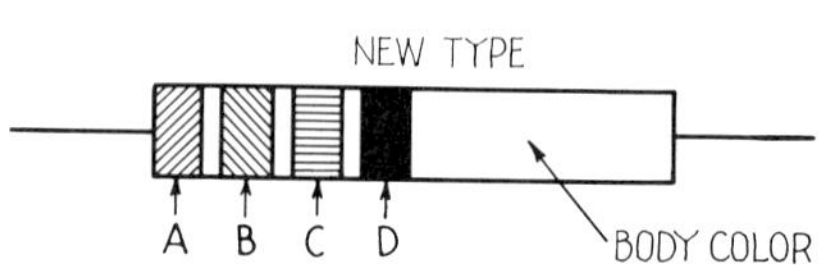

For new type only, body color indicates type of resistor as follows: black—composition, noninsulated; any color other than black, tan preferred—composition, insulated; dark brown—wire-wound, insulated.

An example will indicate the use of the new type of resistor coding. Assume the following with colors are given: *A*, red; *B*, green; *C*, orange; *D*, silver.

Decoding, we get

A	*B*	*C*	*D*
2	5	1,000	10%

The resistor has a value of 25 times 1,000, or 25,000 ohms, and a tolerance of ±10 percent.

E.I.A. COLOR CODE FOR FLEXIBLE RESISTORS

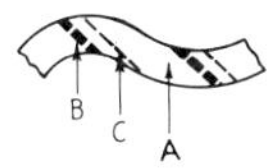

The same color code holds for flexible resistors as for carbon resistors. For flexible resistors the first digit is the body color (*A*). The second digit is the thick thread color (*B*). The multiplier is the thin thread color (*C*).

Ohm's Law and Its Derivatives

Where E = volts, I = amperes, and R = ohms,

$$I = \frac{E}{R} \qquad E = I \times R \qquad R = \frac{E}{I}$$

Where E = volts, I = amperes, R = ohms, and W = watts,

$$W = I^2 \times R \qquad I = \sqrt{\frac{W}{R}} \qquad R = \frac{W}{I^2}$$

$$W = I \times E \qquad I = \frac{W}{E} \qquad E = \frac{W}{I}$$

$$W = \frac{E^2}{R} \qquad E = \sqrt{W \times R} \qquad R = \frac{E^2}{W}$$

Electronic Units

m = milli = 1/1,000
μ = micro = 1/1,000,000
n = nano = 1/1,000,000,000
p = pico = 1/1,000,000,000,000
f = femto = 1/1,000,000,000,000,000

Note: pico is the newer name for micromicro farads ($\mu\mu$).

Examples:

10 ma = 10 milliamperes
2 μf = 2 microfarads
60 pf = 60 picofarads
= 60 micromicrofarads

Cycles are now often called Hertz, abbreviated Hz

Examples:

60 cycles = 60 Hz
770 kilocycles = 770 KHz
24 megacycles = 24 MHz

Abbreviations used with Hz are:

K = kilo = 1,000
M = mega = 1,000,000
G = giga = 1,000,000,000
T = tera = 1,000,000,000,000

PREFERRED VALUES

The average values for resistors used in diagrams in this book have been expressed in round numbers for the sake of simplicity. Actual resistors in use have been standardized at preferred values. For example, where we speak of a 500k resistor, the actual resistor will be 470k. There are three common listings of preferred values, based on tolerance ratings of 20, 10, and 5 percent.

20 PERCENT	10 PERCENT	5 PERCENT
10	10	10
		11
	12	12
		13
15	15	15
		16
	18	18
		20
22	22	22
		24
	27	27
		30
33	33	33
		36
	39	39
		43
47	47	47
		51
	56	56
		62
68	68	68
		75
	82	82
		91

The listed numbers are the basic numbers. Resistors will have these values or values that are multiples of 10 of these numbers. For example, a preferred value resistor may be one of 33, 330, 3,300, 33k, 330k ohms, etc., or 3.3 megohms, etc.

Most manufacturers use resistors with a 10 percent tolerance rating. In service work, use the nearest value in the 10 percent column for replacement of a resistor whose value is given in round numbers.

E.I.A. COLOR CODE FOR CAPACITORS (MMF)

Basic Reference Chart

COLOR	1ST FIGURE (A)	2ND FIGURE (B)	MULTIPLIER (C)	TOLERANCE (D)
Black	0	0	1	20%
Brown	1	1	10	1%
Red	2	2	100	2%
Orange	3	3	1000	2½% or 3%
Yellow	4	4	10000	
Green	5	5		5%
Blue	6	6		
Violet	7	7		
Gray	8	8		
White	9	9		10%
Gold			0.1	
Silver			0.01	10%

Molded Mica Capacitors

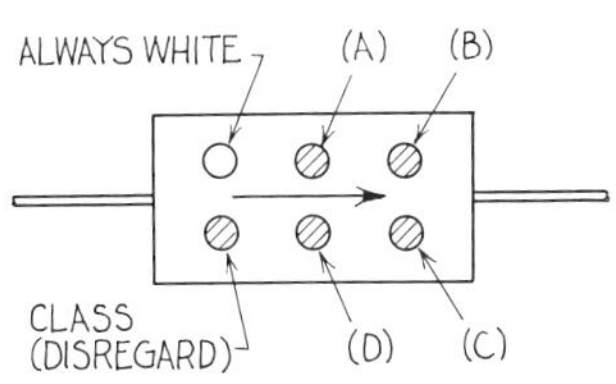

First dot is always white. This indicates a molded mica capacitor. Direction for reading indicated by arrow or equivalent marking.

Ceramic Capacitors

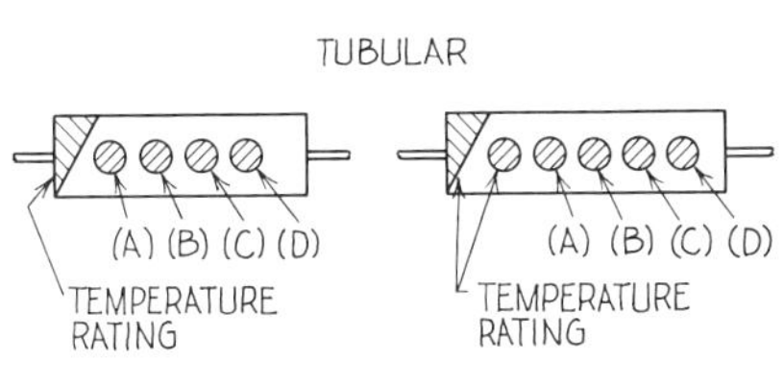

Tubular capacitors read from the end color to the right. End color identifies inside lead. Leads may be axial or radial.

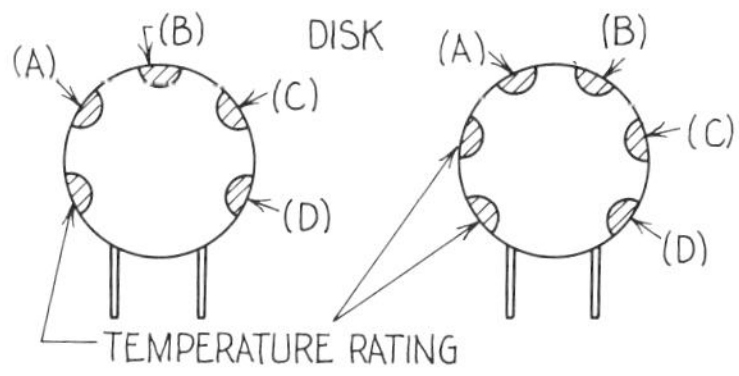

Disk capacitors are read from left to right with leads held downward.

For replacement purposes in tuned circuits, use ceramic capacitors with the same temperature markings as the original.

GRAPHIC SYMBOLS

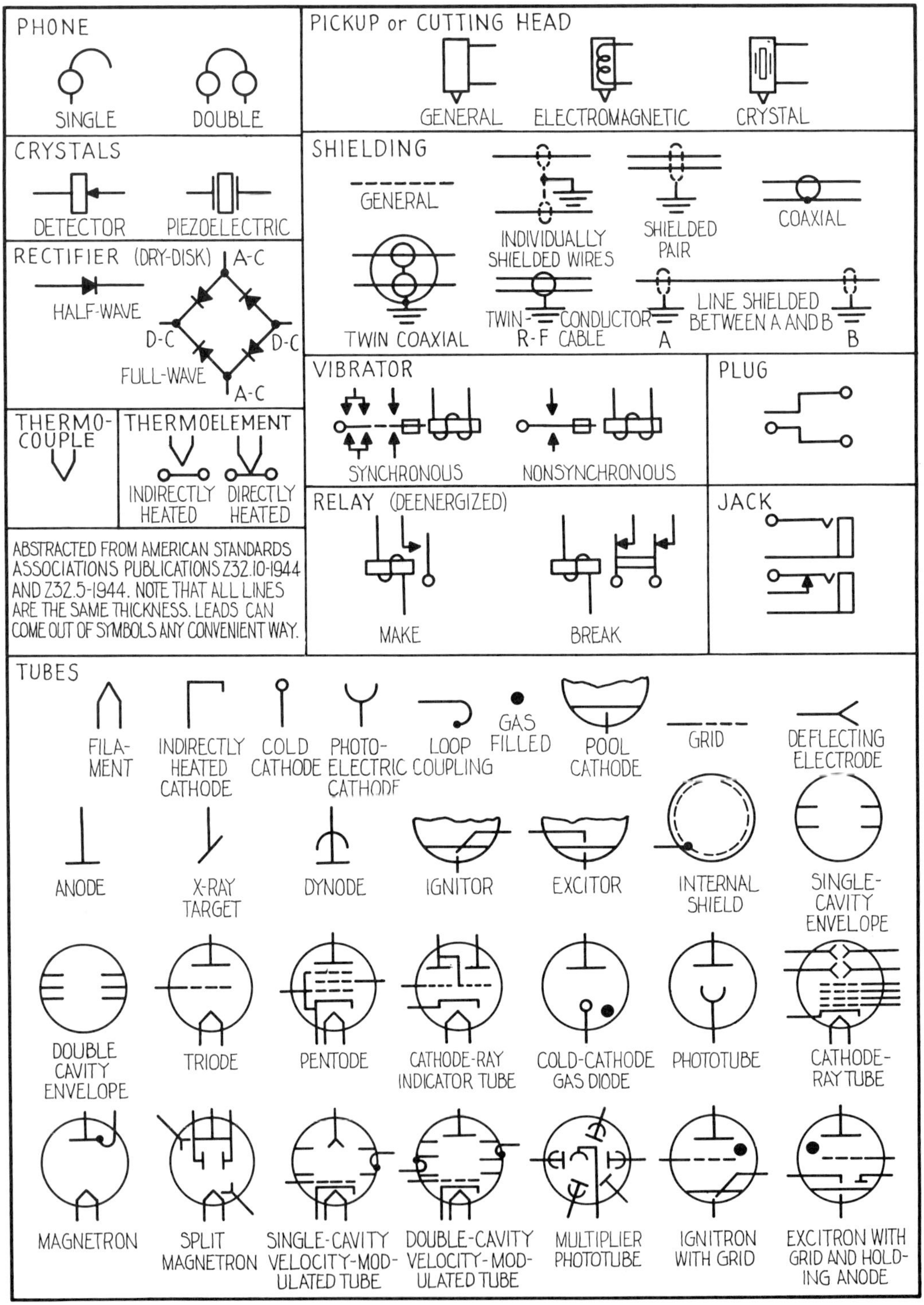

GRAPHIC SYMBOLS

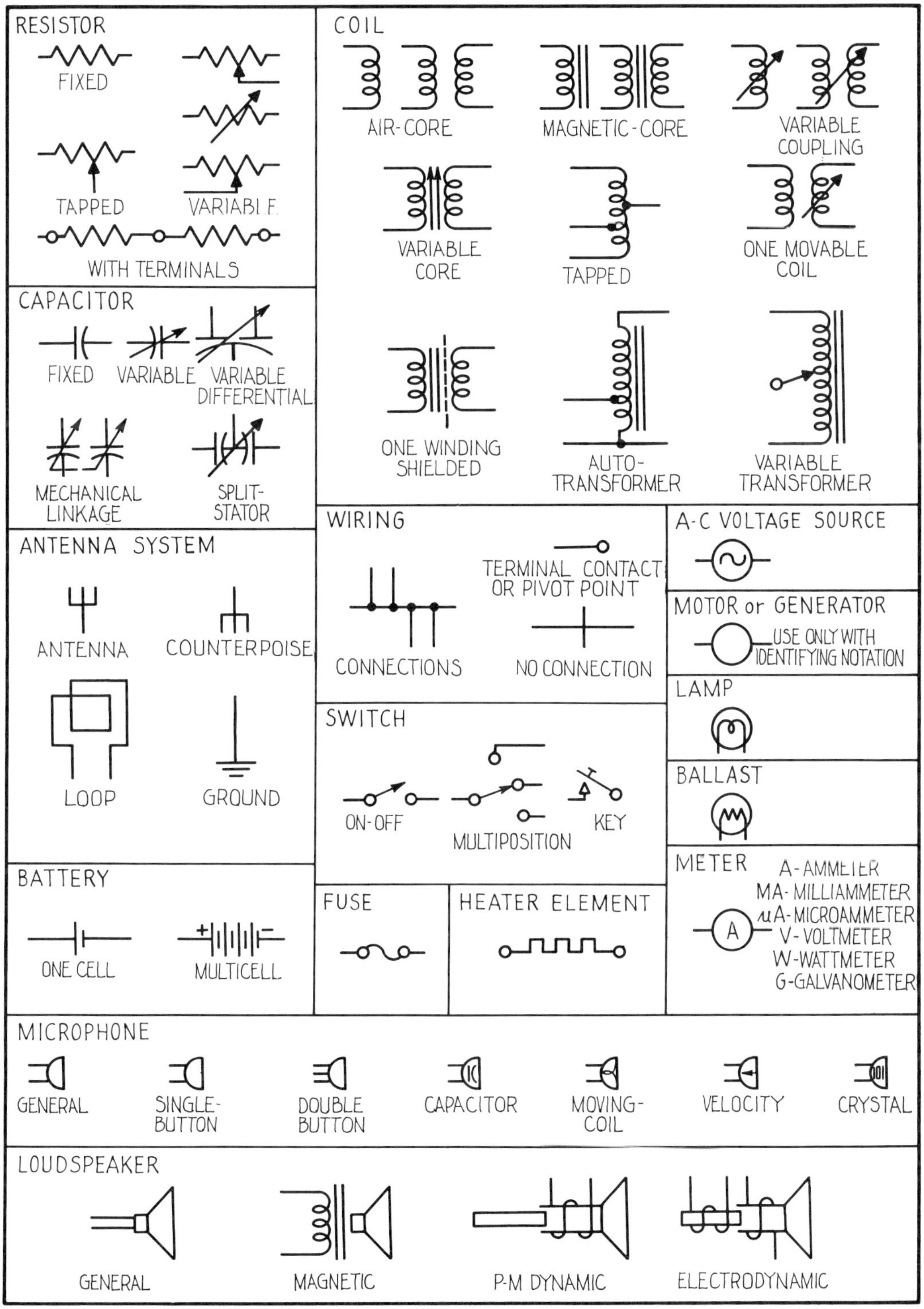

TUBE PRONG NUMBERING SYSTEM

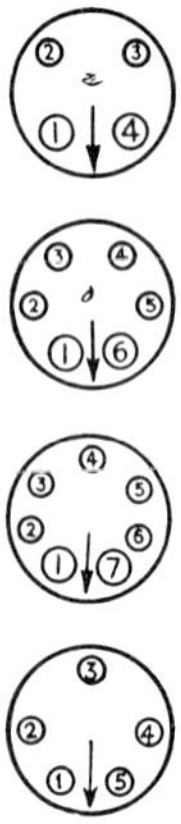

The four-prong tube has two large and two small prongs. The large prongs (1 and 4) are filament connections. The six-prong tube has two large heater prongs (1 and 6). The seven-prong tube has two large prongs (1 and 7) for the heater connections.

All prongs in the five-prong tube are the same size. Prong 3 has more separation than the others. The heater prongs are 1 and 5.

There are two eight-prong bases, the octal and lock-in types. In both cases, numbering begins from the left of the key slot and continues in a clockwise direction.

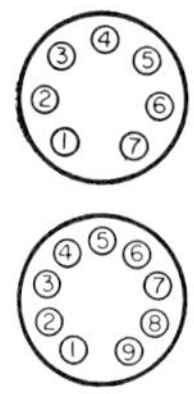

In the miniature seven-prong and nine-prong tube bases, numbering begins at the left of the wide space and continues in a clockwise direction.

BASE DIAGRAMS FOR TRANSISTORS

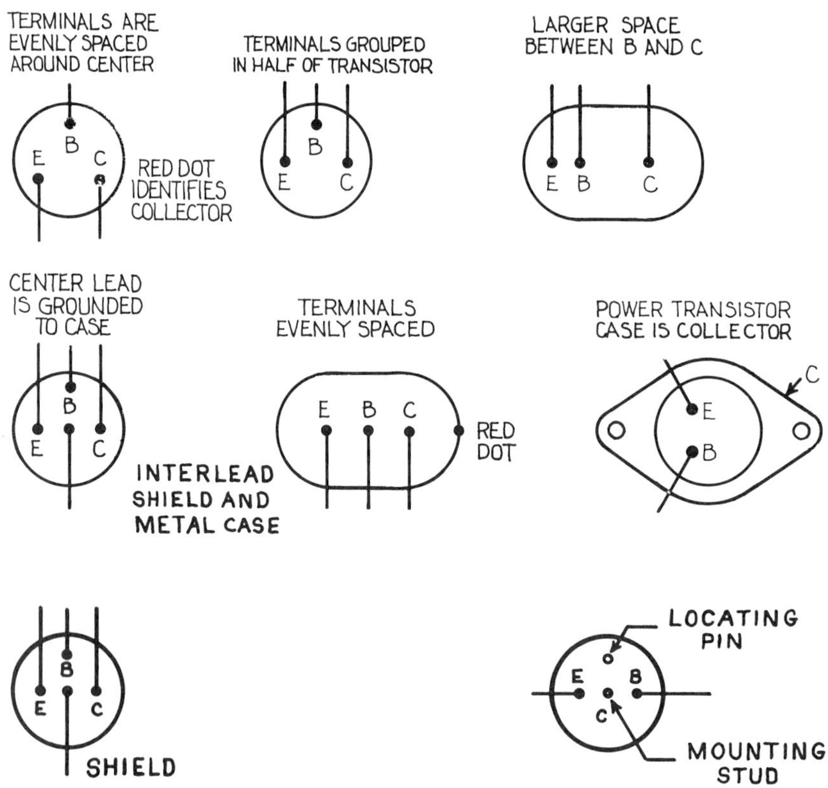

INDEX